19th International Conference on Electron Dynamics in Semiconductors, Optoelectronics and Nanostructures 2015 (EDISON 19)

Journal of Physics: Conference Series Volume 647

Salamanca, Spain
29 June – 2 July 2015

Editors:

Tomas Gonzalez
Maria Jesus Martin-Martinez
Javier Mateos

ISBN: 978-1-5108-1474-5
ISSN: 1742-6588

19th International Conference on Electron Dynamics in Semiconductors, Optoelectronics and Nanostructures 2015 (EDISON 19)

Journal of Physics: Conference Series Volume 647

Salamanca, Spain
29 June – 2 July 2015

IOP
Institute of Physics

Printed from e-media with permission by:

Curran Associates, Inc.
57 Morehouse Lane
Red Hook, NY 12571

Some format issues inherent in the e-media version may also appear in this print version.

Printed by Curran Associates, Inc. (2016)

For permission requests, please contact the Institute of Physics at the address below.

Institute of Physics
Dirac House, Temple Back
Bristol BS1 6BE UK

Phone: 44 1 17 929 7481
Fax: 44 1 17 920 0979

techtracking@iop.org

Additional copies of this publication are available from:

Curran Associates, Inc.
57 Morehouse Lane
Red Hook, NY 12571 USA
Phone: 845-758-0400
Fax: 845-758-2634
Email: curran@proceedings.com
Web: www.proceedings.com

Table of contents

Volume 647

19th International Conference on Electron Dynamics in Semiconductors, Optoelectronics and Nanostructures (EDISON'19)
29 June to 2 July 2015, Salamanca, Spain

Papers

PREFACE

The 19th International Conference on Electron Dynamics in Semiconductors, Optoelectronics and Nanostructures (EDISON'19) was held at the Hospedería Fonseca (Universidad de Salamanca, Spain), on 29 June - 2 July, 2015, and was organized by the Electronics Area from the University of Salamanca.

The Conference is held biannually and covers the recent progress in the field of electron dynamics in solid-state materials and devices. This was the 19th meeting of the international conference series formerly named *Hot Carriers in Semiconductors* (HCIS), first held in Modena in 1973. In the edition of 1997 in Berlin the name of the conference changed to *International Conference on Nonequilibrium Carrier Dynamics in Semiconductors*, keeping the same acronym, HCIS; and finally in the edition of Montpellier in 2009 the name was again changed to the current one, *International Conference on Electron Dynamics in Semiconductors, Optoelectronics and Nanostructures* (EDISON). The latest editions took place in Santa Barbara, USA, in 2011 and Matsue, Japan, in 2013.

Research work on electron dynamics involves quite different disciplines, and requires both fundamental and technological scientific efforts. Attendees to the conference come mostly from academic institutions, belonging to both theoretical and experimental groups working in a variety of fields, such as solid-state physics, electronics, optics, electrical engineering, material science, laser physics, etc. In this framework, events like the EDISON conference become a basic channel for the progress in the field. Here, researchers working in different areas can meet, present their latest advances and exchange their ideas.

The program of EDISON'19 included 13 invited papers, 61 oral contributions and 73 posters. These contributions originated from scientists in more than 30 different countries. The Conference gathered 140 participants, coming from 24 different countries, most from Europe, but also with a significant participation from Japan and USA.

The two topics receiving more abstracts correspond to fields attracting a lot of attention and research activity in the last years: electron dynamics in graphene and related materials and devices, and THz phenomena in nanostructures. Other topics like coherent dynamics in ultrafast optical phenomena or quantum processing, and semiconductor-based spintronics, had also an important presence in the program.

Thanks are given to the members of the International and Scientific Program Committees for their valuable and qualified assistance in the selection of invited speakers and the review of the submitted contributions; and also to those attendees who assisted us in the review process of the papers included in this volume.
We would like also to thank the institutions and sponsors that contributed to support the conference. Firstly the University of Salamanca, one of the oldest in the world, commemorating within three years, in 2018, the 8th century of its foundation in 1218. EDISON'19 can be considered as one more of the events taking place to celebrate such centennial. The Salamanca City Council also contributed in the organization of the conference. We acknowledge as well the support from the Spanish Ministry of Economy and Competitiveness, and from our sponsors and exhibitors, Oxford Instruments and Rohde&Schwarz.
Finally, we would like to thank all participants and authors of the proceedings for their support and contributions to the conference.

We devote this volume to the memory of Prof. Daniel Pardo, our dear boss in Salamanca for many years, who passed away in 2013. He was the pioneer of our activity in the field of the conference and would have enjoyed a lot the celebration of EDISON in Salamanca.

The Editors
Tomás González, María Jesús Martín-Martínez, and Javier Mateos
Universidad de Salamanca, Spain

Organizing Committee (University of Salamanca)

Tomás González (Chair)

María J. Martín (Scientific Secretary)
Susana Pérez (Local Management)
Enrique Díez
Sergio García
Óscar García
Jose Manuel Iglesias
Ignacio Íñiguez-de-la-Torre
Javier Mateos
Yahya Meziani
Elena Pascual
Raúl Rengel
Vincent Talbo
Beatriz G. Vasallo
J. Enrique Velázquez

Scientific Program Committee

I. Cirac	Max Planck Institute of Quantum Optics	Germany
A. di Carlo	University of Rome	Italy
L. Eaves	University of Nottingham	U.K.
D. K. Ferry	Arizona State University	U.S.A.
T. Gonzalez	University of Salamanca	Spain
K. Hirakawa	University of Tokyo	Japan
A. P. Jauho	Technical University of Denmark	Denmark
K. Kalna	University of Swansea	U.K.
T. Kuhn	University of Münster	Germany
T. Otsuji	University of Tohoku	Japan
G. Platero	ICMM-CSIC	Spain
R. Sobolewski	University of Rochester	U.S.A.
A. M. Song	University of Manchester	U.K.
L. Varani	University of Montpellier	France

International Advisory Committee

Journal of Physics: Conference Series **647** (2015) 011003 doi:10.1088/1742-6596/647/1/011003

SPONSORS

VNiVERSiDAD D SALAMANCA
CAMPUS OF INTERNATIONAL EXCELLENCE

ROHDE & SCHWARZ

OXFORD INSTRUMENTS
The Business of Science®

GOBIERNO DE ESPAÑA
MINISTERIO DE ECONOMÍA Y COMPETITIVIDAD

Peer review statement

All papers published in this volume of *Journal of Physics: Conference Series* have been peer reviewed through processes administered by the proceedings Editors. Reviews were conducted by expert referees to the professional and scientific standards expected of a proceedings journal published by IOP Publishing.

Graphene-InSe-graphene van der Waals heterostructures

S. A. Svatek[1], G. W. Mudd[1], Z. R. Kudrynskyi[1], O. Makarovsky[1], Z. D. Kovalyuk[2], C. J. Mellor[1], L. Eaves[1], P. H. Beton[1], and A. Patanè[1]

[1] School of Physics and Astronomy, The University of Nottingham, Nottingham, NG7 2RD (UK)

[2] Institute for Problems of Materials Science, The National Academy of Sciences of Ukraine, Chernivtsi, 58001 (Ukraine)

amalia.patane@nottingham.ac.uk

Abstract. We exploit the broad-band transparency of graphene and the favorable energy band line-up of graphene with *n*-type InSe thin films to create graphene-InSe-graphene heterostructures with high photosensitivity at room temperature. We use a semiclassical model to describe the photoresponse and its dependence on the electron transit time through the InSe layer.

1. Introduction

Van der Waals (vdW) heterostructures are made by stacking thin exfoliated layers of two-dimensional crystals such as graphene, hexagonal boron nitride, and the transition metal dichalcogenides [1-2]. Because the component layers are held together by weak van der Waals forces, both the structural integrity and the electronic and optical properties of the individual atomic planes are maintained. These new materials are not only of fundamental interest, but have the potential for exploitation in novel electronic and optoelectronic devices.

Here we present vdW crystalline heterostructures based on graphene and the direct-band gap metal chalcogenide InSe, an exfoliable and stable semiconductor that extends the current library of vdW crystals. Recent reports of bendable photodetectors [3], large-scale image sensors [4], *p-n* LED junctions [5], and FETs [6] have demonstrated the potential of InSe for future technologies. In particular, our recent work [7] on graphene/InSe heterostructures has demonstrated that the weak optical absorbance of graphene makes this single atomic carbon layer a particularly attractive option for use as a transparent electrical contact compared to conventional metals [8]. Furthermore, the work function of graphene can be adjusted by the electric field effect, an important feature for adjustable band alignment at the interface with a vdW crystal.

We focus on a device architecture where the InSe and two graphene layers are "vertically" stacked. In this graphene/InSe/graphene heterostructure, the small separation (~ 100 nm) between the graphene electrodes leads to a short transit time for the electrons photogenerated in the optically active InSe, thus enabling high photoresponsivity at room temperature.

2. Samples and device fabrication

Our γ-InSe crystals were grown using the Bridgman method from a polycrystalline melt of $In_{1.03}Se_{0.97}$. The primitive unit cell contains three InSe layers each of which has a thickness of 8.320 Å and consists of four covalently bonded monoatomic sheets in the sequence Se-In-In-Se; along the *c*-axis, the primitive unit cell has a lattice constant of $c = 24.961$ Å and, within each *a*-*b* plane, atoms form hexagons with lattice parameter $a = 4.002$ Å (Figure 1). In its bulk form, InSe is a semiconductor with direct band gap energy ($E_g = 1.26$ eV at $T = 300$K) that can be increased due to quantum confinement by reducing the number of layers in the crystalline sheet [9]; also, InSe contains native donors due to In-interstitial atoms [10] and it retains a relatively high electron mobility (10^{-2}-10^{-1} m^2/Vs) [8] over an extended range of InSe thicknesses down to a few nanometers (~ 10 nm).

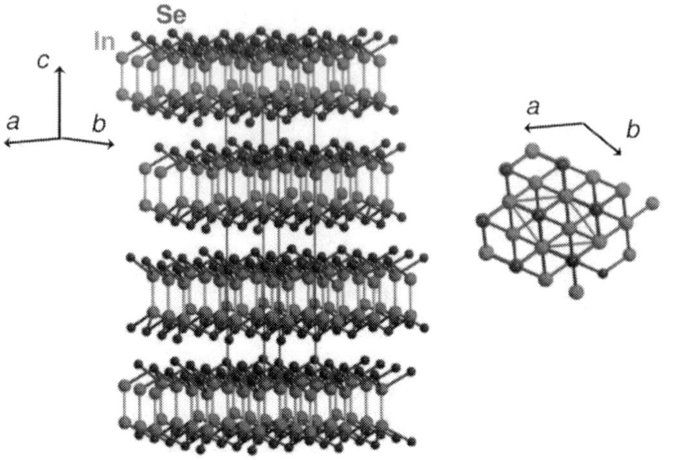

Figure 1. Crystal structure of γ-InSe van der Waals crystals.

The graphene/*n*-InSe/graphene heterostructures were fabricated using large area CVD graphene grown on copper. The main steps in the fabrication of these heterostructures are shown in Figure 2 and differ from those previously used to form vertical tunnel devices with exfoliated graphene [11].

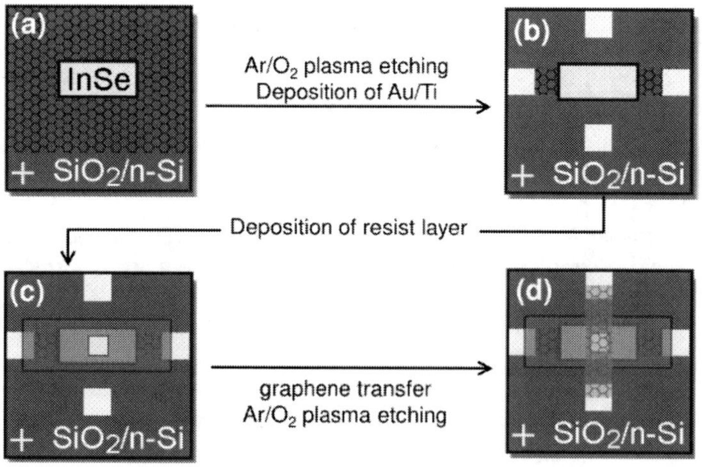

Figure 2. Main steps in the fabrication of graphene/*n*-InSe/graphene heterostructures.
a) An InSe flake is deposited on CVD graphene.
b) The graphene layer is patterned and etched to leave a graphene strip under the InSe flake. Au/Ti contact pads are then deposited.
c) An isolation layer, AR-N, is formed to cover the InSe flake and the lower graphene contact. A window is formed in the AR-N layer on the surface of the InSe.
d) The top graphene contact strip is defined by etching.

In our approach, following deposition of Au/Ti alignment marks on a SiO_2/Si substrate (SiO_2 layer thickness of 300 nm), CVD graphene was transferred by a wet transfer technique and cleaned. InSe flakes were deposited by exfoliation and a flake of thickness *t* was selected using optical microscopy (Figure 2a). The graphene layer was patterned and etched in an Ar/O_2 plasma to leave a graphene strip, which provided a continuous connection to the bottom of the InSe flake. Au/Ti contact pads (110 nm/15 nm) were then deposited at each end of the lower contact strip. In the same step Au/Ti pads were also formed to provide contacts to the upper graphene strip, which was deposited in a later fabrication step (Figure 2b). An isolation layer was then formed by exposing a negative resist, AR-N, using electron beam lithography; this covered the lower graphene contact including the edges of the InSe flake, to avoid the formation of electrical shorts between upper and lower graphene strips. A window with an area of approximately 2 μm × 2 μm was formed in the AR-N layer on the top surface of the InSe to facilitate the mechanical contact with the upper graphene layer (Figure 2c). Finally, a second layer of graphene was transferred; in this case the supporting PMMA film, which was added to the graphene/copper before etching in $FeCl_3$, was not removed, but instead used as the resist layer in a further stage of electron beam lithography in which the upper graphene contact strip was defined by etching in an Ar/O_2 plasma. The resulting upper contact runs at right angles to the lower strip to form a graphene/*n*-InSe/graphene vertical heterostructure and overlaps the Au/Ti contacts formed at an earlier stage of the process (see Figure 2d).

3. Results and discussion

Figure 3a illustrates the schematic layered structure and an optical image of a graphene-*n*-InSe-graphene device. Here the top- (g_1) and bottom-graphene (g_2) layers overlap an InSe nanoflake of thickness t = 130 nm and serve as source and drain to the InSe conducting channel. As shown in Figure 3b, the dependence of the dark current, I, on the bias voltage, V_s, applied between the two graphene electrodes is non-linear. A non-linear V_s-dependence is also observed for the photocurrent ΔI (Figure 3c).

The non-linearity and asymmetry of the current-voltage characteristics is not observed in planar graphene-*n*-InSe-graphene devices [7] and may arise from a spurious contact resistance at the InSe/graphene interface caused by surface defects introduced during the fabrication process. In fact, the upper contact to InSe is formed through mechanical transfer of CVD graphene onto a hole within an isolation layer (AR-N), which is formed from a negative resist. In this case the top InSe surface is coated with resist, which is subsequently removed during the development process. In addition, the upper graphene layer is transferred by floating the PMMA/graphene on water after a cleaning cycle, and then submerging the target substrate under the PMMA/graphene in order to scoop it up. In this process there is scope for residual resist layers to remain on the InSe surface; also, the InSe layer is immersed in water; both these processes may compromise the top graphene/InSe interface. Nevertheless, the implementation of these fabrication processes to realize a working vdW heterostructure device represents significant progress and offers a route to the device scalability using large area CVD graphene.

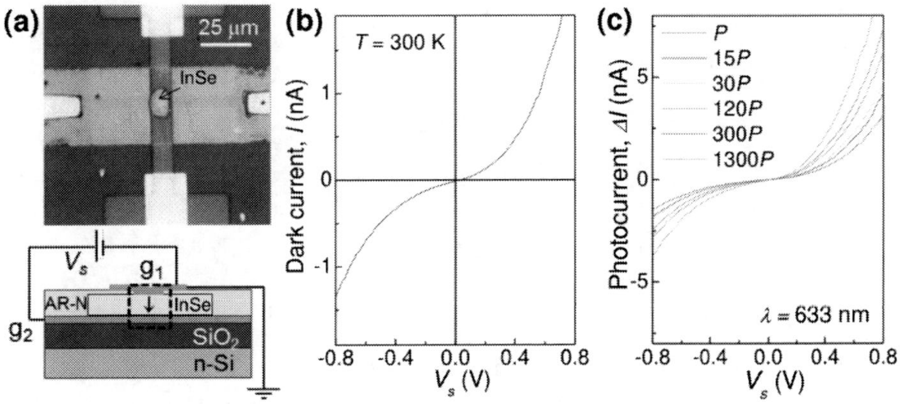

Figure 3. a) Schematic structure and optical image of a graphene-*n*-InSe-graphene vertical device. The graphene layers, g_1 and g_2, are separated by an InSe flake of thickness t = 130 nm. b) Dark current, I, and c) photocurrent, ΔI, versus applied bias V_s at T = 300 K. The photocurrent is measured with a laser beam of wavelength λ = 633 nm ($P = 10^{-13}$ W).

As shown in Figure 4, our heterostructure exhibits a stable and reproducible photoresponsivity R with values of up to 10^4 A/W at $\lambda = 633$ nm and $V_s = 1$V, and response to light at powers P as low as $\sim 10^{-13}$ W over an area of the InSe flake of $\sim 4\ \mu m^2$. Also, a power law relation of the form $R = CP^{-n}$, with $n = 2/3$ where C is a constant, provides a good empirical fit to the data (see lines in Figure 4).

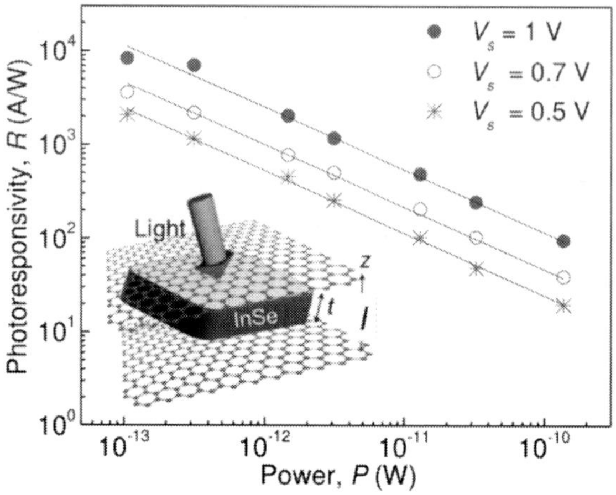

Figure 4. Photoresponsivity versus laser power at $T = 300$ K and $\lambda = 633$ nm, and various applied voltages V_s. The lines describe the empirical power law, i.e. $R = CP^{-n}$, where $n = 2/3$ and C is a constant that depends on the applied voltage V_s.

To model the photoresponsivity, we consider an n-type InSe channel of width w, thickness t, and length l, with ohmic (graphene) contacts across its thickness t (inset of Figure 4). When light of power P and energy $h\nu$ larger than the band gap energy of InSe falls on the surface of the InSe flake, electron-hole pairs are generated. The densities of photo-created electrons and holes are equal and, at steady state, they can be expressed in terms of the minority carrier (hole) lifetime, τ_l, and the rate of carrier generation by light G, *i.e.* $\Delta n = \Delta p = G\tau_l$ [12], where $G = (t\alpha P/h\nu)(wtl)^{-1}$ and α is the absorption coefficient of InSe at the energy $h\nu$. Since the mobility μ of electrons is much larger than that of holes, we express the photocurrent ΔI in terms of the dominant electron current, $\Delta I = \Delta n e v_d w l$, where $v_d = t/\tau_t$ is the electron drift velocity and τ_t is the electron transit time across an InSe channel of thickness t. Thus we can express ΔI as

$$\Delta I = (et\alpha P/h\nu)(\tau_l/\tau_t), \quad (1)$$

from which we derive the photoresponsivity R, the external quantum efficiency EQE, and the internal quantum efficiency IQE:

$$R = \Delta I/P = (et\alpha/h\nu)(\tau_l/\tau_t), \quad (2)$$

$$EQE = R(h\nu/e) = (t\alpha)(\tau_l/\tau_t), \quad (3)$$

$$IQE = EQE/t\alpha = \tau_l/\tau_t. \quad (4)$$

From equations 2-4, it can be seen that R, EQE, and IQE are all proportional to the ratio between the minority carrier lifetime and electron transit time, τ_l/τ_t. In particular, a small separation t between the graphene electrodes and/or a larger applied bias V_s should lead to a faster electron transit time, $\tau_t = t^2/\mu V_s$, and hence larger values of R, EQE ($\sim V_s/t$) and IQE ($\sim V_s/t^2$). For $V_s = 1$V, $t = 130$ nm and an electron mobility $\mu = 0.1$ m^{2}V^{-1}s^{-1}, we estimate an electron transit time $\tau_t \sim 1.7\times10^{-13}$ s. Thus a value of

$R \sim 10^4$ A/W, measured at the lowest incident power P, corresponds to $\tau_l/\tau_t \sim 1.5\times10^5$ and $\tau_l \sim 2.6\times10^{-8}$ s for $\alpha = 10^6$ m^{-1} at $h\nu = 1.96$ eV.

As shown in Figure 4, the photoresponsivity depends on the optical power P. The measured decrease of R with increasing P suggests a decrease of τ_l and/or an increase of τ_t. The transit time of photo-induced carriers may be increased by the power due to enhanced scattering. Also, an increasing power can induce Auger recombination processes and increases the carrier recombination rate, thus reducing the carrier lifetime τ_l.

The measured photoresponsivity is significantly larger than that reported for photodetectors based on graphene and/or vdW 2D crystals at similar wavelengths and laser powers, which do not exceed values of 10^3 A/W [13]. Also, we note that although a higher photoresponsivity can be achieved in phototransistors based on graphene and colloidal nanocrystals [14], the performance of such hybrid photodetectors tends to be compromised by a slow (~ 1s) optical response due to the slow escape rate of photogenerated charges confined in the nanocrystals. In contrast, for our structures, the mechanism responsible for the photoresponse does not rely on a charge trapping effect as light generates free carriers. These features of our heterostructures enable a modulation of the dark and photocurrent at shorter (~ms-μs) time scales.

4. Conclusion

In summary, our findings demonstrate that mechanically formed heterojunctions of InSe and graphene have optical and electrical properties with potential for applications in optoelectronics. Our innovative fabrication methods could be extended to other material systems and device architectures. It should be possible to achieve different band alignments by combining InSe with other layered crystals, such as semiconducting gallium and indium chalcogenides (*e.g.* InX and GaX, where X = S, Se, Te). These layered semiconductors have direct band-gaps covering a wide wavelength range from the visible to the near-infra-red (NIR) (0.5-1 μm), and provide a class of 2D compounds for versatile band-gap engineering. In combination with transparent, conducting graphene, these new heterostructures could be used to create miniaturized device structures for 2D electronics and optoelectronics compatible with a wide variety of materials including dielectric, polymer and flexible substrates.

Acknowledgements
We thank Neil Wilson and Alex Marsden (Department of Physics, Warwick University) for the graphene grown by CVD, which was used in these studies. This work was supported by the Engineering and Physical Sciences Research Council (EPSRC), the National Academy of Sciences of Ukraine, and the EU Graphene Flagship Programme.

References

[1] A. K. Geim, I. V. Grigorieva, Nature 2013 **499,** 419-425.
[2] A.C. Ferrari et al., Nanoscale 2015, **7**, 4598.
[3] R. T. Srinivasa et al., Nano Lett. 2014, **14**, 2800.
[4] S. Lei et al., Nano Lett. 2015, **15**, 259.
[5] N. Balakrishnan et al., Adv. Opt. Mater. 2014, **2**, 1064.
[6] W. Feng, W. Zheng, W. Cao, and P. A. Hu, Adv. Mater. 2014, **26**, 6587.
[7] G.W. Mudd et al., Adv. Materials 2015, DOI: 10.1002/adma.201500889.
[8] W. Feng, X. Zhou, W. Q. Tian, W. Zheng, and P.A. Hu, Phys. Chem. Chem. Phys. 2015, **17**, 3653.
[9] G.W. Mudd et al., Adv. Materials 2013, **25**, 5714.
[10] G.W. Mudd et al., Appl. Phys. Lett. 2014, **105**, 221909.
[11] C. R. Dean et al., Nature 2013, **497**, 598.
[12] S.M. Sze, Semiconductor devices, Physics and Technology, 2nd edition, pages 61-62.
[13] F. Xia, H. Wang, D. Xiao, M. Dubey, and A. Ramasubramaniam, Nat. Photon. 2014, **8**, 899.
[14] L. Turyanska et al., Adv. Electr. Mat. 2015, DOI: 10.1002/aelm.201500062.

PROTEOTRONICS: The emerging science of protein-based electronic devices

Eleonora Alfinito[1], **Jeremy Pousset**[2] **and Lino Reggiani**[3]
[1]Dipartimento di Ingegneria dell'Innovazione, Università del Salento, via Monteroni, I-73100 Lecce, Italy
[2]Istituto per la Microelettronica e Microsistemi (IMM), Consiglio Nazionale delle Ricerche, I-73100 Lecce, Italy
[3]Dipartimento di Matematica e Fisica "Ennio De Giorgi", Università del Salento, via Monteroni, I-73100 Lecce, Italy

E-mail: `lino.reggiani@unisalento.it`

Abstract. Protein-mediated charge transport is of relevant importance in the design of protein based electronics and in attaining an adequate level of understanding of protein functioning. This is particularly true for the case of transmembrane proteins, like those pertaining to the G protein coupled receptors (GPCRs). These proteins are involved in a broad range of biological processes like catalysis, substance transport, etc., thus being the target of a large number of clinically used drugs. This paper briefly reviews a variety of experiments devoted to investigate charge transport in proteins and present a unified theoretical model able to relate macroscopic experimental results with the conformations of the amino acids backbone of the single protein.

1. Introduction

Proteotronics was introduced as a new science by the Authors in 2014 [1]. The word originates from the combination of proteomics, introduced in 1996 by Marc Wilkins to indicate the large-scale study of proteins structures and functions, and electronics, introduced by Thomas Edison in 1883, to indicate the development of devices manipulating electrical current to perform useful tasks. The main objective of proteotronics is to propose and develop innovative electronic devices, based on the selective action of specific proteins. Here, we briefly survey the pillars of this emerging science by reporting a series of experiments appropriate to investigate the electrical properties of a given protein and a theoretical approach based on an impedance network protein analogue (INPA) able to provide a unifying microscopic interpretation of the experimental data. For a comprehensive review on the subject readers are sent to Ref. [2]

2. Physical system of interest

Proteins appear like linear chains of lego-like building blocks (the amino acids) able to swiftly (few microseconds) organize in space so as encoded in their sequence. In other words, proteins are polymers constituted by twenty different types of amino acids (the alphabet of proteins) with a sequential length containing $20 \div 1000$ amino acids. Figure 1 shows a segment structure of a typical transmembrane protein, bovine rhodopsin, sensitive to light. Here we are interested in sensing proteins that under an external stimulus (e.g., light, odours, drugs, etc) change their tertiary structure going from a native to an activated state. The conformational change is crucial

Journal of Physics: Conference Series **647** (2015) 012002 doi:10.1088/1742-6596/647/1/012002

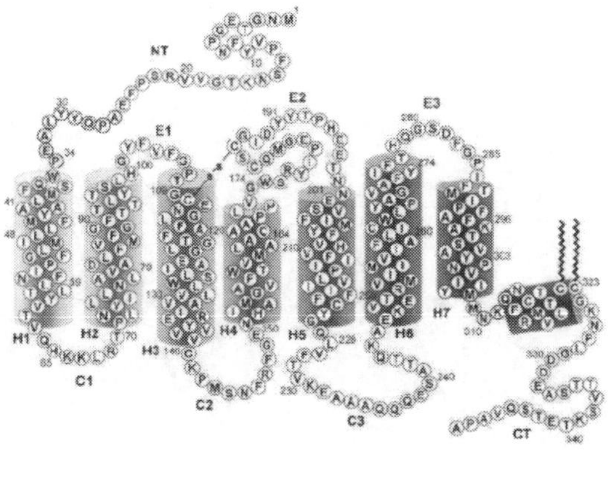

Figure 1. Segment structure of bovine rhodopsin.

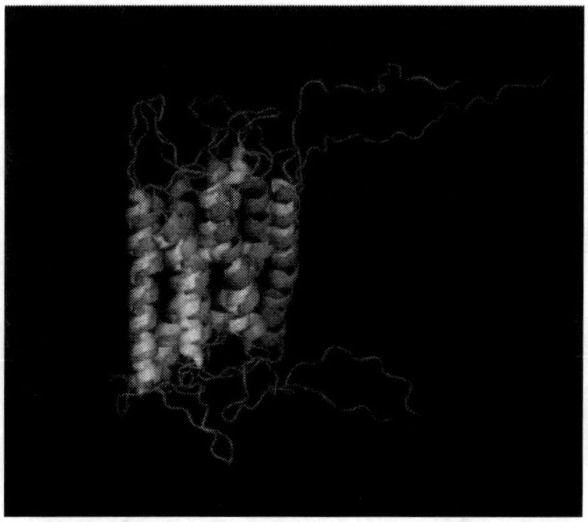

Figure 2. OR-7D4 in its ribbon representation: native state in purple and activated state in green.

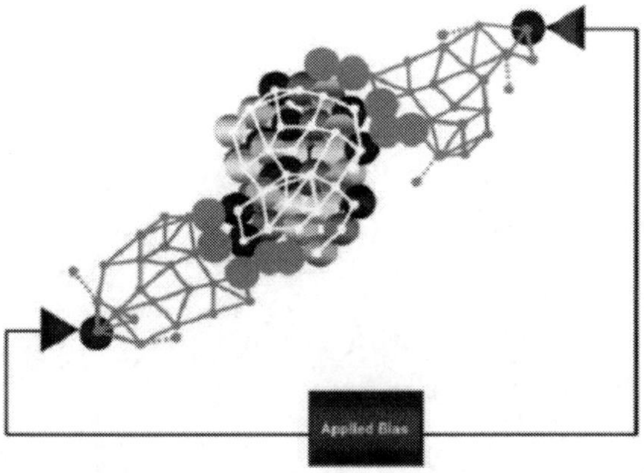

Figure 3. Cartoon of a protein embedded into the analogous impedance network.

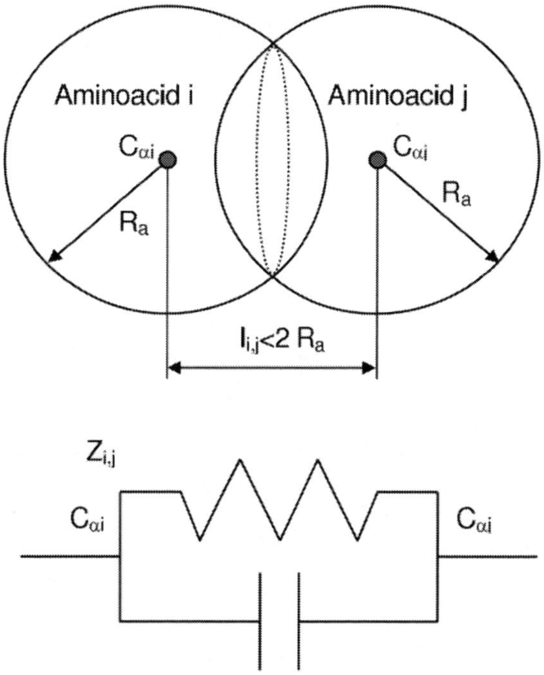

Figure 4. Overlap between amino acids and equivalent-circuit element. $C_{\alpha i}$ identifies the center of the sphere corresponding to the α-carbon atom of the i-th amino acid.

for the external/internal communication of a living cell. Figure 2 reports the ribbon structure of an olfactory receptor (OR), the chimpanzee OR-7D4, in its native (purple) and activated (green) state. In this respect, a sensing protein is the ultimate miniaturization of a sensor. In

Journal of Physics: Conference Series **647** (2015) 012002 doi:10.1088/1742-6596/647/1/012002

nature, the information associated with the conformational change moves to the brain through a biological chain. This abiity to detect the change of an electrical property of the single protein correlated to the conformational change is crucial for the development of a nano-bio-sensor, a significant example of proteotronics [2].

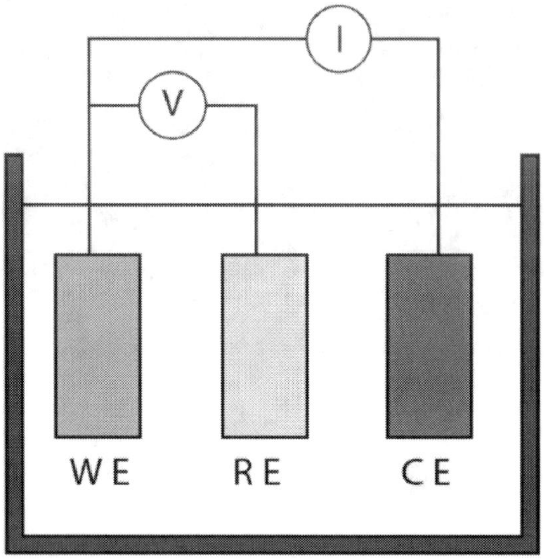

Figure 5. Schematic of a three electrodes electrochemical-cell. WE, RE and CE refer, respectively, to working, reference, and control electrodes.

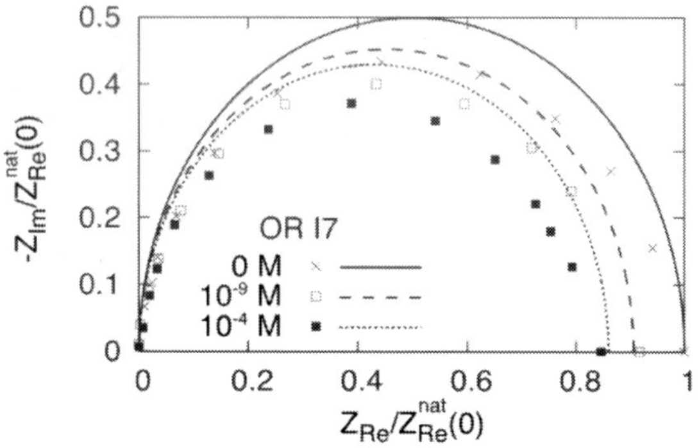

Figure 6. Nyquist plot of the rat OR-I7 in the presence of different concentrations of the specific odorant. Curves refer to INPA calculations, symbols to experiments.

3. Theoretical model

The theoretical model implements an impedance protein network analogue (INPA) where the 3D backbone representation of a single protein is associated with a topological network (graph), consisting of nodes and links, as depicted in Fig. 3. By connecting couples of nodes with the rule that they must be closer than an assigned interaction radius, R_a, the graph turns into an impedance network when an elemental impedance is associated with each link. The elemental impedance is taken as that of an RC parallel circuit, as shown in Fig. 4,

$$Z_{i,j} = \frac{l_{i,j}}{\mathcal{A}_{i,j}} \frac{1}{(\rho^{-1} + \mathrm{i}\epsilon_{\mathrm{i,j}}\,\epsilon_0\omega)} \tag{1}$$

where $\mathcal{A}_{i,j} = \pi(R_a^2 - l_{i,j}^2/4)$, is the cross-sectional area between the spheres of radius R_a centered on the i, j nodes, respectively, $l_{i,j}$ is the distance between these centers, $\rho = 10^{10}$ Ωm the resistivity of each amino acid, with the indicative value following the fit with experiments, (the microscopic mechanism responsible of charge transfer is taken to be of tunneling type), $\mathrm{i} = \sqrt{-1}$ is the imaginary unit, ϵ_0 is the vacuum permittivity and ω is the circular frequency of the applied harmonic voltage. The relative dielectric constant pertaining to the couple of i, j amino acids, $\epsilon_{i,j}$, is expressed in terms of the intrinsic polarizability of the i, j amino acids.

By positioning the input and output electrical contacts on the first and last node, respectively, for a given applied bias the network is solved within a linear Kirchhoff scheme and its global impedance spectrum, $Z(\omega)$, is calculated in the frequency range $10^{-3} \div 10^5$ Hz which is standard for experiments. By construction, besides the small-signal dynamic response, this network produces a parameter dependent static current-voltage (I-V) characteristic determined as:

$$V = Z(0)I \tag{2}$$

To this purpose, we assume that the 3D structures of the protein in its native (without ligand) and activated (with ligand) states are known from the protein data bank available in the literature or from an homology modeling. The role of the medium (lipids) surrounding the

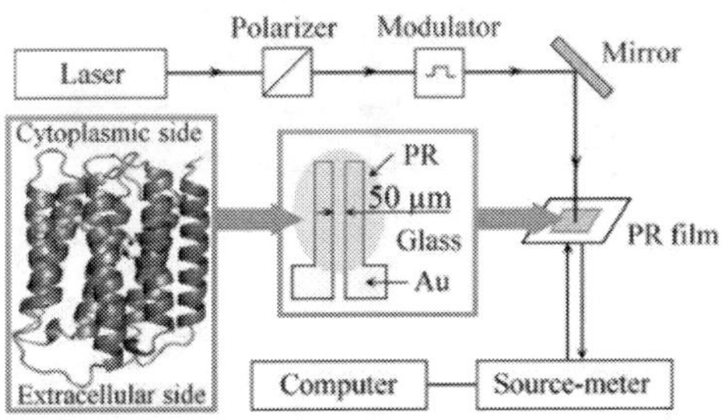

Figure 7. Thin film experimental set-up, sample configuration and 3D structure of pR drawn as a ribbon diagram with the retinal chromophore [3].

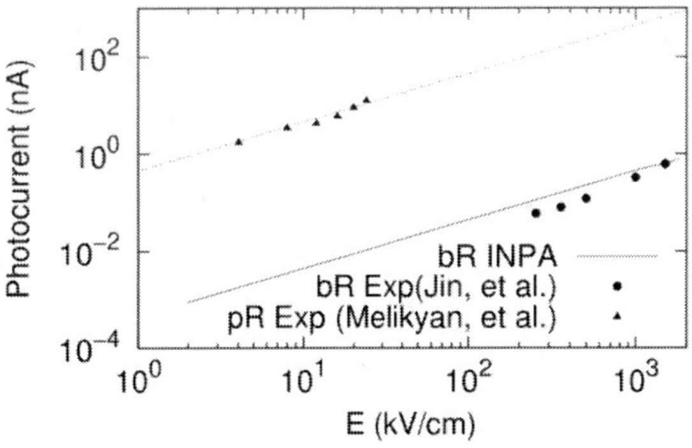

Figure 8. Photocurrent as a function of the applied electric field. Full circles and triangles represent experimental results respectively for bR [3] and pR [4]. Continuous line represents results of simulations carried out within the INPA model, dashed line represents a linear fitting of pR experiments.

protein is here neglected. Indeed, the medium is not involved in the conformational change and so it does not imply a variationn of the global impedance.

4. Results and discussions

This section briefly outlines the main experimental techniques and the associated experimental data available from the literature together with the microscopic modeling carried out with the INPA technique.

Figure 5 shows the schematic of the electrochemical impedance spectroscopy (EIS) carried out on a self assembled monolayer of a given protein with a macroscopic area of a few cm^2 at a small applied voltage in the mV range and in the frequency range $10^{-3} \div 10^5$Hz. Figure 6 reports the impedance of the rat OR-I7 at different odorant concentrations and represented as a Nyquist plot.

Figure 7 reports the set-up for an I-V characteristic carried out on thin-films of proteorhodopsin (pR) in the voltage range $0 \div 100$ V corresponding to average electrc fields in the range $1 \div 100$ kV/cm and in the presence or absence of a visible light. Figure 8 reports the photocurrent of two proteins, bacterio-rhodopsin (bR) and pR belonging to the family of opsins, that are sensitive to visible light [5]. The experiments are successfully described by the INPA model or by a simple linear fit inspired by the theoretical model..

Figure 9 reports the set-up for I-V measurements on monolayers of single protein with a macroscopic surface area of a few cm^2. The contacts are designed to permit the illumination of the sample by visible light. Figure 10 reports a set of I-V characteristics in the range $-1 \div 1$ V for the case of bR together with the theoretical interpretation obtained by the INPA model.

Figure 11 is a schematic of the atomic force microscope (AFM) set-up used for the I-V measurements on protein monolayers properly assembled on a conductive substrate in the range $-10 \div +10$ V. Figure 12 reports a set of results carried out on bacterio-rhodopsin samples for three identations of the tip, 1.2, 2.8 and 4.6 nm from the bottom contact, with the theoretical interpretation performed using the INPA model..

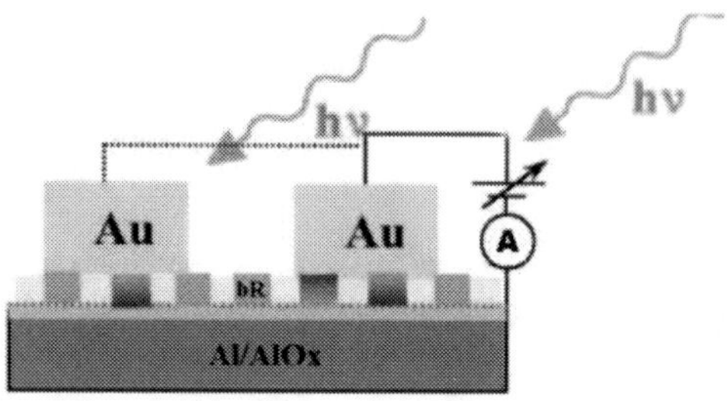

Figure 9. Schematic of the metal-protein-metal nanostructure for I-V measurements in the presence of light.

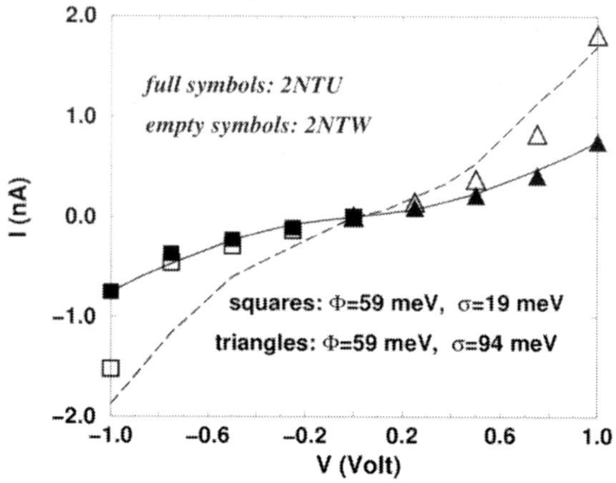

Figure 10. I-V characteristics of bR in the native (2NTU) and activated (2NTW)states. Data were calculated by using a Gaussian distribution of barrier heights between amino acids. The continuous and dashed lines report the experimental data in the native and activated state, respectively.

5. Conclusions and perspectives

Proteotronics is an emerging science aiming at investigating the coupling between the protein world (proteomics) and electronics. In particular, major results can be summarized as follows. (i) The correlation between the 3D structure of proteins and their sensing properties is confirmed from an experimental and a theoretical point of view. (ii) The electrical properties of a single protein is found to be similar to that of a medium-gap semiconductor with Ohmic resistivities estimated in the range of 10^5 Ωcm at room temperature. (iii) The current-voltage characteristics are found to exhibit a superlinear behavior at increasing applied voltages, which is associated with a sequential tunneling mechanism of charge transfer between neighbouring amino acids. Typical barrier heights are in the range $50 \div 300$ meV and barrier width lower than 1 nm. (iv) The electrical change associated with the conformational change following a sensing action of the protein is found to be in the range from a few per cent up to about a factor of two in magnitude. This variation can be positive or negative and, from a theoretical interpretation, is consistent with the change in the spatial location of the amino acids of the single protein. In a conformational change, the variation in the average volume of a protein is estimated to be of a few per cent. (v) A microscopic modeling based on an impedance network protein analogue (INPA) is found to provide a unified interpretation of experiments available from literature. Once the tertiary structure of a protein is given, INPA is useful to visualize the amino acids interconnection through the construction of the contact maps and to predict the electrical properties in a wide range of applied bias. INPA is actually a numerical code based on a Monte Carlo technique and suitable for further implementation to include advances in the knowledge and characterization of proteins. Yet, the scarcety od data and the simplifications in the model reflect the discepancies between numerical and experimental findings which, for the first step of an emerging science, is a plausible result.

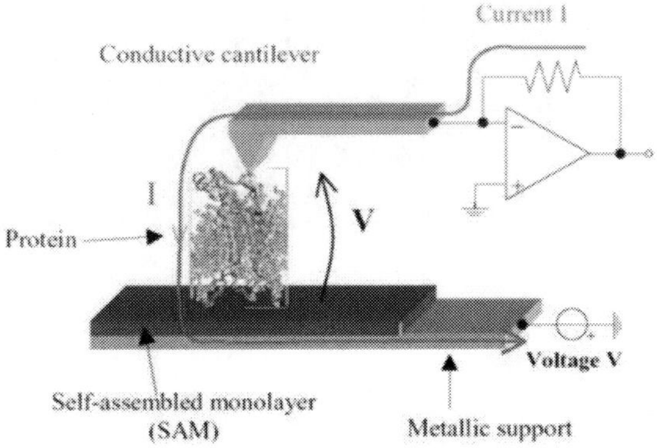

Figure 11. Schematic of the AFM set-up including the detection.

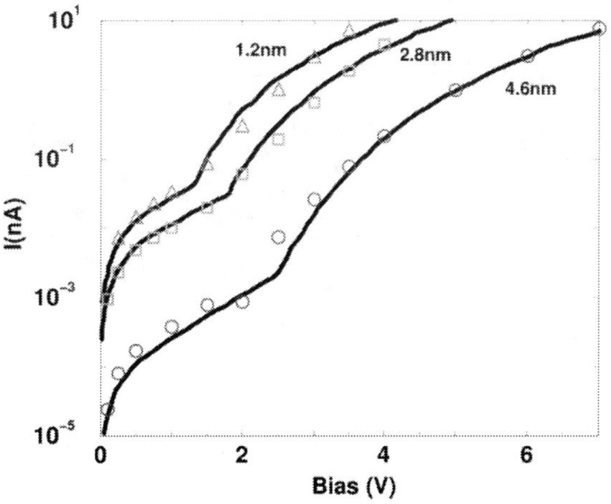

Figure 12. I-V characteristics obtained by simulations with extended contacts at different indentations in nanometers including a leakage current. The thick continuous lines refer to experimental data [6] and the symbols to INPA numerical simulations.

(vi) A first attempt to develop a new generation of nano-bio-sensors using the sensing property, manifested by a conformational change, into an electrical response is reported. A proof of concept on the viability of such an attempt was successfully demonstrated for several sensing proteins (olfactory receptors: rat OR I7, human OR 17-40, light receptors: bacterio rhodopsin, proteo rhodopsin, etc.). As a final comment, we stress that proteotronics can be considered as a branch of *pantatronics*, a new term announced here to underline the historical development and the overwhelming trend of electronics as applied to a multitude of fields of scientific/technological interest (atoms, chemistry, mechanics, molecules, solids, spin, etc.).

References

[1] Alfinito E, Pousset J, and Reggiani L 2015 *Proteotronics: electronic devices based on proteins* Sensors, Lecture Notes in Electrical Engineering Vol. 319, edited by D. Compagnone et al. (Springer International Publishing, 2015), 3.

[2] Alfinito E, Pousset J, and Reggiani L 2015 *Proteotronics: Development of Protein-Based Electrononics* Pan Stanford Publishing Pte. Ltd. Penthouse Level, Suntec Tower 38 Temasek Boulevard Singapore 038988, in press.

[3] Jin Y, Friedman N, Sheve M, He T, and Cahen D 2006 *Bacteriorhodopsin (bR) as an electronic conduction medium: Current transport through bR-containing monolayers* Proceedings of the National Academy of Sciences 103, no. 23, 8601.

[4] Melikyan H, Khishigbadrakh B-E, Babajanyan A, Lee K, Choi A-R, Lee J-H, Jung K-H, and Friedman B 2011 *Proteorhodopsin characterization based on metalinsulatormetal structure technique* Thin Solid Films 519, 3425.

[5] Alfinito E, Pousset J, Reggiani L and Lee K 2013, *Photoreceptors for a light biotransducers* Nanotechnology24 39551.

[6] Casuso I, Fumagalli L, Samitier J, Padrs E, Reggiani L, Akimov V, and Gomila G 2007 *Nanoscale electrical conductivity of the purple membrane monolayer* Physical Review E 76, no. 4, 041919.

Journal of Physics: Conference Series **647** (2015) 012003 doi:10.1088/1742-6596/647/1/012003

Carrier-carrier and carrier-phonon interactions in the dynamics of photoexcited electrons in graphene

J. M. Iglesias, M. J. Martín, E. Pascual and R. Rengel

Applied Physics Department, University of Salamanca, Plaza de la Merced, S/N, Salamanca 37008, Spain

E-mail: josem88@usal.es

Abstract. In this work we use the Monte Carlo method to simulate the photoexcited carrier relaxation dynamics in suspended monolayer graphene to unveil the role of the different scattering mechanisms involved. The results show a strong activity of carrier-carrier and carrier-optical phonon interactions in the very early stages of the relaxation process driving the thermalization and cooling dynamics. The consideration of the coupling between the electron and phonon population reveals an intense damping in the cooling rates that is dependent on the pumped carrier concentration.

1. Introduction

Due to its outstanding electronic and optical properties[1] graphene has become a promising material for future developments of a wide range of highly efficient photonic and optoelectronic applications. Its gapless linear band structure makes graphene particularly interesting for novel devices working in the THz spectral range[2, 3]. Experiments comprising pump-probe differential transmission spectroscopy with time resolutions in the femtosecond scale [4–6] provide, from a macroscopic interpretation, insight into the far-of-equilibrium hot carrier dynamics that takes place after photoexcitation. This dynamics can be split in intraband and interband processes. The first consists of a very fast thermalization of the carriers driven by the energy-conserving dual carrier scattering which spreads the electron distribution around the highly populated energy level associated with the wavelength of the optical excitation, leading to a very fast depletion thereof, accompanied with a simultaneous cooling starred by the intrinsic and substrate phonons cascade emissions as the main relaxation mechanisms. Electron-hole recombination completes this carrier dynamics towards the full thermodynamic equilibrium.

Here we will study the intraband thermalization and cooling by means of ensemble Monte Carlo simulations. This methodology allows to tackle the electron dynamics from a particle approach: the system is described by a number of superparticles that resembles a group of electrons (or holes), which are simulated one by one during consecutive time steps, thus allowing the tracking of different quantities, such as the average carrier energy, distribution functions in energy and K-space or the scattering mechanisms undergone per electron. The main features of the model are described in previous works.[7, 8] The contributions of the interactions driving the described processes are the transverse-acoustic (TA), longitudinal-acoustic (LA), transverse-optic (TO), longitudinal-optic (LO) phonon branches. In addition, impurity and imperfection (defects) scattering are included; however, they are disregarded in the present work due to their single-particle elastic nature. Due to the strong coupling between carriers in graphene, short range carrier-carrier (e-e) scattering is modelled through a static Coulomb screened potential model[9]. The time step in the simulation is set to 0.5 fs. The scattering probabilities are updated every time step. Screening potentials for impurities and electrons are considered by numerically calculating the Lindhard dielectric function in the static limit[10] from the actual energy distribution function obtained in the simulation in order to grant an adequate description of the polarizability under the initial far-from-equilibrium state. The graphene sample is considered to be suspended. Finally, the effect of the hot electron-hot phonon coupling is also incorporated. The population of each phonon mode is updated every time step on a gridded q-space accounting for an histogram of emissions and absorptions and a decay within a relaxation time approximation for lifetimes $\tau_{op} = 2.5$ ps[11] and $\tau_{ac} = 5$ ps[12].

We perform simulations for various pumping energies and concentrations (resulting of different pump fluences), setting an initial carrier distribution according to the expression $f(\varepsilon) = F(\varepsilon, T, \mu = 0) + g_{max} \exp\left[-(\varepsilon - \varepsilon_{photon}/2)^2/2\sigma^2\right]$, where the first member of the RHS is the Fermi equilibrium distribution for intrinsic graphene ($\mu = 0$) and the second represents the excited carriers arising from optical pumping of a light source with an spectral bandwidth peaked at ε_{photon} and σ accounting for a gaussian Fourier bandwidth limited pulse. Possible puddle states were not considered. The parameter g_{max} is the maximum occupation at $\varepsilon = \frac{1}{2}\varepsilon_{photon}$ and relates the pumped concentration with the radiation fluence and the pulse bandwidth.

2. Results

In the first place, we examine the time evolution of the energy distribution function in a generic case to discern the role of the interactions involving the mentioned relaxation dynamics. Figure 1-a shows the evolution of the carrier energy distribution function vs. time, when only e-e interactions are activated, and phonon scattering is disregarded. The results stand out the fast depletion of the electron population at the pumped energy level, since at $t = 100$ fs there is no trace of the initial gaussian distribution. Under these circumstances, carriers spread around the pumped energy level until reaching a thermal distribution at approximately $t = 250$ fs. Let us now check this evolution if e-e interactions are turned off and phonon scattering is activated (fig. 1-b). Cascade phonon emission is clearly visible as the initial pumped gaussian population peak replicates itself at lower energies with a separation corresponding to the energy of the intrinsic optical phonon (165 meV). Finally, fig. 1-c illustrates the dynamics with the full

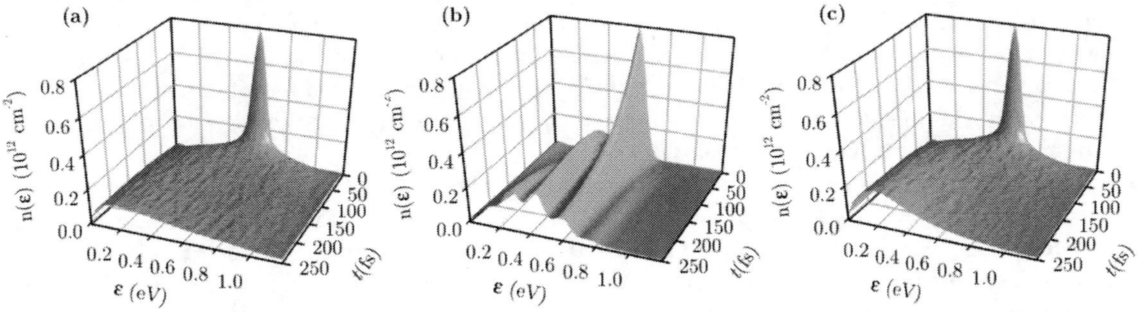

Figure 1. Energy distribution function vs. time obtained from Monte Carlo simulations of (a) the carrier thermalization (phonon scattering turned off), (b) carrier cooling (electron-electron interaccions turned off) and (c) full carrier dynamics during the first 250 fs. Photon energy is $\hbar\omega = 1.2$ eV and $n_s = 5 \times 10^{12}$ cm^{-2}.

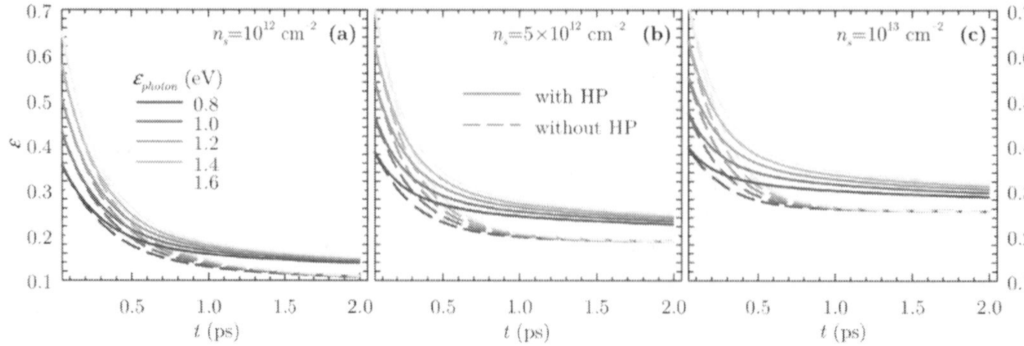

Figure 2. Time evolution of the average system energy considering equilibrium and hot phonon (HP) populations as a function of the pumped carrier density for excited concentrations (a) 10^{12} cm^{-2}, (b) 5×10^{12} cm^{-2}, and (c) 10^{13} cm^{-2}

intraband interactions, where e-e scattering and inelastic phonons interplay between each other: the depletion rate at the excitation energy obeys that of the "e-e only" case, while carriers tend to occupy lower energy levels as time passes as a result of phonon emission. A correct description of hot carrier dynamics involves the consideration of the coupling between charge carriers and the interacting phonons. In particular, hot electrons are expected to emit phonons that, in turn would be out of equilibrium (hot) as well. As a result, the cooling of the carriers is related to that of the phonons, which end up reaching a population distribution according to the lattice temperature thereof. To examine this effect, we have analysed a situation where the phonon population is considered to be in equilibrium and another case for which a hot phonon population (coupled to phonon emission and absorption by carriers) is taken into account. Fig. 2 shows the average system energy, $\varepsilon(t) = \sum_0^N \varepsilon_i(t)/N$ where $\varepsilon_i(t)$ is the energy of each superparticle at a given time and N is the total number of superparticles. In the first place, we observe that the decrease of the system energy is biexponential. Without considering the coupling between carrier and phonon population, all the curves for given n_s tend to overlap. For the lowest electron density ($n_s = 10^{12}$ cm^{-2}) the average energy continues decaying at 2.0 ps but for $n_s = 5 \times 10^{12}$ and $n_s = 10^{13}$ cm^{-2} the average energy reaches its minimum at around 1.8 and 1.4 ps respectively. When hot phonons are included in the dynamics, the initial decay rate is less pronounced as phonon emission instantaneously stimulates its re absorption, pushing carriers back towards higher energies and slowing down the process. Also, the later section of the curves show higher values and the average energies for different pumping wavelengths do not overlap for this time scale showed in the figure, remaining at higher values those of carriers excited with shorter wavelengths. This discrepancy between the hot-phonon and phonon at equilibrium cases is more evident with higher electron density.

One of the most relevant quantities provided by the Monte Carlo method is the scattering count over time, which sheds light on the role of each interaction driving carrier relaxation dynamics. In the fig. 3 we show the time evolution of the average number of intrinsic phonon and e-e scatterings per unit time undergone by a single carrier as a function of the pumping energy. For both scattering mechanisms the initial activity grows linearly with the pumping energy; however the rate of e-e scatterings is one order of magnitude higher. In all cases, e-e interaction drops faster in the first picosecond due to the reduction of the coupling of the excited population; the trend for both intrinsic phonon and e-e scattering is to increase with the photon energy, being this phenomenon more evident at increasing carrier densities. In the latest 500 fs of the sampled time, the scattering rates tend to become independent of the pumping energy,

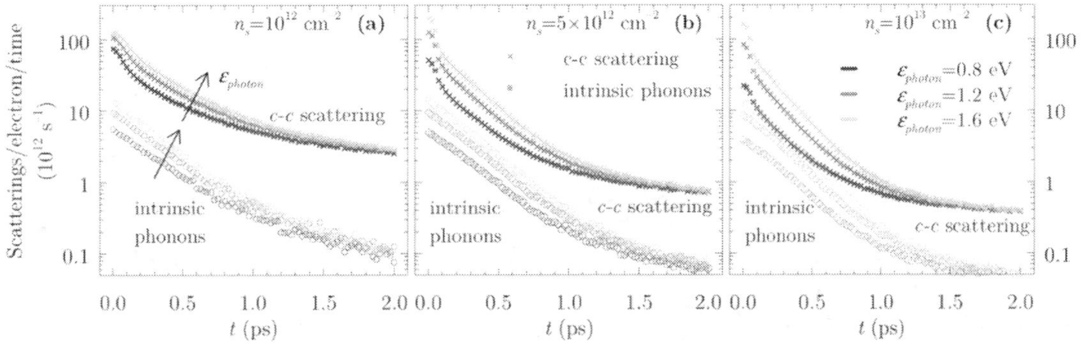

Figure 3. Time evolution of the average number of scatterings per unit time as a function of the pumping energy for (a) $n_s = 10^{12}$ cm^{-2} (b) $n_s = 5 \times 10^{12}$ cm^{-2} and (c) $n_s = 10^{13}$ cm^{-2}

since the system tends to the same quasy-equilibrium distribution functions. In this last section of the plot the carriers with more scattering activity correspond to the situations where the quasi-Fermi surface is smaller (reduced stimulated carrier concentrations).

3. Conclusions

We present a Monte Carlo model for the study of the early stages of the ultrafast intraband relaxation dynamics in monolayer graphene. The role of the most relevant scattering mechanisms (intrinsic phonons and e-e coulomb interactions) ruling this dynamics are disentangled along the time under investigation. The results show that high-energy electron stimulated heated phonon population results in a lower cooling rate of the carriers. Carrier-carrier scattering is the most active interaction and exhibits the fastest drop in the first tens of femtoseconds as a consequence of the depletion of the excited energy levels.

Acknowledgments

This work was supported by research project TEC2013-42622-R from the Ministerio de Economía y Competitividad.

References

[1] Novoselov K S, Geim A K, Morozov S V, Jiang D, Katsnelson M I, Grigorieva I V, Dubonos S V and Firsov A A 2005 *Nature* **438** 197–200
[2] Sensale-Rodriguez B, Yan R, Kelly M M, Fang T, Tahy K, Hwang W S, Jena D, Liu L and Xing H G 2012 *Nat Comms* **3** 780
[3] Ryzhii V, Dubinov A A, Otsuji T, Mitin V and Shur M S 2010 *J. Appl. Phys.* **107** 054505
[4] Sun D, Divin C, Mihnev M, Winzer T, Malic E, Knorr A, Sipe J E, Berger C, de Heer W A, First P N and Norris T B 2012 *New Journal of Physics* **14** 105012
[5] Winnerl S, Göttfert F, Mittendorff M, Schneider H, Helm M, Winzer T, Malic E, Knorr A, Orlita M, Potemski M, Sprinkle M, Berger C and de Heer W A 2013 *Journal of Physics: Condensed Matter* **25** 054202
[6] Tielrooij K J, Song J C W, Jensen S A, Centeno A, Pesquera A, Zurutuza Elorza A, Bonn M, Levitov L S and Koppens F H L 2013 *Nat Phys* **9** 248–252 ISSN 1745-2481
[7] Rengel R and Martín M J 2013 *J. Appl. Phys.* **114** 143702
[8] Rengel R, Pascual E and Maríin M J 2014 *Applied Physics Letters* **104** 233107
[9] Li X, Barry E A, Zavada J M, Nardelli M B and Kim K W 2010 *Applied Physics Letters* **97** 082101
[10] Hwang E H and Sarma S D 2007 *Physical Review B* **75**
[11] Wang H, Strait J H, George P A, Shivaraman S, Shields V B, Chandrashekhar M, Hwang J, Rana F, Spencer M G, Ruiz-Vargas C S and et al 2010 *Appl. Phys. Lett.* **96** 081917 ISSN 0003-6951
[12] Bonini N, Garg J and Marzari N 2012 *Nano Letters* **12** 2673–2678

The potential for sensitivity enhancement by the thermoelectric effect in carbon-nanotube and graphene Tera-FETs

M Bauer[1], M Andersson[2], A Zak[2], P Sakalas[3], D Čibiraitė[4], A Lisauskas[1,4], M Schröter[3], J Stake[2] and H G Roskos[1]

[1]Department of Physics, Goethe-University Frankfurt, Max-von-Laue-Strasse 1, DE-60438 Frankfurt, Germany
[2]Department of Michrotechnology and Nanoscience, Chalmers University, Kemivägen 9, SE-41296 Gothenburg, Sweden
[3]Chair for Electron Devices and Integrated Circuits, Technical University Dresden, Helmholtzstrasse 18, DE-01062 Dresden, Germany
[4]Department of Radiophysics, Vilnius University, 9 Saulėtekio, LT-10222 Vilnius, Lithuania

E-mail: `m.bauer@physik.uni-frankfurt.de`

Abstract. We report on terahertz (THz) measurements with graphene field-effect transistors with integrated antennas (Tera-FETs) and lay special emphasis on thermoelectric contributions to the detected THz photoresponse. Graphene Tera-FETs with integrated broad-band bow-tie antennas were fabricated in a CVD-based growth process and were successfully applied for detection at 600 GHz with optical NEPs down to 515 pW/$\sqrt{}$Hz. While rectification of THz radiation by (distributed) resistive mixing of charge-density waves induced in the gated transistor channel region is well known, significant additional contributions to the detected signal have experimentally been observed and hot-carrier thermoelectric effects have been identified as a possible origin of these signals. We also observe similar signal contributions in carbon-nanotube transistors.

1. Introduction

Detectors for terahertz (THz) radiation employing field-effect transistors (Tera-FETs) have been implemented in various material systems for more than one decade (for an overview see, e.g., [1]). The detection principle is based on the self-mixing properties of charge-density (plasma) waves induced in the transistor channel by an applied THz signal [2]. The rectification mechanism is not limited by conventional cut-off frequencies of the transistors and has successfully been demonstrated up to several THz at room temperature (e.g., [3]). Besides optimization of the transistor geometry, integration of antenna structures for radiation coupling is a crucial step to realize high-sensitive THz detectors. In combination with novel material systems, e.g., carbon-based materials such as graphene and carbon-nanotubes (CNTs), Tera-FETs can also serve as powerful tools for the investigation of physical properties, such as transport properties and carrier dynamics of the materials.

In this contribution, we report on the detection of THz radiation with Tera-FETs using graphene or CNTs as the transistor channel. We fabricated sensitive CVD-grown graphene Tera-FETs with integrated bow-tie antennas [4] and characterized them at 600 GHz, where in addition to the self-mixing signal by plasma wave rectification, we also observe signal contributions which we attribute to a hot carrier photothermoelectric effect arising due to the specific device geometry. We observe similar contributions in other material systems, particularly pronounced in CNT Tera-FETs, which offers the potential for significant sensitivity enhancement of carbon-based Tera-FET detectors.

2. Self-mixing of plasma waves in Tera-FETs

The detection principle of Tera-FETs is based on nonlinear self-mixing of radiation-induced charge density oscillations (plasma waves) in the gated transistor channel region [2]. In most cases, in particular at room temperature, strong damping leads to a decay of the charge density waves before they can reach the end of the channel - they are overdamped ($\omega\tau \ll 1$). Nonetheless, efficient rectification due to (distributed) resistive self-mixing is present in this case and has been shown to yield high-sensitive THz detectors, which can compete with or even exceed the performance of other established detector technologies in this frequency regime, e.g., Schottky-diode-based, pyroelectric, or acousto-optic detectors. The work presented in this contribution is focusing on the case of non-resonant mixing. In Tera-FETs using CNTs as the active transistor channel, however, the transport of charge carriers through the channel can be ballistic and resonances at the p-n-p-junctions of the gated channel region have been observed [5].

Efficient detection by self-mixing of plasma waves in a FET is only possible when certain boundary conditions for the transistor are fulfilled. Most important, the coupling of the incoming high frequency radiation to the transistor contacts must be asymmetric with respect to the source and drain terminal [2]. Otherwise, plasma waves would be launched into the channel from both sides and the signals generated by self-mixing of the waves would cancel out. Asymmetric coupling can be introduced by various means, e.g., by creating an asymmetric transistor geometry or by using a double-transistor layout with parallel signal read-out at the virtual AC ground in the symmetry plane [3]. Another often employed realization of full asymmetric coupling is the introduction of a shunting capacitance between, for example, the gate and drain terminals. In this way, the applied THz signal will be concentrated with full field strength at the source terminal, while the drain terminal forms an AC ground. Under the above considerations, the self-mixing voltage response ΔU measured between drain and source can roughly be predicted from measurements of the DC conductance σ versus gate voltage U_{g} and the applied THz amplitude U_{a} reaching the transistor by [6]

$$\Delta U \propto \frac{U_{\mathrm{a}}^2}{4}\frac{1}{\sigma}\frac{\mathrm{d}\sigma}{\mathrm{d}U_{\mathrm{g}}}. \tag{1}$$

3. Thermoelectrics

In measurements of the photoresponse of Tera-FET detectors, additional contributions to the detected voltage (current) signal can be observed in various material systems. It has been suggested before [7], that the observed signal contributions can be of thermoelectric origin stemming from the specific device geometry. The required asymmetric coupling conditions for the incident radiation discussed above lead to a strong localization of the electric field at one side of the transistor channel. This can result in a local heating of charge carriers. For capacitive gate-drain coupling, for example, only the source side of the channel is heated while the drain side (AC ground) remains at the temperature of thermal equilibrium.

A local difference in the electronic temperature in a material gives rise to a voltage, which is known as the thermoelectric or Seebeck effect. We attribute the additional signals in our measurement to this effect. We observe especially pronounced contributions in graphene and CNT transistors. It has been shown in a number of studies that in graphene, after initial relaxation of excited carriers, electron-lattice interactions can be strongly suppressed due to high

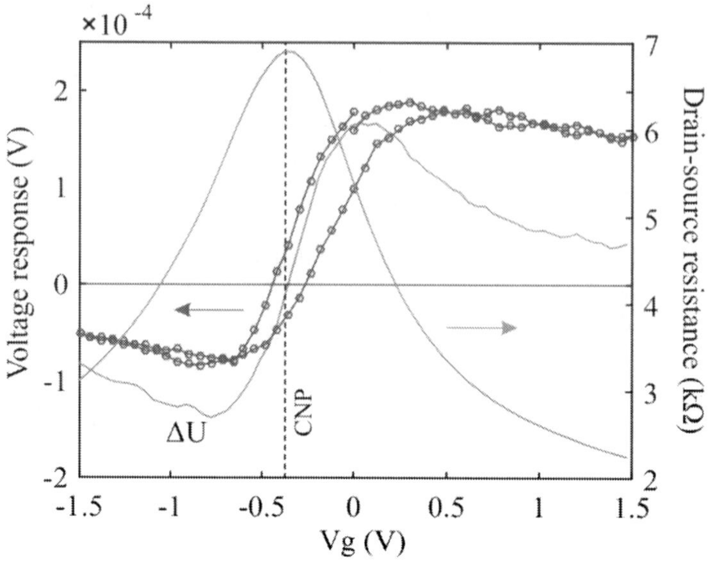

Figure 1. Photoresponse of a CVD-grown graphene Tera-FET at 600 GHz (blue line/symbols, left axis). The gray line (right axis) is the DC resistance of the device. The red line shows the anticipated THz response ΔU due to resistive self-mixing for an arbitrary $U_a \sim 25$mV.

optical phonon energies and long acoustic phonon scattering times, leading to a distribution of hot carriers even under weak electrical driving [8, 9]. In addition, due to the different nature of charge carriers in graphene (electrons or holes) depending on the applied gate voltage, the resulting thermoelectric effect can be of positive or negative sign, thus, enhancing or decreasing the detector's self-mixing response signal.

Figure 1 shows measurements of the photoresponse of a graphene Tera-FET performed at 600 GHz on the left axis (device geometry and measurement setup similar to the one in [4]). The transistor was made of single-layer graphene grown in a CVD process and transferred to a Si/SiO_2 substrate. On the right axis of the figure, the light gray line shows the drain-source resistance measured at DC. From the resistance, the predicted self-mixing response ΔU of the Tera-FET can be calculated using Eq. (1). The result (for an arbitrary applied amplitude U_a for reasons of visualization) is plotted in the figure as a solid red line. Comparing the anticipated and the measured voltage response of the detector, the contributions of the thermoelectric signal can clearly be seen. In particular, at the charge-neutral-point (CNP) of the graphene, no self-mixing signal should be present, since the derivative of the channel conductivity vanishes at this point. The Tera-FET, however, shows a large signal at this gate bias and in a previous study[4] it has been shown that even a low difference in the electronic temperature of ~ 1 K can account for such a thermoelectric contribution to the photoresponse. It can also be observed that as expected, depending on the type of charge carriers, on the right side of the CNP, the contribution to the absolute self-mixing signal is additive while on the left side it is negative. We also characterized the sensitivity of our detectors and determined an optical noise-equivalent power (NEP) of $515\text{pW}/\sqrt{\text{Hz}}$ [4], which is almost comparable to other advanced THz detector technologies, showing that CVD-grown graphene can be used to fabricate sensitive Tera-FET detectors for the THz frequency range.

In measurements performed with transistors realized in various other material system, namely silicon CMOS, GaN and CNTs (data not shown here), we observe similar signal contributions to the self-mixing response. For CMOS detectors, the effect is only visible at cryogenic temperatures, but even for GaN it is observable at room temperature. Here, the thermoelectric contribution is of opposite sign compared to the self-mixing signal. A phase change in the photoresponse is observed. Moreover, we observe a strongly pronounced effect in the investigated CNT transistors at room temperature. Remarkably, no phase change is present in this case meaning that the thermoelectric signal adds positively to the self-mixing. We qualitatively explain this behaviour by using Motts formula for the calculation of the Seebeck coefficient:

$$S = -\int_{E_C}^{\infty} g(E) f(E, E_F)(1 - f(E, E_F)) \frac{E - E_F}{k_B T} \mathrm{d}E \tag{2}$$

where $f(E, E_F)$ is the Fermi function, E_F the Fermi energy, k_B Boltzmann's constant, T is temperature and $g(E)$ is the density of states. The latter is constant for two-dimensional materials but is proportional to $1/\sqrt{E}$ for one-dimensional systems. In this case, Evaluation of Eq. (2) yields a sign reversal in the Seebeck coefficient. Therefore, the resulting thermoelectric current or voltage also shows a reversed sign. Hence, the contribution to the total measured photoresponse is of the same phase as the self-mixing signal, thus, increasing the measured photoresponse. The observation of this theoretical prediction in our measurements is another strong indication that the additional signal contributions are indeed of thermoelectric origin. Although the effect shows up as a side effect in our graphene and CNT detectors, with proper device design THz detectors could be optimized to realize a sensitivity enhancement of the self-mixing effect by the hot carrier thermoelectric effect.

4. Conclusion

We fabricated Tera-FET detectors in a set of materials, in particular graphene and CNTs. Besides efficient detection by plasma wave self-mixing, contributions to the detected signal, which cannot be explained by the plasmonic mixing picture, have been observed in all investigated material systems. We attribute these signals to a photo-thermoelectric effect, arising from the required asymmetric radiation coupling. We discuss the influence of these contributions on the self-mixing effect and show that the strength of the thermoelectric effect is especially pronounced in graphene and CNTs and there can add positively to the plasmonic self-mixing. Hence, these materials offer the potential to strongly enhance or exceed the self-mixing response and could be optimized for the realization of a new set of THz detectors, exploiting the thermoelectric effect. So far, no detailed quantitative analysis of the thermoelectric effect in Tera-FETs is available and is subject to further in-depth investigation.

Acknowledgement

The authors would like to acknowledge fabrication of the CNT transistors by RF Nano Corp., CA, USA.

References

[1] Otsuji T and Shur M 2014 *IEEE Microw Mag* **15** 43–50
[2] Dyakonov M and Shur M 1996 *IEEE T Electron Dev* **43** 380–387
[3] Boppel S *et al.* 2012 *IEEE T Microw Theory* **60** 3834–3843
[4] Zak A *et al.* 2014 *Nano Lett* **14** 5834–5838
[5] Zhong Z, Gabor N M, Sharping J E, Gaeta A L and McEuen P L 2008 *Nat Nanotechnol* **3** 201–205
[6] Sakowicz M *et al.* 2011 *J Appl Phys* **110** 054512
[7] Vicarelli L, Vitiello M S, Coquillat D, Lombardo A, Ferrari A C, Knap W, Polini M, Pellegrini V and Tredicucci A 2012 *Nature Mat* **11** 865–871
[8] Song J C W *et al.* 2011 *Nano Lett* **11** 4688–4692
[9] Gabor N *et al.* 2011 *Science* **334** 648–652

Efficient THz emission from the grating coupled AlGaN/GaN heterostructure on sapphire substrate

I Kašalynas , R Venckevičius, J Laužadis, V Jakštas, E Širmulis, K Požela, and G Valušis

Center for Physical Sciences and Technology, A. Gostauto 11, LT-01108 Vilnius, Lithuania

E-mail: irmantas.kasalynas@ftmc.lt

Abstract. In this work the emission spectra of the AlGaN/GaN heterostructures with large area metallic grating electrodes were obtained over the frequency range 10-500 cm^{-1} and at 300-400 K. Comparative analysis revealed that the emission spectrum below 10 THz was described by the black body radiation model. Indistinct radiation modified with 2DEG plasma resonances was observed only for 7 µm period grating couplers below 3 THz. Measurements of the absolute optical power from the grating-coupled AlGaN/GaN emitter indicated the possibility to achieve an electrical-to-optical conversion efficiency of up to 10^{-3}.

1. Introduction

Properties of electron plasma and lattice oscillations have been widely investigated in polar semiconductors and heterostructures such as GaAs and InGaAs [1], and GaN and AlGaN alloys [2-3]. Recently thermally stimulated selective THz emission from GaAs and AlGaAs semiconductors have been experimentally observed at frequencies of coupled oscillations of free electron plasma and lattice optical phonons [4]. If such an emitter can be made efficient, then it will be a potential candidate for the development of compact room temperature THz imaging system [5].

In this work, the THz emission spectra of the grating-coupled AlGaN/GaN heterostructures with two-dimensional electron gas (2DEG) were investigated. Metal grating couplers with the period from 6 to 150 µm were positioned on top of the AlGaN/GaN structures employing Schottky-type and Ohmic-type contacts. The emission spectra were acquired over the frequency range 10-500 cm^{-1} (0.3-15 THz) with the accuracy of up to 1 cm^{-1} at sample temperature up to 500 K. Measurement of the emission power with an absolute THz power meter ("Thomas Keating") demonstrated that the electrical-to-optical conversion efficiency over the whole frequency range can be up to 0.1 %.

2. Fabrication of samples

The AlGaN/GaN heterostructures with an Al respect to a Ga content up to 30 % were grown in MOCVD reactor on the sapphire substrate. 2DEG density in the range of $n_{2DEG} = (4 - 5)\ 10^{12}\ cm^{-2}$ at the depth of 20 nm was obtained; and the mobility was of 2000 $cm^2/(V{\cdot}s)$. Grating-type electrodes with the metal line width of 3-4 µm and the period up to 150 µm covering an area of 2x2 mm^2 were fabricated using a standard ultraviolet (UV) photolithography procedure and e-beam evaporation of thin metal alloys. Two Ohmic contacts one at each side of the grating was fabricated to provide electric current flow through the 2DEG channel in parallel with the grating vector. Ohmic contacts were formed of a Ti/Al/Ni/Au stack annealed at 830 °C in a N_2 ambient for 30 s, while Schottky contacts were obtained depositing a Ni/Au metal [6].

Journal of Physics: Conference Series **647** (2015) 012005 doi:10.1088/1742-6596/647/1/012005

Electro-thermally stimulated emission spectra of the grating-coupled AlGaN/GaN heterostructures were acquired with a Fourier transform spectrometer (FTS) with a vacuum option sensitive to the polarization of an incident beam. The THz radiation spectrum of the bare AlGaN/GaN heterostructures was also obtained under the same conditions and later used as reference. The emission power was measured with an absolute THz power meter ("Thomas Keating") using the calibrated THz filters.

3. Results

The thermally stimulated emission of the grating-coupled AlGaN/GaN structures is shown in figure 1. Note the results were plot in semi-log scale. Experimental data were fitted to the radiation spectrum of the black body accounting the transmission performance of the optical components of the FTS such as the beam splitter, optical filters and mirrors. Fitting parameters were the body temperature and the emissivity of the grating coupler that varied in the range of 40-57 % for all the samples. In addition the emission spectrum of the AlGaN/GaN structure without the grating coupler was measured and the result is shown in Fig. 1i. Analysis of the radiation spectra revealed that the electro-thermally stimulated THz emission could be nicely described by the radiation spectrum of the black body in a case of grating-coupled structures with large period (see Figs. 1 b-g). And only AlGaN/GaN structures with the grating period of 7 μm independent of the Schottky- or the Ohmic-type grating demonstrated significant deviation from the black body radiation spectrum in the frequency range of 50-250 cm^{-1}.

The frequency of the 2DEG plasma resonances in a conductive channel has been earlier described [7-8]. For a case of the GT1 and GT8 samples with the grating period of 7 μm, the fundamental and higher order modes of the 2DEG plasma were roughly estimated and expected at a frequency of 0.1-1 THz (3.3-33 cm^{-1}). To observe the influence of the plasmonic part, normalized spectra of the grating-coupled I_{GT} to the bare AlGaN/GaN structure I_{HEMT} radiation were obtained by the $(I_{GT} - I_{HEMT})/I_{HEMT}$. The results are shown in figure 2. Indistinct radiation, influenced by minor modes of 2DEG plasma resonance in AlGaN/GaN heterostructures with the grating period of 7 μm, was noticeable in the frequency range of 50-100 cm^{-1}. Moreover, the shape of the normalized spectra was independent on the grating type if one compares the results shown in Fig. 2a and Fig. 2d.

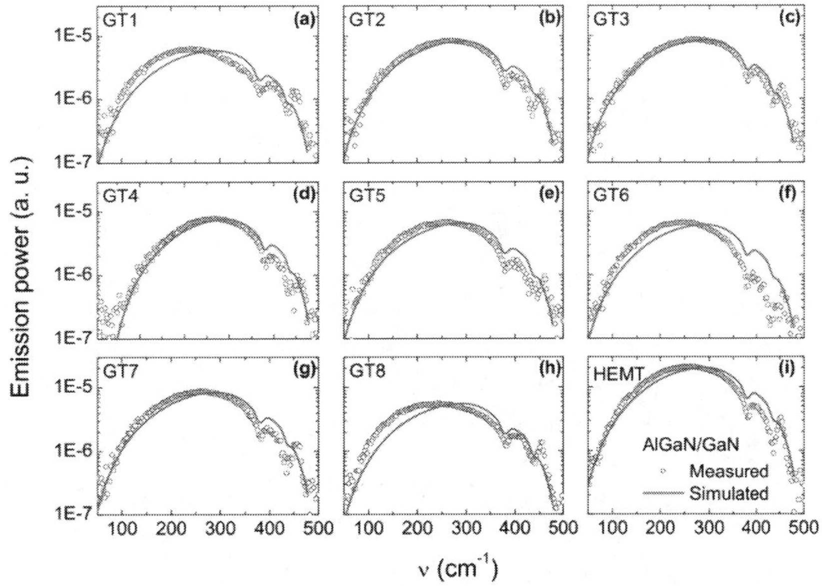

Figure 1. Measured (points) and simulated (dashed lines) emission spectra of the grating-coupled AlGaN/GaN heterostructure with the Schottky-type (a-g) and Ohmic-type (h) electrodes arranged periodically every 7 μm (a, h) and 15 μm (b), 27 μm (c), 34 μm (d), 38 μm (e), 50 μm (f), and 150 μm (g). The emission spectrum of the bare AlGaN/GaN sample (i).

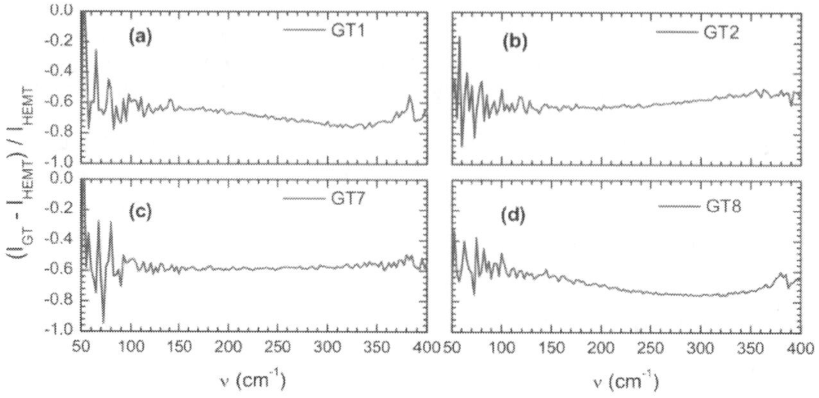

Figure 2. Normalized emission spectra of the grating-coupled AlGaN/GaN heterostructures with the Schottky-type (a-c) and Ohmic-type (d) electrodes arranged periodically every 7 μm (a,d) and 15 μm (b) and 150 μm (c).

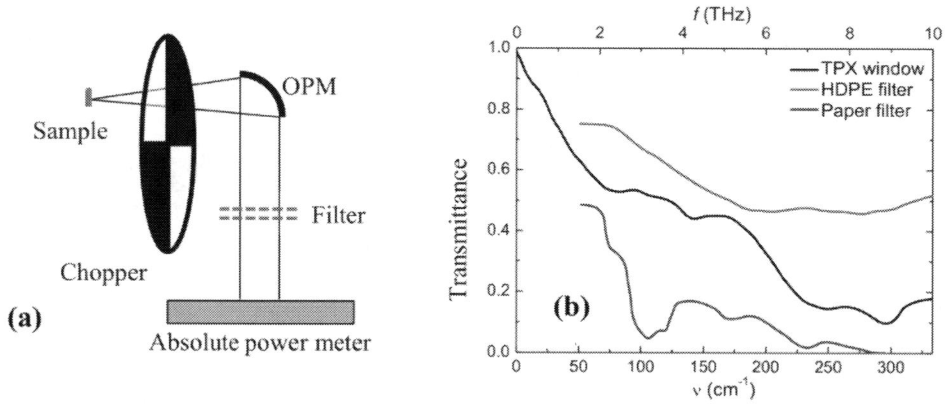

Figure 3. (a) Setup for absolute power measurement. (b) Transmittance of the optical filters located between the Off-axis Parabolic Mirror (OPM) and the absolute power meter.

Table 1. Emission performance of the grating-coupled AlGaN/GaN heterostructures

	Sample 1	**Sample 2**
Applied Voltage (V)	7.0	4.7
Electric Current (mA)	38.2	39.6
Electric Power (mW)	267.4	186.12
Electric Power per solid OPM angle (mW/sr)	4.3	3.0
Absolute power with the TPX window (μW) / Efficiency (%)	26 /0.6	12 / 0.4
Absolute power with the TPX and HDPE (μW) / Efficiency (%)	6 /0.14	2 / 0.07
Absolute power with the TPX and paper (μW)	0 (noise)	0 (noise)

The most important parameters of the THz emitter are the emission power and the conversion efficiency from electrical to optical power. The experimental setup for absolute power measurements is shown in figure 3a. The radiation emitted into free space was collimated with the off-axis parabolic mirror (OPM) and redirected to the calibrated power meter through a TPX window used to filter out extrinsic radiation above 10 THz (see figure 3b). The high-density polyethylene (HDPE) and paper filters were also used to manipulate the radiation spectrum of the THz emitter. The transmittance spectrum of each filter was measured with the FTS and it is shown in figure 3b.

Measurements of the emission power are summarized in table 1. The electrical power used to stimulate thermal radiation of the heterostructures was obtained measuring the applied dc-bias voltage and the current. The solid angle of the rays cone, collected from the THz emitter with a 50-mm focal length and 25-mm diameter OPM, was $\Omega = 0.20$ sr. It was assumed the isotropic radiation model neglecting differences of power emission into the substrate and air. In such a way the OPM redirected 0.016 part of the radiated power to the sphere angle of 4π and only a small part of the supplied electrical power was directed to the absolute power meter. The Sample 1 and 2 was identically processed but the contact resistance influencing the total current flow in the circuit was found different. Collected total radiation power from the THz source based on grating-coupled AlGaN/GaN structures was up to 26 μW. This is the value obtained before the HDPE filter was put in. Additional usage of the HDPE and paper filters allowed us to filter out all background from the visible and IR bands (above 600 cm^{-1} and 100 cm^{-1}, respectively) which might be acquired with the absolute power meter. Thus, a conservative value of the emission power with the HDPE filter was measured up to 6 μW indicating the electrical-to-optical conversion efficiency of about 10^{-3} in the frequency range of 0-10 THz.

4. Conclusion

In summary, we have studied the spectra of the electro-thermally stimulated emission of AlGaN/GaN heterostructures with different grating coupler in the frequency range 10-500 cm^{-1}. Comparative analysis revealed that the shape of the emission spectra below 10 THz could be nicely described by the black body radiation model. Only heterostructures with a grating couplers of 7 μm period have demonstrated deviation from the black body radiation spectrum at the frequency range of 50-250 cm^{-1}. In addition indistinct radiation modified with 2DEG plasma resonances were observed below 3 THz. The electrical-to-optical conversion efficiency for the electro-thermally stimulated THz source was demonstrated up to 0.1 %.

References

[1] Ibáñez J, Tarhan E, Ramdas A, Hernández S, Cuscó R, Artús L, Melloch M, and Hopkinson M 2004 Direct observation of LO phonon-plasmon coupled modes in the infrared transmission spectra of n-GaAs and n-$In_xGa_{1-x}As$ epilayers *Phys. Rev. B* **69** 075314

[2] Talwar D.N 2010 Direct evidence of LO phonon-plasmon coupled modes in n-GaN *Appl. Phys. Lett.* **97**, 051902

[3] Rahbany N, Kazan M, Tabbal M, Tauk R, Jabbour J, Brault J, Damilano B, and Massies J 2013 Measurement of the effect of plasmon gas oscillation on the dielectric properties of p- and n-doped $Al_xGa_{1-x}N$ films using infrared spectroscopy *J. Appl. Phys.* **114**, 053505

[4] Požela K, Širmulis E, Kašalynas I, Šilėnas A, Požela J, and Jucienė V, 2014 Selective thermal terahertz emission from GaAs and AlGaAs *Appl. Phys. Lett.* **105**, 091601

[5] Kašalynas I, Venckevičius R, and Valušis G, 2013 Continuous wave spectroscopic terahertz imaging with InGaAs bow-tie diodes at room temperature *IEEE Sensors Journal* **13**, pp.50-54

[6] Jakštas V, Kašalynas I, Šimkienė I, Strazdienė V, Prystawko P and Leszczynski M 2014 Schottky diodes and high electron mobility transistors of 2DEG AlGaN/GaN structures on sapphire substrate *Lith. J. Phys.* **54** 227–32

[7] Dyakonov M and Shur M, 1993 Shallow Water Analogy for a Ballistic Field Effect Transistor: New Mechanism of Plasma Wave Generation by dc Current *Phys. Rev. Lett.* **71** p.2465-2468.

[8] Hirakawa K, Yamanaka K, Grayson M and Tsui D. C, 1995 Far-infrared emission spectroscopy of hot two-dimensional plasmons in Al0.3Ga0.7As/GaAs heterojunctions Appl. Phys. Lett. **67**, p.2326-2328.

0.69 THz room temperature heterodyne detection using GaN nanodiodes

C. Daher[1], J. Torres[1], I. Íñiguez-de-la-Torre[2], P. Nouvel[1], L. Varani[1], P. Sangaré[3], G. Ducournau[3], C. Gaquiere[3], J. Mateos[2], T. González[2]
[1] IES, Université de Montpellier - CNRS UMR 5214, F-34000, Montpellier, France
[2] Departamento de Física Aplicada, Universidad de Salamanca, 37008 Salamanca, Spain
[3] IEMN, Université de Lille 1 - CNRS UMR 8520, F-59100, Lille, France

E-mail: carlos.daher@ies.univ-montp2.fr

Abstract. GaN-based asymmetric nanodiodes have been used as heterodyne detectors up to 0.69 THz in free-space configuration where Intermediate Frequency bandwidth has been measured up 13 GHz. Monte Carlo simulations, used to estimate nano-device intrinsic conversion losses of 27 dB have confirmed these results.

1. Introduction

Heterodyne detection is a very convenient way to detect terahertz (THz) radiations. Indeed, heterodyne detection allows the use of conventional microwaves systems (such as spectrum analyzers) and simplifies the signal post-processing by down-converting the high-frequency radiation f_{RF} (usually named RF-signal) into a signal with an intermediate frequency (IF) falling into the microwave range [1]. State-of-the-art of electronic components for heterodyne detection of high frequency radiation are mainly achieved by Schottky diodes [2]. Since these components are based on vertical transport, they require advanced materials, challenging fabrication processes and critical technology aspects to operate at THz frequencies. Based on a different concept, the asymmetric nanodiodes have been proposed to obtain planar devices with non-linear, diode like, current voltage (I-V) characteristic [3]. Moreover, the use of high-mobility semiconductors make possible to achieve THz operations. Even if the use of unipolar GaN-nanochannels as THz mixers is fairly new, it presents a lot of room for improvements to make them competitive with Schottky technology. In a previous paper, through a systematic numerical study based on Monte Carlo (MC) simulations, we have presented the performances of such GaN nanodiodes as direct and heterodyne detectors in the 0.300 THz frequency range [4]. The aim of this paper is to present, over a wider frequency range, a systematic experimental study of the performances of GaN nanochannels as heterodyne detectors with a particular attention to their dependence on bias conditions and frequency of operation. In particular, we show for the first time that GaN nanochannels can be used as heterodyne detectors at high frequencies as 0.690 THz.

2. Experimental details

2.1. Sample description

The sample used is based on 2DEG unipolar nanochannels with a broken symmetry created by etching in the AlGaN/GaN heterojunction two symmetrical L-shaped trenches (see [5] for technological details). In order to reduce the impedance of our nanodevice, 32 nanochannels were disposed in parallel without the need of any interconnection. In this way, the extrinsic parasitic elements can be limited: this is an important feature at these frequencies. We emphasize that the simplicity of the technological process used for the fabrication of these nanodiodes is remarkable, since it only involves the etching of insulating recess lines on a semiconductor surface. Another key aspect of these nanodevices is their planar geometry, which provides important advantages for the free-space radiation coupling over traditional diodes (Schottky barrier) used in current THz systems. Each nanochannel of our studied device has a length L and a width W such as L = W = 500 nm. In order to use this device in free-space configuration, planar bow-tie THz antenna was monolithically integrated.

2.2. Experimental setup

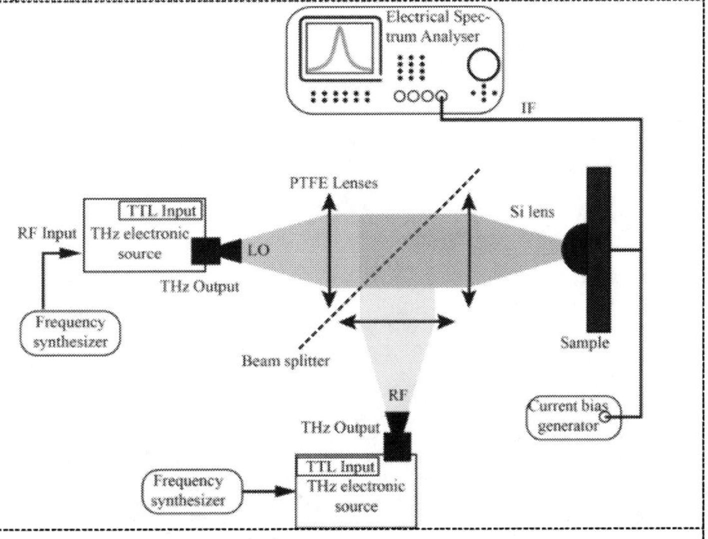

Figure 1. Heterodyne detection set-up using two electronic sources.

To highlight the ability of these devices to mix two THz radiations in the 0.300 THz frequency range and in the 0.650 THz frequency range – figure 1. We have used each time two electronic sources from Virginia Diodes Inc. (VDI®) which frequencies fall into the same range. As this kind of electronic sources are composed of a multiplication chain, the measurement frequency points have been verified to be spurious free (eg monochromatic signal). The RF input of the source is connected using a coaxial cable with subminiature version A (SMA) connectors to a frequency synthesizer, which delivers the microwave signal to be multiplied by the source. The latter will then radiate a THz frequency signal. For example, to reach 0.650 THz, we need a THz source with a multiplying factor of 48, and an input frequency of 13.54 GHz provided by the frequency synthesizer. In the 0.300 THz frequency range we fixed the RF frequency at 0.290 THz, and we varied the LO frequency from 0.290 to 0.330 THz, which means that the Intermediate Frequency varies between 0 and 40 GHz which is the bandwidth of our electrical spectrum analyzer. In the 0.650 THz frequency range the RF frequency was fixed at 0.660 THz, and the LO frequency was varied from 0.660 to 0.673 THz, which means that the Intermediate Frequency varies between 0 and 13 GHz which represents the cut-off frequency of the FR4 sample holder. This frequency was simulated using CST® microwave studio program. This holder was especially designed for 50 Ω impedance matching with the input impedance of our electrical spectrum analyzer. The antenna of the device was connected via gold-microbondings to printed HF-lines to SMA connector. The two THz radiations are collimated and focused using PTFE lenses, whereas a polyethylene terephthalate beam-splitter helps to spatially-superimpose the two radiations. Finally, a silicon hemispherical lens of 6 mm diameter, is placed at the rear side of our sample to improve the focalization on the device, to reduce the free-space coupling losses and to limit interferences inside the Si-bulk layer. A bias-tee is connected to the SMA connector of the sample in order to provide the DC-bias to the device, and to measure the IF with the help of a Rohde & Schwartz® FSU electrical spectrum analyser. In all cases the experiments were performed in free space and at room temperature configurations. In order to understand the physical mechanisms of mixing (in terms of output power), a semi-classical Monte Carlo (MC) simulation self-consistently coupled with a Poisson solver is used as previously explained in [6].

3. Results and discussion

The IF power measured with our device, as a function of frequency is presented in figure 2 for the RF and LO signals falling into the 0.300 THz range (a) and 0.650 THz range (b), for different bias current. In both cases, a strong attenuation around 13 GHz is observed: this effect cannot be attributed to the cut-off frequency of our device. Indeed, important figures of merit to be taken into account are the conversion losses (CL) of our mixer.

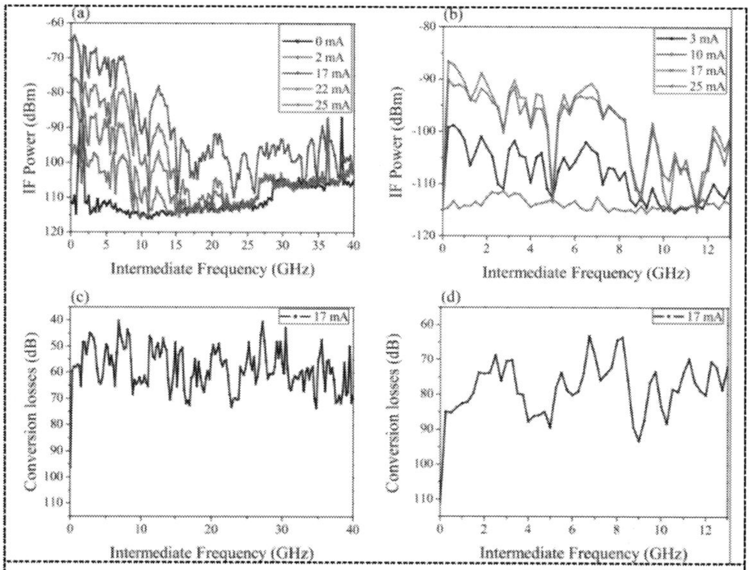

Figure 2. Experimental IF power measured for different bias current, with RF and LO belonging to the frequency ranges (a) 0.300 THz and (b) 0.650 THz. (c) and (d) show the corresponding conversion losses for I = 17 mA.

These have been determined as the ratio of the power at the IF to the power at the f_{RF} subtracting the losses induced by connecting elements and HF-lines, that were simulated using CST microwave studio and present a cutoff frequency around 13 GHz. Consequently, the CL can be plotted according to the IF – Figure 2 (c) and (d) – in the 0.300 and 0.650 THz frequencies ranges, respectively for the best bias conditions of I = 17 mA. The CL of the whole integrated device are 58 ± 2 dB in the 0.300 THz frequency range and increase by an additional 20 dB at higher frequencies to achieve 78 ± 3 dB in the 0.650 THz frequency range. These values do not represent the intrinsic CL of the device since the losses due to radiation coupling with the antenna (especially in the 0.650 THz frequency range) and bondings-wires are difficult to measure. Since CL are mainly constant in each given frequency range,

Journal of Physics: Conference Series **647** (2015) 012006 doi:10.1088/1742-6596/647/1/012006

one can conclude that the bandwidth of our mixer is at least of 40 GHz, i.e. the frequency limit of our electrical spectrum analyzer.

4. Conclusion

We have investigated the potentialities of AlGaN/GaN based nonlinear nanochannels for THz heterodyne detection. The room-temperature heterodyne detection up to 0.690 THz has been demonstrated for the first time with such a device. As a consequence their practical implementation as mixers of high-power sub-mm wave sources seems feasible. Moreover recent studies of similar GaN-based nanodiodes [6], have demonstrated the possibilities as Gunn oscillators to be employed as active elements of THz emitters. As a consequence, that opens the possibility to develop a fully integrated emitter/detector submillimeter-wave system working at room temperature.

5. References

[1] M C Teich 1980 *NASA Conference Publication* vol. 1 no. 2138 p. 1–10
[2] H W Hübers 2008 *IEEE Journal of Selected Topics on Quantum Electronics* vol. 14 p. 378–391
[3] A M Song et al. 2003 *Applied Physics Letters* vol. 83 no. 9 p. 1881–1883
[4] I Íñiguez-de-la-Torre et al. 2014 *IEEE Trans. on terahertz sci. and tech.* vol. 4 no. 6 p. 670–677
[5] P Sangaré et al. 2013 *Journal of Applied Physics* vol. 113, no. 3
[6] A Íñiguez-de-la-Torre et al. 2012 *Journal of Applied Physics* vol. 111 p. 113-705

Enhancement of sub-terahertz detection by drain-to-source biasing on strained silicon MODFET devices

Y.M. Meziani[1], S. Morozov[2,4], J.A. Delgado Notario[1], K. Maremyanin[2], J.E. Velázquez[1], K. Fobelets[3]

[1] Departamento Físca Aplicada, Universidad de Salamanca, 37008 Salamanca, España.
[2] Institute for Physics of Microstructures of Russian, Academy of Sciences, 603950, GSP-105, Nizhny Novgorod, Russia.
[3] Department of Electrical and Electronic Engineering, Imperial College, London SW7 2AZ, UK.

E-mail: meziani@usal.es

Abstract. We report on non resonant detection of sub-terahertz radiation (148-353 GHz) using strained silicon modulation field effect transistor with different gate lengths. The devices were excited at room temperature by a Backward Wave Oscillator (BWO) source. Enhancement of the photoresponse signal by drain-to-source bias was observed. Increasing drain-to-source voltage leads to asymmetry between the boundary conditions at the source and drain contacts.

1. Introduction

Development of terahertz sensors compatible with mainstream CMOS technology is of big interest due to it's capabilities for different terahertz applications like security, imaging, and inspection [1, 2, 3]. Terahertz rays are located in the spectral region $0.1 - 10$THz (3mm $- 30\mu$m, $3 - 300$cm^{-1}) between the microwave and the infrared portion of the electromagnetic spectrum [4]. In early 1990's, Dyakonov and Shur [2] theoretically demonstrated the possibility of using sub-micron field effect transistors as detectors of terahertz radiation by means of the oscillations of plasma waves in the channel. Those devices present many advantages: low cost, small size, room temperature operation and possible tuning by mean of gate bias. Experimental and theoretical investigations have been conducted in semiconductor devices, demonstrating their capabilities for detection of terahertz radiation. Experimental non resonant detection (broadband detection) was demonstrated using different types of transistors like GaAs/AlGaAs FETs [5, 6] and Si-MOSFET [7, 8]. A responsivity of 80 kV/W and a NEP (Noise Equivalent Power) of 300 pW/Hz$^{1/2}$ as well as imaging at 0.65 THz were reported using an array of Si-MOSFET processed by 0.25 μm CMOS technology [9]. Recently, asymmetric double grating gates devices based on GaInAS/InP HEMTs have shown a record of responsivity and NEP of around 20 kV/W and 0.48 pW/Hz$^{1/2}$) [10, 11].
In the present paper, we have studied the detection of sub-terahertz radiation by using Strained

[4] Also with: Lobachevsky State University of Nizhni Novgorod, 23 Gagarin Avenue, Nizhny Novgorod,603950, Russia

Silicon MODFET at room temperature and under drain-to-source bias. Two Si-MODFET devices were excited by a BWO at different frequencies, the response exhibits a non-resonant feature with a good signal to noise ratio. For the first device, with better performance, an abrupt increase of the responsivity was observed. This was attributed to the asymmetry between the boundary conditions at the source and drain contacts.

2. Strained Silicon MODFET and experimental setup

The epistructure of the MODFETs used in this work was grown by molecular beam epitaxy (MBE) on a thick relaxed SiGe virtual substrate grown by low-energy plasma-enhanced chemical vapor deposition (LEPECVD) over a p-doped conventional Si wafer (Fig. 1). The final Ge molar concentration in the virtual substrate was $x_{Ge} = 0.45$. The device had a 9 nm tensile strained (in terms of bi-axial deformation) Si channel, sandwiched between two heavily doped SiGe electron supply layers to generate a high carrier density in the strained-Si quantum well [12]. The ohmic contacts were not self aligned. Table 2 shows the geometry of two transistors with different gate lengths (100 nm and 250 nm). More details on the epistructure can be found in ref. [13].

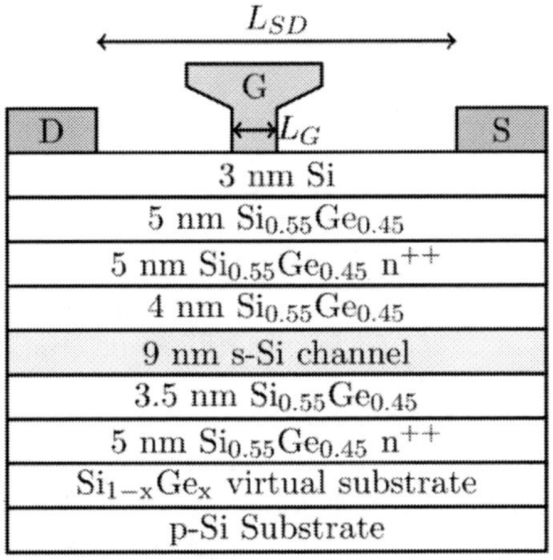

Table 1. Epistructure of the MODFET

	L_{SD} (μm)	L_G (nm)	W_G (μm)
Device 1	2	250	80
Device 2	1	100	80

Table 2. Geometry parameters of devices.

Two BWOs were used for the detection experiments. The first one with tunable frequency from 142 to 248 GHz and the second one with frequencies 286 to 353 GHz. The incoming radiation intensity was modulated by a mechanical chopper at 1.29 kHz and coupled to the device via the contact metalization pads. The induced photoresponse signal was measured by using a lock-in amplifier technique. More description of the experimental setup can be found in [7].

3. Results and discussions

The photoresponse signal measured for the device 1 (table 2) with $L_G = 250$ nm is shown in figure 1 versus gate bias at different frequencies from 208 to 353 GHz. The signal was measured at room temperature and at fixed $V_{DS} = 0V$. Good level of signal to noise ratio was obtained with a maximum around the threshold voltage ($V_{th} \simeq -0.9V$). The latter is related to the non-resonant detection caused by over-damped plasma waves that generate into space-charge waves. The intensity of the signal however, do depend on the incoming radiations frequencies. The maximum is obtained at 353 GHz and the minimum at 252 GHz. This behavior can be attributed to the fluctuation of the emitted power from the BWO sources for different frequencies and also to a better coupling of the incoming radiation to the device at the specific frequency in

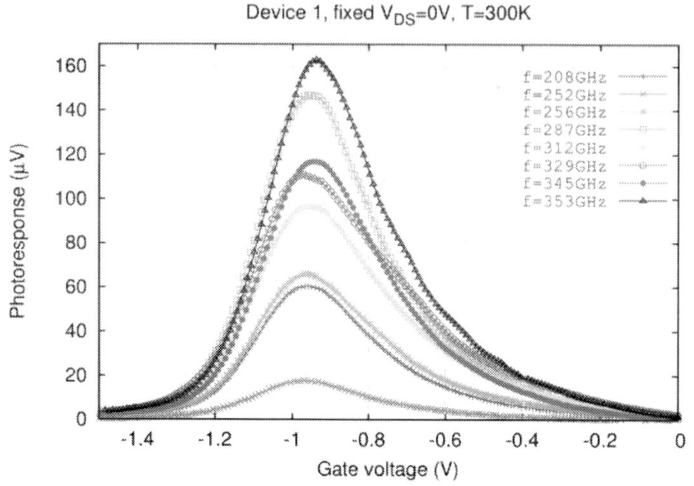

Figure 1. Photoresponse signal vs. gate bias at different incident radiations frequencies from 208 to 353 GHz. It was obtained for device 1 at room temperature and at fixed V_{ds}=0V.

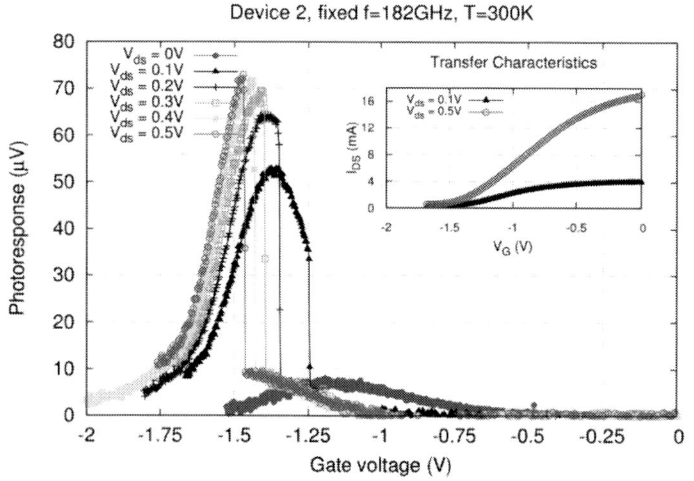

Figure 2. Photoresponse signal vs. gate bias under different drai-to-source biasing (V_{ds} from 0 to 0.5 V). Experiment performed on Device 2 at room temperature and under electromagnetic radiation at 182 GHz. Inset shows the transfer at V_{ds}=0 and 0.5 V

agreement with the theory. Tuning the frequency is obtained by increasing the cathode voltage of the Magnet (1-5 kV), it's difficult to maintain the emission power constant for all frequencies and it fluctuates from 1 to 10 mW [14].

For device 2 (table 2), the photoresponse signal is shown in figure 2. Inset of figure 2 shows its transfer characteristic for two values of the drain-to-source voltage V_{ds}=0.1 and 0.5 V, for which value of $V_{th} \simeq -1.4$V was extracted. At a fixed $V_{ds} = 0.1$V, the maximum photoresponse was observed around -1.2 V with an intensity around 10μV. When V_{ds} was increased to higher voltages (up to 0.5 V), the intensity of the signal was increased by several orders with an abrupt increase of the signal. The estimated responsivity R_v was calculated using the following equation [10]:

$$R_v = \frac{S_t \Delta U}{S_d P_t} \tag{1}$$

where ΔU is the measured photoreponse signal, P_t is the total power of the THz radiation in the detector plane (the power was $\simeq$0.15 mW lower than the emitted one from BWO because of the attenuation in atmosphere), S_t is the spot size of the radiation beam ($\simeq$3 mm), and S_d is the active area of the detector. A responsivity of $\simeq$ 2 kV/W and 20 kV was obtained for

V_{ds}=0 and 0.5 V, respectively. This clearly shows an increase of the responsivity by two orders of magnitude induced by V_{ds}. Similar effect was observed by Lu et al. [15] on a GaAlAs/GaAs HEMT related to the asymmetry of the boundary condition induced by V_{ds}. The gate-to-drain capacitance C_{GD} decreases as function of V_{ds} and the gate-to-source capacitance (C_{GS}) increases and it saturates for higher values of V_{ds} [15]. This behavior induces a greater asymmetry degree in the boundary conditions of the plasma waves and greatly enhances the responsivity of the detector. The abrupt increase of the photoresponse may be attributed to carrier transfer to a parasitic channel above the upper supply layer, but further investigation needs to be carried out to fully understand the phenomenon.

Conclusion

We report on enhancement of the photoresponse signal as well as the responsivity of the detector by means of drain-to-source bias. An abrupt increase of the signal was observed and an enhancement of the responisivity by two orders was demonstrated. This was related to the the asymmetry between the boundary conditions at the source and drain contacts.

Acknowledgments

Devices were kindly supplied by Dr. T. Hackbarth. Authors from Salamanca University acknowledge the financial support from the MINECO/FEDER through the Project No. TEC2012-32777. Y.M. Meziani acknowledge the support from the RIEC Nation-Wide Collaborative Research Project, Japan.

References

[1] Dyakonov M and Shur M S 1993 *Phys. Rev. Lett.* **71** 2465.
[2] Dyakonov M and Shur M S 1996 *IEEE Trans. Electron Dev.* 43 380.
[3] Meziani Y M et al. 2011 *Semicond. Sci. Technol.* **26** 105006.
[4] Siegel P. *IEEE Trans Microwave Theory Tech* **50(3)** 910-928.
[5] Knap W, Deng Y, Rumyantsev S, L J-Q, Shur MS, Saylor CA, et al. 2002 *Appl Phys Lett* **80(18)** 3433-5. http://dx.doi.org/ 10.1063/1.1473685.
[6] Knap W, Kachorovskii V, Deng Y, Rumyantsev S, L J-Q, Gaska R, et al. 2002 *J Appl Phys 2002* **91(11)** 9346?53. http://dx.doi.org/10.1063/1.1468257.
[7] Meziani YM, Lusakowski J, Dyakonova N, Knap W, Seliuta D, Sirmulis E, et al. 2006 *IEICE Trans Electron* **E89-C** 993-998.
[8] Tauk R, Teppe F, Boubanga S, Coquillat D, Knap W, Meziani YM, et al. 2006 *Appl Phys Lett* **89(25)** 253511. http://dx.doi.org/10.1063/1.2410215.
[9] Lisauskas A, Pfeiffer U, jefors E, Bolvar PH, Glaab D, Roskos HG. 2009 *J Appl Phys* **105(11)** 114511. http://dx.doi.org/10.1063/1.3140611.
[10] Kurita Y, Ducournau G, Coquillat D, Satou A, Kobayashi K, Boubanga Tombet S, Meziani YM, Popov V V, Knap W, Suemitsu T and Otsuji T 2014 *Appl Phys Lett.* **104(25)** 251114.
[11] Popov V V, Fateev D V, Ivchenko E L, Ganichev S D 2015 *eprint arXiv:1505.06847*
[12] Rumyantsev S L, Fobelets K, Veksler D, Hackbarth T, Shur MS 2008 *Semiconductor Science and Technology* **23(10)** 105001.
[13] Meziani Y M, García-García E, Velázquez-Pérez J E , Coquillat D , Dyakonova N, Knap W, Grigelionis I and Fobelets K 2013 *Solid-State Electronics* **83** 113
[14] Booske J H, Dobbs R J, Joye C D, Kory C L, Neil G R, Park G S, Park J, and Temkin RJ 2011 *IEEE Transactions on Terahertz Science and Technology* **1**, 54-75
[15] Lü J Q and Shur M S 2001 *Appl. Phys. Lett.* **78** 2587.

Investigation of possibility of VLWIR lasing in HgCdTe based heterostructures

S V Morozov[1,2], V V Rumyantsev[1,2], A M Kadykov[1,3], A A Dubinov[1,2], A V Antonov[1,2], K E Kudryavtsev[1,2], D I Kuritsin[1,2], N N Mikhailov[4,5], S A Dvoretskii[4], F Teppe[3] and V I Gavrilenko[1,2]

[1] Institute for Physics of Microstructures of Russian Academy of Sciences, 603950, Nizhny Novgorod, Russia
[2] Lobachevsky State University of Nizhny Novgorod, 603950, Nizhny Novgorod, Russia
[3] UMR CNRS 5221, GIS-TERALAB, Université Montpellier II, 34095, Montpellier, France
[4] A.V.Rzhanov Institute of Semiconductor Physics, Siberian Branch of Russian Academy of Sciences, 630090 Novosibirsk, Russia
[5] Novosibirsk State University, 630090, Novosibirsk, Russia

E-mail: more@ipmras.ru

Abstract. The optical properties of a number of $Hg_{1-x}Cd_xTe$ bulk epilayers (x = 0.152 – 0.23) and heterostructures with quantum wells (QW) based on narrow gap HgCdTe are examined aiming to reveal the prospects of such structures for laser development in long wave infrared and very long wave infrared ranges. Experimental evidence of long wavelength superluminescence, i.e. amplification of spontaneous emission, at 8.4 μm in narrow gap HgCdTe bulk epitaxial film at 100 K is reported. Employing heterostructures with QW is demonstrated to be promissory for furthering the radiation wavelength to 10 - 30 μm range.

HgCdTe material has been intensively studied for more than half a century due to its paramount importance for infrared optoelectronics. Development of IR sensors has driven the HgCdTe growth techniques, in particular molecular beam epitaxy (MBE), which nowadays allows producing high quality epitaxial structures with in situ control of layers composition and thickness. Besides bulk epilayers with high uniformity and low residual carrier density ($\sim 10^{14}$ cm^{-3}) MBE can also provide HgCdTe based heterostructures with quantum wells (QW) energy spectrum of which can be tuned by changing QW width and Cd content in it. A number of peculiar fundamental properties [1, 2] were demonstrated in such systems as well as pronounced long-wavelength photoconductivity (PC) up to THz range [3, 4]. However, processes of light emission from narrow gap HgCdTe were less investigated; in particular, spectra of interband photoluminescence (PL) from HgCdTe based structures in very long wavelength infrared range (VLWIR) are scarce in literature. Stimulated emission was investigated only in wavelength region up to 5.3 μm [5, 6]. In this work we have examined the optical properties of a number of $Hg_{1-x}Cd_xTe$ bulk epilayers and heterostructures with QW based on narrow gap HgCdTe aiming to reveal the prospects of such structures for laser development in LWIR and VLWIR ranges.

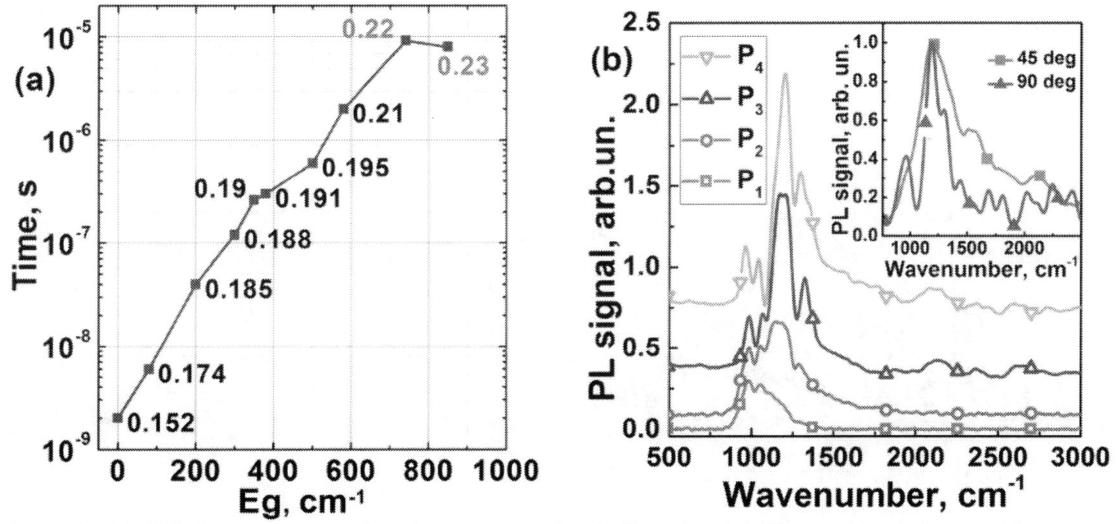

Figure 1. (a) PC decay time vs. bandgap energy for bulk epitaxial films with different Cd content x (see labels) at 4.2K. Pumping wavelength was in the range of $\lambda_{exc} = 11 \div 17$ μm. (b) Spectra of PL collected from the edge of the sample with x = 0.22 (in-plane direction) at different photon densities in pumping pulse: P_1 – 1.1×10^{12}, P_2 – 2×10^{14}, P_3 – 2.6×10^{14}, P_4 – 2.7×10^{15}. Beam diameter was approximately 1 cm. Insert represents spectra of PL signal collected from different directions (relative to structure growth axis) at $P \sim 1.4\times10^{15}$ photons in pumping pulse. T = 100K.

All structures under study were grown on semi-insulating GaAs(013) substrates with ellipsometric control of the layer content and thickness [7]. The "active" parts of the structures were grown on a relaxed CdTe buffer layer about 5 μm thick. Typical residual carrier density in active layer is about $n = 5\cdot10^{14}$ cm^{-3}. PL spectra studies were carried out in a closed-cycle cryostat optically connected to a Fourier transform infrared spectrometer Bruker Vertex 80v, operating in step-scan mode. Details on experiment can be found in [8].

All structures were firstly characterized by PC spectra and kinetics measurements [3, 4]. Figure 1(a) summarizes the PC decay time data, i.e. information on carrier lifetimes in structures under study. One can see that PC decay time grows with x up to ~ 10 μs at x = 0.22 and then slightly decreases. The radiative lifetime is known to decrease with the bandgap [9]; hence, such behavior, i.e. carrier lifetimes increase with the bandgap indicates that recombination is non-radiative. Review of literature data suggests that the dominating mechanism of carrier recombination is Shockley-Read-Hall (SRH) process [9-11]. However, according to recent calculation of carrier recombination rates [11] lifetime of 9.2 μs measured in our bulk epilayer (x = 0.22) practically coincides with the calculated radiative lifetime for n-doped (5×10^{14} cm^{-3}) $Hg_{1-x}Cd_xTe$ with x = 0.224 at low temperatures (20 K). Thus, one can suggest that in epilayers under study carrier recombination is mostly controlled by radiative processes when $x \geq 0.22$ at least at low temperatures. This is confirmed by further decrease in carrier lifetime when x increases to 0.23 (see Figure 1(a)). Moreover, domination of radiative process over SRH recombination can be enhanced when excitation intensity is increased. At high enough excitation level (in the case of PL measurements, for example) recombination via trap centers tends to saturate while rates of radiative recombination increase and thus radiative recombination readily dominates in solid solutions with x = 0.21 – 0.23 at low temperature [4, 8]. As a result, this allowed observation of interband PL on record wavelength of 26 μm [8] in bulk $Hg_{1-x}Cd_xTe$ with x = 0.189. PL measurements in such structures could be the first step in the development of light emitting devices for the spectral range 20 – 30 μm, which is practically inacceptable for A_3B_5 based devices such as quantum cascade lasers, due to strong two-phonon absorption.

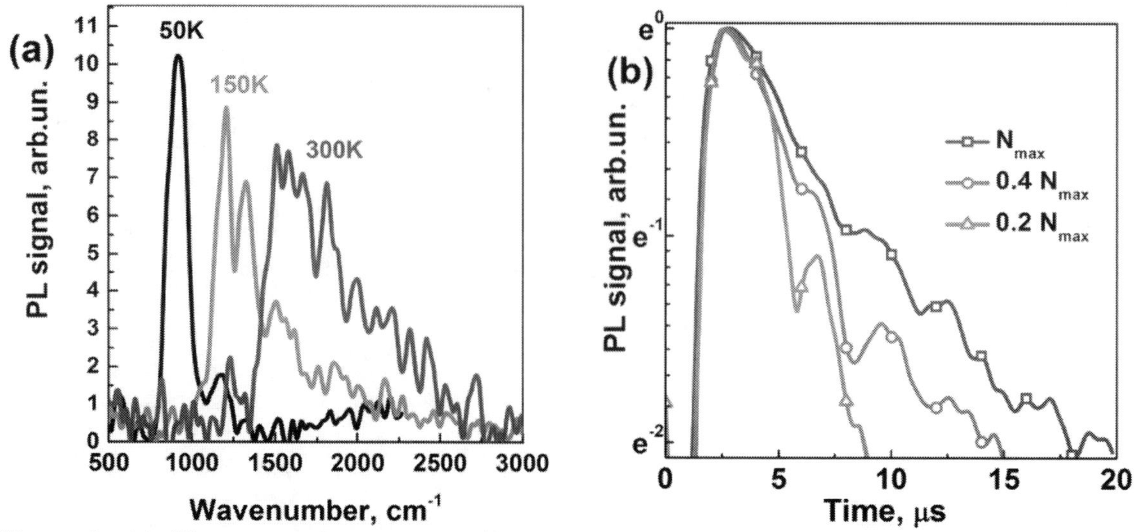

Figure 2. (a) PL spectra from narrow (5 nm) HgTe QW sandwiched between two layers of $Hg_{0.47}Cd_{0.53}Te$ at different temperatures (b) Kinetics of interband PL from wide (30 nm) $Hg_{0.814}Cd_{0.186}Te$ QW sandwiched between two layers of $Hg_{0.3}Cd_{0.7}Te$ under different excitation levels N. ($N_{max} = 10^{16}$ photons in 10 ns pulse)

However, the next question to be answered is whether the amplification of long-wavelength radiation can be achieved in such structures. In order to reveal the possibility of light amplification in samples under study the PL spectra transformation under intense optical excitation was examined. For this experiment the bulk epilayer with x = 0.22 was chosen since it was the narrowest gap structure that according to calculation contained the dielectric waveguide for radiation wavelength corresponding to interband transitions. PL radiation was collected from the edge of the sample since in-plane direction of the epilayer is thought to be optimal for amplification of radiation. Indeed, at high enough excitation level spectra of PL radiation propagating along the film plane is considerably narrower as compared to radiation propagating in 45^0 angle relative to growth direction (see the insert in figure 1(b)). Figure 1(b) demonstrates the genesis of a narrow line of amplified PL on broad band of spontaneous emission as the intensity of pumping grows. Graph P_1 represents PL spectra under pulsed excitation with ~10^{12} photons in 130 fs pulse, which proved to be almost identical to PL spectra measured under continuous wave excitation with P = 300 mW. As the pumping power grows PL line broadens at first due to filling of conduction and valence band by non-equilibrium carriers (compare graphs P_1 - P_2). Then a narrow line, which is superimposed on the broad background band of spontaneous emission, appears at 1170 cm^{-1} and rapidly grows with pumping intensity (compare graphs P_2 - P_3) finally slightly shifting to 1196 cm^{-1} (8.4 μm) (graph P_4). At excitation power P_4 the full width at half maximum (FWHM) of PL line is almost 3 times smaller as compared to power P_2.

At this point it should be noted that such experiments were not performed earlier apparently because Auger recombination was believed to be absolutely detrimental for light emission processes in such narrow gap material. However, it is worth mentioning some recent papers that give an optimistic view concerning the negative influence of Auger recombination in narrow gap MBE epilayers. Firstly, calculations of recombination rates that exploit the full band structure obtained in hybrid pseudopotential tight-binding Hamiltonian model predict longer Auger time than those calculated with widely used analytical expressions [11]. Secondly, at high carrier densities Auger recombination rates are affected by Coulomb screening that should also increase the Auger recombination time [9]. Though at rather high temperature as 100K the Auger recombination lifetime is of the same order of magnitude as spontaneous radiative recombination time according to [11] somewhat of Auger mitigation is indirectly confirmed by our experiments as well.

Nevertheless, the FWHM and intensity of superluminescent line demonstrate saturation at higher excitation. This is due to plasmonic contribution to dielectric function in the active layer of the structure. High carrier density formed by intense excitation increases plasma frequency bringing it closer to the PL radiation frequency. This leads to considerable decrease in dielectric function in the active layer, which has dramatic effect on the light confinement. Estimations show that the waveguide in the structure under study almost disappears at $n = 2\times10^{18}$ cm^{-3}. Thus, at high density of non-equilibrium carriers, the amplification of radiation is weak and electron-hole recombination takes place mostly via fast non-radiative processes. However, further amplification can be achieved in structures with separate light and carrier confinement. Indeed, the negative influence of excess carriers on the light confinement can be reduced by placing QWs inside the core layer of the dielectric waveguide. QWs will accumulate the non-equilibrium carriers providing the same gain but the change of dielectric function in thin QW layers will have small impact on waveguide properties of the structure.

QW are also expected to be beneficial for additional suppression of non-radiative processes [12]. Unlike bulk material the energy spectrum of QW structures is characterized by close effective masses of electrons and holes that should result in mitigation of the Auger recombination. This is indirectly confirmed by low temperature quenching of PL in QW structures as compared to bulk samples [13]. PL signal at room temperature is easily detected not only in wide QW but also in rather narrow wells (figure 2(a)). Moreover, SRH recombination can also be suppressed by saturating the trap centers with intense pumping. Time resolved PL measurements in QWs show that the carrier relaxation time grows with excitation intensity and can be as high as 5 μs under high excitation (figure 2(b)). This infers that carrier density of several 10^{11} cm^{-2} required for amplification [14] can be achieved under reasonable level of excitation. We calculated mode localization and modal gain coefficients for several optimized designs of epitaxial HgCdTe structures aimed at obtaining stimulated emission in VLWIR range. These structures are supposed to contain 5 HgTe QW in core layer of waveguide. Calculations show [15] that these QW would provide amplification coefficient of several 10 cm^{-1} at non-equilibrium carrier concentration of $1 - 4\times10^{11}$ cm^{-2} (for $\lambda \sim 15 - 30$ μm).

The work was supported by the Russian Foundation for Basic Research (grants #13-02-01154, #15-02-05154, #15-42-02249), President Program of Support for Leading Scientific Schools of Russia (Scientific school #1214.2014.2) and the Russian Academy of Sciences. The work was done using equipment of Center "Physics and technology of micro- and nanostructures" at IPM RAS.

References

[1] Zholudev M, *et al.*, *Phys. Rev.* B, 2012. **86**(20) 205420.
[2] Orlita M, *et al.*, *Nat. Phys.*, 2014. **10**(3) 233-238.
[3] Rumyantsev V V, *et al.*, *Semicond. Sci. Technol.*, 2013. **28**(12) 125007.
[4] Rumyantsev V V, *et al.*, *Semicond.*, 2013. **47**(11) 1438-1441.
[5] Arias J M, *et al.*, *Semicond. Sci. and Tech.*, 1993. **8**(1S) S255.
[6] Bleuse J, *et al.*, *J. Cryst. Growth*, 1999. **197**(3) 529-536.
[7] Dvoretsky S, *et al.*, *J. Electron. Mater.*, 2010. **39**(7) 918-923.
[8] Morozov S V, *et al.*, *Appl. Phys. Lett.*, 2014. **104**(7) 072102.
[9] Jóźwikowski K, *et al.*, *J. Appl. Phys.*, 2012. **112**(3) 033718.
[10] Rogalski A, *Rep. Prog. Phys.*, 2005. **68**(10) 2267-2336.
[11] Krishnamurthy S, *et al.*, *J. Electron. Mater.*, 2006. **35**(6) 1369-1378.
[12] Jiang Y, *et al.*, *J. Appl. Phys.*, 1991. **69**(10) 6869.
[13] Morozov S V, *et al.*, *Appl. Phys. Lett.*, 2014. **105**(2) 022102.
[14] Morozov S V, *et al.*, *Semicond.*, 2012. **46**(11) 1362-1366.
[15] Morozov S V, *et al.*, *THz Emission of Narrow- Gap HgCdTe Films and HgTe/CdTe Quantum Wells Structures*, in *38th International Conference on Infrared, Millimeter and Terahertz Waves, Mainz on the Rhine, Germany, September 1– 6 2013*. 2013, Fraunhofer IPM and University of Kaiserslautern Mainz on the Rhine, Germany. Mo6-5.

Terahertz excitations in HgTe-based field effect transistors

A. Kadykov[1,2], C. Consejo[1], F. Teppe[1], W. Desrat[1], L. Viti[3], M. S. Vitiello[3], D. Coquillat[1], S. Ruffenach[1], S. Morozov[2], S. Kristopenko[2], M. Marcinkiewicz[1], N. Dyakonova[1], W. Knap[1], V. Gavrilenko[2], N. N. Michailov[4], S.A. Dvoretskii[4]

[1]Laboratoire Charles Coulomb (L2C), UMR 5221, GIS-TERALAB, CNRS-Université de Montpellier, Montpellier, F-France
[2]Institute for Physics of Microstructures, Russian Academy of Sciences, Nizhny Novgorod, Russia
[3]NEST, Istituto Nanoscienze-CNR and Scuola Normale Superiore, I-56127 Pisa, Italy
[4]Institute of Semiconductor Physics, Siberian Branch, Russian Academy of Sciences, pr. Akademika Lavrent'eva 13, Novosibirsk, 630090 Russia
Email : Christophe Consejo@univ-montp2.fr

Abstract. We report on Terahertz detection by inverted band structure HgTe-based Field Effect Transistor. Photoconductivity measurements allow for the observation of cyclotron resonance and Shubnikov-de-Haas-like oscillations. However, an unexpected peak was observed at the critical magnetic field value for which zero mode Landau Levels are crossing. Therefore, this specific feature of TeraHertz photoconductivity spectra can tentatively attributed to this magnetic field driven topological phase transition.

1. Introduction

The present main challenges in nano-electronics are to reduce power consumption and to extend field effect transistor (FET) operating frequency in more intriguing frequency domains, by reducing their size while still retaining or improving the electron mobility. For these purposes, classical semiconductor technologies are coming to their limits and new materials are recently emerging as valuable alternatives to overcome these restrictions. In this perspective, topological insulators[1] are extremely promising. Whereas their bulk is insulator, their edges are indeed composed of a set of dissipation-less states having Dirac dispersion. Therefore, because their charge carriers are massless and protected from backscattering by the intrinsic topology of the semiconductor band structure, these materials might be ideal to simultaneously achieve high electron mobility and low power consumption. Moreover, it was shown in different materials that using plasma wave effects in 2D electron channels, nanometer sized FETs can operate as efficient resonant or broadband Terahertz (THz) detectors, mixers, phase shifters and frequency multipliers at frequencies far beyond their fundamental cut-off frequency[2]. Therefore, using plasma wave phenomena in 2D topological insulator-based FETs turns out to be an extremely promising way to take up the present-day nano-optoelectronics challenges.

In this work, we study an inverted band structure HgTe-based FET by THz photoconductivity as a function of quantizing magnetic field. We show that THz detection by nonlinearities in the FET channel allows for the observation of the magnetic field-induced quantum phase transition between non-trivial inverted band structure and trivial QH regime. These experimental results pave the way towards THz topological FETs.

2. Results and discussion

We have studied an 8.3 nm wide [013]-oriented HgTe QWs embedded in between $Cd_xHg_{1-x}Te$ barriers with x = 0.77. The sample was grown by molecular beam epitaxy (MBE) with ellipsometric control on semi-insulating GaAs [013] substrates with relaxed CdTe buffers[3]. Its band structure was inverted. The sample has symmetric In-doped (n-type) areas around the QW layer.

Devices were fabricated using low-temperature positive electron beam lithography (EBL). The layered structure was etched using a plasma of CH_4:H_2 in a reactive ion etching (RIE) facility. The PMMA mask was patterned with EBL in shape of a rectangular 12 μm x 5 μm area, placed at the center of a 100 μm x 100 μm etched square. In order to realize a top-gate FET geometry, an 120 nm thick SiO_2 layer was sputtered on the sample surface and a 6 μm wide, 6 μm long gate (G) electrode was patterned at the center of the S-D distance and a 10/100 nm Ti/Au contact thermally evaporated on it.

Photoconductivity experiments were carried out at frequencies of 292 GHz and 660 GHz generated by a multiplied Schottky diode. The THz beam was modulated by a mechanical chopper at 215 Hz and guided to the sample by stainless steel pipes. The transistor was placed at the center of a superconducting coil and cooled to 4 K. We set constant 3.8 μA current in the channel. The detection signal appearing between source and drain under illumination was measured with a standard lock-in technique as a function of the magnetic field at fixed gate voltage from 0 to 750 mV.

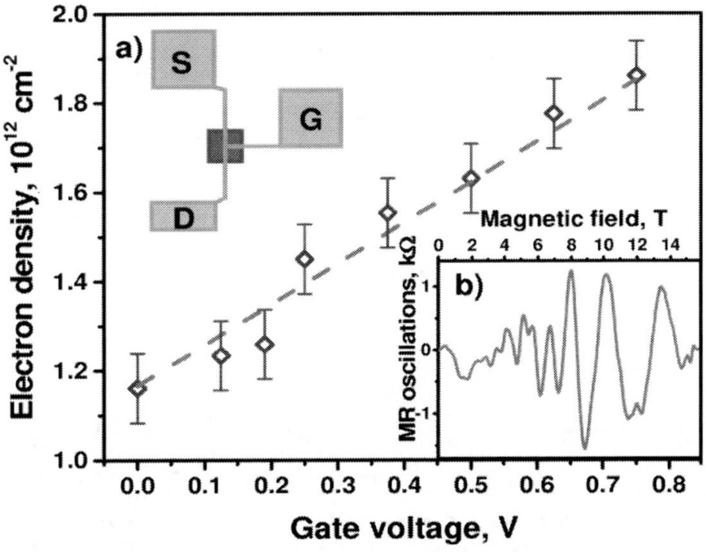

FIG. 1. Measured electron density at 4 K as a function of the swing voltage, extracted from Shubnikov-de Haas oscillation frequencies. Insert (a) is a picture of the field effect transistor, when insert (b) represents an example of measured SdH oscillations up to 16 T at Vg = 750 mV.

The FET gate was leaky but allows to slightly controlling the carrier density. Figure 1 shows the electron concentration tuning with the gate bias, extracted from Shubnikov-de Haas (SdH) oscillation frequencies. Electron density varies from approximately 1.1 to 1.8 x 10^{12} cm^{-2}. Insert (a) shows a picture of the FET used in this work. Insert (b) illustrates an example of SdH curve obtained up to 16 T at 4 K and V_g = 750 mV.

Figure 2 (a) shows the photoconductivity spectra as a function of magnetic field at 292 and 660 GHz. At low fields, one can see broad cyclotron resonances (CR) shifting with incident frequency. This assumption is in agreement with absorption spectra reported in Figure 2 (d) also showing CR lines roughly at the same positions. At approximately 6 T, a well pronounced peak is observed in photoconductivity data, although no trace of it is present in the transmission spectra.

One can notice however that the photon energy used in these experiments (1.2 and 2.7 meV) are few times larger than the heat sink thermal energy (0.35 meV) but smaller than the avoided crossing energy (4.3 meV), thus can hardly induce optical transition between these two bands. That is the reason why we have no other absorption lines in transmission spectra (Figure 2 (d)).

Furthermore, photoconductivity datas obtained on Van der Pauw configuration sample without gate, only show CR line and SdH-like oscillations, without any other visible resonance (Results are not shown here). Interestingly, the position of the resonance at 6 T, therefore only observed in FET configuration, is independent on incident frequency and carrier concentration, and corresponds to a critical magnetic field B_c for which anticrossing of zero-mode LLs is expected.

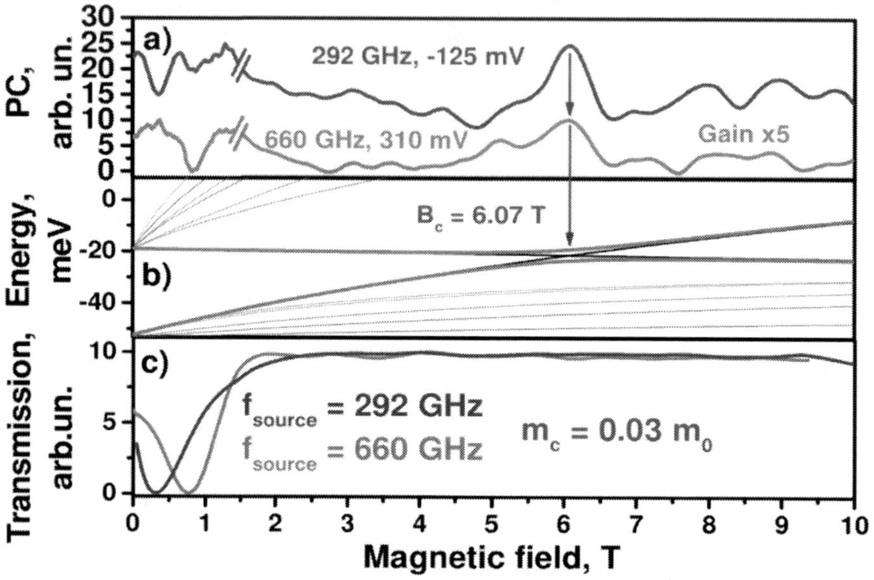

FIG. 2. Comparison between photoconductivity spectra (a and b), Landau levels energy versus magnetic field (c) and transmission spectra (d).

In HgTe inverted band structure QWs, the lowest Landau level (LL) of the conduction band contains a pure heavy hole state, when the highest LL of the valence band has a more electronic character and shifts to higher energies in magnetic field. This leads to a crossing of these two peculiar LLs for a finite magnetic field depending on the QW thickness. The observation of such a LL crossing is first clear indication for the occurrence of an inverted band structure. Therefore, at this critical magnetic field, the Dirac mass parameter M is tuned from negative to positive values, corresponding to a quantum phase transition between inverted, with non-trivial topology, and non-inverted band structure. The crossing of Landau levels for an inverted band structure was demonstrated experimentally by Schultz et al.[4]. However, it was shown in more details in[5] by magneto-transmission Fourier-transform spectroscopy that instead of crossing, these zero-mode LLs are anticrossing[6] because of bulk inversion asymmetry (BIA) of HgCdTe crystal. As measured experimentally and confirmed with the band structure calculations, for an 8.3 nm wide QW, the LL avoided crossing occurs at approximately 6 T.

Kane model calculation of valence and conduction bands splitted into LLs, taking into account BIA, is plotted in Figure 2 (b) as a function of magnetic field. One can see the anticrossing of the lower level of conduction band and the upper level of valence band at $B_c = 6$ T.

On the contrary of transmission spectroscopy, in the photoconductivity experiments, elementary excitations are detected by measuring the photo-induced nonlinear change of the resistance, which monitors exactly the electronic system nonlinearities. This provides chances for exploring unique natures of excitations unable to be investigated by conventional absorption spectroscopy[7]. Furthermore, THz photoconductivity in field effect transistors has been proven since more than 10 years in many materials, to be a powerful tool to probe the 2D electron gaz nonlinearities[8]. In usual FET, the THz electric field modulates electron density in the channel through the gate contact, and modulates carrier velocity by the source side. Thus, THz photoconductivity signal reflects a change of channel's conductivity. Under external magnetic field, if the incident radiation frequency coincides with the cyclotron frequency, electrons are efficiently heated, inducing a significant change of conductivity, giving rise to the typical photovoltaic signal. The position of the observed cyclotron resonance peak is as expected shifting to higher values with increasing incident frequency. The observed peak shifting neither with incident frequency nor with carrier concentration is measured at around 6T. This magnetic field value corresponds to the well-known peculiarity of the band structure in these QWs, i.e. the anticrossing of the two zero mode Landau levels. Therefore, we assume that this photoconductivity signal at approximately 6 T is linked with the magnetic field driven topological phase transition.

However, in our sample the Fermi level is much higher than the anticrossing position and the system is not in the quantum spin Hall regime. Therefore, topological edge states are not present and the observed peak cannot be attributed to helical edge states conductivity. Moreover, because the zero mode LLs are below the Fermi level and do not participate to the conduction, the electrons of the anticrossing region of the band structure shouldn't affect directly the channel's conductivity. Therefore, deeper theoretical work must be performed to understand how electron-electron interactions can allow the conductivity at the Fermi level to be affected by lower Landau levels.

3. Conclusion

In conclusion, we demonstrated THz photoconductivity response in inverted band structure HgTe FETs, showing a well pronounced peak at magnetic field of 6 T. Its position is independent on incident frequency and carrier concentration, and corresponds to a critical magnetic field for which particular anticrossing of zero-mode Landau levels is expected. Therefore, we assume that this resonance is related to the magnetic field-induced topological phase transition appearing in HgTe inverted band structure QWs.

Acknowledgments

This work was supported by the CNRS through LIA TeraMIR project and by the Russian Academy of Sciences, the non-profit Dynasty foundation, the Russian Foundation for Basic Research (Grants 13-02-00894, 15-02-08274), "COST action MP 1204" and by the Italian Ministry of Education, University, and Research (MIUR) through the program FIRB-Futuro in Ricerca 2010, RBFR10LULP, "Fundamental research on terahertz photonic devices".

Reference

[1] C. L. Kane, J. E. Moore 2011 *Physics World* **24** 32

[2] W. Knap, S. Rumyantsev, M.S. Vitiello, et al. 2013 *Nanotechnology* **24** 214002

[3] S. Dvoretsky, N. Mikhailov, Y. U. Sidorov, V. Shvets, S. Danilov, B. Wittman, and S. Ganichev 2010 *J. Electron. Mater.* **39** 918

[4] M. Schultz, U. Merkt, A. Sonntag, U. Rössler, R. Winkler, T. Colin, P. Helgesen, T. Skauli, and S. Lovold 1998 *Phys. Rev. B* **57** 14772

[5] M. Zholudev, F. Teppe, M. Orlita, et al.2012 *Phys. Rev. B* **86** 205420

[6] M. S. Zholudev, F. Teppe, S. V. Morozov, M. Orlita, C. Consejo, S. Ruffenach, W. Knap, V. I. Gavrilenko, S. A. Dvoretskii, N. N. Mikhailov 2015 *JETP Letters* **100** 790

[7] S. Holland, Ch. Heyn, D. Heitmann, E. Batke, et al. 2004 *Phys. Rev. Lett.* **93** 186804

[8] M. I. Dyakonov, M. S. Shur 1993 *Phys. Rev. Lett.* **71** 2465

Journal of Physics: Conference Series **647** (2015) 012010 doi:10.1088/1742-6596/647/1/012010

Relaxation and coherent oscillations in the spin dynamics of II-VI diluted magnetic quantum wells

F. Ungar[1], **M. Cygorek**[1], **P. I. Tamborenea**[1,2] **and V. M. Axt**[1]

[1] Theoretische Physik III, Universität Bayreuth, 95440 Bayreuth, Germany

[2] Departamento de Física and IFIBA, FCEN, Universidad de Buenos Aires, Ciudad Universitaria, Pab. I, C1428EHA Buenos Aires, Argentina

Abstract. We study theoretically the ultrafast spin dynamics of II-VI diluted magnetic quantum wells in the presence of spin-orbit interaction. We extend a recent study where it was shown that the spin-orbit interaction and the exchange sd coupling in bulk and quantum wells can compete resulting in qualitatively new dynamics when they act simultaneously. We concentrate on $Hg_{1-x-y}Mn_xCd_yTe$ quantum wells, which have a highly tunable Rashba spin-orbit coupling. Our calculations use a recently developed formalism which incorporates electronic correlations originating from the exchange *sd*-coupling. We find that the dependence of electronic spin oscillations on the excess energy changes qualitatively depending on whether or not the spin-orbit interaction dominates or is of comparable strength with the sd interaction.

Ultrafast spin dynamics in semiconductors is attracting nowadays a great deal of attention. In a recent article we explored the interplay between the exchange sd interaction (EXI) and the spin-orbit interaction (SOI) in II-VI diluted magnetic semiconductors (DMS) [1]. In that study we found that the EXI and the SOI can be tuned to overrun each other or to compete on an equal level in realistic bulk and quantum well systems. Importantly, we found that the inclusion of the SOI introduces oscillations in the spin dynamics which are completely absent when only the EXI is relevant. In the present article, we characterize systematically the decay and the periods of oscillations seen in the spin dynamics in quantum wells, as functions of the excess energy of the electron population in the conduction band (mean energy of a Gaussian occupation of spin-polarized photoexcited electrons). We employ a microscopic density-matrix theory that models on a quantum-kinetic level the spin dynamics, taking into account the exchange-induced correlations and the localized character of the Mn spins [2, 3]. For the sake of brevity we shall only sketch the model here and refer the reader to Refs. [1] and [3] for a complete description.

We consider conduction band electrons in Mn-doped II-VI DMS coming from low-intensity optical excitations. We work in the regime of electron densities n_e much lower than the Mn density n_{Mn}, in which the Mn spin variables are nearly stationary [3, 4, 5, 6]. The spin dynamics is described by $\langle \mathbf{s}_{\mathbf{k}}^{\perp}\rangle(t)$ and $\langle s_{\mathbf{k}}^{\parallel}\rangle(t)$, the mean electronic spin components, perpendicular and parallel to the Mn magnetization, respectively, corresponding to the conduction-band state **k**.

In this study we concentrate on $Hg_{1-x-y}Mn_xCd_yTe$ quantum wells, since this alloy offers great flexibility in the control of the Rashba SOI. This control is achieved thanks to the strong dependence of the band gap on the doping fraction $x + y$ [7]. With this material, Rashba coefficients of the order of $\alpha_R \approx 10\,\mathrm{ps}^{-1}\mathrm{nm}$ can be obtained for realistic quantum well specifications [1]. Throughout the paper we shall assume the Mn magnetization to be perpendicular to the quantum well. Furthermore, we take a Gaussian electron occupation

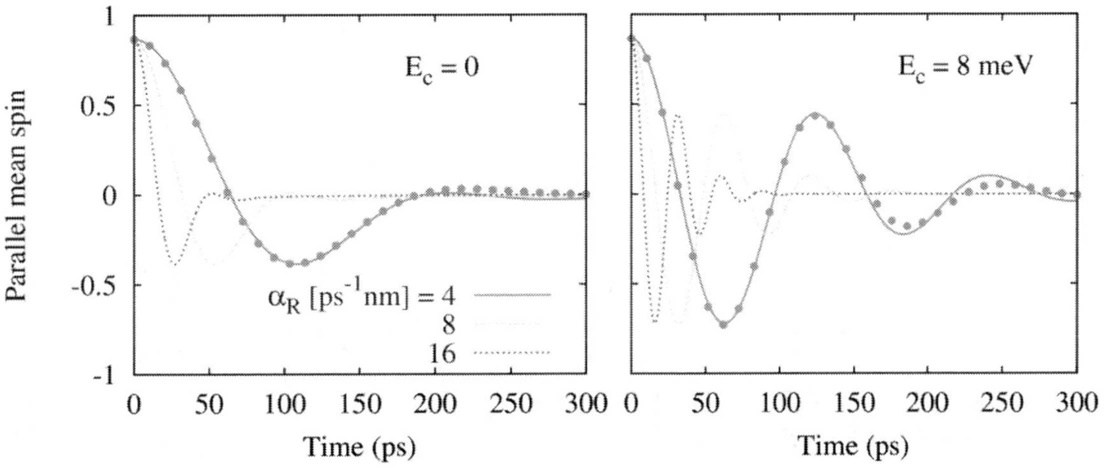

Figure 1. Electron spin dynamics in a $Hg_{1-x-y}Mn_xCd_yTe$ quantum well with Rashba interaction and no exchange sd coupling for a Gaussian electron distribution centered at E_c. Lines: simulations for different values of α_R expressed in units of ps^{-1}nm. Red dots: simple fits to the first oscillation of the curves with $\alpha_R = 4\,\text{ps}^{-1}\text{nm}$ (see text for details).

centered at an energy E_c above the conduction-band edge with a standard deviation of $E_s = 3$ meV and initial spin-polarization rotated 30° with respect to the Mn magnetization.

It is instructive to examine first the spin dynamics resulting only from the Rashba spin-orbit interaction (i.e., no exchange sd coupling is accounted for). In this case the Mn magnetization does not enter the dynamics. For later comparison we used, however, the above described initial condition where the direction of the initial electronic spin is related to direction of the Mn magnetization. Figure 1 shows the time evolution of the summed parallel spin component, $\langle s^{\|}\rangle(t) = \sum_{\mathbf{k}}\langle s^{\|}_{\mathbf{k}}\rangle(t)$, for three different values of the Rashba coupling constant α_R. The cases $E_c = 0$ (Gaussian occupation centered at the band edge) and $E_c = 8\,\text{meV}$ are shown. We see that the Rashba interaction produces well-defined oscillations and decay. Note that without the EXI, the time evolution of the total spin is given by coherent precessions of individual electron spins around the $\mathbf{k}$-dependent magnetic Rashba field, which are collectively responsible for the decay. Red dots in both panels of Fig. 1 are fits to the initial oscillations of the $\alpha_R = 4\,\text{ps}^{-1}\text{nm}$ evolutions, done with a function $f(t) \propto \exp[-(t/\tau)^2]\cos(2\pi t/T)$. For a Gaussian electron distribution with spins pointing in the growth direction, we find from Eq. (17) of Ref. [1]:

$$\langle s^{\|}(t)\rangle = C\int_0^\infty dk\; k\, \exp\left[-\frac{\left(\hbar^2k^2 - 2m^*E_c\right)^2}{(2m^*E_s)^2}\right]\cos(2\alpha_R kt), \tag{1}$$

where m^* is the effective mass and C is a constant determined by the initial value of the total spin. The integral in Eq. (1) is close to the (half-sided) Fourier transform of a function with a single central peak indicating in time regime a damped oscillation with roughly the peak frequency. An initially exponential decay of $\langle s^{\|}(t)\rangle$ would require a Lorentzian decay in the energy domain. However, the function in Eq. (1) decays much faster for large k explaining why the initial behavior of $\langle s^{\|}(t)\rangle$ is much better approximated by a Gaussian than by an exponential. Indeed, Fig. 1 reveals that the Gaussian fit is almost perfect at early times but worsens somewhat later. Applying the Gaussian fit to a number of different cases we find that for given E_c both $\tau \propto \alpha_R^{-1}$ and $T \propto \alpha_R^{-1}$ hold to a very good approximation. A similar behavior is observed for $|\langle \mathbf{s}^{\perp}\rangle(t)| = |\sum_{\mathbf{k}}\langle \mathbf{s}^{\perp}_{\mathbf{k}}\rangle(t)|$, whose initial evolution can be well fitted with

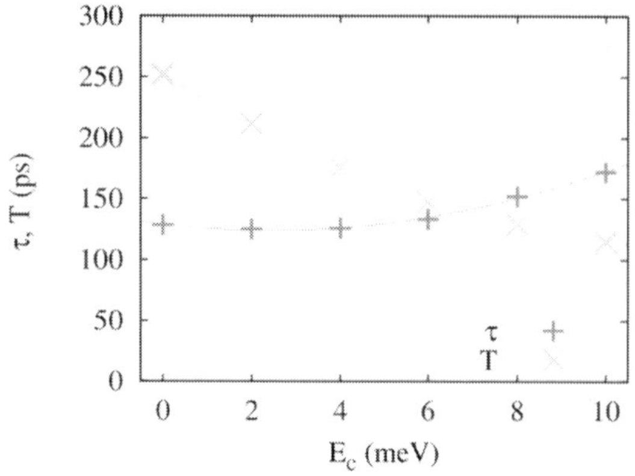

Figure 2. Gaussian electron spin relaxation time, τ, and period of the oscillations, T, in a $Hg_{1-x-y}Mn_xCd_yTe$ quantum well with Rashba interaction $\alpha_R = 4\,\mathrm{ps}^{-1}\mathrm{nm}$ and no exchange sd coupling as a function of the excess energy E_c. Symbols: full calculation; lines: parabolic fit.

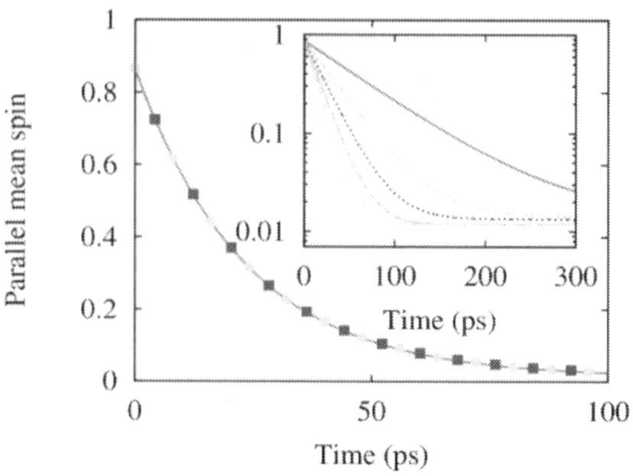

Figure 3. $\langle s^{\parallel}(t)\rangle$ with exchange sd and no Rashba interaction. $x_{\mathrm{Mn}} = 0.3\%$, $S = 0.1$, $E_c = 0$ (red line), $4\,\mathrm{meV}$ (green circles), $8\,\mathrm{meV}$ (blue squares). Inset: $E_c = 0$, $S = 0.1$, various x_{Mn}: 0.1% (red), 0.2% (green), 0.3% (blue), 0.4% (magenta), 0.5% (cyan).

$f(t) \propto \exp\left[-(t/\tau)^2\right]\cos\left(2\pi t/T\right) + 1$, with the same values of τ and T as for $\langle s^{\parallel}\rangle(t)$. We found that τ and T can be precisely fitted as functions of E_c with parabolas, as seen in Fig. 2. The decrease of T and the increase of τ with rising E_c reflect the fact that the effective Rashba field is k dependent becoming stronger for larger k.

Let us now look at the spin dynamics under the influence of the EXI only (no Rashba SOI). For the EXI coupling constant we take the value $J_{\mathrm{sd}} = 26.8\,\mathrm{meV\,nm}^3$ [1]. The effects of the EXI on the carrier spins can be controlled via the Mn concentration, x_{Mn}, and the initial net Mn magnetization, $S = |\langle \mathbf{S}\rangle|$ [1]. The dynamics of the electron spin component parallel to the Mn magnetization is shown in Fig. 3. As found previously, it is approximately described by an exponential decay to an in general non-zero equilibrium value with a decay time [cf. Eq.(19) of Ref. [3]] $\tau^{\parallel} = (J_{sd}^2 n_{Mn} m^*/\hbar^2 d)\langle S^2 - S^{\parallel 2}\rangle$, where d is the width of the quantum well and $\langle S^2 - S^{\parallel 2}\rangle$ is a second moment of the spin-$\frac{5}{2}$ Mn system. In particular, $\tau^{\parallel}$ is linear in x_{Mn} and independent of the excess energy E_c. Note that no oscillations appear in the time evolution of the parallel spin component when only the EXI is present. The perpendicular component decays to zero with a slightly different rate while it precesses around the Mn magnetization [3].

We now come to the combined effects of the Rashba SOI and EXI. Figure 4 shows the spin dynamics for $x_{\mathrm{Mn}} = 0.3\%$, $S = 0.1$, $\alpha_R \approx 4\,\mathrm{ps}^{-1}\mathrm{nm}$, and three different values of E_c. Note the semilog scale chosen to better visualize the long time behavior. This set of parameters defines a "strong sd" and "weak Rashba" situation. Accordingly, the initial decay is exponential (not Gaussian like in Fig. 1), with an E_c-independent decay rate, like in Fig. 3. However, at later times fairly regular oscillations appear, with essentially constant, E_c-dependent amplitude, due to the presence of the Rashba interaction. We note that the frequency of the oscillations depends only slightly on E_c, becoming higher for higher E_c, which indicates a slight dependence on the Rashba mechanism. The frequency of the oscillations (period of about 35 ps) is close to the precession frequency about the net Mn magnetic field, which is independent of E_c. The oscillations do not decay because all electrons precess with nearly the same frequency governed mainly by the Mn magnetization and thus do not dephase. The fact that the amplitude of the oscillations does not

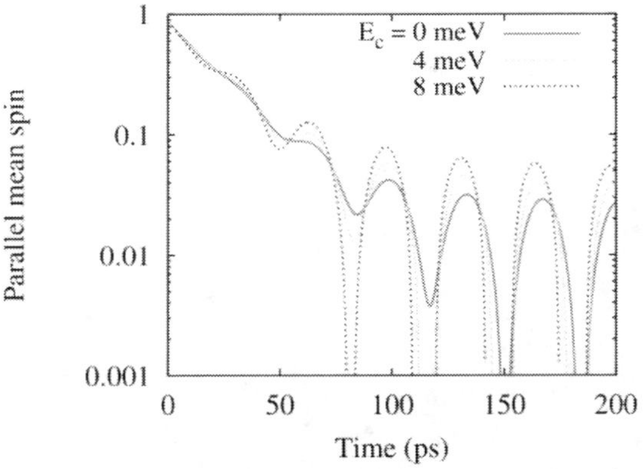

Figure 4. Evolution of the mean parallel electron spin under exchange sd and Rashba interactions. Parameters: $x_{\mathrm{Mn}} = 0.3\%$, $S = 0.1$, $\alpha_R \approx 4\,\mathrm{ps}^{-1}\mathrm{nm}$, and three values of E_c.

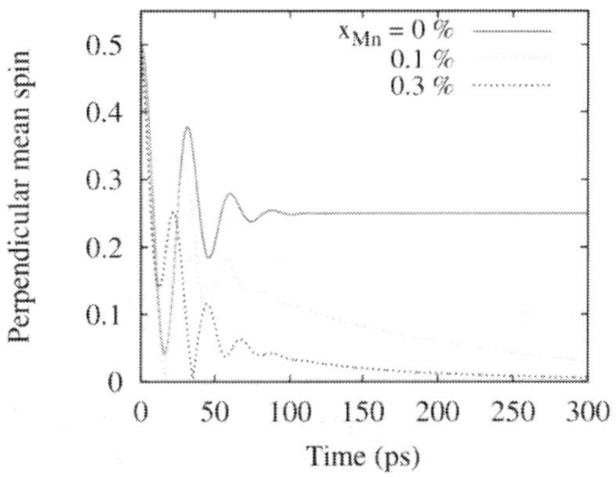

Figure 5. Evolution of the mean perpendicular electron spin component under exchange sd and Rashba interactions. Parameters: $\alpha_R = 16\,\mathrm{ps}^{-1}\mathrm{nm}$, $E_c = 8\,\mathrm{meV}$, $S = 0.1$, and three values of x_{Mn}.

decay with time and the near independence of the period on E_c distinguish qualitatively these oscillations from the ones observed with Rashba coupling alone. Also note that again there is a saturation value different from zero as seen above in the sd-only case. However, a new feature produced by the presence of the Rashba interactions is that the perpendicular spin component does not go to zero as in the sd-only evolution (not shown for brevity).

Finally, Fig. 5 shows the time evolution of the mean perpendicular spin component under EXI and SOI for fixed Rashba constant $\alpha_R = 16\,\mathrm{ps}^{-1}\mathrm{nm}$, $E_c = 8\,\mathrm{meV}$, and three different values of x_{Mn}. This figure shows the effect of an increasing EXI coupling in the presence of a strong Rashba coupling on the perpendicular spin component. We see that as the Mn concentration increases, starting from a Rashba-only situation in which the equilibrium value $\langle s_{\mathbf{k}}^{\perp} \rangle$ is half its initial value [1], a decay to zero sets in. At the same time, the frequency of the oscillations increases and their amplitude goes down.

In conclusion, we have studied theoretically the effects of the Rashba spin-orbit interaction in II-VI diluted-magnetic-semiconductor quantum wells. We characterized the dependence of the spin dynamics on the excess energy of a Gaussian population of electrons in the conduction band. Our findings provide qualitative signatures that could aid experimentalists in distinguishing the relative importance of spin-orbit and exchange interactions in DMS quantum wells.

We gratefully acknowledge the financial support of the Deutsche Forschungsgemeinschaft (grant No. AX17/9-1), the Universidad de Buenos Aires (UBACyT 2011-2014 No. 20020100100741), and CONICET (PIP 11220110100091).

References

[1] Ungar F, Cygorek M, Tamborenea P I and Axt V M 2015 *Phys. Rev* B **91** 195201
[2] Thurn C and Axt V M 2012 *Phys. Rev.* B **85** 165203
[3] Cygorek M and Axt V M 2015 *Semicond. Sci. Technol.* **30** 085011
[4] Cygorek M and Axt V M 2014 *Phys. Rev.* B **90** 035206
[5] Thurn C, Cygorek M, Axt V M and Kuhn T 2013 *Phys. Rev.* B **88** 161302(R)
[6] Thurn C, Cygorek M, Axt V M and Kuhn T 2013 *Phys. Rev.* B **87** 205301
[7] Kossut J 1988 *Diluted Magnetic Semiconductors* (*Semiconductors and Semimetals* vol 25) ed Furdyna J K and Kossut J (San Diego: Academic Press)

Journal of Physics: Conference Series **647** (2015) 012011 doi:10.1088/1742-6596/647/1/012011

Electron and hole in-plane g-factors in single InAs quantum rings

R. Kaji[1], T. Tominaga[1], Y.-N. Wu[2], S.-J. Cheng[2], and S. Adachi[1]

[1] Department of Applied Physics, Hokkaido University, Sapporo, Japan

[2] Department of Electrophysics, National Chiao Tung University, Hsinchu, Taiwan

E-mail: r-kaji@eng.hokudai.ac.jp

Abstract. The electron and hole in-plane g-factors were studied in individual InAs/GaAs quantum rings (QRs). From the magneto-photoluminescence measurements under a transverse magnetic field, we evaluated the in-plane g-factors of electron and hole spins by rotating the sample systematically along the crystal growth axis. The experimental results indicate that the in-plane and the out-of-plane anisotropies in hole g-factor are larger than those of electron g-factor, and the value of the hole in-plane g-factor varies largely from QR to QR while the electron g-factor is almost a constant value. From the model calculation considering the effects of shape anisotropies and uniaxial stress, we examined the correlation between the in-plane g-factors and the degree of valence band mixing. Further, the experimentally obtained trend in g-factors was in agreement qualitatively with theoretical consideration.

1. Introduction

Spin confined in the semiconductor nanostructures is one of the prospective candidates for the basis in coming quantum information technology. The confinement of carrier wave functions yields a long spin coherence time via the strong suppression of spin relaxation processes originated from the spin-orbit interaction, and indeed, quite longer spin coherence times compared to that reported in the bulk or quantum wells have been achieved in the quantum dots (QDs). [1] Thus, the spin manipulation in these systems has gained considerable attention from both scientific and engineering viewpoints. Since the hole spin is expected to have a longer coherence time than the electron spin in QDs due to a weaker interaction with lattice nuclei, [2, 3] an arbitrary manipulation of hole spin is especially favored.

One of the key parameters for spin manipulation is carrier g-factor. Several works have focused on developing techniques for the control and the evaluation of electron and/or hole g-factors. These techniques include the appropriate adoption of materials and structures in III-V bulk semiconductors, [4] the application of bias voltages to In(Ga)As QDs, [5, 6] and the utilization of nuclear spin polarization in InAlAs QDs. [7] Since In(Ga)As and GaAs are widely-used III-V materials, the electron and hole g-factors in their QDs have been studied intensively. [8]

In this study, we focus on the electron and hole g-factors (g^e, g^h) in the individual InAs/GaAs quantum rings (QRs). Semiconductor QRs exhibit not only the discrete energy levels like QDs, but also the distinctive properties such as the Aharanov-Bohm effect. [9, 10] In the previous work, we have evaluated the longitudinal components of (g^e, g^h), where the hole g-factor showed the correlation to the degree of valence band mixing. [11] In order to obtain the comprehensive knowledge of g-factors in QRs, we investigated the transverse components of (g^e, g^h) in the same InAs/GaAs QR sample.

2. Sample and Setup

A self-assembled InAs/GaAs QR sample grown on an undoped (100) GaAs substrate [12] was studied under the transverse magnetic fields ($|B_\perp| \leq 5$ T) at 6 K. The average inner and outer radii and height of QR were found to be 10 nm, 40 nm, and 1.5 nm, respectively, from AFM measurements on a reference uncapped layer. These values were considered in the model calculations described below. Photoluminescence (PL) spectra from single QRs were detected by a triple grating spectrometer and a liquid N_2-cooled Si-CCD. The energy resolution to determine the spectral peaks was ~5 μeV. A cw-Ti:sapphire laser tuned to the transition energy at the wetting layer of QRs (λ ~864 nm) was employed for excitation, and its polarization was set to temporal averaging nonpolarization in order to avoid the additional effects depending on the initial population of carrier spins. [13] The polarization of PL signal was determined by a rotating half-wave plate and a linear polarizer inserted in the detection path.

3. Results and Discussion

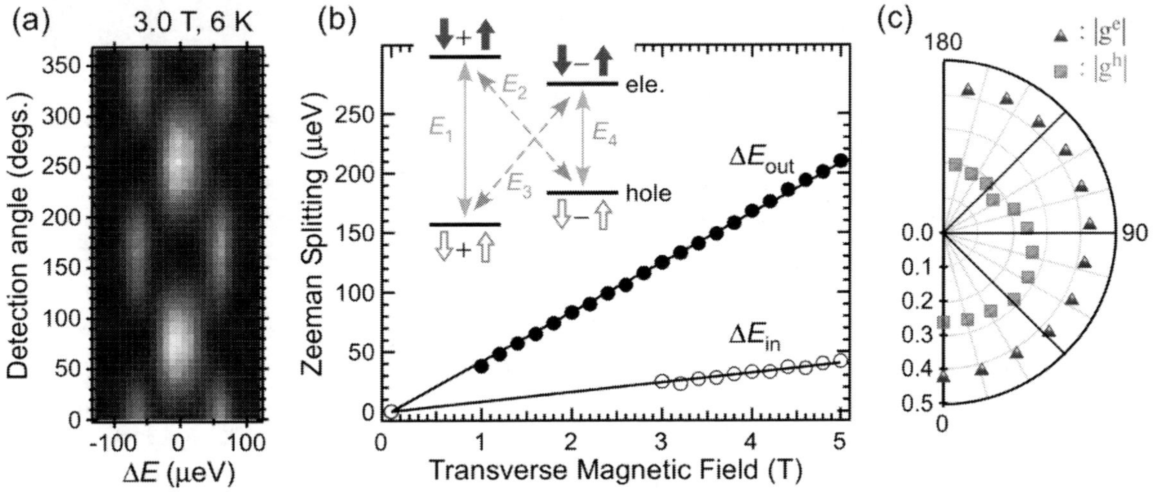

Figure 1. (a) 2D plot of the charged exciton PL as a function of the detection angle. The horizontal axis was converted to the energy difference (ΔE) from E_0(=1.3272 eV). (b) Energy splittings of the outer and inner peaks as a function of $B_\perp$. Inset shows the energy diagram of electron (hole) spins depicted as solid (open) arrows. (c) Polar plot of $|g^e_\perp|$ (triangles) and $|g^h_\perp|$ (squares) in a typical individual QR.

First, we explain briefly the evaluation of the electron and hole in-plane g-factors ($g^e_\perp$, $g^h_\perp$). As shown in the inset of Fig. 1(b), the transverse magnetic field $B_\perp$ induces the spin-state mixing, and thus, the four transitions (E_1, E_2, E_3, E_4) with linearly polarized emission and absorption become optically-active. The magnitudes of ($g^e_\perp$, $g^h_\perp$) can be deduced from the Zeeman splittings of the outer peaks $\Delta E_{\rm out}=E_1 - E_4$ and the inner peaks $\Delta E_{\rm in}=E_2 - E_3$ whose proportional coefficients are given by the sum or difference of ($|g^e_\perp|$, $|g^h_\perp|$). Figure 1(a) shows the PL spectra of the charged exciton from a typical single QR as a function of the detection angle under $B_\perp$=3.0 T. The spectra indicate the four peaks with cross-linear polarizations between the outer (E_1, E_4) and inner peaks (E_2, E_3), and the Zeeman splitting $\Delta E_{\rm out}$ ($\Delta E_{\rm in}$) depicted as solid (open) circles in Fig. 1(b) increases linearly with increasing $B_\perp$. From the line fittings, the in-plane g-factors were found to be $|g^e_\perp|=0.43 \pm 0.01$ and $|g^h_\perp|=0.26 \pm 0.01$.

Further, we investigated the in-plane anisotropies of ($|g^e_\perp|$, $|g^h_\perp|$) by rotating the sample around the crystal growth axis (z-axis). Figure 1(c) shows the polar plot of ($|g^e_\perp|$, $|g^h_\perp|$). While $|g^e_\perp|$ (triangles) shows the isotropic nature, $|g^h_\perp|$ (squares) varies depending on the direction of $B_\perp$ and shows the anisotropic feature. To quantify the degree of anisotropy in ($|g^e_\perp|$, $|g^h_\perp|$), we introduced an anisotropic parameter $\alpha^i = (|g^i_{\rm max}| - |g^i_{\rm min}|)/\overline{|g^i_\perp|}$ (i= e, h), where $|g^i_{\rm max(min)}|$ is the maximum (minimum) value of $|g^i_\perp|$ and

$\overline{|g^i_\perp|}$ is the average value over the measurements under the different direction of $B_\perp$. The anisotropic parameters of ($|g^e_\perp|$, $|g^h_\perp|$) in this QR were found to be $\alpha^e = 0.035$ and $\alpha^h = 0.470$, respectively.

We performed the same procedure for the several QRs in the same sample and obtained the following findings; $\overline{|g^h_\perp|}$ varies largely from QR to QR ($0.04 \leq \overline{|g^h_\perp|} \leq 1.03$) compared to $\overline{|g^e_\perp|}$ ($0.34 \leq \overline{|g^e_\perp|} \leq 0.50$), and α^h is larger than α^e for most of the measured QRs. In addition, g^h shows a large out-of-plane anisotropy, that is, a large difference between the transverse and the longitudinal components ($\overline{|g^h_\perp|}_{\rm ave} = 0.58$, $|g^h_\parallel|_{\rm ave} = 1.50$) as contrast with the small out-of-plane anisotropy of g^e ($\overline{|g^e_\perp|}_{\rm ave} = 0.44$, $|g^e_\parallel|_{\rm ave} = 0.60$). [11] The observed differences in (g^e, g^h) are attributed to the differences in the symmetry of their Bloch functions and/or in the mixing with the other energy bands.

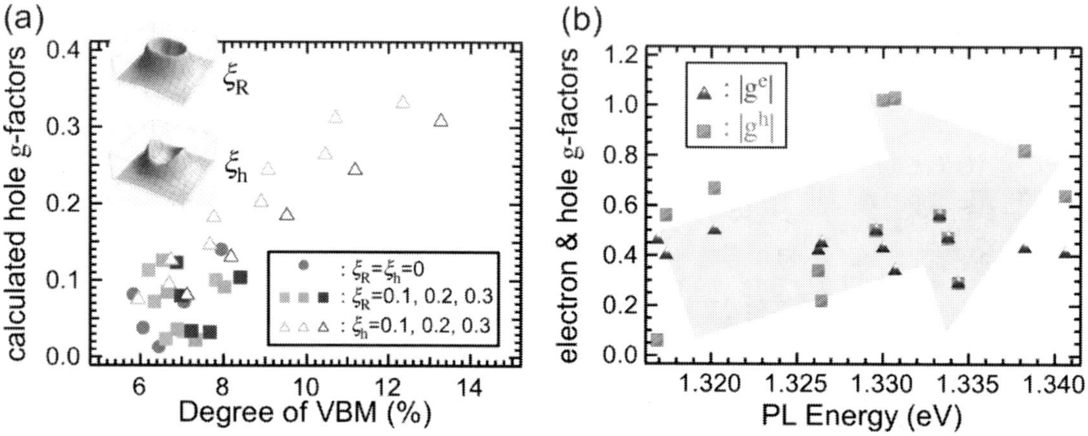

Figure 2. (a) The calculated hole g-factors in x-direction as a function of β. Various symbols indicate the implementation of the shape anisotropies in height and radius (ξ_h, ξ_R) and the external stress in x-direction. The QR size was fixed to the measured mean size. (b) Experimentally obtained $\overline{|g^e_\perp|}$ (triangles) and $\overline{|g^h_\perp|}$ (squares) as a function of $E_{\rm PL}$ of the corresponding individual QRs. $\overline{|g^h_\perp|}$ indicates a certain degree of dependence on $E_{\rm PL}$ while $\overline{|g^e_\perp|}$ is in almost constant over the whole range.

Next, we consider the effect of band mixing, especially in the valence band mixing (VBM), on the observed ($|g^e_\perp|$, $|g^h_\perp|$). The VBM, which is the band mixing between the heavy holes (HHs: $\varphi_{\pm 3/2}$) and the light holes (LHs: $\varphi_{\pm 1/2}$), changes the pure HHs to the "*mainly HHs*" defined as $|\tilde{\varphi}^{\pm}\rangle = (|\varphi_{\pm 3/2}\rangle + |\beta| e^{\pm i 2\theta_s} |\varphi_{\mp 1/2}\rangle)/\sqrt{1+\beta^2}$, where β and θ_s are the degree of VBM and the coupling phase, respectively. One of the remarkable effects originated from VBM is the modulation of g^h, which has been reported for the longitudinal component. [11] The in-plane g-factors of pure HHs and LHs can be written as $g^{\rm HH}_\perp = 3q_\perp$ and $g^{\rm LH}_\perp = 4\kappa_\perp + 10q_\perp$, respectively ($q_\perp$, $\kappa_\perp$: in-plane component of Luttinger parameter). Although $g^{\rm HH}_\perp$ in bulk semiconductor is negligible due to the relation $q_\perp \ll \kappa_\perp$, non-zero values of hole in-plane g-factors have been observed commonly in the self-assembled QDs and QRs. The reason is that the VBM is enhanced in these self-assembled structures, and thus, the state we could access is not the pure HH but the *mainly HH*. In general, the origin of the VBM is attributed to the confinement of carrier spins and the anisotropies in the shape, in the interfacial properties, [14] and in the distribution of the residual strain. [15] Since our QRs were originated from self-assembled QDs formed with the lattice mismatch to the barrier component, the residual strain was the most expected source of VBM. Note that the VBM does not affect so much on g^e even in the QDs and QRs.

Figure 2(a) shows the calculated hole g-factor in x-direction (g^h_x) as a function of the degree of VBM, β, for asymmetric strained QRs. The simulation was carried out using Luttinger-Kohn four band $\boldsymbol{k} \cdot \boldsymbol{p}$ theory [16] for volcano-shaped QRs with the consideration of shape anisotropies in height and radius of ring (parametrized by the parameters ξ_h and ξ_R, respectively). [17] To highlight the VBM effects, a controlled external uniaxial stress was applied to the QRs along the shape-elongation axis, and it was varied from $\sigma = -0.2$ GPa (compressive) to $\sigma = +0.2$ GPa (tensile) to tune the degree of VBM and to examine the corresponding changes of g-factor, as shown in Fig. 2(a). Other parameters used in the calculation were adopted by considering the materials of QR and barrier layers. From the calculation, the following was found; increasing the shape anisotropies ($\xi_{\rm h}$, $\xi_{\rm R}$) and the magnitude of the tensile stress, the degree of VBM and resultant $g_x^{\rm h}$ increase as overall trend since they enhance the anisotropy in the system of our model. Although the magnitude of calculated $g_x^{\rm h}$ are smaller than the experimentally obtained $|g_\perp^{\rm h}|$ at this stage, all calculated points agree with simple consideration. Further, we found that even without the effects of shape anisotropies and the external stress, the quantum confinement could induce a significant VBM about 6.5% in our small QRs.

To confirm the theoretical considerations, we plotted the observed g-factors ($|g_\perp^{\rm e}|$, $|g_\perp^{\rm h}|$) as a function of the PL energy ($E_{\rm PL}$) in Fig. 2(b). $E_{\rm PL}$ can be used as a measure of the confinement strength of carrier wave functions. Although the large variation in a whole range of $E_{\rm PL}$ obscures the trend, $|g_\perp^{\rm h}|$ seems to increase with increasing $E_{\rm PL}$, and this point is in contrast with the case of $|g_\perp^{\rm e}|$; it seems to keep a constant value. The increase in the confinement strength yields the increases in both $E_{\rm PL}$ and β. Thus $|g_\perp^{\rm h}|$ is expected to increase corresponding to the change in $E_{\rm PL}$, and indeed, this simple consideration agrees qualitatively with experimental observation. More detailed discussion including the quantitative comparison between experimental and theoretical results will been seen in future works.

4. Conclusion

The electron and hole in-plane g-factors in individual self-assembled InAs/GaAs QRs were studied. The evaluated hole in-plane g-factor varies largely from QR to QR, and the in-plane and the out-of-plane anisotropies in hole g-factor were observed; these features were contrast with the case of electron g-factor. From the model calculation, the effects of in-plane anisotropies and the confinement of carrier spins on the degree of VBM and the resultant change in hole in-plane g-factor were investigated. Although more detailed discussions are necessary for the quantitative agreement between the experimental and theoretical results, our observations will serve a useful information for the hole spin manipulation in nanostructures.

Acknowledgments

We are grateful to S. Suraprapapich and C. W. Tu of the University of California, San Diego, for providing the quantum ring samples. This work has been supported by JSPS KAKENHI Grant Number 26800162.

References

[1] Greilich A, Oulton R, Zhukov E A, Yugova I A, Yakovlev D R, Bayer M, Shabaev A, Efros Al L, Merkulov I A, Stavarache V, Reuter D and Wieck A 2006 *Phys. Rev. Lett.* **96**, 227401
[2] Testelin C, Bernardot F, Eble B and Chamarro M 2009 *Phys. Rev.* B **79**, 195440
[3] Kaji R, Ohno S, Hozumi T and Adachi S 2013 *J. Appl. Phys.* **113** 203511
[4] Kosaka H, Kiselev A A, Baron F A, Kim K W and Yablonovitch E 2001 *Electron. Lett.* **37** 464
[5] Nakaoka T, Tarucha S and Arakawa Y 2007 *Phys. Rev.* B **76** 041301(R)
[6] Bennett A J, Pooley M A, Cao Y, Sköld N, Farrer I, Ritchie D A and Shields A J 2013 *Nat. Comm.* **4** 1522
[7] Kaji R, Adachi S, Sasakura H and Muto S 2007 *Appl. Phys. Lett.* **91** 261904
[8] Bayer M, Kuther A, Forchel A, Gorbunov A, Timofeev V B, Schafer F and Reithmaier J P 1999 *Phys. Rev. Lett.* **82** 1748
[9] Lorke A, Luyken R J, Govorov A O and Kotthaus J P 2000 *Phys. Rev. Lett.* **84** 2223
[10] Bayer M, Korkusinski M, Hawrylak P, Gutbrod T, Michel M and Forchel A 2003 *Phys. Rev. Lett.* **90** 186801
[11] Kaji R, Hozumi T, Hachiyama Y, Tomii T, Sasakura H, Jo M and Adachi S 2014 *Appl. Phys. Express* **7** 065002
[12] Suraprapapich S, Panyakeow S and Tu C W 2007 *Appl. Phys. Lett.* **90** 183112
[13] Ohno S, Adachi S, Kaji R, Muto S and Sasakura H 2011 *Appl. Phys. Lett.* **98** 161912
[14] Krebs O and Voisin P 1996 *Phys. Rev. Lett.* **77** 1829
[15] Léger Y, Besombes L, Maingault L and Naruette H 2007 *Phys. Rev.* B **76** 045331
[16] van Bree J, Silov A Yu, Koenraad P M, Flatté M E and Pryor C E 2012 *Phys. Rev.* B **85** 165323
[17] Fomin V M, Gladilin V N, Klimin S N, Devreese J T, Kleemans N A J M and Koenraad P M, 2007 *Phys. Rev.* B **76** 235320

Journal of Physics: Conference Series **647** (2015) 012012 doi:10.1088/1742-6596/647/1/012012

Optical control of exciton and spin states in a quantum dot by excitation with twisted light

T Kuhn[1], **D E Reiter**[1,2] **and G F Quinteiro**[3]

[1] Institut für Festkörpertheorie, Universität Münster, 48149 Münster, Germany
[2] Imperial College London, South Kensington Campus, London SW7 2AZ, UK
[3] Departamento de Física and IFIBA, Universidad de Buenos Aires, Buenos Aires, Argentina

E-mail: `Tilmann.Kuhn@uni-muenster.de`

Abstract. We discuss how the excitation of a quantum dot can benefit from the special properties of twisted light beams. First we address the question of the appropriate theoretical description of the light-matter coupling in this system, i.e., of the most appropriate gauge. Then we discuss the optical transitions induced by different types of twisted light beams showing that both the envelope and the spin state of the exciton can be well controlled.

1. Introduction

Highly inhomogeneous light beams carrying a phase singularity at the beam center – also called vortex beams or twisted light (TL) – attract much attention for their interesting effects in different areas of physics [1]. Theoretical studies in solid-state physics have predicted, for instance, that TL can induce electric currents in quantum rings [2] and new electronic transitions (forbidden for plane waves) in quantum dots (QDs) [3, 4]. In this contribution we will discuss how the excitation of a QD by different types of TL beams can be used to control both the envelope and the spin parts of the generated excitons.

2. Gauge-invariant formulation of the twisted-light–matter interaction

We study the excitation of a QD by a TL Bessel beam with the vector potential given in cylindrical coordinates (r, φ, z) by

$$\mathbf{A}^{(\ell,\sigma)}(\mathbf{r},t) = \frac{1}{2}A_0\left[J_\ell(q_r r)\mathbf{e}_r + i\sigma J_\ell(q_r r)\mathbf{e}_\varphi - i\sigma\frac{q_r}{q_z}J_{\ell+\sigma}(q_r r)\mathbf{e}_z\right]e^{i[(q_z z-\omega t)+(\ell+\sigma)\varphi]} + \text{c.c.} \quad (1)$$

Here, A_0 is the amplitude, ω the frequency, q_r and q_z the wave vectors in longitudinal and radial direction, respectively, and $\mathbf{e}_r, \mathbf{e}_\varphi, \mathbf{e}_z$ are unit vectors. J_ℓ is a Bessel function with the integer ℓ characterizing the topological charge of the beam, $\sigma = \pm 1$ describes the handedness of the circular polarization (CP), and c.c. denotes the complex conjugate. Equation (1) is an exact solution of the vectorial Helmholtz equation and it satisfies the Coulomb and radiation gauge, i.e., $\text{div}\mathbf{A}(\mathbf{r},t) = 0$ and $U(\mathbf{r},t) = 0$, U being the scalar potential. For beams with large beam waist, i.e., $(q_r/q_z) \ll 1$, the z-component can often be neglected, which corresponds to the paraxial approximation. Here we will consider tightly focused beams, in which $q_r \approx q_z$, and therefore keep the full vector potential. For simplicity, we will refer to ℓ as the orbital angular momentum (OAM), although this strictly holds only in the paraxial limit.

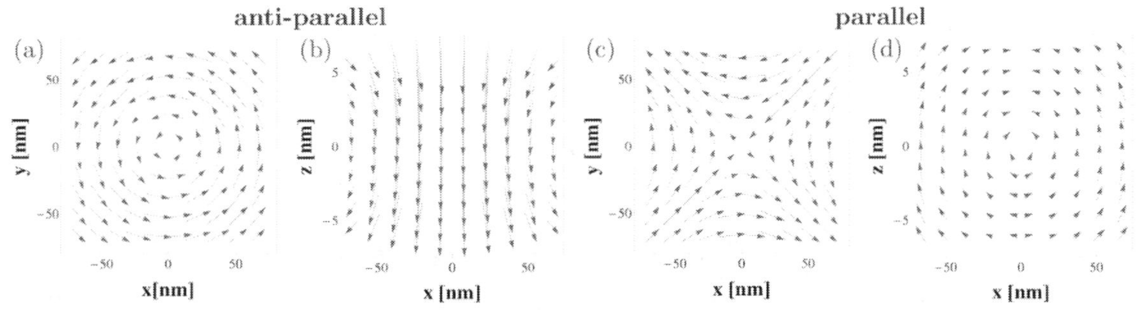

Figure 1. Electric field profiles of tightly focused TL Bessel beams with OAM $\ell = 1$ traveling along the z-direction. (a),(b): CP $\sigma = -1$ (anti-parallel class); (c),(d): CP $\sigma = +1$ (parallel class). (a),(c): profiles in the xy-plane; (b),(d): profiles in the xz-plane.

The general form of the light-matter interaction is given in terms of a Hamiltonian $\sim \mathbf{p}\cdot\mathbf{A}(\mathbf{r}, t)$ with the momentum operator $\mathbf{p}$. This form has the drawback that it depends on the gauge. Although the results are of course gauge independent as long as exact calculations are being performed, this does not hold anymore if approximations are made. The light-matter coupling in QDs is typically treated in terms of the dipole coupling by the Hamiltonian $-\mathbf{d}\cdot\mathbf{E}(\mathbf{r}_0, t)$, where $\mathbf{d}$ is the dipole operator and $\mathbf{E}(\mathbf{r}_0, t)$ is the electric field at the position of the QD. Being described in terms of the electric field, it is obviously gauge invariant. However, if the QD is placed close to the center of a TL beam – which is necessary to benefit from the peculiarities of this beam – the field is strongly inhomogeneous over the QD and it even often vanishes directly at the center, such that this form cannot be used. We have recently shown [5] that a gauge (the TL gauge) exists where the coupling can still be written in terms of the electric field, however, in a form different from the dipole coupling. For a TL beam with given ℓ and σ the coupling Hamiltonian in the TL gauge then reads

$$H_{\text{TL-matter}} = \frac{1}{|\ell|+1} erE_r^{(\ell,\sigma)}(\mathbf{r}, t) + \frac{1}{|\ell+\sigma|+1} ezE_z^{(\ell,\sigma)}(\mathbf{r}, t) \qquad (2)$$

with $\mathbf{E}^{(\ell,\sigma)} = -\partial\mathbf{A}^{(\ell,\sigma)}/\partial t$ and e being the elementary charge. In [5] we have shown that this is always a good approximation for beams in which $\mathrm{sign}(\ell) = \mathrm{sign}(\sigma)$, which we call the parallel class. In the anti-parallel class $[\mathrm{sign}(\ell) \neq \mathrm{sign}(\sigma)]$ in general the magnetic terms cannot be neglected. This different behavior can be traced back to the electric field profiles, which are shown in Fig. 1 for the example of TL beams with $\ell = 1$ and $\sigma = \pm 1$ traveling along the z-axis. In both cases the in-plane field is described by the function J_1 and therefore has a vortex at the beam center. The in-plane field of the parallel class [Fig. 1(c)] can be well approximated by a gradient of a scalar potential, which is obviously not true for the field profile in the anti-parallel case [Fig. 1(a)]. However, while in the parallel class the electric field close to the beam center is dominated by the in-plane part (the z-component involves the function J_2), in the anti-parallel class the z-component (involving the function J_0) dominates [Fig. 1(b)]. It is essentially homogeneous in that region and therefore can also be written as a gradient field. Hence, if there is a coupling to the z-component, this usually dominates and the TL gauge is valid also here.

3. Twisted-light–induced transitions in quantum dots

We will analyze the transitions that can be induced by exciting a QD, which is placed close to the beam axis, by different types of TL beams. We assume that the lateral size of the QD is smaller than q_r^{-1} and the height is smaller than q_z^{-1}, such that the lowest orders in the expansions in

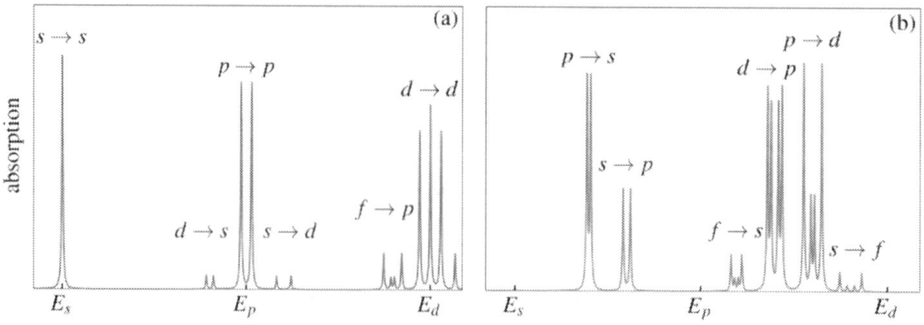

Figure 2. Qualitative shape of a QD absorption spectrum as a function of the photon energy for the case of excitation by the in-plane electric field of a TL beam (a) with $\ell = 0$ and (b) with $\ell = \pm 1$. The labels denote the initial (VB) and final (CB) shell of the transition.

$(q_r r)$ and $(q_z z)$ are sufficient. The electron and hole wave functions in a QD can be written as a product of an envelope function $\phi^{e,h}(\mathbf{r})$ and the lattice-periodic Bloch functions of conduction band (CB) and valence band (CB). Typically the CB Bloch functions are of s-type with total angular momentum $J^e = \frac{1}{2}$ and $J^e_z = \pm\frac{1}{2}$, i.e., $|1/2, +1/2\rangle = |s\uparrow\rangle$ and $|1/2, -1/2\rangle = |s\downarrow\rangle$, while the VB Bloch functions with total angular momentum $J^h = \frac{3}{2}$ can be separated into heavy hole (HH) states with $J^h_z = \pm 3/2$ and light hole (LH) states with $J^h_z = \pm 1/2$ given by

$$\begin{aligned} |3/2, +3/2\rangle &= 2^{-1/2}|(p_x + ip_y)\uparrow\rangle, \quad |3/2, +1/2\rangle = 6^{-1/2}\left(|(p_x + ip_y)\downarrow\rangle - 2|p_z\uparrow\rangle\right), \\ |3/2, -3/2\rangle &= 2^{-1/2}|(p_x - ip_y)\downarrow\rangle, \quad |3/2, -1/2\rangle = 6^{-1/2}\left(-|(p_x - ip_y)\uparrow\rangle - 2|p_z\downarrow\rangle\right). \end{aligned} \tag{3}$$

For the calculation of the TL-induced transition rates – and thus of the absorption spectrum – the matrix elements of the interaction Hamiltonian (2) between VB and CB states are required. Since the electric field varies slowly over an elementary cell and the Bloch functions of CB and VB have opposite parity, the transition matrix elements can be separated into a product of the bulk dipole matrix element $\mathbf{d}_{cv}$ and envelope integrals $\mathbf{G}^{eh}$ with

$$d_{cv} = \langle s|x|p_x\rangle = \langle s|y|p_y\rangle = \langle s|z|p_z\rangle, \qquad \mathbf{G}^{eh} = \int \phi^{e*}(\mathbf{r})\mathbf{E}(\mathbf{r})\phi^h(\mathbf{r})d^3r. \tag{4}$$

We model the QD by harmonic confinement potentials for electrons and holes, where we assume that the confinement in z-direction (growth direction) is much stronger than in x- and y-direction. The energy levels of electrons and holes are then grouped into shells denoted by s, p, d, f ... We assume a slight in-plane asymmetry, such that the states in higher shells are split, and characterize the transition by the initial (VB) and final (CB) shell.

Let us start by considering a beam with $\ell = 0$ and $\sigma = \pm 1$. In such a beam the in-plane components of the field dominate. Its radial dependence is given by $J_0(q_r r)$ and is therefore essentially homogeneous over the QD. Hence the transitions induced by this beam are the same as for plane waves. Both in the case of transitions from the HH and the LH band excitons with $J_z = J^e_z + J^h_z = \pm 1$ are excited. A qualitative absorption spectrum is plotted in Fig. 2(a). We find strong peaks when the VB and CB shells agree and weaker peaks when the the VB and CB shell index differs by two (in general by an even number). The presence of these peaks is due to a slight difference between electron and hole wave functions of the same shell. Similar spectra have been obtained with more refined exciton models and are also seen in experiments [6].

As the first genuine TL beam we consider a beam with $\ell = \pm 1$ and $\sigma = \ell$, i.e., from the parallel class. Also here the in-plane components dominate; however, now their radial dependence is given by $J_1(q_r r)$, which in the region of the dot can be approximated by $(q_r r)$, and the angular dependence is $e^{\pm i\varphi}$. At the beam center there is a vortex with vanishing electric and magnetic field. The electric field has odd parity, such that only transitions between shells differing by one (in general by an odd number) can be excited. Such a spectrum is qualitatively shown in Fig. 2(b). Since only in-plane dipole matrix elements are involved, the spectrum is again qualitatively the same for HH and LH excitons. Thus we find that with this kind of TL beams excitons which are usually dark due the symmetry of their envelope functions can be generated.

Finally we consider a beam with again $\ell = \pm 1$ but now $\sigma = -\ell$, i.e., a beam belonging to the anti-parallel class. In such a beam the radial dependence of the in-plane components is $\sim J_1(q_r r)$, while that of the z-component is $\sim J_0(q_r r)$. Therefore the latter one is much larger and, as soon as the dipole matrix element has a non-vanishing z-component, this will give rise to the dominant transition. While for HH excitons there are no such transitions, LH excitons with angular momentum $J_z = 0$ can be excited [see Eq. (3)]. Since the integrals over the envelope functions are the same as in the case of the beam with $\ell = 0$, the spectrum will again look similar to Fig. 2(a). We thus find that this kind of TL beam can be used to generate excitons with the same envelope structure as in the case of plane waves, but with different spin configuration.

4. Discussion and conclusions

We have shown that TL beams, due to their more complex spatial structure compared to plane waves, provide additional flexibility which can be used to excite specific transitions in QDs: (i) beams with dominant in-plane components and $\ell \neq 0$ can excite transitions from the p- to the s-shell and vice versa; (ii) beams with dominant z-component can excite LH excitons with $J_z = 0$. Let us comment in some more detail on this latter aspect.

For different applications in quantum information processing using the switching either of the electron spin in a charged QD [7] or of the spin of a magnetic impurity in a QD [8] LH excitons have been shown to be useful or even necessary for the desired operation. This is due to the possibility to optically address excitons with $J_z = 0$. With plane waves this requires cleaving the sample and exciting from the side, in addition to an excitation from the top, which is still necessary to address excitons with $J_z = \pm 1$. TL therefore provides the big advantage of addressing all transitions with co-propagating beams. QD structures, in which the lowest exciton transition is of LH type, have indeed recently been fabricated [9] using a suitable strain engineering. Therefore the combination of these structures with TL excitation opens up new ways for spin control phenomena in QD systems.

Acknowledgments

GFQ acknowledges financial support from the Argentine research agency Consejo Nacional de Investigaciones Científicas y Técnicas (CONICET) through the *Programa de Becas Externas*. DER gratefully acknowledges financial support from the German Academic Exchange Service (DAAD) within the P.R.I.M.E. program.

References

[1] Andrews D L 2008 *Structured light and its applications: An introduction to phase-structured beams and nanoscale optical forces* (New York: Academic Press)
[2] Quinteiro G F and Berakdar J 2009 *Optics Express* **17** 20465
[3] Quinteiro G F and Tamborenea P I 2009 *Phys. Rev.* B **79** 155450
[4] Quinteiro G F and Kuhn T 2014 *Phys. Rev.* B **90** 115401
[5] Quinteiro G F, Reiter D E and Kuhn T 2015 *Phys. Rev.* A **91** 033808
[6] Huneke J *et al.* 2011 *Phys. Rev.* B **84** 115320
[7] Pazy E *et al.* 2003 *Europhys. Lett.* **62** 175
[8] Reiter D E, Kuhn T and Axt V M 2011 *Phys. Rev.* B **83** 155322
[9] Huo Y H *et al.* 2014 *Nat. Phys.* **10** 46

Ultrahigh responsivity of optically active, semiconducting asymmetric nano-channel diodes

Y Akbas[1], A Stern[1], L Q Zhang[2], Y Alimi[2], A M Song[2], I Iñiguez-de-la-Torre[3], J Mateos[3], T González[3], G W Wicks[1] and Roman Sobolewski[1]

[1]University of Rochester, Rochester, NY 14627-0231, USA
[2]University of Manchester, Manchester, M13 9PL, UK
[3]Universidad de Salamanca, E-37008 Salamanca, Spain

E-mail: roman.sobolewski@rochester.edu

Abstract. We present our research on the fabrication and optical characterization of novel semiconducting asymmetric nano-channel diodes (ANCDs). We focus on optical properties of ANCDs and demonstrate that they can be operated as very sensitive, single-photon–level, visible-light photodetectors. Our test devices consisted of 1.2-µm-long, ~200- to 300-nm-wide channels that were etched in an InGaAs/InAlAs quantum-well heterostructure with a two-dimensional electron gas layer. The ANCD I–V curves were collected by measuring the transport current both in the dark and under 800-nm-wavelength, continuous-wave–light laser illumination. In all of our devices, the impact of the light illumination was very clear, and there was a substantial photocurrent, even for incident optical power as low as 1 nW. The magnitude of the optical responsivity in ANCDs with the conducting nano-channel increased linearly with a decrease in optical power over many orders of magnitude, reaching a value of almost 10,000 A/W at 1-nW excitation.

1. Introduction

Room-temperature nano-devices based on quantum confinement and/or ballistic nonlinearities are intrinsic nanostructures, not simply scaled-down conventional circuitry, and are gaining wide-spread research attention. Among the nano-devices, one of most popular are the asymmetric nano-channel diodes (ANCDs), also referred to as self-switching diodes (SSD), first proposed by Song *et al.* [1]. ANCDs, contrary to conventional diodes, do not rely on energy-barrier concepts to achieve rectification, but rather their nonlinear I–V characteristics result from the carrier transport in an asymmetric nano-channel. The ANCD planar geometry allows for a flexible design and easy integration as a multi-element sensor or with either optical nano-concentrators or THz coupling antennas. Based on Monte Carlo (MC) simulations, ANCDs are expected to be efficient THz generators [2], which have been demonstrated to be viable THz detectors [3].

Typically, carrier transport in ANCDs is confined to a two-dimensional electron gas (2DEG) layer in order to take advantage of the 2DEG ultrahigh mobility and in this way, minimize the carrier transient time. Depending on the device's dimensions and/or fabrication process and the level of its control, there are two basic types of SSDs: "normally OFF" devices with a channel open/depleted at zero bias and "normally ON" devices, where the channel is always conducting. These two types exhibit quite different I–V curves, as shown in figures 1(a) and 1(b), where MC simulations [4] of two ANCDs of the same geometry are presented but with different values for the surface charge σ.

Journal of Physics: Conference Series **647** (2015) 012013 doi:10.1088/1742-6596/647/1/012013

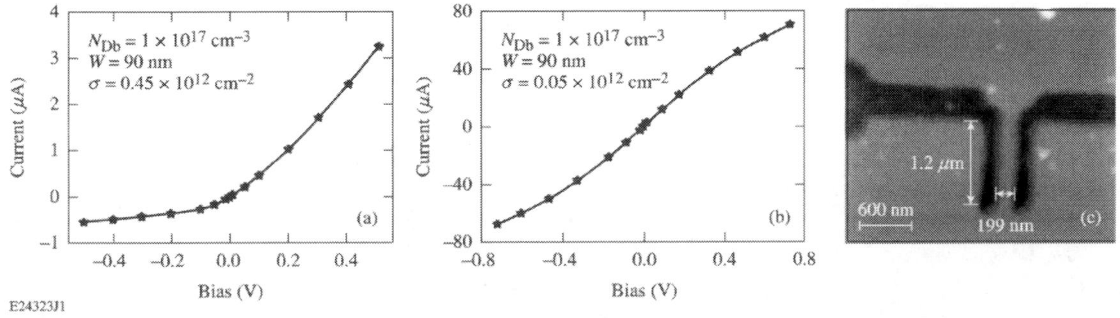

Figure 1. Monte Carlo (MC) results for I–V curves of an asymmetric nano-channel diode (ANCD) with a channel width W = 90 nm and two different values of the surface charge σ, leading to (a) a depleted channel and diode-like characteristics, (b) a conducting channel (effective width of 80 nm at zero dc bias) and nonlinear characteristics, and (c) an atomic force microscope image of an actual device.

Despite the popularity of SSDs, extensive fundamental and applied research is still needed to fully understand their carrier transport dynamics through the asymmetric nano-channel, as well as their performance in both the THz and optical radiation ranges. Our work presented here focuses on photoresponse properties of ANCDs, and we demonstrate that especially normally ON devices possess a very strong, intrinsic internal photoelectron gain mechanism, making them attractive as possible ultrafast optical detectors with single-photon sensitivity.

2. Device fabrication and experimentation

Our tested devices consisted of ~1.2-μm-long and ~200- to 300-nm-wide channels [see figure 1(c)] and were fabricated on an InGaAs/InAlAs quantum-well heterostructure grown on an InP wafer. The fabrication started with the formation of mesa structures by wet chemical etching with an $H_3PO_4/H_2O_2/H_2O$-based solution. Ohmic contacts were then formed by thermal evaporation of 50 nm of Au/Ge/Ni alloy, followed by a 200-nm Au layer. Finally a nano-channel was defined using electron-beam lithography and wet chemical etching.

The ANCD I–V curves were collected by measuring the transport current for the voltage-source biasing condition, both in the dark and under light illumination. For optical excitation, we used 800-nm-wavelength continuous-wave (cw) radiation generated by a commercial, non-mode-locked Ti:sapphire laser.

3. Experimental results and analysis

Figure 2 presents a family of dc I–V characteristics of a normally OFF device, measured in the dark (black dots) and with a single-μW level of light illumination (see figure caption) focused on the device. We note that the curve measured in the dark resembles well the one presented in figure 1(a). The unbiased ANCD is clearly in the OFF state and, when biased, a diode-like characteristic is well reproduced. A leakage current at negative bias is ascribed to the relatively large channel width of this ANCD sample, while the change of the slope near 3-V bias is caused, according to MC simulations, by velocity saturation associated with electron scattering from Γ into L satellite valley.

The impact of the light illumination is very strong, as it is illustrated in the top left inset in figure 2, where we plot a family of photocurrent characteristics versus the bias voltage. The observed behaviour very strongly points to optical gating of the nano-channel. In fact, the first quadrant (positive voltage and positive photocurrent) of the inset graph closely resembles the characteristics of a field-effect transistor (FET), except that here the various curves of the graph correspond to different optical intensities with a threshold optical power of about 0.9 μW. Optical gating in FETs has been recently studied [5]; however, here we have a much simpler structure—an asymmetric nano-channel—and the effect is very strong, even as the incident light power changes in a very narrow range from 1 to 1.4 μW.

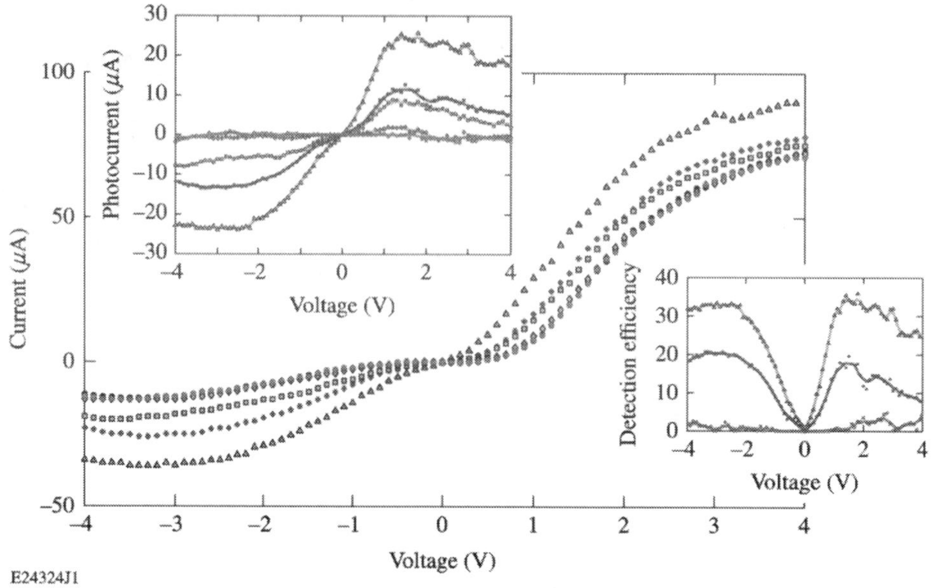

E24324J1

Figure 2. Direct current I–V characteristics for a normally OFF device (channel width of ~200 nm) in darkness (black dots) and under 800-nm-wavelength cw laser illumination. The top left inset shows the photocurrent's dependence on the bias voltage, collected as the current difference between the I–V curve under a given illumination and the one measured in the dark. The bottom right inset presents the detection efficiency's (DE's) dependence on the bias voltage, calculated for a given curve as the ratio of the photocurrent to incident laser power and expressed in units of (electron/photon). The legend for all presented traces is as follows: yellow triangles 1.4 μW; violet crosses 1.27 μW; light blue squares 1.05 *m*W; green diamonds 0.91 μW; and red stars 0.78 μW.

The optical gating is caused by photogenerated holes that move to locations below and/or beside the channel and are likely to partly compensate negative surface charges at the walls of the trenches. This effect reduces the channel depletion in the analogous way to a gate voltage in the FET, further opening the 2DEG channel and, consequently, leading to a large optical responsivity. The latter conclusion is in full agreement with the detection efficiency (DE) data presented in the bottom right inset in figure 2. The DE value increases with the incident light power and the ANCD structure exhibits an intrinsic gain, i.e., a DE >> 1 resulting from the optical gating of the channel of the SSD.

A family of I–V characteristics, this time collected for a normally ON device, is shown in the top right inset in figure 3 and was measured under light illumination conditions similar to those in figure 2. We note that now the I–V curve measured in the dark (black dots) has an S-like shape and resembles the one presented in figure 1(b). The unbiased diode is clearly in the ON state, and when biased, the current is in the mA rather than the μA range. The observed nonlinearity (transition toward saturation) comes, as indicated earlier, from the *Γ–L* electron scattering. Figure 3 (main panel) shows that the maximum optical responsivity (red dots), expressed in A/W (see also the left bottom panel), increases linearly (red solid line) with decreasing optical power over many orders of magnitude with only very slight tapering at the lowest light power levels, reaching a value of almost 10,000 A/W, comparable to the gain of avalanche-type, single-photon detectors.

Optical gain in normally ON ANCDs is consistent with a model proposed for photoconductive gain in high-electron-mobility transistors (HEMTs) [6]. In this model, the band bending present in the 2DEG of the HEMT captures photoexcited electrons, which transit the channel of the device, while photoexcited holes are pushed away from the 2DEG and become trapped in the substrate or in surface states on the sidewalls of the channel. The value of the photoconductive gain is the ratio of the hole trapping time to the electron transit time. The gain is dependent on the optical power because the density of photogenerated charges affects the potential (as in the case of the normally OFF device) that separates the photogenerated holes from the 2DEG, thereby affecting the hole trapping time. The subpicosecond electron transit time of our ANCDs translates into a very large optical gain.

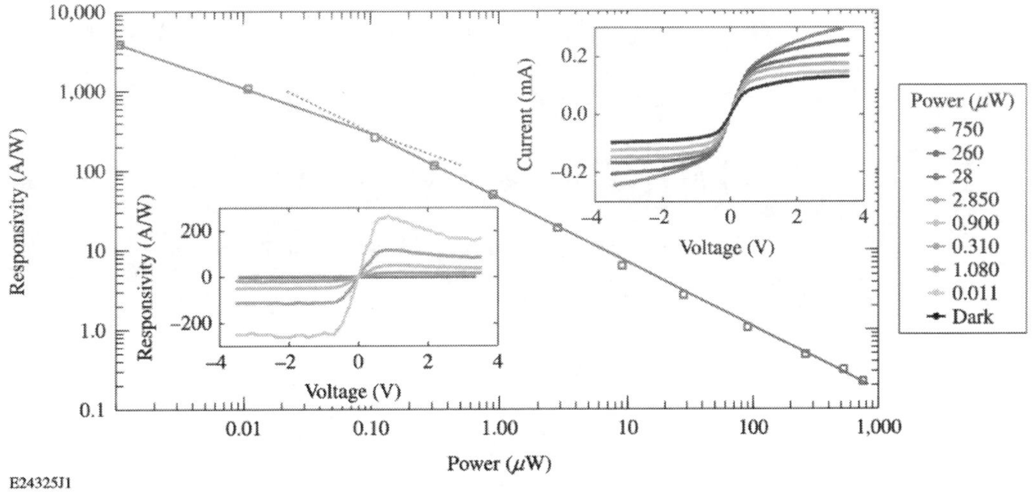

Figure 3. Maximum optical responsivity's (red dots; solid line is a linear fit) dependence on the incident optical power for a normally ON ANCD (channel width ~300 nm). The top right inset shows the I–V characteristics collected in the dark (black dots) and under 800-nm cw light illumination for power levels in the range of 750 to 0.11 μW. The bottom left inset presents examples of the responsivity versus the bias voltage.

4. Conclusions

We have shown that ANCDs, originally intended for THz applications, have also very interesting photoresponse properties. In normally OFF devices, where the nano-channel width is defined by the depletion layers and electric-field–controlled, optical illumination plays a role in optical gating, analogous to the gate in FET structures. On the other hand, the physics of the photoresponse gain mechanism in normally ON ANCDs arises from a dramatic difference between a subpicosecond transient time of electrons travelling in the 2DEG nano-channel layer and the microsecond lifetime of holes, optically excited and, subsequently, pushed toward the substrate. When cooled (to minimize the dark current), our ANCDs should be practical photon counters.

Acknowledgment

The work in Rochester was supported in part by a University of Rochester Research Grant and by ARO Grant #W911NF-15-1-0356.

References

[1] Song A M, Missous M, Omling P, Peaker A R, Samuelson L and Seifert W 2003 *Appl. Phys. Lett.* **83** 1881
[2] Xu K Y, Wang G and Song A M 2008 *Appl. Phys. Lett.* **93** 233506
[3] Balocco C, Kasjoo S R, Lu X F, Zhang L Q, Alimi Y, Winnerl S and Song A M 2011 *Appl. Phys. Lett.* **98** 223501
[4] Mateos J, Vasallo B G, Pardo D and González T 2005 *Appl. Phys. Lett.* **86** 212103
[5] Tripathi S and Jit S 2011 *J. Appl. Phys.* **109** 053102
[6] Romero M A, Martinez M A G and Herczfeld P R 1996 *IEEE Trans. Microw. Theory Tech.* **44** 2279

Electrostatic enhancement of light emitted by semiconductor quantum well

A Krokhin[1], A Neogi[1], A Llopis[1], M Mahat[1], L Gumen[2], S Pereira[3] and I Watson[4]

[1] *Department of Physics, University of North Texas, Denton 76203, USA*
[2] *Universidad Popular Autónoma del Estado de Puebla, 21 Sur, #1103, 72160, Mexico*
[3] *CICECO, University of Aveiro, 3810-193 Aveiro, Portugal*
[4] *SUPA, Institute of Photonics, University of Strathclyde, Glasgow, United Kingdom*

E-mail: Arkadii.Krokhin@unt.edu

Abstract. Carrier dynamics in metal-semiconductor structures is driven by electrodynamic coupling of carriers to the evanescent field of surface plasmons. Useful modifications in electron and hole dynamics due to presence of metallic inclusions show promise for applications from light emitters to communications. However, this picture does not include contributions from electrostatics. We propose here an electrostatic mechanism for enhancement of light radiated from semiconductor emitter which is comparable in effect to plasmonic mechanism. Arising from Coulomb attraction of e-h pairs to their electrostatic images in metallic nanoparticles, this mechanism produces large carrier concentrations near the nanoparticle. A strong inhomogeneity in the carrier distribution and an increase in the internal quantum efficiency are predicted. In our experiments, this manifests as emission enhancement in InGaN quantum well (QW) radiating in the near-UV region. This fundamental mechanism provides a new perspective for improving the efficiency of broadband light emitters.

1. Introduction

Metallic inclusions in semiconductor structures play the role of environment which may strongly affect the rate of electron-hole recombination. Depending on the final density of photonic states the environment may either decrease or increase the recombination rate. In the former case energy of the radiated photon fits the photonic band gap[1] and in the latter case recombination occurs faster due to Purcell effect [2]. During the last two decades several experimental manifestations of Purcell effect have been observed in semiconductor quantum wells (QW's) covered by thin metallic film [3, 4, 5]. The level of enhancement strongly depends on how exactly the frequency ω of the emitted lights matches the resonant frequency of plasmonic resonance ω_s. If the mismatch between ω and ω_s increases the effect of plasmonic enhancement becomes negligible. This frequency-matching condition limits application of plasmonic mechanism for light enhancement in solid-state light emitters with broadband spectrum. Emitters that radiate blue or near-UV light require coating by silver thin film, since silver exhibits plasmonic resonance in this region of the spectrum. However, because of its fast oxidation silver cannot be widely used in commercial devices. Here gold is relatively immune to corrosion but its resonant frequency lies in the red-IR region. Thus, gold cannot be used in the whole visible spectrum. One more disadvantage of plasmonic mechanism is its relatively high Joule losses which accompany

Journal of Physics: Conference Series **647** (2015) 012014 doi:10.1088/1742-6596/647/1/012014

propagation of surface plasmons in metal. Here we report a new non-resonant and non-dissipative electrostatic mechanism of light enhancement.

The proposed mechanism is based on electrostatic attraction of electrons and holes to their electrostatic images in metallic nano-inclusions. Neutral metal equally attracts electrons and holes, causing the carriers to drift towards, and concentrate near, the nano-inclusion. The rate of e-h recombination depends not only on the density of states but also on the local concentration of carriers, $n(\mathbf{r})$. In the regions with higher concentration the wave functions of electrons and holes strongly overlap that increases the probability of recombination. This, however, does not necessary lead to higher intensity of light if most of the recombination processes are non-radiative. In direct band-gap heterostructures the radiative recombination occurs mostly through the bimolecular mechanism which is quadratic in the carrier concentration, $\Gamma_r = Bn^2$. In contrast, the Shockley-Read-Hall mechanism of non-radiative recombination is characterized by a linear rate, $\Gamma_{nr} = An$ [6]. Thus, faster increase of Γ_r with concentration is beneficial, since it leads to higher internal quantum efficiency of the emitter,

$$\eta = Bn^2/(An + Bn^2). \tag{1}$$

To demonstrate the new mechanism of enhancement we studied photoluminescence from InGaN/Gan multi-QW shown in Figs. 1 and 2. This QW radiates blue light with maximum intensity near $\lambda = 416$ nm. The QW was infiltrated by gold nano-particles. The plasmonic mechanism of enhancement is ruled out since gold nanospheres of diameter from 30 to 35 nm used in our experiment have plasmonic resonance at approximately 2λ.

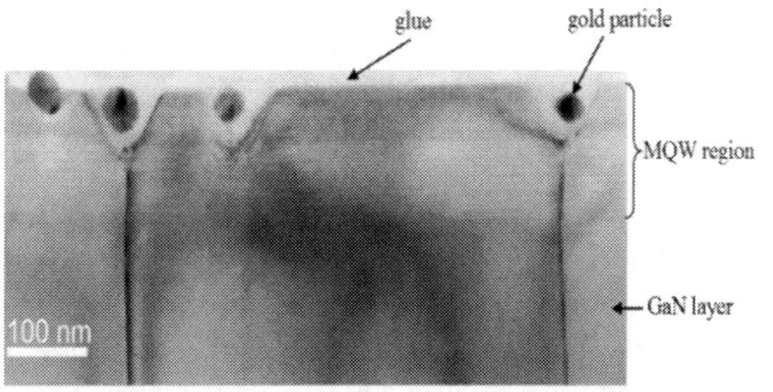

Figure 1. TEM image of InGaN/GaN multi-quantum well with gold particles of ~ 30 size inside the pits.

Figure 2. A pit with several gold nanoparticles.

2. Sample preparation and experimental results for the emission spectra

The size of the NP plays a principal role here, determining the level on enhancement. The electrostatic drift of the carriers towards the nanoparticles becomes noticeable if the ratio $\gamma = (e^2/\varepsilon R)/kT$ of electrostatic energy to thermal energy becomes $\gamma \sim 1$. For the effect to be observable at room temperature, this condition requires that the typical size $2R$ of the NP be $2R < 50$ nm given that the dielectric constant of the QW is $\varepsilon \sim 10$.

We measured the PL enhancement from the same material heterostructures of InGaN/GaN which were previously used in Ref. [4, 5] to demonstrate the plasmonic mechanism of enhancement and recently in Ref. [7] to demonstrate the electrostatic mechanism. In nitride heterostructures threading dislocations end off with inverted pits at the surface provide a natural and invasive way to incorporate metallic nanoparticles inside a QW heterostructure, see Fig. 1.

In order to obtain a pit size of $\sim$ 50 nm, we grew a $In_{0.09}Ga_{0.91}N/Ga$ 14-layer QW with a well thickness of 2.5 nm, barrier thickness of 7.5 nm, and a 15-nm cap layer above the final QW. This light emitting heterostructure was grown by MOCVD under conventional conditions along the (0001) axis without intentional doping [8]. The V-shaped pits act as a potential barrier for non-radiative recombination center. The Au nanoparticles of 30-35 nm suspended in water or DMSO solution were trickled into the pits using a combination of sonication, density of fluid and flow rate. Mechanical contacts between the nanoparticles and the surface of the QW occur randomly at very few points, as it can be seen in Figs. 1 and 2. Near these points of contact a depletion layer may be formed due to Schottky barrier. It repels carriers and in principle may reduce the attraction due to image charges. However, since the area of possible mechanical contacts is negligible as compared to the area of a nanoparticle, the Coulomb attraction remains the dominant force of interaction.

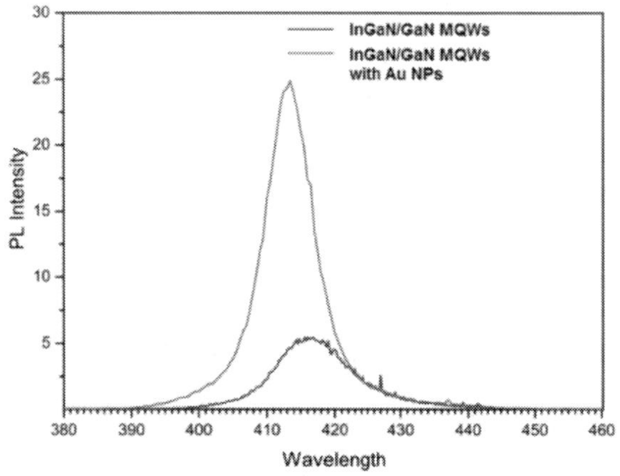

Figure 3. Photoluminescence spectrum of the multi-QW measured at room temperature without (blue curve) and with (red curve) gold nanoparticles.

The PL spectra of the multi-QW with and without gold NPs are shown in Fig. 3 with a maximum photoluminescence intensity at $\lambda = 413$ nm and $\lambda = 416$ nm, respectively. There is roughly a six-fold enhancement in the PL emission with gold at room temperature. As mentioned before, a gold nanoparticle does not serve as plasmonic source of enhancement for an InGaN QW due to having almost two times lower localized surface plasmon energy. Instead, the Coulomb attraction of the free carriers to their image charges in the nanoparticle leads to enhancement. Image charges, being of equal magnitude and opposite sign to the image-inducing carrier, cause both electrons and holes to drift towards the surface of the nanoparticle that increases the electron-hole and exciton recombination rate. Excitons, while being electrically neutral are also attracted to a neutral metal surface. The origin of this electrostatic mechanism is best demonstrated by Thomson's theorem which states that the equilibrium distribution of charges on the surface of an isolated conductor corresponds to the minimum of the potential energy [9]. A direct consequence of this theorem is that a neutral conductor attracts *any* given set of electric charges. Therefore, electron-hole pairs and excitons within the QWs shown in Fig. 1 are attracted towards the metal nanoparticles. The attraction of a single carrier located at a distance r from a neutral metallic sphere of radius R decays fast far away from the sphere, $eE(r) \sim 1/r^5$. However, close to the surface, it grows as $eE(r) \sim 1/(r-R)^2$. This Coulomb force leads to a permanent drift of the carriers. The local concentrations of electrons is obtained from the non-linear diffusion equations in the presence of the electrostatic force $e\mathbf{E}(\mathbf{r})$.

$$\dot{n}_e = D_e \nabla^2 n_e + \frac{eD_e}{kT} \nabla \cdot (\mathbf{E} n_e) - n_e(A - Bn_h) + g. \tag{2}$$

The equation for holes is obtained by replacement $e \Leftrightarrow h$. Here $D_{e,h}$ are the diffusion coefficients for electrons and holes, and g is the density of e-h pairs produced by the pumping laser per second. The electrostatic field $\mathbf{E}$ in the drift term in Eq. (2) is obtained from the Poisson equation where the real charges and also their electrostatic images in metal must be included.

The strong contribution of the drift term to the e-h kinetics can be demonstrated by an asymptotic steady-state solution if we linearize Eq. (2) in the vicinity of the nanosphere, $r - R \ll R$:

$$n_{e,h}(r) \propto \exp[\gamma R/4(r-R)], \;\; \gamma = (e^2/\varepsilon R)/kT. \tag{3}$$

Here the parameter γ defines the width of the region with higher concentration of carriers. Since it scales as $\gamma \sim 1/T$ the effect of enhancement becomes stronger at lower temperatures. Thus, the electrostatic drift leads to an exponential singularity of the carrier concentration at the surface. This strong non-physical singularity appears since the asymptotic solution (3) is not self-consistent. Once Debye screening, which strongly reduces attraction to the image charges, is taken into account the singularity becomes much weaker. Moreover, Debye screening leads to the effect of saturation of emission enhancement at high pumping power [7]. It is worth mentioning that electrostatic enhancement is not accompanied by Joule losses, unlike the plasmonic enhancement, which inevitably leads to dissipation of energy.

The proposed electrostatic mechanism successfully explains the enhanced spectra in Fig. 3. At the same time the emission spectrum of the QW coated by high-quality gold film does not exhibit enhancement at all [4, 5]. It is obvious that in this case the carriers in the QW are also attracted towards their electrostatic images. However, in the plane geometry this attraction pushes electrons and holes towards the metal film, without changing the average in-plane separation between them. It turns out that the increase of the concentration only along a single coordinate (the one which is perpendicular to the quantum confinement) is not sufficient to produce the effect of enhancement. In this case the image charges may only reduce the effective width of the QW that gives rise to the blue shift of the spectrum, which was not explained in Refs. [4, 5]. Gold nanospheres, being practically point attractors, lead to both effects, therefore the blue shift and emission enhancement are clearly seen in Fig. 3.

3. Conclusions

We propose a mechanism of light emission enhancement from semiconductor heterostructures which is due to attraction of carriers to neutral metallic nano-inclusions. Unlike the well-known plasmonic mechanism, it is free from the frequency-matching condition, thus, is may be used to increase the performance of broad-band solid-state light emitters. Due to its electrostatic nature the new mechanism is non-dissipative and universal, i.e. *any* metal can be used and this does not affect the level of enhancement. Finally, it does not require high-quality coating since any irregularities in the shape of metallic inclusions increase electrostatic image-charge forces.

References

[1] Noda S, Fujita M and Asano T 2007 *Nature Photon.* **1** 449
[2] Purcell E M 1946 *Phys. Rev.* **69** 681
[3] Hecker N E, Höpfel R A, Sawaki N, Maier T. and Strasser G 1999 *Appl. Phys. Lett.* **75** 1577
[4] Neogi A, Lee C W and Everitt H O 2002 *Phys. Rev. B* **66** 153305
[5] Okamoto K, Niki I, Shvartser A, Narukawa Y, Mukai T and Scherer A 2004 *Nature Mat.* **3** 601
[6] Dai Q, Schubert M F, Kim M H, Kim J K, Schubert E F, Koleske D D, Crawford M H, Lee S R, Fischer A J, Thaler G and Banas M A 2009 *Appl. Phys. Lett* **94** 111109
[7] Llopis A, Lin J, Pereira S M S, Trindade T, Martins M A, Watson I M, Krokhin A A and Neogi A, 2013 *Phys. Rev. B* **87** 201304(R)
[8] Deatcher C J, Liu C, Pereira S, Lada M, Cullis A G, Sun Y J, Brandt O and Watson I M 2003 *Semicond. Sci. Technol.* **18** 212
[9] Landau L D, Lifshitz E M and Pitayevskii L P 1984 *Electrodynamics of Continuous Media* 2nd Ed (Oxford: Pergamon Press)

Terahertz emission from lithium doped silicon under continuous wave interband optical excitation

A V Andrianov[1,5], A O Zakhar'in[1], R K Zhukavin[2], V N Shastin[2,3] and N V Abrosimov[4]
[1] Ioffe Institute, St Petersburg, 194021, Russia
[2] Institute of Physics of Microstructures, Nizhny Novgorod, 603600, Russia
[3] N I Lobachevskii Nizhny Novgorod State University, 603950, Russia
[4] Leibniz Institute for Crystal Growth, Berlin, 12489, Germany

E-mail: alex.andrianov@mail.ioffe.ru

Abstract. We report on experimental observation and study of terahertz emission from lithium doped silicon crystals under continuous wave band-to-band optical excitation. It is shown that radiative transitions of electrons from 2P excited states of lithium donor to the 1S(A1) donor ground state prevail in the emission spectrum. The terahertz emission occurs due to capture of nonequilibrium electrons to charged donors, which in turn are generated in the crystal as a result of impurity assisted electron-hole recombination. Besides the intracentre radiative transitions the terahertz emission spectrum exhibits also features at about 12.7 and 15.27 meV, which could be related to intraexciton transitions and transitions from the continuum to the free exciton ground state.

1. Introduction

During last two decades, terahertz (THz) sources of electromagnetic radiation with frequencies between 0.1 and 10 THz have attracted attention for many applications such as remote sensing, bio-chemical detection, communication, and medical diagnostics. One of the possible schemes of a relatively simple THz emitter is based on optical transitions between energy levels of shallow impurities in semiconductors. The intracenter THz radiative transitions appear at the energy relaxation of nonequilibrium carriers created in the allowed band, e.g. at the impact ionization of impurities in the electric field [1-4] or at the photoionization of impurities by infrared laser radiation [5]. THz lasing on intacenter optical transitions in germanium and silicon was demonstrated in [2, 5]. It has recently been shown that the impurity related THz radiation can also appear under interband optical excitation of semiconductors doped with shallow centres [6]. Such a THz photoluminescence (PL) was detected in a number of materials [6-8]. The THz PL is due to features of electron-hole recombination involving impurity states. Such a recombination results in the formation of charged impurity centres and free carriers in the allowed band. The subsequent capture of free carriers to charged centres is accompanied by THz radiation [7, 8] in similar way as it occurs under electrical breakdown of impurity centres [1-4]. The THz PL in semiconductors can be quite intense and it can find applications in THz technologies. For example, it might be more convenient to use the band-to-band optical

[5] To whom any correspondence should be addressed.

pumping to excite intracenter radiative transitions when the impurity breakdown requires high electric fields (e.g. in Si, where impurity breakdown threshold field exceeds 1000 V/cm [3]).

THz emitters based on impurity optical transitions in silicon are of interest because they can be directly integrated with silicon electronics. Lithium in silicon demonstrates quite a few unusual properties comparing with other donor impurities. One can mention an ability of lithium to form complexes with others impurities, which exist in the crystal. The ionization energy of Li donor in Si is substantially lower than that for others donors, for example, formed by V group elements. The electronic structure of the Li donor ground state is also unusual. There is so called "inverted" in comparison with V group donors arrangement of the sublevels of the Li donor ground state [9]. Lithium in Si is one of possible candidates for a Qbit in a quantum computer with the information readout based on intracenter optical transitions [10]. In view of said above, it is of interest to investigate intracenter THz luminescence associated with Li donors in Si.
This paper reports the detection and study of THz PL from Si:Li crystals.

2. Experimental details

2.1. Samples under the test
The experiments were curried out on silicon single crystals doped with lithium to a level of 10^{16} cm^{-3}. The crystals were doped during the growth from melt by the method of growth from a pedestal (a variant of float-zoning technique). According to measurements of optical absorption in the middle infrared range, the concentration of oxygen was below 10^{16} cm^{-3}. The samples for measurements were prepared in the form of chemically-mechanically polished 1.5-mm-thick 5x7-mm plates.

2.2. THz measurements
The Si:Li samples were placed in a helium optical cryostat with a controlled temperature, which was optimized for the THz spectral domain. A CW semiconductor laser emitting at λ=660 nm with maximal output power of about 45 mW was used as a source of the interband photoexcitation. The intensity of photoexcitation did not exceed 1 W/cm^2. The main measurements of THz PL were performed in "transmission" geometry. Some control experiments were also performed in "backscattering" geometry. The results of these measurements were almost the same. The spectral measurements were done with using a step-scan Fourier spectrometer for spectral range of 5-350 cm^{-1}, described elsewhere [4]. The spectral resolution in most cases was 5 cm^{-1} (0.62 meV). A THz radiation signal was measured with liquid helium cooled Si bolometer by synchronous detection method at the laser-radiation modulation frequency by a mechanical interrupter (80 Hz).

3. Results and discussion
The characteristic spectrum of the THz PL from Si:Li crystal observed at liquid helium temperatures is shown on the figure 1a. It is seen that the emission lines with maxima at 24.71 meV (5.98 THz) and 19.63 meV (4.75 THz) dominate in the spectrum. The difference between the energies of these lines is 5.08 meV. This value is very close to well-known energy difference of 5.11 meV between the $2P_{\pm}$ and $2P_0$ levels of the donor impurity in silicon, which was thoroughly calculated and confirmed experimentally for various donor centres in Si (see Ref. [9]). Besides that the position of the 24.71-meV emission line is in good agreement with the position of the line of the transition between the $2P_{\pm}$ excited state and $1S(A_1)$ sublevel of the ground state of Li donor in silicon, which was observed in far infrared absorption experiments [9]. It allows us to attribute the intense emission lines at 24.71 and 19.63 meV to intracenter transitions $2P_{\pm}\rightarrow 1S(A_1)$ and $2P_0 \rightarrow 1S(A_1)$ in Li donor, respectively (see figure 1b).
The character of the THz PL spectrum indicates that the lines of radiative transitions from 2P states to the lowest sublevel $1S(E+T_2)$ of the ground state of Li donor are significantly suppressed as compared to the lines of transitions to the $1S(A_1)$ sublevel.

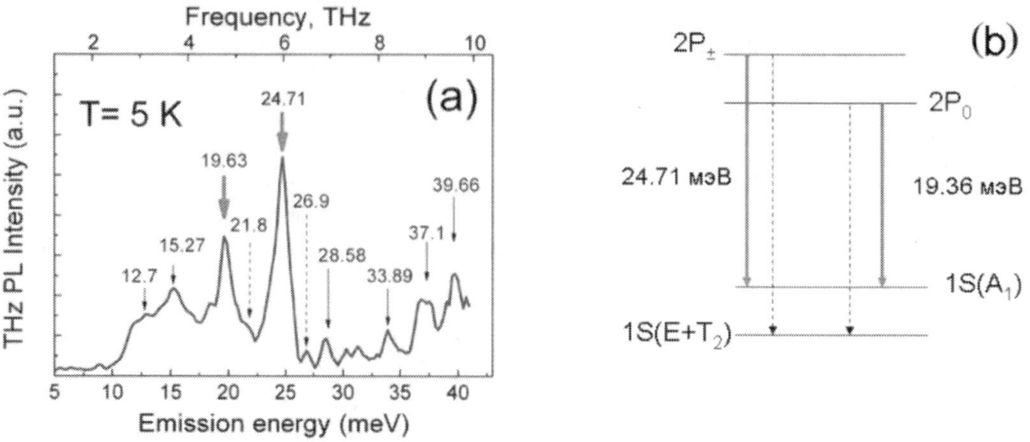

Figure 1. a) Typical THz PL spectrum at I_{exc}~ 1W/cm^2. Arrows point to the positions of THz emission lines. The labels show the energies in "meV". b) Scheme of main intracenter radiative transitions.

Weak line at 26.9 meV and a feature at the energy of ~21.8 meV, which looks as a high-energy shoulder of the line of the $2P_0 \rightarrow 1S(A_1)$ transition, marked by dashed arrows in the figure 1, are most likely related to transition to the $1S(E+T_2)$ ground state of Li donor from the $2P_\pm$ and $2P_0$ excited states, respectively. Strong suppression of the THz emission associated with transitions to the $1S(E+T_2)$ state as compared to emission caused by transition to $1S(A_1)$ state can be attributed to the effect of reabsorption of the radiation. Indeed, due to the total internal reflection only a small fraction of the radiation is directly emitted from the excitation region. The main fraction of radiation, multiply reflected from the surface of the crystal, travels inside it and undergoes absorption. At helium temperatures, the lowest sublevel $1S(E+T_2)$ of the donor ground state is populated much higher than the sublevel $1S(A_1)$. Therefore, the absorption of THz radiation on the lines corresponding to the $2P \rightarrow 1S(E+T_2)$ transition is much higher than on the lines corresponding to the $2P \rightarrow 1S(A_1)$ transitions. We have to note that, in the experiments on THz electroluminescence (EL) under impurity breakdown in Si:P [3], similar regularity was observed. The authors have observed significant (almost complete) suppression of the emission associated with transitions to the lowest sublevel of the donor ground state as compared to the emission associated with transitions to higher lying sublevels of the donor ground state. But this feature was not explained in [3]. We believe that this feature of the THz EL spectra in Si:P can also be explained by the THz radiation reabsorption effect.

Weak emission lines in the THz PL spectrum of silicon with lithium in the energy range from 27.5 to 40 meV (see figure 1a) can be attributed to intracenter optical transitions in donors caused by Li-O complexes. The lines with maxima at 28.58 and 33.89 meV are due to optical transitions to the ground state $1S(A_1)$ of the Li-O donor from the $2P_0$ and2 $P_\pm$ excited states, respectively. Wider line with the maximum at about 37.1 meV can be attributed to the superposition of transitions from $4P_\pm$ and $3P_\pm$ states to the $1S(A_1)$ state, and the 39.66 meV line is possibly due to transitions of electrons from the conduction band to the $1S(A_1)$ state. Such an interpretation is in good agreement with the data on the spectrum of energy levels of the Li-O donor in silicon [9].

The THz PL spectrum (figure 1a) also exhibits relatively intense features at energies of 12.7 and 15.27 meV. It is hard to assign these features to intracenter transitions in the lithium donors. On the other hand, it is known that in the energy range of 10-14 meV the absorption associated with $1S \rightarrow 2P$ transitions in free excitons in Si is observed [11]. Furthermore, the binding energy of free exciton in Si is known to be about 15 meV [12]. The capture of free carriers to excitons can be accompanied by THz radiation similar to radiation at the capture of carriers into ionized impurity centres. Therefore, we tentatively attribute the 12.7- and 15.27-meV THz emission lines to intraexcitonic radiative transitions and transitions from continuum to the free exciton ground state.

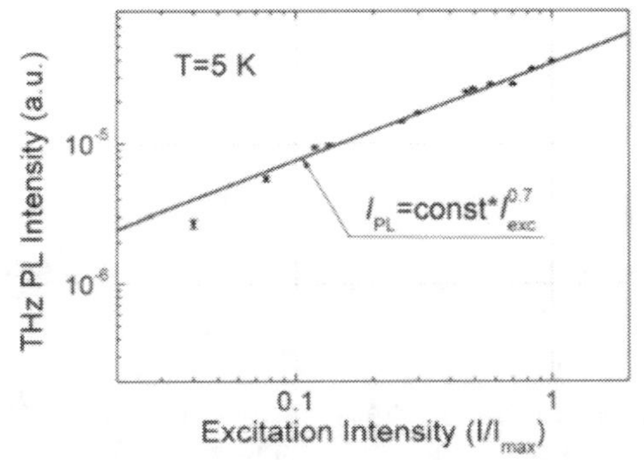

Figure 2. The total PL intensity versus the excitation intensity. I_{max}=1 W/cm^2.

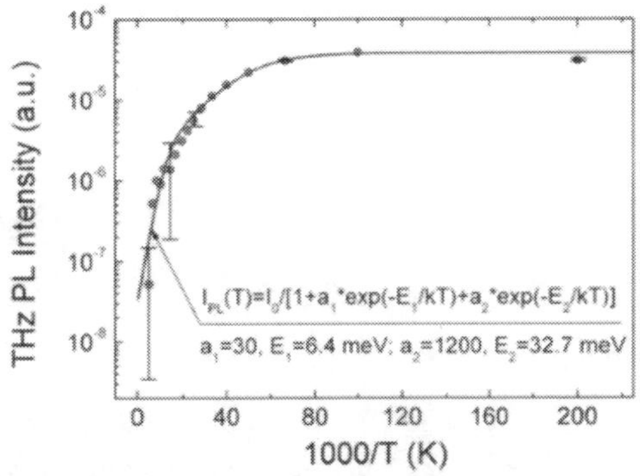

Figure 3. Temperature dependence of the THz PL. The solid curve is a result of fitting of the data.

It was revealed that the THz PL intensity is sublinear versus the excitation intensity (see figure 2). At the excitation intensity above 0.1 W/cm2 this dependence can be well approximated by the power law of $I_{PL}\sim I_{exc}^{0.7}$. This fact is important and it indicates that the main THz radiation is formed as a result of recombination of nonequilibrium holes with electrons on neutral donors. An analysis, similar to that is given in [8], shows that a sublinear dependence of the THz PL intensity on the excitation intensity should be expected in the case of this mechanism of THz emission.

The THz PL decreases with increase in temperature. Figure 3 shows the temperature dependence of the total intensity of the THz PL in Arrhenius coordinates. As is seen the data can be well approximated by usual activation equation with two temperature quenching energies: E_1=6.4 meV and E_2=32.7 meV. The energy of 32.7 meV is consistent with the binding energy of Li donor in Si [9]. Therefore, the main temperature quenching of the THz PL can be attributed to the temperature dissociation of neutral Li donors. The energy of 6.4 meV is in agreement with binding energy of $2P_{\pm}$ excited state [9]. Most likely this quenching energy reflects a thermal release of electrons from the 2P donor states since transitions from 2P states give significant contribution to the observed THz PL.

4. Conclusion

THz luminescence from Si:Li crystals has been detected and studied under interband optical excitation. It is shown that the main THz PL is caused by intracenter transitions from 2P excited states to the $1S(A_1)$ sublevel of the ground state of the Li donor. The experimental data indicate that the main radiation is formed as a result of recombination of nonequilibrium holes with electrons on neutral donors. The THz PL spectrum also exhibits lines, which are possibly due to intraexcitonic radiative transitions and transitions from continuum to the ground state of excitons.

References

[1] Koenig S H and Brown R D 1960 *Phys. Rev. Lett.* **4** 170
[2] Gousev Yu P et al 1999 *Appl. Phys. Lett.* **75** 757
[3] Lv P-C et al 2004 *Appl. Phys. Lett.* **85** 22
[4] Andrianov A V, Zakhar'in A O, Yassievich I N and Zinov'ev N N 2004 *JETP Lett.* **79** 365
[5] Pavlov S G et al 2000 *Phys. Rev. Lett.* **84** 5220
[6] Andrianov A V , Zakhar'in A O, Ivanov Yu L and Kipa M S 2010 *JETP Lett.* **91** 96
[7] Zakhar'in A, Andrianov A, Egorov A and Zinov'ev N 2010 *Appl. Phys. Lett.* **96** 211118
[8] Zakhar'in A O, Bobylev A V and Andrianov A V 2012 *Semiconductors* **46** 1135
[9] Jagannath C, Grabowski Z W and Ramdas A K 1981 *Phys. Rev.* B **23** 2082
[10] Smelyanskiy V N, Petukhov A G and Osipov V V 2005 *Phys. Rev.* B **72** 081304(R)
[11] Labrie D, Thewalt M L W, Booth I J and Kirczenow G 1988 *Phys. Rev. Lett.* **61** 1882
[12] Green M A 2013 *AIP Adv.* **3** 1112104

Journal of Physics: Conference Series **647** (2015) 012016 doi:10.1088/1742-6596/647/1/012016

Relativistic Doppler reflection as a probe for the initial relaxation of a non-equilibrium electron-hole plasma in silicon

Mark D Thomson[1], **Fanqi Meng**[1], **Bo E Sernelius**[2] **and Hartmut G Roskos**[1]
[1] Physikalisches Institut, Goethe-University, D-60438 Frankfurt am Main, Germany
[2] Department of Physics, Chemistry and Biology, Linköping University, 58183 Linköping, Sweden

E-mail: thomson@physik.uni-frankfurt.de, roskos@physik.uni-frankfurt.de

Abstract. This paper reviews the status of investigations of the relativistic Doppler reflection of a broadband terahertz pulse at a counter-propagating plasma front of photo-excited charge carriers in undoped silicon. When a THz pulse with 20-THz bandwidth impinges onto a moving plasma front with a carrier density in the range of 10^{19} per cm^3, one observes a spectral up-shift, which is, however, much less pronounced than expected from simulations assuming a Drude plasma characterized by a single carrier relaxation time τ of the order of 15-100 fs. Qualitative agreement between simulations and experiments can be achieved if τ is chosen to be less than 5 fs. In order to explore carrier relaxation in more detail, optical-pump/THz-probe experiments in the conventional co-propagation geometry were performed. If the pump-probe delay is long enough for monitoring of the equilibrium value of the scattering time, τ ranges from 200 fs at low carrier density to 20 fs in the 10^{19}-cm^{-3} density range. For small (sub-picosecond) pump-probe delay, the data reveal a significantly faster scattering, which slows down during energy relaxation of the charge carriers.

1. Introduction

Early on in the history of relativity, the Doppler frequency shift arising upon reflection of an electromagnetic wave at a fast-moving mirror was investigated as an illustrative example of relativistic effects [1]. If the mirror is a moving plasma front, the same frequency shift is expected independent of whether the charge carriers participate in the motion or are immobile [2, 3]. In the latter case, no energy can be transferred from the mirror to the wave, while a transfer of kinetic energy from the mirror occurs if the particles, which constitute the plasma, propagate with the front. A situation with a plasma front at relativistic speed and static or slow-moving charge carriers can be obtained by the optical inter-band excitation of a semiconductor material, which has to be a volume material in order to enable the propagation effect, while the carriers themselves can be spatially confined (e.g., in quantum structures) or unconfined.

In this contribution, we investigate the relativistic Doppler reflection of THz pulses at counter-propagating photo-ionization fronts in undoped silicon at room temperature (see Fig. 1). Unlike in gaseous media, an amplified laser pulse can readily generate a charge-carrier plasma in silicon which has the density and homogeneity to act as an effective mirror for a focused THz pulse. The relatively weak absorption of the pump light by the silicon sample resulting from the indirect character of the inter-band transitions enables a sufficient long interaction time of the plasma front and the THz pulse of about 140 fs [4]. The reflection coefficient strongly depends on the carrier density. Below a critical density N_{cr} (under-dense plasma), only a small portion of the incoming wave is reflected. Good reflectivity requires over-dense-plasma conditions, which are achieved if all angular frequencies ω_i of the incoming THz pulse are below a critical angular frequency, i.e. $\omega_i \leq \omega_{cr}$. The critical angular frequency is connected with the angular plasma frequency at the critical density according to $\omega_{cr}^2 = \omega_{pl}^2/\Gamma = (N_{cr}e^2)/(\varepsilon_r\varepsilon_0 m_{eff}\Gamma)$. Here, Γ is the up-shift factor which determines how the angular frequency ω_i of the input wave transforms upon reflection in the laboratory frame, $\omega_r = \Gamma\omega_i$. For an estimate of the expected up-shift, one can assume a lossless (collisionless) plasma. In this case, Γ only depends on the ratio $\beta = U/c$ of the velocity U of the ionization front and the velocity c of the wave in silicon propagating in the opposite direction: $\Gamma = (1+\beta)/(1-\beta)$. For the material parameters of Si relevant for our experiments (photoexcitation at a wavelength of 775 nm), the up-shift factor has a value of about eight [4]. Absorption losses and pump-pulse depletion effects reduce the up-shift factor [4]. In the experiments, we achieve a plasma frequency $\omega_{pl}/2\pi$ of up to 30 THz (half of this value in the experiments shown in Fig. 2) [5, 6], which suggests that over-dense conditions can be reached for frequencies $\omega_i/2\pi$ of the incoming waves of up to 11 THz (5.5 THz in Fig. 2), and up-shifted signals should be observable above 80 THz (above 40 THz in the data of Fig. 2).

2. Measurement of the relativistic Doppler up-shift in highly photo-excited silicon

The Doppler up-shift is measured with optical pulses of a 1-kHz Ti:Al_2O_3 amplifier laser whose 150-fs pulses are spectrally broadened in an Ar-gas-filled hollow-core fibre and compressed with chirped mirrors. Most of the 30-fs-long pulses is used to excite the silicon sample, the remainder is employed for the generation of THz pulses by two-colour photo-ionization of air [7-11], respectively their detection by sum-frequency generation in a 500-μm-thick ZnTe crystal (sensitivity up to 150 THz) [5, 6, 11, 12].

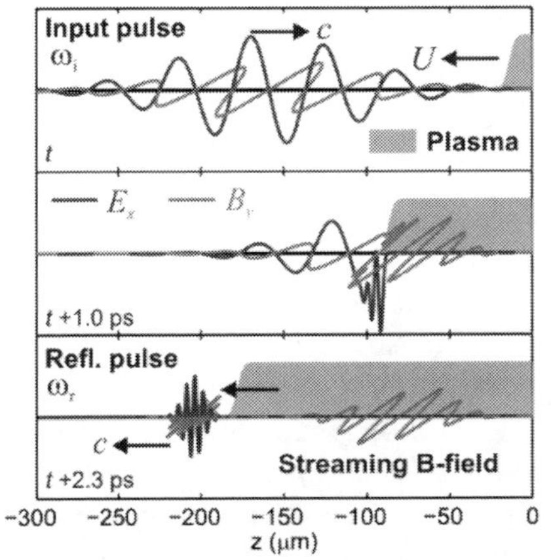

Figure 1. Relativistic Doppler reflection in the over-dense case. A THz wave propagating to the right encounters an ionization front moving to the left. Upon interaction, a substantial part of the energy of the THz pulse is lost to the plasma by excitation of a stationary, so-called streaming magnetic field. Pump depletion and carrier relaxation (which would dampen the streaming magnetic field) are neglected. From [4].

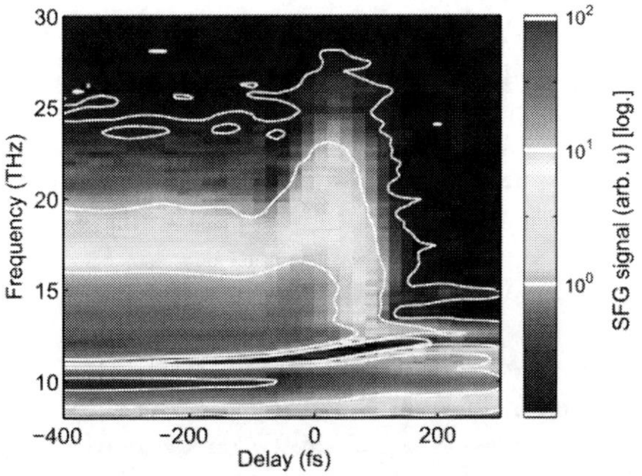

Figure 2. Measured frequency up-shift occurring around time delay zero, when the THz pulse encounters a counter-propagating plasma front. The spectrum at negative time delay (pump pulse has not arrived yet) is that of the unperturbed THz pulse. For positive time delay beyond 100 fs, when the pump pulse is completely absorbed, the THz pulse is strongly attenuated by the static plasma. From [5].

Figure 2 displays the key result of the reflection measurements. If the arrival of the THz pulse, which has measureable signal up to 25 THz (see Fig. 2 for delay times below -100 fs), is properly timed such that it encounters the moving plasma front, one observes newly created frequency components up to 28 THz. Compared with Drude model calculations [4, 5], this degree of up-shift and the intensity of the up-shifted signal are lower than expected, even if one assumes a value of the

Journal of Physics: Conference Series **647** (2015) 012016 doi:10.1088/1742-6596/647/1/012016

scattering time τ as low as 15 fs [5]. A reduction of the up-shift factor and of the signal intensity can be qualitatively explained with a scattering time of less than 5 fs (data to be published in [13]).

In the Drude model, τ is a momentum relaxation time to which elastic and inelastic scattering events of the charge carriers contribute. Phonon emission in silicon is not effective [6], hence carrier-carrier scattering is the main candidate for momentum relaxation. In the literature, little is found on Drude scattering in silicon at the relevant excitation densities and for non-equilibrium conditions. It is worth remarking that the scattering time as estimated above for our single-crystalline sample appears to approach that of amorphous silicon [14]. The experiments described in the following are carried out in order to shed more light on the carrier dynamics.

3. Measurement of the equilibrium Drude scattering times of highly photo-excited silicon

The measurements are performed as optical-pump/THz-probe measurements, but now in co-propagation geometry (see Fig. 3). The data are again analyzed with a Drude ansatz. High-quality fits with the Drude model necessitate that the Drude response has been measured over a broad frequency range around $\omega = \omega_{pl}$, the region where the spectral changes of the Drude response are most pronounced. For this reason, we employ in this study white-light THz/mid-IR pulses with a bandwidth of more than 150 THz combined with a suitable sum-frequency detection technique, which via analysis of the spectra allows us to reliably extract scattering times even shorter than 10 fs [6, 7].

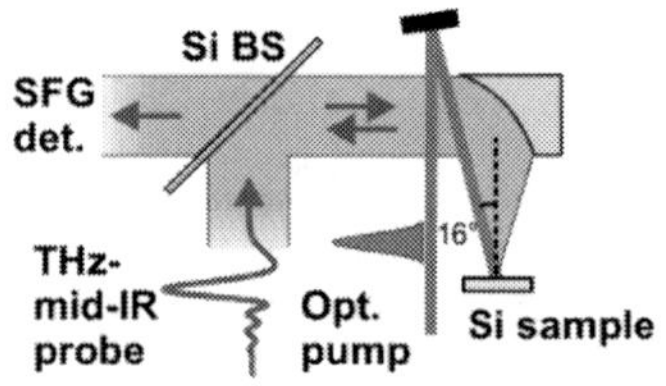

Figure 3. Excitation of an undoped silicon sample by 150-fs pulses at 775 nm, and front-side reflective probing by THz/mid-IR pulses with a bandwidth of 150 THz (pulse duration of about 100-150 fs due to dispersive broadening). Si BS: Silicon beam-splitter; SFG: Sum-frequency generation for the detection of the THz pulses. From [6].

Our study reveals that τ exhibits a time evolution during the initial energy relaxation of the photo-excited charge carriers. Equilibrium values are measured if the time delay between the maxima of the pump and probe pulses is 1 ps and longer. Figure 4 displays THz/mid-IR reflection data for a delay time of 1 ps. The spectra of the reflected pulses can be reproduced very well by a Drude analysis if care is taken to treat the depth-distributed character of the reflection process properly [6]. For an excitation density ranging from 0.35 to 2.1×10^{19} electron/hole pairs per cm^3, and a plasma frequency ranging from 15 to 31 THz (effective reduced mass of Si: 0.15 m_0), the scattering time is found to decrease from 31 fs at the lowest density to 20 fs at the highest.

Figure 4 shows these data (open circles) together with additional results (full circles) for low excitation densities obtained by us with conventional time-domain THz spectroscopy (0.1-3 THz). The experiments cover the density regime up to 3.3×10^{17} cm^{-3}, the pump-probe delay in these measurement is 5 ps [6]. τ is found to decrease from 200 fs at low excitation density to 90 fs at 3.3×10^{17} cm^{-3}.

The curves in Fig. 4 are the results of generalized Drude model calculations based on the theoretical determination of scattering rates in a thermalized plasma as outlined in [6, 15]. The solid curve traces the excitation density dependence of the total scattering rate. The theoretical prediction reproduces the measured data both with regard to the density trend and the absolute values well. According to theory, carrier momentum scattering at low densities is dominated by energy relaxation of the carriers via deformation-potential interaction with acoustic phonons. Interaction with optical phonons is weak as expected. With increasing density, the acoustic relaxation channel becomes still more effective because of the increasing carrier velocity, before phase-space constraints reduce its effectiveness. At a density of 5×10^{17} cm^{-3} (mean carrier distance of about 10-15 nm), a cross-over occurs above which electron-hole scattering is the dominant momentum relaxation process.

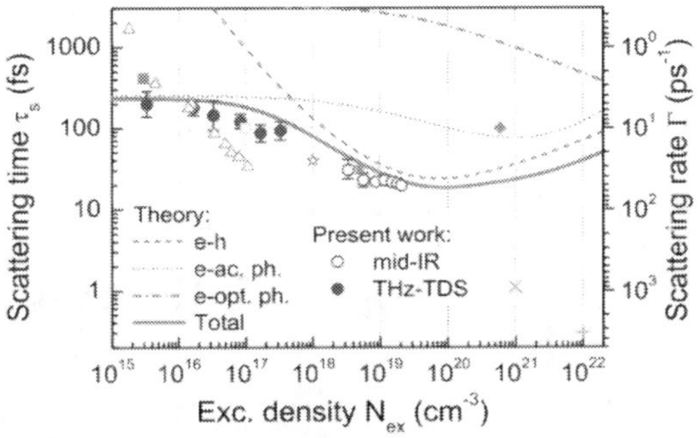

Figure 4. Measured (symbols) and calculated (full line) Drude scattering time of a thermalized electron-hole plasma as a function of electron/hole pair density. The symbols in colours other than blue are from the literature, the reader is referred to [6]. The broken lines show the contributions from various relaxation channels. From [6].

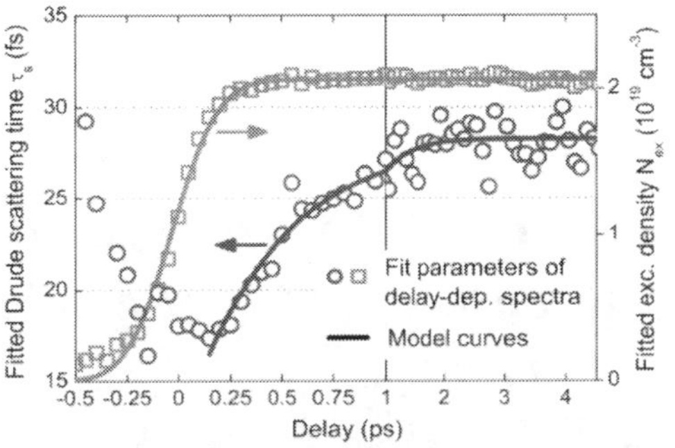

Figure 5. Drude scattering time and excitation density as a function of pump-probe delay. The scattering time exhibits a rise which is delayed with respect to that of the excitation density. The lines are fits with an exponential step function. From [6].

4. Subpicosecond evolution of the Drude scattering time for highly photo-excited silicon

If one extends the THz/mid-IR reflection measurements to shorter delay times, and performs a Drude-model analysis of the resultant reflection spectra, one obtains the data shown in Fig. 5. Compared with the build-up of the carrier density, τ exhibits a delayed rise with a time constant of 450 fs. This value is close to that attributed to the initial energy relaxation of the charge carriers [16].

5. Conclusion

We have demonstrated a frequency up-shift of THz pulses reflected off a counter-propagating front of photo-excited charge carriers in silicon. The degree of up-shift is strongly limited by extremely fast carrier momentum relaxation of less than 5 fs at our excitation densities in the 10^{19}-cm^{-3} range. Additional measurements show that the scattering time gradually increases to a density-dependent value of 20-30 fs as the charge carriers thermalize first among each other and then with the lattice. Practical exploitation of the Doppler up-shift will require much reduced electron-hole scattering.

Acknowledgments

The authors acknowledge funding by the Deutsche Forschungsgemeinschaft.

References

[1] Pauli W 1958 *Theory of relativity* (Pergamon, New York)
[2] Semenova V I 1972 *Sov. J. Radiophys. Quantum Electron.* **15** 1375
[3] Lampe M, Ott E and Walker J H 1978 *Phys. Fluids* **21** 42
[4] Thomson M D, Tzanova S M and Roskos H G 2013 *Phys. Rev.* B **87** 085203
[5] Meng F, Thomson M D and Roskos H G 2014 *Phys. Rev.* B **90** 155207
[6] Meng F, Thomson M D, Sernelius B E and Roskos H G 2015 *Phys. Rev.* B **91** 075201
[7] Thomson M D, Blank V and Roskos H G 2010 *Opt. Express* **18** 23173
[8] Thomson M D, Kress M, Löffler T and Roskos H G 2007 *Laser and Photonics Rev.* **1** 349
[9] Kress M, Löffler T, Eden S, Thomson M D and Roskos H G 2004 *Optics Lett.* **29** 1120
[10] Kress M et al. 2006 *Nature Phys.* **2** 327
[11] Blank V, Thomson M D and Roskos H G 2013 *New J. Phys.* **15** 075023
[12] Thomson M D, Blank V and Roskos H G 2013 *EPJ Web of Conferences* **41** 09011
[13] Thomson M D, Meng F and Roskos H G 2015 *Proc. SPIE*
[14] Esser A et al. 1990 *Phys. Rev.* B **41** 2879
[15] Sernelius B E 1991 *Phys. Rev.* B **43** 7136
[16] Zielbauer J and Wegener M 1996 *Appl. Phys. Lett.* **68** 1223

Electron capture processes in quantum dots due to one-and two-phonon assisted transitions: The role of optical phonon confinement

A L Vartanian, K A Vardanyan, V N Mughnetsyan and A A Kirakosyan
Department of Solid State Physics, Yerevan State University,
1, A.Manoogian, 0025 Yerevan, Armenia

E-mail: karvard@ysu.am

Abstract. Electron capture induced by carrier heating in spherical quantum dot–quantum well (QD-QW) structure is studied theoretically. The capture rate (CR) in one- and two-polar-optical-phonon-mediated capture processes has been studied by taking into account the phonon confinement. We have derived the analytic expressions for carrier CRs which can be conveniently applied to practical calculations for the spherical quantum dot systems. The numerical results of the CR as function of dot radius, lattice temperature and electron density in GaAs/AlAs/vacuum and CdSe/ZnS/H2O QD systems are obtained and discussed.

1. Introduction

The understanding of ultrafast carrier dynamics in QDs and, particularly, the processes involving carrier capture into the dots and intradot relaxation, have been under extensive theoretical and experimental studies during the past decade. Two different capture and relaxation processes have been considered; via carrier–carrier interaction (Auger processes) or via carrier–phonon coupling [1-3]. Most of the nanoscale microcrystals have a spherical shape. In many cases, these spherical QDs are composed of a spherical core of one (core-well) material embedded in a matrix of another (shell-barrier) material. The inhomogeneous nature of nanostructures leads to strong modifications of the electronic properties as well as the phonon spectrum. New types of confined (LO1, LO2) interface (IO1, IO2) and surface (SO1, SO2) optical modes can occur in QD quantum wells [4]. We have investigated the effect of phonon confinement on one and two phonon capture processes in a spherical QD-QW structure, in the framework of the effective-mass approximation.

2. Results and discussion

Within the framework of perturbation theory, the rates of electron capture from an initial state $|k, p_1, l_1\rangle$ with energy $E_{kp_1l_1}$ to the QD state $|t_3, p_3, l_3\rangle$ with energy E_0 via one- and two-phonon processes can be, respectively, written as [5]

$$R_0^{One} = \frac{\sqrt{2m}}{4\hbar^2} \sum_{t_3} \sum_{\alpha smn} (N_{\alpha smn} + 1) Q_{k_{01},s}^{t_3,0}(\alpha smn) \frac{f\left(E_{k_{01}}^{(1)}\right)}{\sqrt{E_{k_{01}}^{(1)}}} \left(1 - f(E_0)\right), \tag{1}$$

$$R_0^{Two} = \frac{\sqrt{2m}}{4\hbar^2} \sum_{t_3 p_1} \sum_{\substack{\alpha smn,\\ \alpha' s' m' n'}} (N_{\alpha smn}+1)(N_{\alpha' s' m' n'}+1) \sum_{l_1} \left| Y(sm, s'm', p_1 l_1) \right|^2 \frac{f\left(E_{k_{02}}^{(2)}\right)}{\sqrt{E_{k_{02}}^{(2)}}} \times$$

$$\times \left| \sum_{t_2} \left(\frac{Q(\alpha sn, \alpha' s' n'; t_3 0, t_2 s', k_{02} p_1)}{E_{k_{02}} - E_{t_2, s', -m'} - \hbar\omega_{\alpha smn}} + \frac{Q(\alpha' s' n', \alpha sn; t_3 0, t_2 s, k_{02} p_1)}{E_{k_{02}} - E_{t_2, s, -m} - \hbar\omega_{\alpha smn}} \right) \right|^2 (1 - f(E_0)), \quad (2)$$

where $f(E_k)$ is the Fermi-Dirac distribution function, $N_{\alpha smn}$ is the distribution function of the α type (LO1, LO2, IO1, IO2, SO1, SO2) optical phonons with quantum number set (s,m,n), frequency $\omega_{\alpha smn}$ and the electron-LO or- IO(SO) phonon radial coupling functions $\Gamma_{sn}^{\alpha}(r)$ [4],

$$E_{k_{01}}^{(1)} = E_0 + \hbar\omega_{\alpha smn} \equiv \frac{\hbar^2 k_{01}^2}{2m}, \qquad E_{k_{02}}^{(2)} = E_0 + \hbar\omega_{\alpha smn} + \hbar\omega_{\alpha' s' m' n'} \equiv \frac{\hbar^2 k_{02}^2}{2m}, \quad (3)$$

$$Y(sm, s'm', i_1 j_1) = \int_0^{\pi} d\theta \sin\theta \int_0^{2\pi} d\varphi Y_{sm}(\theta,\varphi) Y_{s'm'}(\theta,\varphi) Y_{i_1 j_1}(\theta,\varphi), \quad (4)$$

$$Q_{t_1 i_1}^{t_3 i_3}(\alpha sn) = \int_0^{\infty} F_{t_3 i_3}^{*}(r) \Gamma_{sn}^{\alpha}(r) F_{t_1 i_1}(r) r^2 dr, \quad Q(\alpha sn, \alpha' s' n'; t_3 i_3, t_2 i_2, t_1 i_1) = Q_{t_1 i_1}^{t_2 i_2}(\alpha sn) Q_{t_2 i_2}^{t_3 i_3}(\alpha' s' n'), \quad (5)$$

$F_{ti}(r)$ are the electron radial wave functions of the stationary states of the discrete and continuous spectrum [6]. Expressions (1) and (2) allow us to present the phonon system in electron capture rate calculations more realistically, because in nanostructures the existence of boundaries between the constituting materials and/or to the vacuum introduces a coupling of the longitudinal and transverse

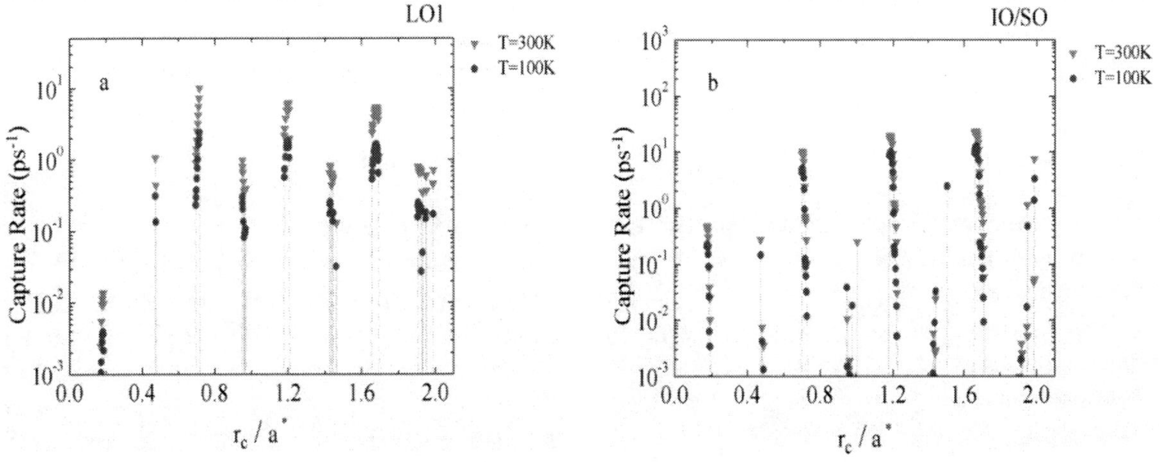

Fig. 1. Dependence of CR on CdSe/ZnS/H_2O QD radius at $T = 100K$ and $T = 300K$ for fixed electron density of $10^{17} cm^{-3}$ via emission of (a) LO1 and (b) IO/SO phonons.

phonon modes and leads not only to the formation of new phonon modes (IO and SO) but also changes the potential field created by bulk LO phonons. The latter is represented by the $\Gamma_{sn}^{\alpha}(r)$ functions. The calculated electronic CRs in CdSe/ZnS/H_2O QDs due to LO1and IO/SO phonon modes as a function of dot radius for temperatures of T=300K and T=100K and carrier density $n=10^{17} cm^{-3}$ are shown in the Fig. 1a and b, respectively. These results exhibit some interesting features. First, the CR of an electron from the barrier region to the quantum dot ground state due to the emission of confined or interface optical phonons exhibits strong resonances versus dot radius. The graphs exhibit

Journal of Physics: Conference Series **647** (2015) 012017 doi:10.1088/1742-6596/647/1/012017

clear and pronounced peaks at approximately regular intervals. This is because the capture is extremely efficient at resonance ($E_k = E_0 + \hbar\omega_{\alpha smn} = 0$), while it is more and more unlikely when the energy of the impinging electron increases mainly because $f(E_k)$ becomes exponentially small. Second, the comparison of the CRs due to LO1 and IO phonon scatterings shows that the interaction

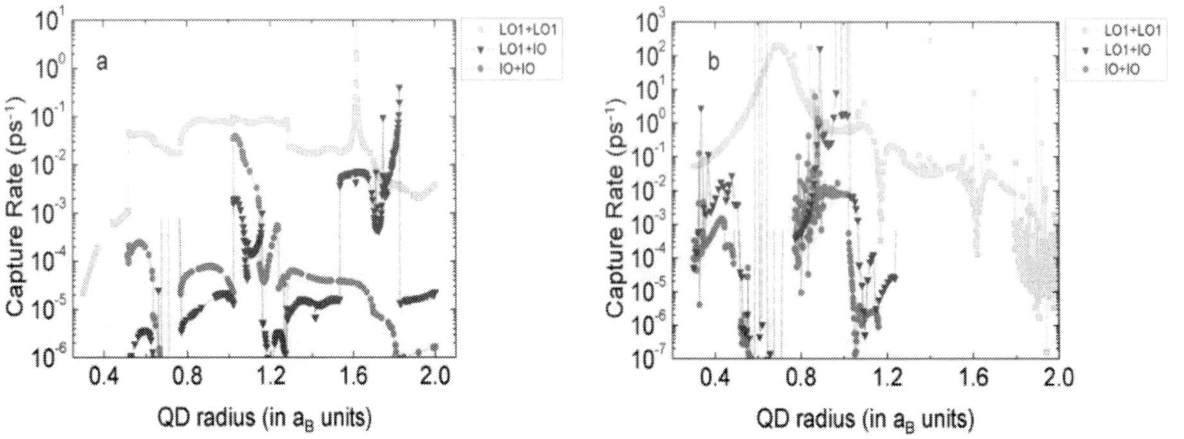

Fig. 2. Two-phonon (LO1+LO1, LO1+IO, IO+IO) capture rate dependence on GaAs/AlAs QD radius (in units of effective Bohr radius) at $T = 300K$ and for fixed density $10^{17} cm^{-3}$: a. caused by discrete intermediate states, b. caused by continuous intermediate states.

with IO phonons leads to a one order of magnitude larger CR than the electron–LO1phonon coupling for the same conditions. The electron capture time defined as $\tau = 1/R_0$ for the coupling with LO1 phonons has the same order of magnitude as that obtained by Magnusdottir et al. in GaAs QDs without taking into account phonon confinement [7]. In contrast to the case of CdSe/ZnS/H_2O QDs, the dependence of the electron CR on the radius of GaAs QDs exhibits wider areas with a high rate of electron capture both with [5] and without [7] the phonon confinement effect. Moreover, it has been shown the appearance of these bands is not caused by phonon confinement [5]. Only the increase of the height of the resonances in the CR with increasing quantum dot radius is assumed to be due to the spatial confinement of the phonon modes. A capture with the emission of two phonons may occur in the states of the quantum dot when the E_b binding energy satisfies the $\hbar\omega_{ph} < E_b < 2\hbar\omega_{ph}$ condition. In Fig. 2 the dependences of the two-phonon CR due to emission of LO1+LO1, LO1+IO and IO+IO phonon pairs as function of QD radius is shown. Here we consider the electron scattering processes in the cases where a discrete (Fig. 2a) or continuous (Fig. 2b) electronic states are involved as intermediate states.

The CR for LO1+LO1 processes with discrete intermediate states is almost constant for QDs with radii from $0.8a_B$ to$1.25a_B$ and constitutes about $10^{-1}ps^{-1}$. At larger values of the radius the CR decreases and becomes less than $10^{-2}ps^{-1}$ in case of radius value $2a_B$. Note, however, that when the radius becomes about $1.6a_B$ (which is surely the radius separating two bands), the capture rate sharply increases more than $1ps^{-1}$. The dependence of the CR on the radius of QD in the emission of phonons LO1+LO1 shows that in processes involving continuous intermediate states the capture rate can get over $1ps^{-1}$ when $0.5a_B \le r_c \le 1.2a_B$. Subsequent increase of QD radius leads to a reduction of the CR reaching up to $10^{-2}ps^{-1}$ when the radius becomes about $2a_B$.The dependence of the CR on the radius of QD in the emission of phonons LO1 + IO shows that in processes involving both continuous and discrete intermediate states the CR over the entire range of the radius less (rather several orders) than

$10^{-2}ps^{-1}$. The values for the CR of the order of $10^{-2}ps^{-1}$are obtained only for some narrow intervals of the quantum dot radius. This picture for the CR is also conserved in the emission of phonons IO + IO, the only difference is that in processes involving continuous intermediate states CR is of the order of $1ps^{-1}$, when $0.9a_B \le r_c \le 1a_B$.

The two-phonon CR dependencies on the lattice temperature and electron density in the barrier region for the quantum dot with radius $1.45a_B$ are presented in Figures 3 and 4, respectively. As can

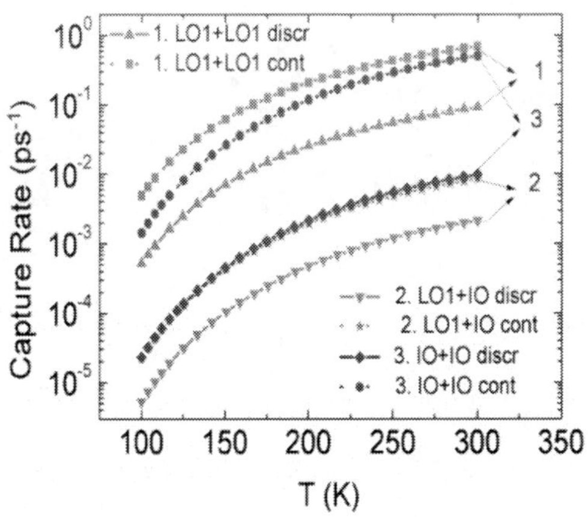

Fig.3. Two-phonon capture rate dependence on temperature at $a = 1.45a_B$ and $n = 10^{17}cm^{-3}$.

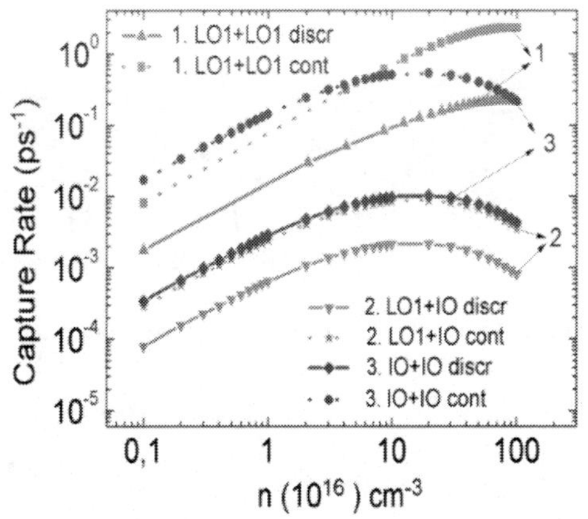

Fig.4. Two-phonon capture rate dependence on carrier density at $a = 1.45a_B$ and $T = 300K$.

be seen, with increasing temperature the CR increases, which, as in the single-phonon capture case, is due to the increase of the average number of phonons. Compared to one-phonon capture, the behavior of the CR does not change depending on the electron density. The only difference is that in the single-phonon CR maximum occurs in concentrations 10^{16} cm^{-3} while the two-phonon capture maximum is observed at higher concentrations.

In conclusion, the capture rates in one- and two-polar-optical-phonon-mediated capture processes have been studied by taking into account a more realistic model for phonon system. We have derived the analytic expressions for carrier capture rates which can be conveniently applied to practical calculations for the quantum dots.

Acknowledgements
This work was supported by the State Committee of Science MES RA (Projects No.13-1C196, 13IE-038, 13RF-093, 13YR-1C0044) and the ANSEF grant (condmatth-3821) New York, USA.

References
[1] Borri P, Schneider S, Langbein W, and Bimberg D 2006 *J. Opt. A, Pure Appl. Opt.* **8** S33
[2] Piwonski T, O'Driscoll I, Houlihan J, Huyet G, and Manning R 2007 *Appl. Phys. Lett.* **90** 122108
[3] O'Driscoll I, Piwonski T, Schleussner C-F, Houlihan J, Huyet G, and Manning R J 2007 *Appl. Phys. Lett.* **91** 071111
[4] Zhang L, Xie H-J, and Chen C-Y 2002 *Phys. Rev. B* **66** 205326
[5] Vardanyan K A, Vartanian A L, Mughnetsyan V N, Dvurechenskii A V, and Kirakosyan A A 2015 *Physica E* **66** 268
[6] Tkach M, Holovatsky V, Voitsekhivska O, Mykhalyova M, and Fartushynsky R 2001 *Phys. Stat. Sol. (b)* **225** 331
[7] Magnusdottir I, Uskov A, Bischoff S, Tromborg B, and Mørk J 2002 *Journ. Appl. Phys.*, **92** 5982

Journal of Physics: Conference Series **647** (2015) 012018 doi:10.1088/1742-6596/647/1/012018

3D self-consistent percolative model for networks of randomly aligned carbon nanotubes

S Colasanti[1], V Deep Bhatt, A Abdellah and P Lugli
Institute for Nanoelectronics
Technical University Munich (TUM), Munich, 80333, Germany

E-mail: simone.colasanti@nano.ei.tum.de

Abstract. A numerical percolative model for simulations of random networks of carbon nanotubes is presented. This algorithm takes into account the real 3D nature of these networks, allowing for a better understanding of their electrical properties. The nanotubes are modeled as non-rigid bendable cylinders with geometrical properties derived according to some statistical distributions inferred from the experiments. For the transport mechanisms we refer to the theory of one-dimensional ballistic channels which is based on the computation of the density of states. The behavior of the entire network is then simulated by coupling a SPICE program with an iterative algorithm that calculates self-consistently the electrostatic potential and the current flow in each node of the network. We performed several simulations on the resistivity of networks with different thicknesses and over different simulation domains. Our results confirm the percolative nature of the electrical transport, which are more pronounced in films close to their percolation threshold.

1. Introduction

Random carbon nanotube (CNT) networks are nowadays very attractive thanks to the combination of interesting electrical properties, low cost fabrication processes and the possibility to deposit them over flexible substrates. Among the possible applications, these networks are studied as transparent electrodes, thin-film transistors and several types of sensors [1]. Despite this great interest, the main transport mechanisms are still not fully understood. The presented model is based on a stochastic algorithm that can generate non-rigid solid objects in a three-dimensional space, emulating with high fidelity the typical fabrication processes involved (i.e. spray-coating). This represents a crucial step, since with the available two-dimensional models all the nanotubes are projected into one plane, making the number of junctions between different CNTs highly overestimated. An appropriate formalism for the electrical transport in one-dimensional channel has been introduced, together with an iterative algorithm that calculates self-consistently all the electrical quantities in every node of the network. Our work is intended to overcome some of the flaws regarding the modeling of these networks and provide a tool for the controlled design of devices based on this technology.

.

[1] To whom any correspondence should be addressed.

Journal of Physics: Conference Series **647** (2015) 012018 doi:10.1088/1742-6596/647/1/012018

2. Computational models

The software is composed of two main programs, one for the generation of the structure and one for the actual simulation, plus a tool for the visualization of the results. The electrical simulation is carried on by a SPICE software which executes at each call a netlist generated by our program. Everything is embedded in a self-consistent framework, which will update at each iteration the netlist to simulate.

2.1. Network generation

For the generation of the structure we referred to our previously developed simulation tool [2]. Every nanotube is represented as a bendable cylinder with radius and length that are randomly chosen according to a certain statistical distribution, which is inferred from AFM images of the investigated samples [3]. During this step also the electronic properties will be determined by randomly choosing the two indexes of the chiral vector. The number of nanotubes to be generated is determined by a specific density of CNTs/μm^2 that is chosen accordingly to the used experimental setup [4].

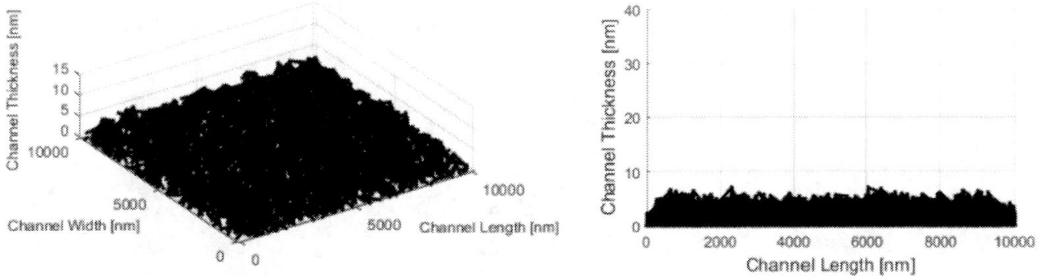

Figure 1. 3D view (left) and side view (right) of a CNT network generated over a 10x10 μm^2 area. The density of nanotubes is around 18 CNTs/μm^2.

In our approach, CNTs are placed sequentially, controlling each time if there are any intersections between the incoming nanotube and the already existing network; if an intersection occurs, the new nanotube will be bent of an angle usually smaller than 10 degrees; if the angle is greater than the maximum allowed, the nanotube will slide until it finds a resting point over the substrate or over another nanotube. The result is a film in a three-dimensional space, which is depicted in figure 1. The values of thickness and surface coverage are in good agreement with the measurements, as we already reported elsewhere [3].

2.2. Network simulation

The resulting set of tube segments and tube junctions must be then converted into a netlist and the overall electrical behavior is computed by coupling our program with a SPICE software. Each segment is represented by a voltage-dependent resistor, whose value is dynamically calculated within a self-consistent algorithm. After a first guess on the value of the resistor, the netlist is solved and the potential at each node of the entire network is calculated. The value of every resistor will be then updated according to the new applied potential and the process will be iterated until the difference between two consecutive steps is less than a previously set tolerance. The expression of the resistance follows equation (1).

$$R(V) = R_0(V)\left(1 + \frac{L}{\lambda_{EFF}}\right) \tag{1}$$

where R_0 represents the so-called quantum resistance, L is the length of the tube segment and λ_{EFF} is the effective mean free path, which contains the effects due to phonon scattering events [5]. The quantum resistance is calculated as the ratio between the applied voltage and the current that flows through the nanotube. The latter is solved according to the theory of one-dimensional conducting channels [6] and follows the equation

$$I(V) = \frac{q}{\hbar} \int_{-\infty}^{+\infty} dE\, D_{\varepsilon}(E) \frac{\gamma_1 \gamma_2}{\gamma_1 + \gamma_2} [f_1(E,V) - f_2(E,V)] \tag{2}$$

where q is the electronic charge, ħ is the reduced Planck constant, $D_\varepsilon(E)$ is the density of states per unit of energy, γ_1 and γ_2 are the broadening functions at the contacts. From (2) we can see that the current is a function of the electrostatic potentials imposed at the two contacts of the nanotube (this dependence is implicit in the Fermi functions $f_1(E,V)$ and $f_2(E,V)$ at the two nodes), which is the reason why an iterative approach for the solution of the system is needed. In equation (2) the variable that discriminates the different properties of the nanotubes (i.e. the chirality) is the DOS, which is computed according to the numerical model developed in [7].

3. Results and discussion

We performed different studies on the resistivity of different networks in terms of film thickness, showing the percolative nature of the transport. In figure 2 we report the current flow through a network of a certain area, and the same plot for two subsets obtained by cutting half of the network along its width.

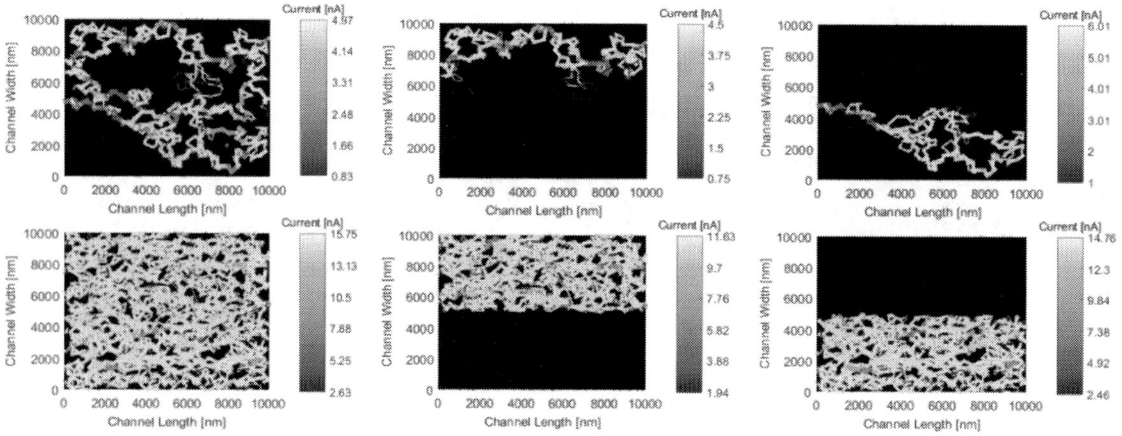

Figure 2. Plot of the current flow through a film of 12 CNTs/μm^2 (top) and a film of 24 CNTs/μm^2 (bottom). From left to right, the entire network, the network with the bottom half cut out and the network with the top half cut out. While for the dense film the density of current is almost constant in the three different cases, for the sparse network it instead changes considerably.

As it stands out, the percolative behavior is much more evident in films with a low density of nanotubes. When the density increases, the network tends to become more uniform, and the film approaches a behavior similar to the one of a bulk material. In figure 3a we show the value of the normalized resistivity versus the aspect ratio of the network. When the film is close to its percolation threshold, it is very likely that a cut could eliminate fundamental percolation paths, increasing dramatically the resistivity of the network. On the other hand, if the width is greater than its initial value, more paths will be formed, causing a serious drop of the resistivity. These effects are much less influential on more dense networks, where the resistivity seems to be less affected by the variations of the aspect ratio of the film.

A similar analysis has been done on the resistivity of films with different densities, as is shown in figure 3b. Consistently with our previous findings, we observed two different behaviors on the resistivity, depending on the density of the nanotubes. After a certain threshold, an increase of the density does not affect anymore the resistivity of the film, which tends to saturate to a certain value. As we previously mentioned, CNT networks could be employed as transparent electrodes. Since in this particular application a trade-off exists between the resistivity and transparency of the film, this analysis could be useful for a proper design of the electrode. When such a threshold is actually reached, an increase of the film thickness will not bring any appreciable benefit to the conductivity of the network, but will drastically reduce the transparency of the film.

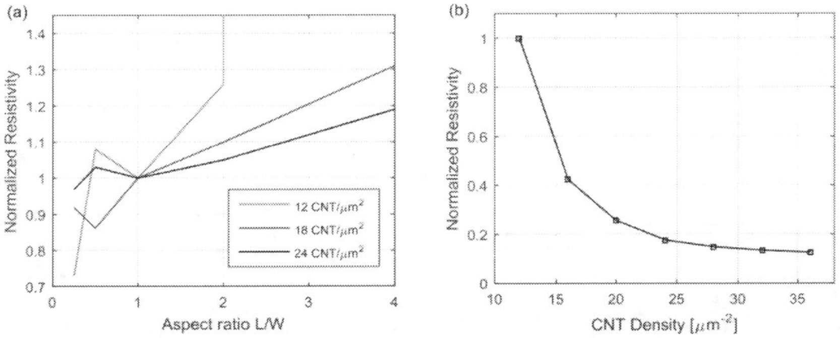

Figure 3. (a) Normalized resistivity versus the aspect ratio of the simulated area for three different densities. (b) Normalized resistivity versus the density of the nanotubes.

4. Conclusion

In conclusion, we report the implementation of a simulation tool based on a percolative model. Accurate morphologies have been generated in a three dimensional space, reducing the gap between the real structures and the one resulting from 2D models previously presented in the literature. We implemented an accurate model for the transport through the single nanotubes and we included it in a self-consistent framework for the simulation of the entire network. The results obtained by our simulations are in agreement with the percolation theory. We observed that the percolative behavior is much more evident for low densities of nanotubes, and it tends to vanish when the networks are far from their percolation threshold.

Acknowledgments

This work was partly supported by the European Commission under grant agreement PITN-GA-2012-317488-CONTEST and the TUM Graduate School.

References

[1] Hu L, Hecht D S and Grüner G 2010 Carbon nanotube thin films: fabrication, properties, and applications. *Chem. Rev.* **110** 5790–844

[2] Colasanti S, Deep Bhatt V and Lugli P 2014 3D Modeling of CNT Networks for Sensing Applications *2014 10th Conference on Ph. D. Research in Microelectronics and Electronics (PRIME)* pp 3–6

[3] Colasanti S, Robbiano V, Loghin F C, Deep Bhatt V, Abdellah A, Cacialli F, and Lugli P 2015 Experimental and Computational Study on the Temperature Behavior of CNT Networks *2015 15th IEEE Int. Conf. Nanotechnol.*, in press.

[4] Abdellah A, Abdelhalim A, Horn M, Scarpa G, and Lugli P 2013 Scalable spray deposition process for high-performance carbon nanotube gas sensors *IEEE Trans. Nanotechnol.* **12** (2) 174–181

[5] Pennington G and Goldsman N 2005 Low-field semiclassical carrier transport in semiconducting carbon nanotubes *Phys. Rev. B* **71** 205318

[6] Datta S 2005 *Quantum Transport Atom to Transistor* (New York: Cambridge University Press)

[7] Mintmire J and White C 1998 Universal Density of States for Carbon Nanotubes *Phys. Rev. Lett.* **81** 2506–9

Comparison of phonon scattering in nanowire field effect transistors with Si, GaAs and InGaAs cores using the NEGF formalism

A Price and A Martinez

College of Engineering, Swansea University, UK

E-mail: A.C.PRICE.625036@swansea.ac.uk

Abstract.

Using the Non-equilibrium Green's Function (NEGF) formalism, the impact of electron-phonon scattering on the performance of different core nanowire field effect transistors (NWFETs) has been investigated. Three core materials have been considered: Si, GaAs and InGaAs. The effective mass approximation has been used, with masses extracted from tight-binding simulations. The I_D-V_G characteristics at low and high drain bias are shown. It was found that at low drain bias, scattering caused a 86%, 72% and 50% percentage reduction in the current at high gate bias in the Si, GaAs and InGaAs core, 2.2×2.2 nm^2 cross-section NWFETs respectively. The phonon-limited mobility and percentage tunnelling have also been calculated.

1. Introduction

Gate-all-around nanowire field effect transistors (NWFETs) are a likely contender to succeed FinFETs. They have strong electrostatic integrity and good scaling properties, with 6 nm channel lengths achievable. The introduction of III-V channel materials such as GaAs and InGaAs are of interest due to their high mobility. In this work, ballistic and scattering simulations of n-channel Si, GaAs and $In_{0.3}Ga_{0.7}As$, 2.2×2.2 nm^2 and 4.2×4.2 nm^2 cross-section NWFETs with a 6 nm channel length are studied. The source and drain of the NWFETs are doped at 10^{20} cm^{-3} and the channel is undoped. A schematic of the 2.2×2.2 nm^2 cross-section NWFET can be seen in figure 1. All the relevant scattering mechanisms involving acoustic, optical and polar optical phonons have been included. The scattering parameters were taken from ref. [1].

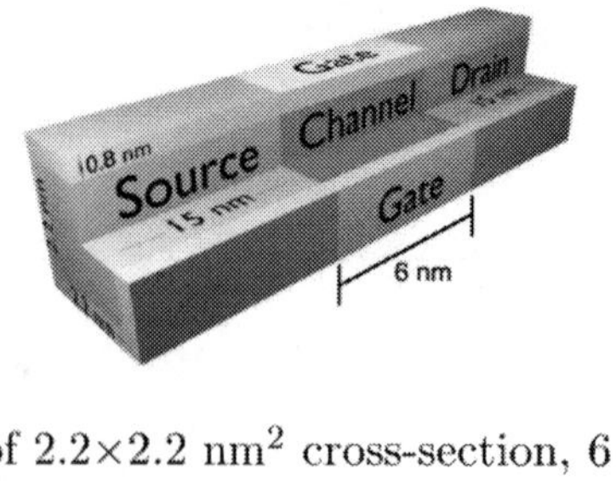

Figure 1. Schematic of 2.2×2.2 nm^2 cross-section, 6 nm channel NWFET.

2. Theory

This study uses the Non-equilibrium Green's Function (NEGF) formalism. Within the NEGF formalism one-particle electronic Green's functions, G, are used. The retarded Green's function, G^r, is given by $G^r(E) = [E1 - H - \Sigma^r(E)]^{-1}$ where H is the Hamiltonian of the system. Within the effective mass approximation the Hamiltonian is given by

$$H_\nu(\vec{r}) = -\frac{\hbar^2}{2}\nabla_i\left(\frac{1}{m_\nu}\right)_{ij}\nabla_j + V(\vec{r}) \quad (1)$$

At these scales bulk masses are not valid, so new confinement effective masses have been calculated using the results of tight-binding simulations, which capture the dependence of the bandstructure on the cross-section of the nanowire [2].

3. Results

In this section, the I_D-V_G characteristics at low and high drain bias are presented. The sub-threshold slope, percentage reduction in the current due to scattering and the percentage tunnelling are also studied. At the end of this section, the phonon-limited mobility is briefly discussed.

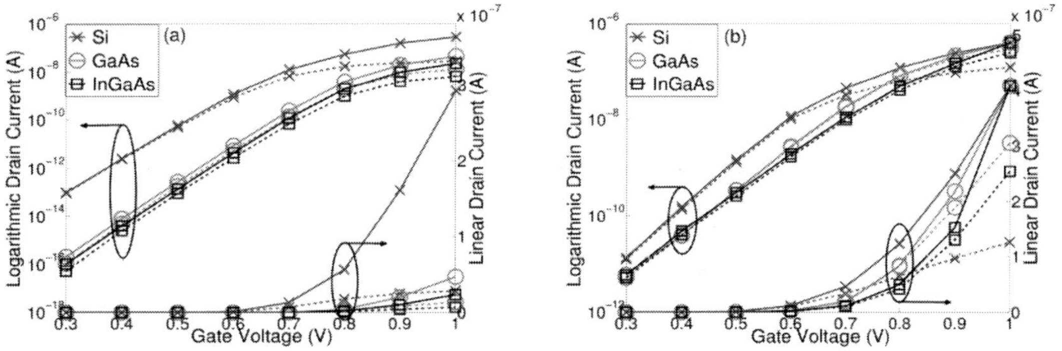

Figure 2. I_D-V_G characteristics at low drain bias, V_D = 1 mV, for Si, GaAs and InGaAs for the (a) 2.2×2.2 nm^2 (b) 4.2×4.2 nm^2 cross-section NWFETs. The ballistic simulations are denoted by the solid lines and the scattering simulations by the dotted lines.

Figure 2 shows the I_D-V_G characteristics at low drain bias for all the simulated devices. For both cross-sections, the highest current is in the Si devices. For the 2.2×2.2 nm^2 cross-section devices, the scattering sub-threshold slope is 76, 69 and 71 mV/dec for the Si, GaAs and InGaAs cores respectively. For the 4.2×4.2 nm^2 cross-section devices, the sub-threshold slope is 99, 109 and 112 mV/dec for scattering in the Si, GaAs and InGaAs cores respectively. The ideal value for the sub-threshold slope is approximately 60 mV/dec. A higher sub-threshold slope indicates worse electrostatic control.

In the 2.2×2.2 nm^2 cross-section, there is a 86%, 72% and 50% percentage reduction in the current due to scattering at high V_G for the Si, GaAs and InGaAs cores respectively. For the 4.2×4.2 nm^2 cross-section devices, there is a 58%, 40% and 37% percentage reduction in the current due to scattering at high V_G for the Si, GaAs and InGaAs cores respectively. The greater percentage reduction in the current for 2.2×2.2 nm^2 NWFETs is due to the strength of electron-phonon coupling increasing with decreasing cross-section.

For the 2.2×2.2 nm^2 cross-section devices at both low and high drain bias, the current is much lower in the III-V NWFETs. However, in the 4.2×4.2 nm^2 cross-section NWFETs the current in the III-V devices is much closer to that of the Si device, approaching the same value for the ballistic on-current. The much lower current in the III-V 2.2×2.2 nm^2 NWFETs is because of the strong confinement. In bulk GaAs and InGaAs, the order of valleys in increasing energy is Γ, L then X. For the III-V 2.2×2.2 nm^2 cross-section devices the low mass Γ-valley becomes raised in energy such that it's higher than the heavier L and X-valleys, whereas in the III-V 4.2×4.2 nm^2 cross-section NWFETs, the valleys have the same order as bulk. The elevation the Γ-valley results in much lower current in the III-V 2.2×2.2 nm^2 devices. The greater degradation of the sub-threshold slope with increasing cross-section for the III-V NWFETs is because of the lower mass Γ-valley beginning to contribute more to the transport as the cross-section increases.

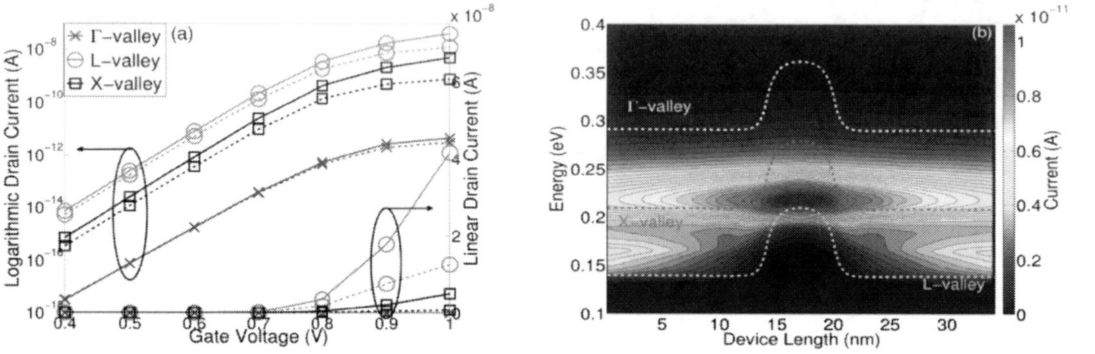

Figure 3. (a) I_D-V_G characteristics for each valley in GaAs and (b) scattering energy-resolved current spectrum at low drain bias, V_D = 1 mV, for GaAs core, 2.2×2.2 nm^2 cross-section NWFET.

Figure 3 (a) shows the I_D-V_G characteristics at low drain bias for each valley in the GaAs core 2.2×2.2 nm^2 cross-section NWFET. Most of the current is in the high mass L and X-valleys, with little current in the low mass Γ-valley. For the 4.2×4.2 nm^2 cross-section NWFET more of the current is in the Γ-valley [3]. This behaviour is also seen in the InGaAs NWFETs. Figure 3 (b) shows the energy-resolved current spectra with the first subband of each valley superimposed, showing the elevation of the Γ-valley. In comparison to the 4.2×4.2 nm^2 cross-section NWFET, the Γ-valley has been shifted a further 0.7 eV from the potential. By increasing the thickness of the oxide the Γ-valley will lower; resulting in much higher current [4].

In order to understand why the current is generally lower in the III-V NWFETs, the percentage tunnelling must be considered. The amount of percentage tunnelling is mostly determined by the effective mass and the size of the potential barrier. For the 2.2×2.2 nm^2 cross-section NWFETs, the Si device has the lowest percentage tunnelling as it has a smaller barrier (due to the poorer electrostatic control) and the main valleys contributing to the transport have smaller effective masses than the III-V devices (as in the 2.2×2.2 nm^2 III-V devices the contribution of the low mass Γ-valley has been reduced as it is elevated). The scattering percentage tunnelling at low drain bias and at high V_G is 11%, 54% and 45% for the Si, GaAs and InGaAs cores respectively. The higher percentage tunnelling in the III-V devices results in a lower total current. The lower tunnelling current in the Si devices contributes to the higher current reduction due to scattering than the III-V NWFETs.

For the 4.2×4.2 nm^2 cross-section devices the scattering percentage tunnelling at low drain bias and high V_G is 3%, 12% and 9% in the Si, GaAs and InGaAs NWFETs respectively. The reduction in the percentage tunnelling with increasing cross-section is because the effective mass decreases with increasing cross-section. A more detailed discussion on the effects of confinement, scattering and percentage tunnelling in GaAs NWFETs can be found in ref. [3].

Figure 4 shows the I_D-V_G characteristics at high drain bias for all the simulated NWFETs. For the 2.2×2.2 nm^2 cross-section devices, the scattering sub-threshold slope is 71, 66 and 63

Journal of Physics: Conference Series **647** (2015) 012019 doi:10.1088/1742-6596/647/1/012019

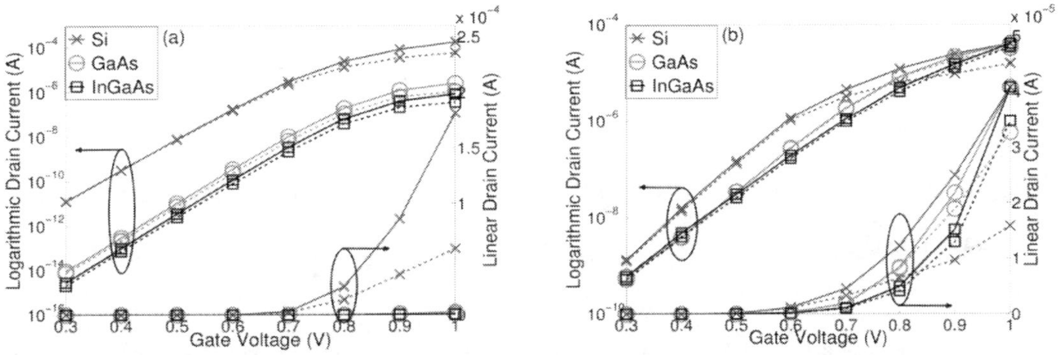

Figure 4. I_D-V_G characteristics at high drain bias, V_D = 0.6 V, for Si, GaAs and InGaAs for the (a) 2.2×2.2 nm^2 (b) 4.2×4.2 nm^2 cross-section NWFETs. The ballistic simulations are denoted by the solid lines and the scattering simulations by the dotted lines.

mV/dec for the Si, GaAs and InGaAs cores respectively. For the 4.2×4.2 nm^2 cross-section devices, the scattering sub-threshold slope is 97, 116 and 121 mV/dec for the Si, GaAs and InGaAs cores respectively.

In the 2.2×2.2 nm^2 cross-section devices, there is a 67%, 57% and 43% percentage reduction in the current due to scattering at high V_G for the Si, GaAs and InGaAs cores respectively. For the 4.2×4.2 nm^2 cross-section devices, there is 61%, 20% and 15% percentage reduction in the current due to scattering at high V_G for the Si, GaAs and InGaAs cores respectively. In comparison to low drain, the high drain characteristics generally show a decrease in the percentage reduction of the current due to scattering. This is because the percentage tunnelling is greater at high drain bias than low drain [3].

The phonon-limited mobility has been calculated at low drain bias for low channel charge. For the 2.2×2.2 nm^2 cross-section NWFETs the highest mobility is in the Si core device. At low channel charge the phonon-limited mobility is 229, 62 and 139 cm^2/Vs for the Si, GaAs and InGaAs NWFETs respectively. However, for the 4.2×4.2 nm^2 NWFETs the InGaAs device has the highest mobility. For the Si, GaAs and InGaAs core devices at low channel charge, the phonon-limited mobility is 546, 1290 and 1887 cm^2/Vs respectively. The low mobility for the III-V 2.2×2.2 nm^2 cross-section devices is due to the elevation of the Γ-valley.

4. Conclusion

Different dimension NWFETs with Si, GaAs and InGaAs cores have been investigated using the NEGF formalism. For both the 2.2×2.2 nm^2 and 4.2×4.2 nm^2 cross-section, the Si core device has the highest current. This is because there is higher percentage tunnelling in the III-V devices. For the III-V 2.2×2.2 nm^2 cross-section devices, the low mass Γ-valley becomes elevated in energy above the heavier L and X-valleys. This results in very low current and mobility in those NWFETs. Scattering was found to cause a large reduction in the current for all the simulated devices.

References

[1] M. Lundstrom, "Fundamentals of carrier transport," *Cambridge University Press* (2000).

[2] Y. M. Niquet, A. Lherbier, N. H. Quang, M. V. Fernandez-Serra, X. Blase and C. Delerue, *Phys. Rev. B* **73**, 165319 (2006).

[3] A. Price and A. Martinez, *J. Appl. Phys.* **117**, 164501 (2015).

[4] A. Price and A. Martinez, *J. Phys. Conf. Ser.* **609**, 012004 (2015).

Ballistic deflection transistors and their application to THz amplification

M Margala[1*], H Wu[2], and Roman Sobolewski[2]

[1]Department of Electrical and Computer Engineering, University of Massachusetts Lowell, One University Avenue, Lowell, Massachusetts 01854, USA
[2]Department of Electrical and Computer Engineering and Laboratory for Laser Energetics, University of Rochester, Rochester, New York, 14627-0231, USA

Email: Martin_Margala@uml.edu

Abstract. We present implementation of recently proposed ballistic deflection transistors (BDTs) as THz amplifiers. BDT is a planar device based on InGaAs/InAlAs/InP heterostructure with quasi-ballistic transport obtained in the two-dimensional electron gas layer that facilitates ultra-short transit time and high performance needed for THz-range circuitry. The BDT performance is optimized through its structural modification and the use of high-k dielectrics. Our time-domain, electrical transient measurements demonstrate sub-THz switching performance of a BDT with a ~1-μm-wide channel. Independently, circuit simulations using experimental parameters of BDTs with a channel width of 430 nm and with the BDTs themselves connected as a multi-stage travelling-wave amplifier, designed for 6-dB gain, predict a 2.7-THz bandwidth with a gain flatness of ±0.3 dB.

1. Introduction

The scaling limit of CMOS technology has given birth to other alternative novel devices. Some devices offer improved speed, some promise unmatched power efficiency, while others are reconfigurable at runtime to offer both versatility and robustness [1]. In addition, advanced nanofabrication techniques and novel material configurations have lead to devices in which the active region is smaller than the mean free path of electrons at room temperature. These devices are ballistic in nature, since the main scattering electrons encounter is with the designed device geometry [2]. In the same context, our group has introduced a ballistic deflection transistor (BDT), which is proposed to work at THz frequencies at room temperature [3]. The BDT exploits the very nature of a two-dimensional electron gas (2DEG) layer, where electrons can travel without virtually any scattering and achieve very high velocities of the order of 10^8 cm/s, what is 2.5 times faster than electron transport in silicon. Given the ultralow capacitive design, due to the fully planar device geometry, initial estimates of the unity-gain current frequency $f_T > 1$ THz for the BDT fabricated with a 430 nm gate width [3].

[*] To whom any correspondence should be addressed.

The main contribution of this work is not only to validate the feasibility of the BDT operation in the THz range through simulations, but also to demonstrate the usefulness of the time-resolved optoelectronic spectroscopy, as a tool for experimental characterization of THz- bandwidth devices.

2. BDT description and operation

The BDT is a six-terminal coplanar structure etched into an InGaAs 2DEG. The device consists of a grounded electron source, left and right gates, and three biased drains. Figure 1 shows a SEM image of the top-view of BDT with the experimental set up. The top-left (V_{LD}) and top-right (V_{RD}) ports are drain ports, bottom-left (V_{LG}) and bottom-right (V_{RG}) ports are gates, top port (V_{TD}) is a bias port that controls gain, and the bottom port (V_{SS}) is the source. Electrons in the high mobility 2DEG are provided by the δ-doped layer, which is separated by the undopped spacer, to reduce the interaction with remote impurities.

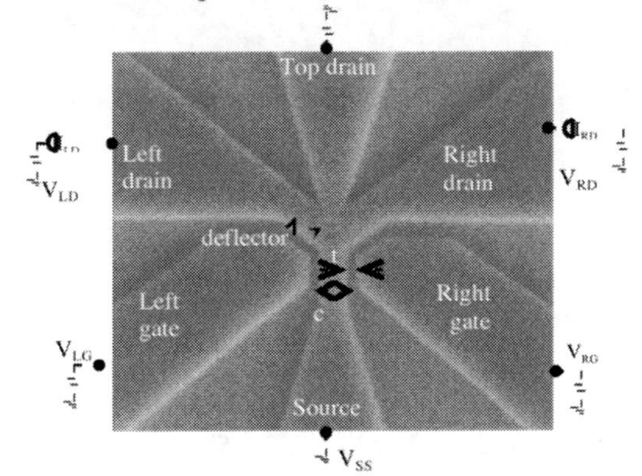

Figure 1. SEM image of top view of BDT with channel width (c) = 240 nm and trench width (t) = 100 nm. Top-left (V_{LD}) and top-right (V_{RD}) ports are drains, bottom-left (V_{LG}) and bottom-right (V_{RG}) ports are gates. Top port (V_{TD}) is a bias port that controls gain, and the bottom (V_{SS}) port is the source.

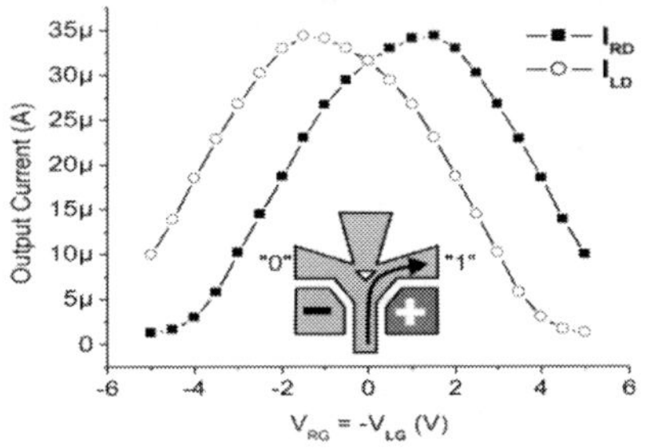

Figure 2. Transfer characteristics of BDT in push-pull gate bias ($V_{RG} = -V_{LG}$). $V_{TD} = V_{LD} = V_{RD} = 1$ V. c = 240 nm, t = 100 nm.

A typical, room temperature transfer curve is shown in figure. 2. The current rises and pinches down with gate voltage. This shape of transfer curve makes a BDT suitable for frequency doubling. A simple operational mechanism of BDT is shown in the inset of figure 2, which exhibits that when the electron current moves along the left drain, transistor registers logic "1" at left drain and logic "0" at right drain and vice-versa. The gate control over the channel can be attributed to three effects: classic channel pinch-off due to the field effect, an electron steering due to the gate bias, and deflection due to the presence of a centre triangular deflector. The steering effect implies that flowing electrons encounter a lateral field directing them towards one output drain or the other, depending on gate bias. Pinch-off can be observed when a strong bias depletes the channel completely. Interaction of electrons with right or left side of the deflector helps electrons in moving towards the corresponding drain. A detailed explanation of the BDT operation can be found in [4] and [5].

3. THz amplification using BDTs

For THz amplifier simulations, we have designed an 8-stage BDT traveling-wave amplifier with 15 BDTs in each stage, as shown in figure 3. BDTs were interconnected with the 185-Ω characteristic impedance coplanar waveguide for both the gate and the drain lines. The length of each differential transmission line was 320 μm, feasible for an on-chip implementation. The actual BDT equivalent circuit parameters were derived based on experimental performance of a BDT transistor with the 430-nm-wide channel, including both capacitive and resistive parasitics. A large number of amplifying stages were needed to achieve a large enough (in our case 3 mS) total transconductance for the assumed total gain of 6 dB.

Simulation results are presented in figure 4 and demonstrate that our amplifier achieved a 2.7-THz bandwidth with a gain flatness of ±0.3-dB. The return loss was better than –10 dB up to 2.5 THz and the gain-bandwidth product reached 10.8 THz. The inset shows the impedance mismatch reduction after every stage until the match was achieved at the stage 8. Although the device representation in our simulations has been based on low-frequency BDT parameters, our studies clearly demonstrate that BDT amplifiers have a high potential to achieve the THz-level performance. The latter observation has been independently supported by the Monte Carlo (MC) simulations [6]. Finally, we stress that in the studied amplifier, its gain was limited by the transconductance of our present-day BDTs and we expect a dramatic improvement by introducing BDT nanostructures with the channel width below 200-nm.

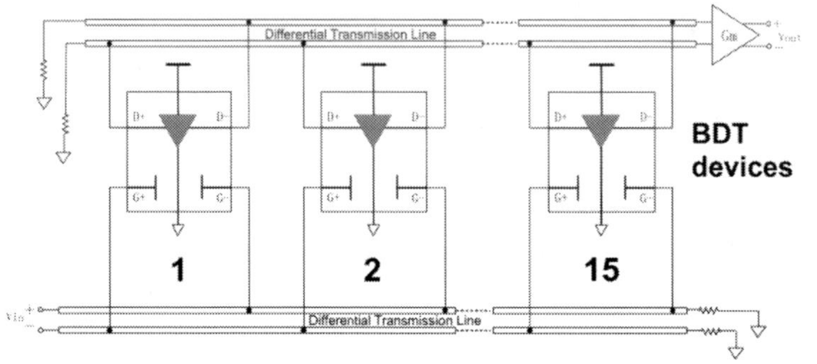

Figure 3. Schematics of a traveling wave THz amplifier based on a sequence of BDTs. The actual, simulated amplifier was a multi-tap structure.

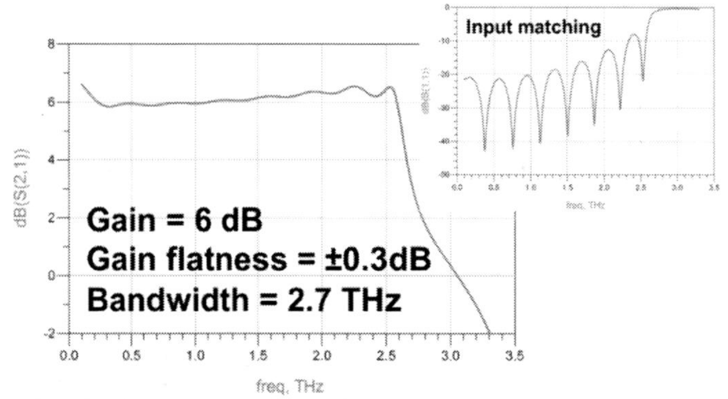

Figure 4. Simulated frequency response of the S_{12} parameter and input matching (inset).

4. Time-domain characterization of BDTs

Figure 5 presents our integrated, "experiment-on-chip" setup [7] that enables picosecond, time-domain, or, equivalently, THz-bandwidth characterization of ultrafast electronic devices, such as a BDT shown schematically at the bottom of the top panel. The top panel shows the laser system with two optical beams (excitation pulses) aimed at the gaps in the coplanar waveguide structure, acting as the optoelectronic transducers. Next, our electrical excitation transients (schematically shown as red pulses) are applied to the BDT gates and the resulting output signals were sampled using electro-optic (EO) transducers at the positions corresponding to the red dots. As a reference, we also time resolved the input electrical transients at the points (blue spots) just before the BDT, also using the EO method [8]. The actual input and output transients are shown in the bottom panel and are indicated by the arrows. The input pulse exhibits a strong, negative reflection from the BDT due to a very large impedance mismatch between the coplanar waveguide and the BDT [7]. Finally we note that the BDT output pulse has a ~230 GHz bandwidth, what is expected from an ~μm-size-channel BDT with semi-ballistic transport, used in this experiment and, in fact, agrees well with MC simulations for a ~800-nm-channel BDT. Based on MC simulations, a scaled-down BDT should achieve sub-ps performance.

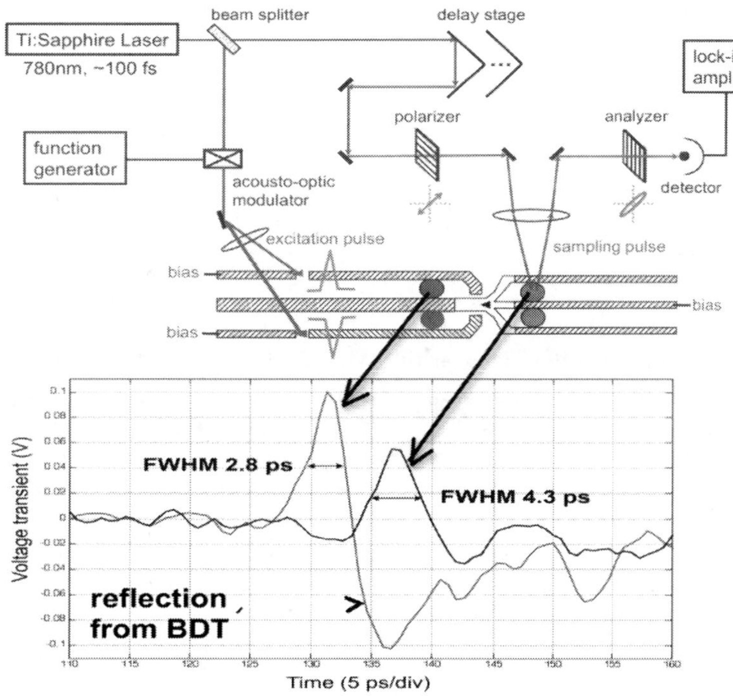

Figure 5. Schematics of an experimental setup (top panel) for OE generation of electrical transients incident on the BDT, measured at the position corresponding to the blue dots, and the EO detection of the output signals (position red dots), transmitted through the device. The bottom panel shows the actual transients recorded using the EO sampling method.

6. Conclusions

We have demonstrated usefulness of a BDT nanostructure for THz frequency range applications at room temperature. Preliminary evaluations, including MC simulations have shown the BDT ultrahigh operating speed and reduced gate capacitance. The latter input was used to build a circuit model working as multi-stage, multi-BDT traveling-wave amplifier. In its simplest implementation, the amplifier achieved THz-level performance with a 2.7-THz bandwidth for a 6-dB gain and the gain-bandwidth product of up to 10.8 THz. In addition, using an optoelectronic time-domain spectroscopy method we have experimentally demonstrated THz switching of BDTs incorporated into coplanar waveguides and excited by picosecond electrical input transients.

Acknowledgment

The authors thank M. Gong for his assistance in experiments. This work is supported by the ARO Grant No. W911NF-12-2-0076.

References

[1] Taur Y *IBM J. Res. Dev.*, **46**, 213, 2002.
[2] Song A M 2004 *Encyclopedia of Nanoscience and Nanotechnology*, **9**, 371.
[3] Diduck Q, Irie H and Margala M 2009 *Int. J. of High Speed Electron. & Sys. (IJHSES)*, **19**, 23.
[4] Kaushal V, Iñiguez-de-la-Torre I, Irie H, Guarino G, Donaldson W R, Ampadu P, Sobolewski R and Margala M 2010 *IEEE Trans. Nanotech.*, **9**, 723.
[5] Kaushal V, Iñiguez-de-la-Torre I and Margala M 2011 *Solid State Electron*, **56**, 120.
[6] Iñiguez-de-la-Torre I, Kaushal V, Margala M, González T and Mateos J 2010 *DRC*, 117.
[7] Irie H and Sobolewski R 2010 *J. Appl. Phys.* **107**, 084315.
[8] Zheng X, Xu Y, Sobolewski R, Adam R, Mikulics M, Siegel M and Kordos P 2003 *Appl. Opt.* **42**, 1726.

EDISON'19 IOP Publishing
Journal of Physics: Conference Series **647** (2015) 012021 doi:10.1088/1742-6596/647/1/012021

Intrinsic Electric Oscillations of Ovonic Devices towards the TeraHertz Limit

E. Piccinini[1], F. Buscemi[2], M. Rudan[1], R. Brunetti[2], C. Jacoboni[2]
[1]DEI Department, University of Bologna, Viale Risorgimento 2, I-40136, Bologna, Italy
[2]FIM Department, University of Modena and Reggio Emilia, Via Campi 213/A, I-41125 Modena, Italy

E-mail: rossella.brunetti@unimore.it

Abstract. The time-dependent response of Ovonic devices to an electric potential ramp signal is analysed by means of an enhanced version of a previously published time-dependent charge-transport model proposed by the authors. Depending on the inevitable parasitics of the system, either stable or oscillating solutions are found according to the position of the load line. The model also allows for speculations on the potential of Ovonic materials in the design of high-frequency oscillating circuits close to the terahertz range.

1. Introduction

The topic of electrical and thermal transport in amorphous chalcogenides has been addressed several times in the last 40 years. A large community of scientists agrees in interpreting the steady-state, reversible switching behaviour of Ovonic materials as connected to hot-carrier phenomena [1, 2]. In particular, for given voltage biases, these semiconductors show the existence of two stable states featuring, on the microscopic scale, different carrier temperatures and, on the macroscopic scale, different resistivities, leading to the typical S-shaped current-voltage characteristic [2, 3, 4]. Most of the analyses published in the literature refer to steady-state conditions. Only recently some researchers pointed out that the investigation of the transient behaviour introduces new issues related to the switching time and to how the bias is applied [5]; others focused on the recovery time after the bias is changed [6].

The present work elaborates on the model by Buscemi *et al.* [7]. Thanks to the simplicity of the physical model, transport equations can be coupled to the circuit equation and used to describe transient situations with a limited computational load.

2. The model

According to a common framework for Ovonic materials, let us consider a homogeneous material, whose electron-state structure is extremely simplified and reduced to two distinct energy levels E_T and E_B, where $E_B = E_T + \Delta E_0$. The former level stands for the energy level of trapped carriers and the latter for that of mobile, or band, carriers. Since the mobility of trapped carriers is negligible, the current density reads $J = q\mu_B n_B F$, where n_B and μ_B are the concentration and the mobility of band carriers and F is the electric field.

The total concentration n of the carriers is fixed, but carriers can be differently distributed between the two levels depending on the transport conditions and on the applied electric field.

In particular, according to a Poole or Poole-Frenkel mechanism, the height of the energy barrier confining carriers in the traps depends on the electric field as $\Delta E_0 - \gamma|F|$, with γ a constant, so that the higher the applied field, the higher the probability of trapped carriers to transfer to band states. Given a field F, we assume that the steady-state band concentration $\tilde{n}_B$ can be expressed by the function [8]:

$$\tilde{n}_B = \frac{n}{1 + \Gamma \exp\left(\frac{\Delta E_0 - \gamma|F|}{kT_e}\right)}, \tag{1}$$

Γ being the ratio of the density of trap over band states, and T_e the electron temperature.
Under time-dependent conditions, the time evolution of the band population is assumed to vary as:

$$\frac{dn_B}{dt} = -\frac{n_B - \tilde{n}_B}{\tau_N}, \tag{2}$$

where τ_N is a relaxation time constant. This assumption makes n_B one of the unknowns of the problem and improves the model in [7], where n_B is instead instantaneously updated to $\tilde{n}_B$, which is a function of T_e.

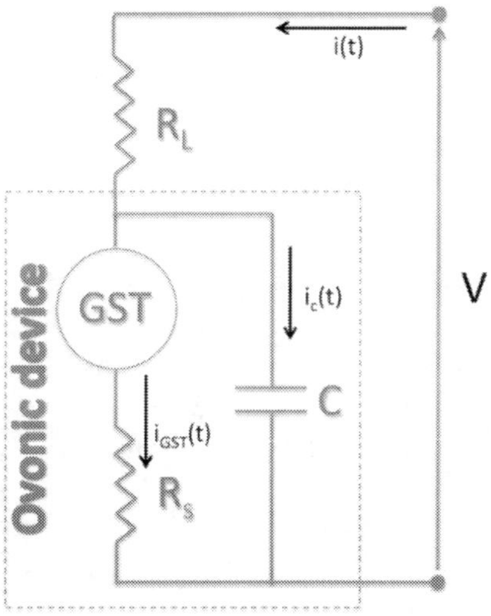

Figure 1. Schematics of the analyzed circuit. The boxed area represents an Ovonic device including the active layer and the parasitics. The capacitor C takes into account all capacitive parasitic effects, independently of their origin.

The time-dependent balance equation for the total energy density ε^{TOT} is written as:

$$\frac{d\varepsilon^{TOT}}{dt} = JF - \left.\frac{d\varepsilon^{TOT}}{dt}\right|_{lattice}, \tag{3}$$

where $\varepsilon^{TOT} = n_T E_T + n_B E_B = n_T E_T + n_B(E_T + \Delta E_0) = nE_T + n_B \Delta E_0$.

Coherently with the simplified model for the electron states, we assume for the energy relaxation due to the lattice [8]:

$$\left.\frac{d\varepsilon^{TOT}}{dt}\right|_{lattice} = n\frac{kT_e - kT_0}{\tau_T}. \tag{4}$$

with τ_T a relaxation time constant for T_e.

In order to exploit the above equations to study a device, we must consider a layer of Ovonic material between electrodes, giving rise to a parasitic capacitance C_S and a contact resistance R_S. Next, a series load resistance R_L and a parallel capacitance C_C due to the circuit are also considered. A schematic diagram of the circuit is sketched in Fig. 1 where the two capacitances have been replaced by the single capacitance $C = C_C + C_S$. The field F is not known *a-priori*, as it is a function of the circuit itself: $F = (V - R_L i)/(\ell + q\mu_B n_B A R_S)$, V being the external potential, ℓ and A being the device length and cross section, respectively.

The circuit theory leads to the following differential equation for the total current i:

$$\frac{di}{dt} = \frac{1}{CR_S R_L}\left[V - F\ell + CR_S\frac{dV}{dt} - i(R_S + R_L)\right], \tag{5}$$

which is coupled to (2) and (3) and solved numerically to find the triplet $\{n_B(t), i(t), T_e(t)\}$, given the external potential $V(t)$.

3. Results and discussion

Usual contact resistances and parasitic capacitances for Ovonic devices are hardly less than 1 kΩ and 1 pF, respectively [5, 6]. Following the literature, in the first part of our analysis we set $R_S = 1$ kΩ and $C = 1$ pF, and consider a large load resistance $R_L > 100$ kΩ, this rendering a current-driven circuit. When a simple bias, like a ramp signal to a voltage plateau V_{max}, is applied, three conditions are possibile, according to where the load line intersects the steady-state $I(V)$ characteristic (Figs. 2 and 3).

If the operating point lies in the lower branch of the $I(V)$ characteristic, a stable electric response is always reached, and the time needed to attain the steady-state condition depends both on the magnitude of the parasitic capacitance and, to a smaller extent, on the ramp slope. Similarly, a stable response is attained also when the load line crosses the upper branch of the $I(V)$ characteristic. In the latter case, a switching phenomenon takes place, and a current overshoot is present before the steady-state condition is reached (see Fig. 2).

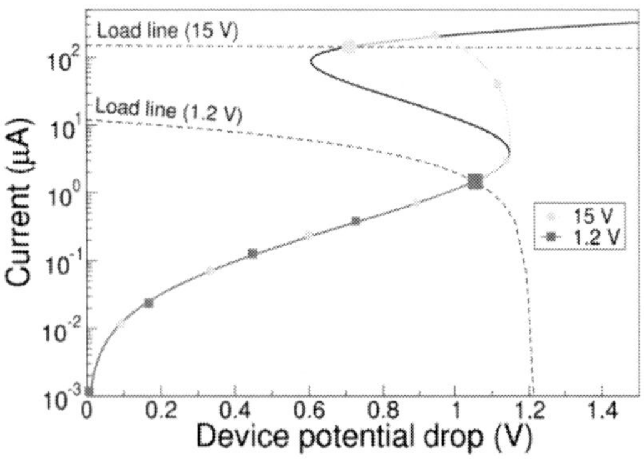

Figure 2. When the load lines intersects the steady-state $I(V)$ characteristic in the upper or lower branch, a stable state is found. The cases $V_{max} = 15$ V and $V_{max} = 1.2$ V, $R_L = 100$ kΩ are shown.

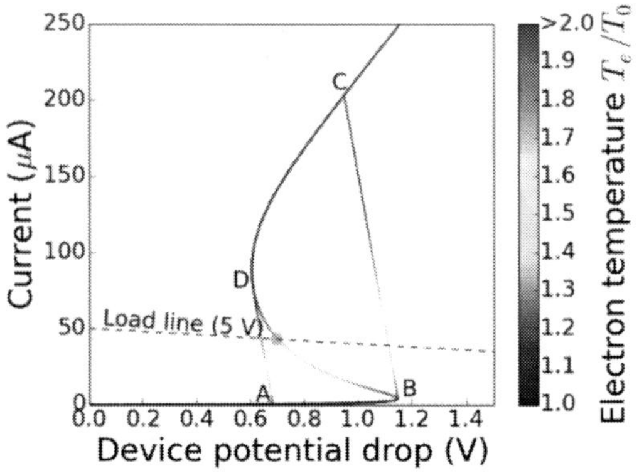

Figure 3. When the load line intersects the steady-state $I(V)$ characteristic in the NDR region ($V_{max} = 5$ V, $R_L = 100$ kΩ), the cycle $ABCD$ is found. Colors denote the electron temperature ratio T_e/T_0.

On the contrary, when the load line and the $I(V)$ characteristic intersect in the negative differential resistance region (NDR), the circuit becomes unstable and oscillations take place. In this case, the analyses of the time evolution of T_e helps in understanding how the cycle sets in. After the switching point is reached, T_e rapidly increases to values much higher than that of the operating point. Consequently, the current quickly jumps to high values. Due to (3), the contribution of energy relaxation is quite large, and eventually T_e jumps back to near-equilibrium values below the operating-point value. Then, the cycle starts again and continues to chase indefinitely the steady-state condition. The oscillating frequency found for these conditions is about 35 MHz. This value is similar to the one found in the literature [9], where the lower-to-upper branch transitions and v.v. are almost instantaneous.

The instability connected to the NDR branch of the $I(V)$ characteristic is a direct consequence of the presence of the parasitic capacitance C. In fact when $C \approx 0$ it is found that the system reaches a stable condition in any case. Apart from any considerations on the parasitics due to probing and connections, we point out that a device with zero intrinsic capacitance cannot exist because of the two contacts necessary to generate the field inside the device.

One should also notice that, during the switching to the upper branch of the characteristic, T_e is always larger than the value it would have on the steady-state curve at the same current, and the opposite condition holds true while the system switches back to the lower branch (Fig.

3). Since such a behaviour is the result of the interplay between the parasitic capacitance and the two relaxation times τ_N and τ_T (here $\tau_N = 0.1$ ps and $\tau_T = 0.28$ ps), we can speculate on the oscillating frequency of the circuit, supposing that we are able to minimise the parasitic capacitance towards its theoretical limit. The smallest capacitance of a 1000 nm$^2\times$40 nm capacitor with an Ovonic material as dielectric can be estimated in the order of 10^{-3} fF.

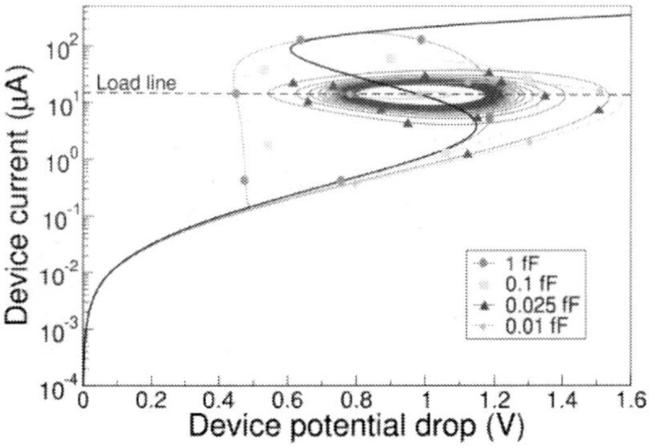

Figure 4. Simulations of the oscillating behaviour when the parasitic capacitance decreases from 1 fF down to 10^{-2} fF. A plateau voltage $V_{max} = 15$ V and a load resistance $R_L = 1$ MΩ have been applied to set the working point in the NDR region.

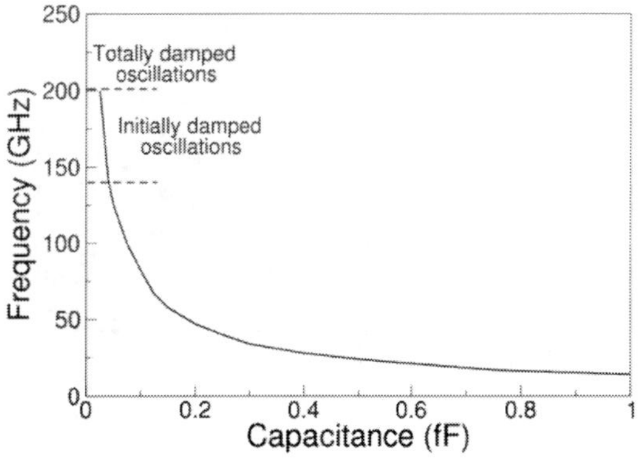

Figure 5. Oscillation frequencies for parasitic capacitances near the theoretical limit. Initially damped oscillations are reported for $0.025 < C < 0.04$ fF, while a totally damped behaviour is found below that limit.

In Fig. 4 we have simulated the circuit responses to a given bias for reduced parasitic capacitances, ranging from 1 fF down to 10^{-2} fF, for an operating point lying in the NDR region. For the "largest" considered capacitances, the usual oscillations between the upper and the lower branches of the characteristic are found. Their frequencies are inversely proportional to C. However, as C decreases, the gap between the actual value of T_e and its "target" value at the same current reduces because the device approaches faster the steady-state conditions. In turn, changes on T_e imply changes of n_B, thus of i_{GST} and F. As a consequence, it may happen that progressively, high-frequency damped oscillations set in. Eventually, for further reductions of C, damping is so effective that the system converges to the operating point.

Performing a Fourier-transform of the voltage drop along the Ovonic device reveals that the maximum frequency that could be achieved is close to the terahertz domain (Fig. 5). This analysis suggests that Ovonic materials could be employed for high-frequency applications in the terahertz range, provided that technological advances do reduce circuit parasitics in such a way as to not overcome the intrinsic capacitance of the device as it happens today.

References

[1] Ielmini D 2008 *Phys. Rev. B* **78** 035308
[2] Piccinini E, Cappelli A, Buscemi F, Brunetti R, Ielmini D, Rudan M, and Jacoboni C 2012 *J. Appl. Phys.* **112** 083722
[3] Jacoboni C, Piccinini E, Buscemi F, and Cappelli A 2013 *Solid-State Electron.* **84** 90
[4] Rudan M, Giovanardi F, Piccinini E, Buscemi F, Brunetti R, Cappelli A, Marcolini G, and Jacoboni C 2013 *J. Comput. Electron.* **12** 666
[5] Wimmer M and Salinga M 2014 *New J. Phys.* **16** 113044
[6] Lavizzari S, Ielmini D, and Lacaita A L 2010 *IEEE Trans. Electron. Dev.* **57** 1838
[7] Buscemi F, Piccinini E, Brunetti R, Rudan M, and Jacoboni C 2014 *Appl. Phys. Lett.* **104** 262106
[8] Buscemi F, Piccinini E, Cappelli A, Brunetti R, Rudan M, and Jacoboni C 2014 *Appl. Phys. Lett.* **104** 022101
[9] Schmidt P E and Callarotti R C 1984 *J. Appl. Phys.* **55** 3144

Bubbles in graphene - a computational study

Mikkel Settnes, Stephen R. Power, Jun Lin, Dirch H. Petersen and Antti-Pekka Jauho

Center for Nanostructured Graphene (CNG),
DTU Nanotech, Technical University of Denmark, DK-2800 Kgs. Lyngby, Denmark

E-mail: `Antti-Pekka.Jauho@nanotech.dtu.dk`

Abstract. Strain-induced deformations in graphene are predicted to give rise to large pseudomagnetic fields. We examine theoretically the case of gas-inflated bubbles to determine whether signatures of such fields are present in the local density of states. Sharp-edged bubbles are found to induce Friedel-type oscillations which can envelope pseudo-Landau level features in certain regions of the bubble. However, bubbles which minimise interference effects are also unsuitable for pseudo-Landau level formation due to more spatially varying field profiles.

1. Introduction

Strain engineering has been proposed as a method to manipulate the electronic, optical and magnetic properties of graphene [1–10].It is based on the close relation between the structural and electronic properties of graphene. An inhomogeneous strain field can introduce pseudomagnetic fields (PMFs), [1, 4, 5] where the altered tight binding hoppings mimic the role of a gauge field in the low energy effective Dirac model of graphene [11, 12]. Guinea *et al* [1] demonstrated that nearly homogeneous PMFs can be generated by applying triaxial strain. One of the most striking consequences of homogeneous PMFs is the appearance of a Landau-like quantization. [1, 8] Scanning tunnelling spectroscopy on bubble-like deformations has observed pseudo-Landau levels (pLLs) corresponding to PMFs stronger than 300 T [2, 3]. Deformations can be induced in graphene samples by different techniques like pressurizing suspended graphene [5, 13] or by exploiting the thermal expansion coefficients of different substrates [3]. As a result, introducing nonuniform strain distributions at the nanoscale is a promising route towards strain engineering. The standard theoretical approach to treat strain effects employs continuum mechanics to obtain the strain field. The strain field can then be coupled to an effective Dirac model of graphene to study the generation of PMFs in various geometries. In most studies, only the PMF distribution is considered as opposed to experimentally observable quantities like local density of states (LDOS). This study calculates the LDOS of such systems without applying periodicity, which can introduce spurious interactions between neighboring bubbles.

2. Model

2.1. Patched Green's function approach

The patched Green's function approach, developed in Ref [14], treats device 'patches' embedded within an extended two dimensional system described by a tight-binding Hamiltonian. This approach allows us to insert a single bubble into an otherwise pristine infinite graphene sheet, and avoids issues such as interferences between a bubble and its periodic images or system

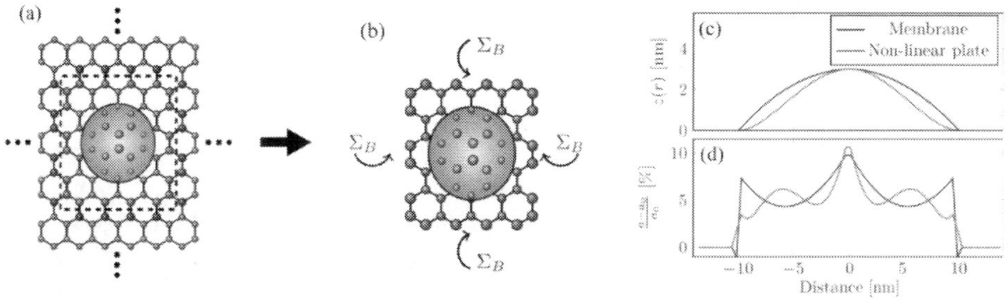

Figure 1. The patched GF method describes the extended graphene sheet away from the bubble device region (a) with a self energy term $\mathbf{\Sigma}_B$ (b). We consider membrane and non-linear plate type bubbles with radially dependent (c) height and (d)strain profiles.

edges. The extended part of the system is treated through a self-energy Σ_B entering the device area. The self-energy is written in terms of Green's functions (GFs) of an infinite, pristine sheet so that it can be calculated using methods taking advantage of periodicity and analytic integrability [15]. The device region Hamiltonian can be tridiagonalized allowing the device region GF to be treated using an adaptive recursive method allowing for efficient calculation of local properties, such as LDOS, everywhere in the device region surrounding a bubble.

2.2. Strain model

We consider two possibilities for the shape of a gas inflated bubble – the *membrane* and *non-linear plate* models. [16] The former is suitable for very large bubbles where bending stiffness can be neglected, whereas the latter is more appropriate for including bending effects near the edges of smaller bubbles. Membrane bubbles therefore have very sharp edges, whereas the edges are smoother in the non-linear plate bubbles. While these continuum models have been found to agree well with experimental shape profiles, more accurate modeling of bubble shapes and strain distributions can be achieved using molecular dynamics simulations. [5, 17, 18]

Using the deformation field $\boldsymbol{u}$ the position of the atom i initially at $\boldsymbol{R}_i^0$ becomes $\boldsymbol{R}_i = \boldsymbol{R}_i^0 + \boldsymbol{u}$. The new bond lengths are afterwards determined as $d_{ij} = |\boldsymbol{R}_i - \boldsymbol{R}_j|$. Hoppings are modified according to $t_{ij} = t_0 \mathrm{e}^{-\beta\left(d_{ij}/a_0 - 1\right)}$, where $t_0 \approx 2.7$ eV is the pristine coupling, $a_0 = 1.42$Å is the carbon-carbon distance and $\beta \approx \partial \log(t)/\partial \log(a)|_{a=a_0} \approx 3.37$ [7]. The strain tensor created by the strain field, $\epsilon_{ij} = \frac{1}{2}(\partial_j u_i + \partial_i u_j + \partial_i z \partial_j z)$, gives rise to a gauge field [11],

$$\boldsymbol{A} = -\frac{\hbar\beta}{2ea_0}\begin{pmatrix}\epsilon_{xx} - \epsilon_{yy} \\ 2\epsilon_{xy}\end{pmatrix}, \tag{1}$$

with the resulting PMF given by $\mathbf{B_s} = \nabla \times \mathbf{A}$. Table 1 shows the height profile, in-plane displacements and PMFs for the two types of bubbles considered here.

3. Results

The LDOS at individual sites, as well as the average DOS, can be quickly calculated using the above techniques. Fig. 2a shows the averaged DOS for membrane (blue) and non-linear plate (red) model bubbles. For the membrane model, we previously distinguished between two different type of oscillations [14]. Sharp peaks, such as those highlighted by the blue circle and triangle, showing an energy dependence, $E_n \sim \sqrt{n}$, are consistent with pLLs arising due to the PMF. In addition, the periodic oscillations at higher energies are Friedel-type oscillations arising due to electron scattering at the sharp edges of the membrane bubble.

Model type	$z(r,\theta)$	$\boldsymbol{u}(r,\theta)=\begin{pmatrix}u_r\\u_\theta\end{pmatrix}$	$B_s(r,\theta)$
Membrane	$h_0\left(1-\frac{r^2}{R^2}\right)$	$\begin{pmatrix}u_0\frac{r}{R}\left(1-\frac{r}{R}\right)\\0\end{pmatrix}$	$\frac{\hbar\beta u_0}{2ea_0R^2}\sin(3\theta)$
Nonlinear plate	$h_0\left(1-\frac{r^2}{R^2}\right)^2$	$\begin{pmatrix}r(R-r)(c_1+c_2r)\\0\end{pmatrix}$	$\frac{\hbar\beta}{2ea_0}\left[(c_1-c_2R)-\frac{32h_0^2r^3}{R^6}\left(1-\frac{r^2}{R^2}\right)\right]\sin(3\theta)$

Table 1. Height profile $z(r,\theta)$, in-plane displacements $\boldsymbol{u}(r,\theta)$ and PMF distributions $B_s(r,\theta)$ for the membrane and non-linear plate bubbles, where R and h_0 are the bubble radius and height respectively, and $u_0=1.136h_0^2/R$, $c_1=1.308h_0^2/R^3$ and $c_2=-1.931h_0^2/R^4$. [16]

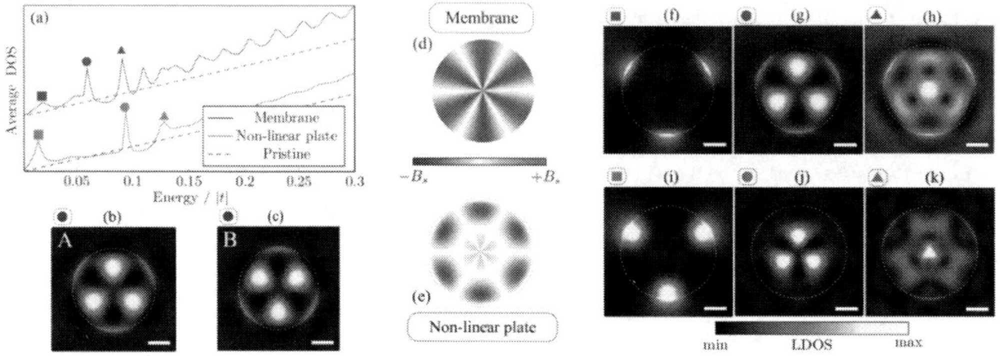

Figure 2. a) shows the averaged DOS with each bubble model with $R=10$nm and $h_0=3$nm, similar in scale to previous theoretical and experimental studies [2,5]. Important peaks in each are highlighted by symbols. b,c) LDOS maps for the membrane model at the circle energy for the A and B sublattices. d,e) PMF distributions for each bubble type f-k) A sublattice LDOS maps for the peaks highlighted in a). The LDOS map scale bar is 5nm.

Before discussing the interplay between the different oscillation types, we note that both vary independently with position throughout the bubble region. The position dependence of the pLLs arises due to the non-uniform PMF distribution, which is plotted in Fig. 2d. This takes maximum amplitudes along the armchair directions which occur every 60°, but is only three-fold symmetric due to a sign change between two consecutive maxima. Unlike real magnetic fields, PMFs conserve time-reveral symmetry by taking opposite signs in the K and K' valleys of graphene. This leads to a strong sublattice polarisation [14, 19, 20], which is clearly visible for the LDOS maps shown in Fig. 2b-c, where the circle LDOS peak from Fig. 2a is localized in different regions on different sublattices. Comparison to Fig. 2d confirm that these correspond to a change in the sign of the PMF. We note that this first "pseudomagnetic peak" is localized along armchair directions where the PMFs are largest and reasonably constant. The position dependence of the Friedel oscillations meanwhile emerges from interference between electrons scattered at different sides of the bubble. The interplay of both oscillation types in membrane bubbles leads to the Friedel type acting as an envelope and quenching the LDOS signature of pLLs in certain regions of the bubble, as is clear for the dark spots in the LDOS map for the higher energy in Fig. 2h. STS measurements taken at such a spot would completely omit this peak due to the enveloping effects of the edge-induced Friedel oscillations.

To detect PMF signatures in gas-inflated bubbles, it may be worth considering softer edge profiles, such as the non-linear plate, which should give rise to weaker Friedel oscillations. This is clear from the averaged curve in Fig. 2a, where higher energy oscillations are considerably suppressed compared to the membrane case. However, here there is also an absence of sharp pLL peaks following a $\sqrt{n}$ distribution, with the possible exception of the red circle peak. This is consistent with the PMF distribution for this bubble type, plotted in Fig. 2e, which has a radial fluctuation in the sign and strength of the PMF. The center of the bubble has a distribution similar to the membrane case, and the central region of the LDOS map in Fig 2j resembles that of the corresponding membrane model peak (Fig 2g). We note also that the Friedel features for the higher energy in Fig. 2k (which does not fit the $\sqrt{n}$ distribution) are more blurred than for the membrane case, as expected for scattering from a less-sharp bubble edge. Thus it seems that bubble shapes which reduce Friedel oscillations also effectively remove pLL effects due to the less uniform PMFs induced by their strain profiles.

Finally, the square symbol energy peak at low energies in both bubble types is a state localized near the bubble edge (Figs. 2f,i). It is not directly related to pseudomagnetic effects, but emerges due to the interface between the pristine region outside the bubble and the strained, perturbed region within. The localized states at this boundary act somewhat like a potential, and induce the scattering which lies behind the Friedel oscillations in these bubbles. These states are far less localized in the non-linear plate bubble than the membrane bubble, due to an edge which is no longer as sharp. This in turn leads to the smoothening and averaging out of the Friedel oscillations that we observed earlier for the non-linear plate bubbles.

4. Conclusions

We studied theoretically the local and averaged densities of states in gas-inflated graphene bubbles embedded in infinite graphene sheets using the patched Green's function approach. We determined that pLL features in sharp-edged bubbles may be hidden by interference effects due to electron scattering at the bubble edges. Softer-edged bubbles were found to display weaker interference effects, however their shape profiles also resulted in PMF distributions unsuitable for pLL formation. Our results suggest that it will be difficult to obtain reliable pLL features in such gas inflated systems, unlike bubbles formed on substrates which often display the triaxial-type strain that is predicted to give a more appropriate, uniform PMF for Landau level formation [1].

This research was supported by Danish National Research Foundation project DNRF58.

References

[1] Guinea F, Katsnelson M I and Geim A K 2009 *Nature Physics* **6** 30–33
[2] Levy N *et al* 2010 *Science* **329** 544–7
[3] Lu J, Castro Neto A H and Loh K P 2012 *Nature communications* **3** 823
[4] Jones G W and Pereira V M 2014 *New Journal of Physics* **16** 093044
[5] Qi Z *et al* 2014 *Phys. Rev. B* **90**(12) 125419
[6] Low T, Guinea F and Katsnelson M I 2011 *Physical Review B* **83** 195436
[7] Pereira V M, Castro Neto A H and Peres N M R 2009 *Physical Review B* **80** 045401
[8] Neek-Amal M, Covaci L, Shakouri K and Peeters F M 2013 *Physical Review B* **88** 115428
[9] Power S R, Gorman P D, Duffy J M and Ferreira M S 2012 *Phys. Rev. B* **86**(19) 195423
[10] Pereira V M and Castro Neto A H 2009 *Phys. Rev. Lett.* **103**(4) 046801
[11] Suzuura H and Ando T 2002 *Phys. Rev. B* **65**(23) 235412
[12] Vozmediano M, Katsnelson M and Guinea F 2010 *Physics Reports* **4** 109–148
[13] Bunch J S *et al* 2008 *Nano Letters* **8** 2458–2462
[14] Settnes M, Power S R, Lin J, Petersen D H and Jauho A P 2015 *Physical Review B* **91** 125408
[15] Power S R and Ferreira M S 2011 *Physical Review B* **83** 155432
[16] Yue K, Gao W, Huang R and Liechti K M 2012 *Journal of Applied Physics* **112** 083512
[17] Neek-Amal M and Peeters F M 2012 *Phys. Rev. B* **85**(19) 195445
[18] Qi Z, Bahamon D A, Pereira V M, Park H S, Campbell D K and Neto A H C 2013 *Nano letters* **13** 2692–7
[19] Moldovan D, Ramezani Masir M and Peeters F M 2013 *Physical Review B* **88** 035446
[20] Schneider M, Faria D, Viola Kusminskiy S and Sandler N 2015 *Phys. Rev. B* **91**(16) 161407

Time-Dependent Modelling of Single-Electron Interferometry with Edge-States

A Beggi[1], **A Bertoni**[2], **P Bordone**[1,2]

[1] Dipartimento di Scienze Fisiche, Informatiche e Matematiche, Università degli Studi di Modena e Reggio Emilia, Via Campi 213/A, I-41125 Modena, Italy
[2] S3, Istituto Nanoscienze-CNR, Via Campi 213/A, I-41125 Modena, Italy

E-mail: andrea.beggi@unimore.it

Abstract. We simulate the time-resolved dynamics of localized electrons in a 2DEG system, where an external magnetic field creates quantum Hall edge states, and properly polarized split gates define a Mach-Zehnder electron interferometer. The carriers travelling inside the Hall channels consist of localized wave packets of edge states: they are propagated numerically by means of a Fourier split-step approach. We find that the energy-dependent scattering process at the quantum point contacts, together with the finite energy distribution of the carriers, have a remarkable effect on the transmission coefficient T of the device. We provide an analytical model to justify the characteristics of T which is in good agreement with the numerical simulations.

1. Introduction

Many recent theoretical and experimental works [1, 2, 3, 4] in the field of electron interferometry rely on edge states (ESs) in the integer quantum Hall regime as perfect 1D channels for coherent carrier transport. Indeed, these channels are highly immune to scattering and decoherence, thus being a good candidate as semiconductor quantum bits. The long-term aim of our work is devising solid-state flying qubits consisting of carriers travelling in ESs, and quantum logic gates based on quantum point contacts (QPCs) and a suitable pattern of surface split-gate designing a network of Hall edge channels. Here, we simulate the coherent propagation in time of electrons in a modulation-doped 2D device, where an applied orthogonal magnetic field creates ESs, and properly polarized split gates define a Mach-Zehnder electron interferometer (MZI). The electrons travelling inside the device are described by localized wave packets (WPs) of ESs. This, as already proposed in the literature [5, 6], is in contrast with the time-independent approach that deals with a single-energy delocalized edge state, thus losing the possibility to describe the explicit dynamics and the single-carrier evolution of the system.

In particular, our study is focused on the energy dependence of the transmission of the QPCs: since our carriers have a finite distribution of energies, this dependence has a remarkable impact on the total transmission coefficient T of the mesoscopic MZI. The results of numerical simulations are supported by a simplified exactly solvable analytical model.

2. Numerical Simulations

The Hall ESs of a 2DEG are chiral delocalized states that follow the boundary between a low-potential and a high-potential region of the device (respectively, allowed and forbidden for electrons). Their wave function has the form $\Psi_n(x, y, k) = \varphi_{n,k}(x)e^{iky}$, with travelling direction y and where the localized wave function $\varphi_{n,k}(x)$ depends on the details of the confining potential profile in the x direction. As a first step, we compute numerically $\varphi_{n,k}(x)$ by solving the effective Hall 1D Hamiltonian (with n and k being the index of the Landau level and the k-vector of

Journal of Physics: Conference Series **647** (2015) 012023 doi:10.1088/1742-6596/647/1/012023

the ES, respectively) in the initialization region, where the split-gates potential is independent on y. Since the state $\Psi_n(x, y, k)$ can not represent localized particles, in order to describe a single-carrier travelling inside the MZI of Fig. 1, we consider a superposition of different ESs (belonging to the first Landau level, i.e. $n = 1$) with gaussian weights, that is:

$$\Psi_0(x, y) = \int dk\, F_\sigma(k)\varphi_{1,k}(x)e^{iky}, \tag{1}$$

where $F_\sigma(k) = \sqrt{\sigma}\,(2\pi)^{-3/4}\, e^{-\sigma^2(k-k_0)^2}e^{-iky_0}$. This wave function is indeed the ES equivalent of a minimum uncertainty WP of plane waves. Therefore, it represents a carrier which is initially localized around $x_0 = -\hbar k_0/eB$ and y_0. It is composed by ES with different momenta around $k = k_0$, and its group velocity is $v_g = \hbar k_0/m^*_B$.

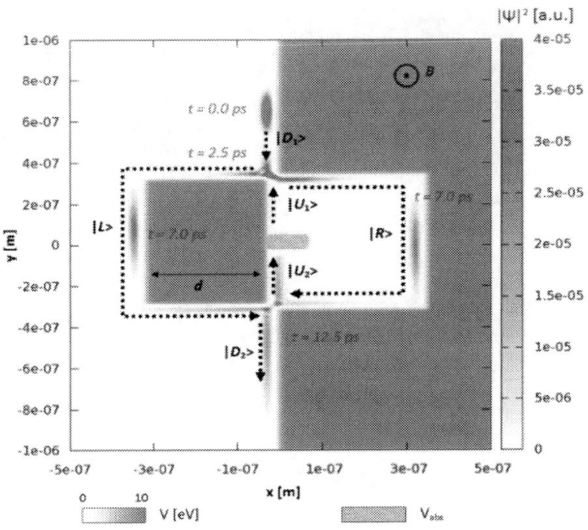

Figure 1. The MZI (in blue) used in our simulations, with snapshots of the electronic wavepacket (in red) at different times t. Dashed lines represent delocalized scattering states. The dimension d is changed in order to control Δl.

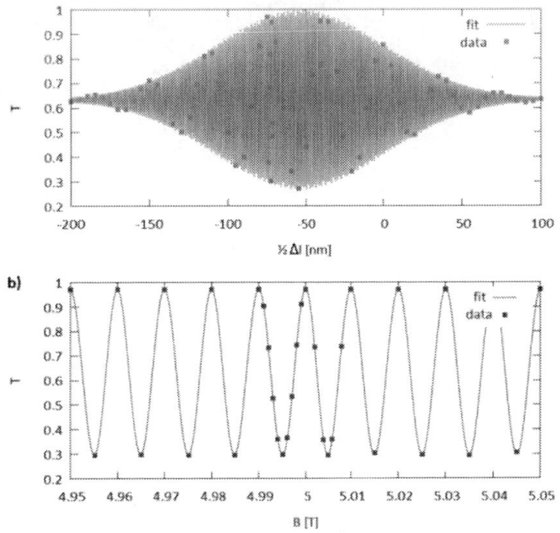

Figure 2. Transmission coefficient T obtained by fitting the results of time-dependent simulations: **a)** at a constant magnetic field $B = 5$ T and variable length mismatch ; **b)** at a constant length mismatch and variable magnetic field (Δl = -75 nm).

Our MZI device operates at filling factor $\nu = 1$. In detail, we model the edges of the split-gate pattern through smooth potential barriers (i.e. Fermi functions), and we tune the gap of both QPCs (the constrictions of the potential V in Fig. 1) in order to have a perfectly half-reflecting behaviour in terms of transmitted and reflected electron density (see Ref.[7] for other details on the device). In our simulations, the electron WP is split by the first QCP into two parts, that travel along the two arms of the device; then they are recollected and interfere at the second QPC, giving a transmitted and a reflected component. The latter is absorbed by a metallic lead, which is modeled with an imaginary potential V_{abs} (yellow rectangle in the center of Fig. 1). Therefore, the norm of the final WF is the transmission coefficient T of the device.

To solve the time-dependent Schrödinger equation including the external magnetic field B, we use a parallel Trotter-Suzuki scheme and the split-step Fourier method[8]. Simulations are performed for WPs of different initial extension, namely $\sigma = 20$ nm, 40 nm, and 60 nm (to obtain an optimal proportion between σ and the dimension of the device).

By changing the value of B or the area of the device (the latter is achieved by tailoring the length of one of the MZI arms), the transmission coefficient T shows Aharonov-Bohm oscillations, as

Initial WP	a) Fit of simulations			b) Analytical model			c) Numerical model		
σ (nm)	T_0	T_1	Σ (nm)	T_0	T_1	Σ (nm)	T_0	T_1	Σ (nm)
20.0	0.737	0.240	41.0	0.710	0.254	39.4	0.760	0.240	39.7
40.0	0.633	0.365	54.5	0.681	0.381	52.5	0.636	0.364	52.7
60.0	0.578	0.422	71.0	0.609	0.435	68.9	0.579	0.421	69.1

Table 1. Parameters for the transmission coefficient $T(\Delta l)$: **a)** from a fit of numerical simulations; **b)** from the analytical model; **c)** from numerical integration of Eq. (3).

reported in Fig. 2, in agreement with Ref[3]. However, when the lengths of the two paths become very different, we observe that the oscillations of the transmission are modulated by a gaussian envelope (see Fig. 2a), which is clearly a direct consequence of the finite-size of the electron WP. This has been already noted in literature[5]: once the length mismatch Δl between the two arms is substantially larger than σ, the two parts of the WF arrive at the second QPC at different times, and cannot interfere.

From a phenomenological point of view, the profile of $T(\Delta l)$ can be fitted with the following expression: $T(\Delta l) = T_0 + T_1 \exp(-\frac{1}{8}(\Delta l + c)^2/\Sigma^2)\cos(k_e \Delta l + \varphi_0)$. The fit parameters for our simulation are reported in columns a) of Table 1, for different dispersions of the initial WP. The visibility[1] $v_{MZI} = T_1/T_0$ is never ideal (i.e. unitary), but it increases as σ grows. Moreover, the amplitude of the gaussian damping of T, which is described by the standard deviation Σ, increases with σ. These effects can be understood by considering that the scattering process at the QPCs is k-dependent.

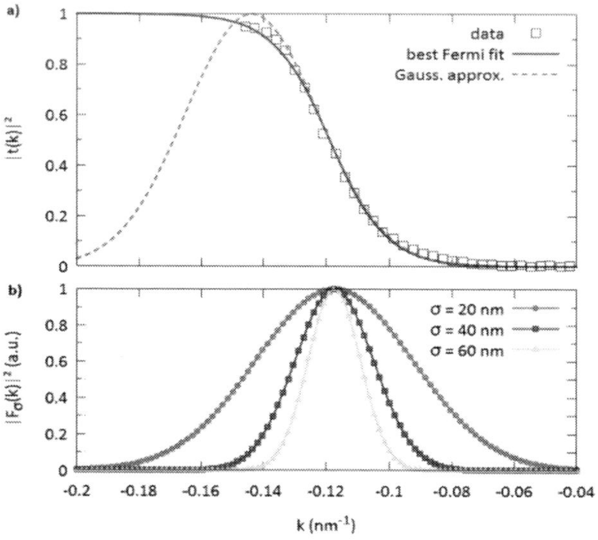

Figure 3. **a)** Transmission coefficient $|t(k)|^2$ for the first QPC: numerical calculations (points), Fermi fit (continuous line) and Gaussian approximation (dashed line). **b)** Gaussian weight distribution $|F_\sigma(k)|^2$ for the simulated WPs.

In order to evaluate these effects, we focus on the scattering process at the first QPC and analyze the reflected and the transmitted components of the WP by projecting them on the local set of ESs to get their spectral composition. Then, by comparing them with the initial WP, the numerical calculation of reflection and transmission probabilities at different k is straightforward: if the reflected WP contains the ES of momentum k with a probability $|F_\sigma^r(k)|^2$, then the reflection probability is $|r(k)|^2 = |F_\sigma^r(k)|^2/|F_\sigma(k)|^2$. A similar calculation can be made for $|t(k)|^2$.

The results of this calculation are shown in Fig. 3a, where the energy-dependence of the scattering amplitude t is evident. We found out that this behaviour is well described by a Fermi function $t(k) = (\exp(+\alpha(k-k_0))+1)^{-\frac{1}{2}}$ (and $r(k) = (\exp(-\alpha(k-k_0))+1)^{-\frac{1}{2}}$ for the reflection coefficient, not shown in the figure). We used the same fitting parameters for r and t in order to satisfy the constraint $|r(k)|^2 + |t(k)|^2 = 1$, and to get a perfect half-reflecting behaviour for $k = k_0$, namely the central wave vector of our WP.

3. Analytical Model

In order to develop an analytical model for the transmission of our MZI, able to take into account the energy dependence of the 2D scattering process in the QPCs, we note that the Fermi function fitting the simulation data of Fig. 3a can be approximated by a Gaussian in the region around the turning point k_0, $(\exp(\alpha(k-k_0))+1)^{-1} \simeq \exp\left(-(\alpha(k-k_0)+\gamma)^2/4\gamma\right)$, with $\gamma = 4\ln(2)$. This approximation is acceptable in the range of k-components of our WPs, at least for $\sigma = 40\,\mathrm{nm}$ and $\sigma = 60\,\mathrm{nm}$ (see Fig. 3b).

We describe the dynamics of the wavepacket inside the MZI with the scattering matrix formalism[1]. The scattering states are represented in Fig. 1. The initial state is described by $|\Psi_0\rangle = \int dk\, F_\sigma(k)\,|D_1(k)\rangle$, while the scattering matrices of the QPCs, which are identical, are given by:

$$S = \begin{pmatrix} r(k) & it(k) \\ it(k) & r(k) \end{pmatrix}. \tag{2}$$

While travelling along the arms of the device, the phase acquired by the wavefunction is described by the matrix $P = \mathsf{diag}\left(e^{i\Delta\varphi}, 1\right)$, where $\Delta\varphi(k) = e\Phi_m/\hbar - k\cdot\Delta l$. Here, Φ_m is the magnetic flux of B through the device. The wavefunction after the scattering process is given by $|\Psi_f\rangle = SPS\,|\Psi_0\rangle$, and the norm of the transmitted part $|\Psi_T\rangle = \left(\frac{1}{2\pi}\int dk\, |D_2(k)\rangle\,\langle D_2(k)|\right)|\Psi_f\rangle$ gives the transmission coefficient T:

$$T = \frac{1}{2\pi}\int dk\, |F_\sigma(k)|^2\, |r^2(k)e^{i\Delta\varphi(k)} - t^2(k)|^2. \tag{3}$$

Using the Gaussian approximation for the transmission and reflection amplitudes, we can integrate analytically the previous expression, to get:

$$T = T_0 - T_1 e^{-\frac{\Delta l^2}{8\Sigma^2}} \cos\left(\frac{e}{\hbar}\Phi_m - k_0\Delta l\right), \tag{4}$$

where T_0 and T_1 – which represent, respectively, the average value of T and the maximum amplitude of its oscillations – are given by $T_0 = 2\frac{\sigma}{\Sigma}\exp\left(-\frac{\gamma\sigma^2}{2\Sigma^2}\right)$ and $T_1 = \frac{1}{2}\frac{\sigma}{\Sigma}$, while the standard deviation Σ is given by $\Sigma = \sqrt{\sigma^2 + \alpha^2/4\gamma}$.

As a consequence, the visibility is given by $v_{MZI} = \frac{1}{4}\exp\left(\gamma\sigma^2/2\Sigma^2\right)$, and we recover the ideal behaviour in the limit for $\sigma \to \infty$. Equation (4) predicts the same trend found in numerical simulations for T. If we used the energy-independent approximation, that is $r(k) = t(k) = 1/\sqrt{2}$, we would have obtained an expression for T which is formally identical to Eq. (4), but with $T_0 = T_1 = 1/2$ and $\Sigma = \sigma$: a result which is not consistent with our numerical simulations. Indeed, the gaussian damping Σ is not proportional to σ, as we observed in the numerical simulations, since there is a correction due to the scattering properties of the QPCs. Therefore, it is sufficient to include the localized nature of the carrier to recover the simulation results. However, in order to quantify the exact extension of this damping, as well as to justify the observed reduction in visibility with respect to the ideal case, we have to take into account the energy-dependent scattering at the QPCs. With this correction, the results of the numerical simulations are well reproduced by the analytical model (see Table I, columns a) and b)). Finally, we mention that a better agreement can be obtained with a direct numerical integration of Eq. (3), at the expense of an exact analytical solution. This result is reported in column c) of Table 1 and detailed in Ref.[7].

References

[1] Weisz E, Choi H K, Sivan I, Heiblum M, Gefen Y, Mahalu D and Umansky V 2014 **344** 1363–1366
[2] Bocquillon E, Freulon V, Berroir J M, Degiovanni P, Plaçais B, Cavanna A, Jin Y and Fève G 2013 **339** 1054–1057
[3] Neder I, Ofek N, Chung Y, Heiblum M, Mahalu D and Umansky V 2007 *Nature* **448** 333–337
[4] Fève G, Mahe A, Berroir J M, Kontos T, Placais B, Glattli D, Cavanna A, Etienne B and Jin Y 2007 *Science* **316** 1169–1172
[5] Haack G, Moskalets M, Splettstoesser J and Büttiker M 2011 *Phys. Rev. B* **84**(8) 081303
[6] Gaury B, Weston J and Waintal X 2014 *Phys. Rev. B* **90**(16) 161305
[7] Beggi A, Bordone P, Buscemi F and Bertoni A 2015 *arXiv preprint arXiv:1504.03837*
[8] Kramer T, Heller E J and Parrott R E *Journal of Physics: Conference Series*

Journal of Physics: Conference Series **647** (2015) 012024 doi:10.1088/1742-6596/647/1/012024

Weak measurement from the electron displacement current: new path for applications

D. Marian[1,2], E. Colomés[1], N. Zanghì[2] and X. Oriols[1]

[1]Departament d'Enginyeria Electrònica, Universitat Autònoma de Barcelona, Bellaterra 08193, Spain

[2]Dipartimento di Fisica dell'Università di Genova and INFN sezione di Genova, Genova 16146, Italy

E-mail: damiano.marian@uab.cat, enrique.colomes@uab.cat, nino.zanghi@ge.infn.it, xavier.oriols@uab.cat

Abstract. The interest on weak measurements is rapidly growing during the last years as a unique tool to better understand and predict new quantum phenomena. Up to now many theoretical and experimental weak-measurement techniques deal with (relativistic) photons or cold atoms, but there is much less investigation on (non-relativistic) electrons in up-to-date electronics technologies. We propose a way to perform weak measurements in nanoelectronic devices through the measurement of the total current (particle plus displacement component) in such devices. We study the interaction between an electron in the active region of a electron device with a metal surface working as a sensing electrode by means of the (Bohmian) conditional wave function. We perform numerical (Monte Carlo) simulations to reconstruct the Bohmian trajectories in the iconic double slit experiment. This work opens new paths for understanding the quantum properties of an electronic system as well as for exploring new quantum engineering applications in solid state physics.

1. Introduction

In the last years an increasing attention has been posed on the utility of weak measurements [1] for exploiting new fascinating quantum phenomena. A weak measurement implies that the measuring apparatus provides a small perturbation on the measured system and that the output result has a large uncertainty. Although quantum mechanics forbids to measure simultaneously the position and the momentum of a quantum particle in a single experiment, weak measurements realized on a large set of identically prepared systems provide well-defined values of momentum and position of the quantum system. Thus, one of the targets of weak measurements techniques has been the Bohmian formulation of quantum mechanics, which explicitly deals with quantum trajectories that have well-defined position and velocity. Therefore this type of measurement has been used, for example, to reconstruct trajectories in recent experiments [2]. In particular, the trajectories reconstructed from weak measurements are exactly the ones predicted by Bohmian mechanics [3, 4]. Up to now, all such experiments are done with (relativistic) photons or cold atoms and, unfortunately, a procedure to perform weak measurements of position and momentum for (non-relativistic) electrons in solid-state devices does not exist. Weak measurements in electronic devices are available in Refs. [5, 6, 7, 8] where a scheme with a quantum point contact weakly coupled to a quantum system was proposed for the

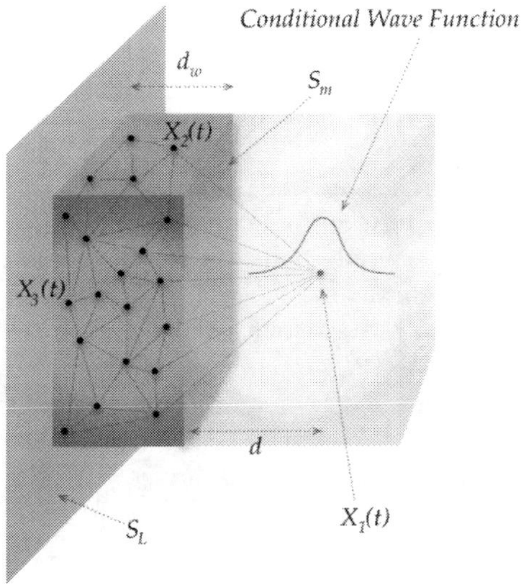

Figure 1. Schematic representation of the system studied numerically. The M electrons in the metal surface interact through Coulomb interaction (red dashed lines) among them and with the electron X_1 in the active region of the device. The capital letters denote the actual (Bohmian) positions of the particles.

amplification of small signals. But this technique is unable to measure the fundamental position and momentum of a quantum electronic state. For this reason, in this work we investigate the feasibility of doing weak measurements from the (high-frequency) displacement current measured by an ammeter in a solid-state electronic device.

2. Displacement current and momentum measurements

First of all, without specifying how the measurement is explicitly implemented, let us discuss mean (ensemble) values of the displacement current and momentum. The mean value of the total electrical current for an (ideal) quantum system $\langle I \rangle$ can be computed straightforwardly as the sum of the (mean) particle current plus the (mean) displacement current on a given surface S_L:

$$\langle I \rangle = \int_{S_L} \langle \mathbf{J}_c(\mathbf{r}, t) \rangle \cdot d\mathbf{s} + \int_{S_L} \epsilon(\mathbf{r}) \frac{d\langle \mathbf{E}(\mathbf{r}, t) \rangle}{dt} \cdot d\mathbf{s}, \tag{1}$$

where $\epsilon(\mathbf{r})$ is the dielectric constant of the material, $\langle J_c(\mathbf{r}, t) \rangle$ the ensemble value of the particle current density and $\langle \mathbf{E}(\mathbf{r}, t) \rangle$ the ensemble value of the electric field on the surface. Identically, $\langle I \rangle$ can be calculated from the Ramo-Schockley-Pellegrini expression [9, 10, 11]. In the particular situation where two metallic surfaces, in which there is no variation of the potential, are placed at distance L_x the total current from the Ramo-Schockley-Pellegrini becomes:

$$\langle I(t) \rangle = \frac{q}{mL_x} \langle p(t) \rangle_\Omega, \tag{2}$$

where $\langle p(t) \rangle_\Omega$ is the mean value of the momentum in the volume Ω of the device. In summary, the measurement of the total current can be interpreted as a measurement of the momentum of the quantum system (only for the particular geometry of the electron device mentioned above).

3. How to treat the system-apparatus interaction?

In this section, we briefly explain the interaction between the quantum system and the rest of the parts of the measuring setup (cables, ammeter, etc) to justify why the measurement of the displacement current can be considered as a weak measurement. In particular, we want to simulate the interaction between an electron in the active region of a device and the electrons composing the metal surface. In figure 1 we depict the studied system: It is considered the interaction between an electron $N = 1$ in the device with M electrons in the metal surface (of area S_m and width d_m), and the interaction between the M electrons in the metals through Coulomb interaction. The many particle Schrödinger equation can not be numerically solved exactly for $N + M$ degrees of freedom due to computer memory limitations; thus, we have to choose a suitable numerical method to have an appropriate approximate solution of the equation. We use the method reported in Ref. [12, 13], which uses the notion of Bohmian *conditional wave function* [3, 14] because it is able to keep track of the dynamical and variable interaction given by the Coulomb potential. Then we solve numerically the following single-particle (conditional) Schrödinger-type equation for the electron in the device (x_1 in figure 1):

$$i\hbar\frac{\partial\psi(x_1,t)}{\partial t} = [H_0 + V]\,\psi(x_1,t), \tag{3}$$

where $V = V(x_1, \mathbf{X}_2(t), ..., \mathbf{X}_{M+1}(t))$ is the conditional Coulomb potential felt by the system and H_0 is its free Hamiltonian. With capital letter, $\mathbf{X}(t)$, we denote the actual positions of the (Bohmian) particles. The M electrons in the metal surface are simulated semi-classically taking into account the interactions with a bath of phonons at room temperature. We assume for simplicity that the electron 1 can move only in the transport direction x.

The output total current in the surface S_L is then computed through the Bohmian trajectories by means of the following equation

$$I_T(t) = \int_{S_L} \epsilon(\mathbf{r})\frac{d\mathbf{E}}{dt}d\mathbf{s} = \sum_{i=1}^{N} \nabla F(\mathbf{X}_i(t))\mathbf{v}_i(t), \tag{4}$$

where the flux F depends on each electron position and $\mathbf{v}_i$ is the Bohmian velocity [15, 16, 17]. We note that in (4) we have inserted only the displacement component of the total current because the electrons are not crossing the surface S_L. By means of numerical simulations performed along the ideas exposed above we show that the measurement of the total current in a large surface S_L turns out to be weak (long-range Coulomb interaction) with almost negligible perturbation of the electron wave function [16, 17].

4. Measurement of the local (Bohmian) velocities

Identically to the last section, it can be demonstrated that the total current measured in a sensing electrode with a *small* surface provides output current only when the electron is very close to the electrode (short-range Coulomb interaction). This latter type of measurement provides a strong measurement of the position [17]. The two measurements just exposed provide a way to reconstruct the Bohmian trajectories in a solid state device [4, 18].

By means of numerical (quantum Monte Carlo) experiments of the system plus measuring apparatus, we test the feasibility of the previous weak measurements. As an example, in figure 2, the electron trajectories in the iconic double-slit scenario are reconstructed from displacement current measurements in a three terminal solid-state device. It can be seen that the trajectories are more dense near the maxima of the interference pattern, while no trajectories are found around the minima: this feature is in accordance with the mentioned experiment for photons [2] and with the general Bohmian theory which states that the trajectories follow the wave function probability density and are distributed according to the modulus squared of the wave function.

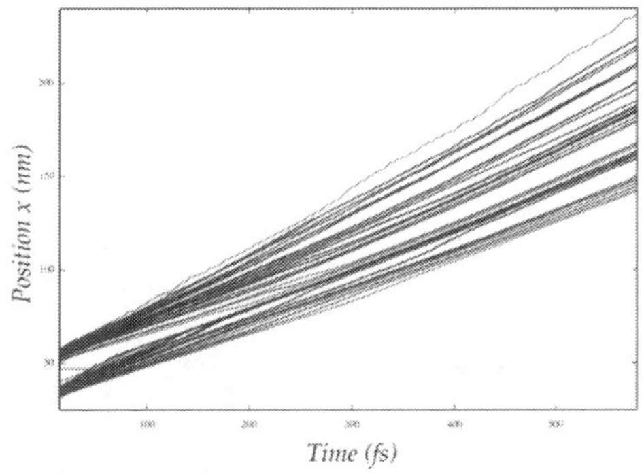

Figure 2. Reconstructed trajectories for an electron from an ensemble of weak measurements of the displacement current. The initial electron wave function is composed of two space separately wave packets that later interfere. It can be seen that the trajectories are more dense around the maxima of the wave function while no trajectories are found around the minima.

5. Conclusions

In this conference, we show how the weak measurements of the displacement current (in multi terminal devices with different geometries) allow the weak measurement of the momentum of a quantum system or its strong measurement of position. From these results, one can expect a new and unexplored path for solid-state quantum engineering applications (quantum tomography, quantum computing, etc.). As an example, the viability of measuring local (Bohmian) velocities from these techniques is shown through a numerical experiment.

Acknowledgment

This work has been partially supported by the "Ministerio de Ciencia e Innovación" through the Spanish Project TEC2012-31330 and by the Grant agreement no: 604391 of the Flagship initiative "Graphene-Based Revolutions in ICT and Beyond". N.Z is supported in part by INFN.

References

[1] Aharonov Y, Albert D Z and Vaidman L 1988 *Phys. Rev. Lett.* **60** 1351
[2] Kocis S, Braverman B, Ravets S, Stevens M J, Mirin R P, Shalm L K and Steinberg A M 2011 *Science* **332** 1170
[3] Dürr D, Goldstein S and Zanghì N 1992 *J. Stat. Phys.* **67** 843
[4] Dürr D, Goldstein S and Zanghì N 2009 *J. Stat. Phys.* **134** 1023
[5] Zilberberg O, Romito A and Gefen Y 2011 *Phys. Rev. Lett.* **106** 080405
[6] Williams N S and Jordan A N 2008 *Phys. Rev. Lett.* **100** 026804
[7] Romito A, Gefen Y and Blanter Y M 2008 *Phys. Rev. Lett.* **100** 056801
[8] Shpitalnik V, Gefen Y and Romito A 2008 *Phys. Rev. Lett.* **101** 226802
[9] Shockley W 1938 *J. App. Phys.* **9** 635
[10] Ramo S 1939 *Proceedings of the IRE* **27** 584
[11] Pellegrini B 1986 *Phys. Rev. B* **24** 5921
[12] Oriols X 2007 *Phys. Rev. Lett.* **98** 066803
[13] Oriols X and Mompart J 2012 *Applied Bohmian Mechanics: From Nanoscale Systems to Cosmology* (Singapore: Pan Standford Publishing).
[14] Dürr D and Teufel S 2007 *Bohmian Mechanics* (Berlin: Springer)
[15] Albareda G, Marian D, Benali A, Yaro S, Zanghí N and Oriols X 2013 *J. Comp. Elect.* **12**(3) 405
[16] Marian D, Zanghì N and Oriols X 2014 *Proc. of IWCE* IEEE 117
[17] Marian D, Zanghì N and Oriols X 2015, To be published somewhere else
[18] Traversa F L, Albareda G, Di Ventra M and Oriols X 2013 *Phys. Rev. A* **87** 052124

Generating sequences of phonon wave packets by optical excitation of a quantum dot

D. E. Reiter[1,2], D. Wigger[1], S. Lüker[1], V. M. Axt[3], P. Machnikowski[4], T. Kuhn[1]

[1] Institut für Festkörpertheorie, Universität Münster, 48149 Münster, Germany
[2] Imperial College London, South Kensington Campus, London SW7 2AZ, UK
[3] Theoretische Physik III, Universität Bayreuth, 95440 Bayreuth, Germany
[4] Department of Theoretical Physics, Wrocław University of Technology, 50-370 Wrocław, Poland

E-mail: Doris.Reiter@uni-muenster.de

Abstract. Phonons can play an active role in controlling solid state systems or in the transport of energy and information. Thus, it is vital to understand generation processes and properties of phonons. By optical excitation of a semiconductor quantum dot, a sequence of phonon wave packets can be created. In this paper we focus on the fluctuation properties of lattice displacement and momentum of these phonon wave packets. For detuned excitation the fluctuations may fall below their respective vacuum values and squeezing occurs.

1. Introduction

While phonons, which are always present in any solid state system, have mostly been regarded as unwanted sources of dissipation or decoherence of the electronic system [1, 2], now the focus shifts towards the active use of phonons. Examples are the phonon assisted state preparation of excitons in quantum dots [3–6] or ultrafast manipulation of lasing processes by strain pulses [7]. Phonons also offer the possibility to study fundamental quantum effects like squeezing in a mechanical system [8–14]. To actively use phonons, it is mandatory to understand how phonons are generated and what the properties of the created phonons are. In this paper we will investigate the properties of acoustic phonons generated by the optical excitation of a semiconductor quantum dot (QD). Longitudinal acoustic (LA) phonons couple strongly to the electronic states of the QD, which makes a QD also a good candidate for a phonon generator. Sequences of phonon wave packets can be generated by switching on a continuous wave (CW) excitation [14, 15]. The properties of these wave packets depend strongly on the excitation conditions like excitation strength and detuning. In previous studies we have analyzed the lattice displacement and its properties [14, 15]. To complement these studies, in this paper we will dicuss the momentum and its fluctuations. This is particularly important for squeezing as displacement and momentum are a pair of conjugate operators that enter the Heisenberg uncertainty principle. From the large variety of excitation conditions we will pick a strong excitation which results in the generation of a long sequence of phonon wave packets. We will consider excitations with three different detunings and discuss the resulting phonon dynamics including the behavior of the fluctuations.

Journal of Physics: Conference Series **647** (2015) 012025 doi:10.1088/1742-6596/647/1/012025

2. Theory

A detailed description of the electron and phonon system can be found in previous publications, e.g. Refs. [14, 15], such that we limit ourself to a brief summary here. We treat the QD as an electronic two-level system consisting of the ground and a single exciton state. The excitation is continuous and switched on smoothly. The switch on is simulated by an error function. The strength of the CW field is characterized by its Rabi frequency $\Omega_{\rm R}$. A resonant excitation refers to the case when the frequency of the light matches the transition energy (including the polaron shift) of the electronic system. We also consider detuned excitations where the detuning Δ between laser frequency and transition energy has a finite value. The LA phonons are treated in harmonic approximation. From the phonon creation and annihilation operators, we calculate the lattice displacement $u(r)$, the lattice momentum $\pi(r)$ and furthermore the squared fluctuations $(\Delta u)^2$ and $(\Delta\pi)^2$, respectively. For simplicity, the QD is taken to be spherical such that all quantities are radially symmetric and only depend on the distance r from the dot center. Squeezing is achieved when the fluctuations fall below their respective vacuum values. To easily identify if the phonons are squeezed we introduce the normalized difference of the fluctuation to the vacuum flucuations as $D_u = ((\Delta u)^2 - (\Delta u)^2_{\rm vac})/(\Delta u)^2_{\rm vac}$ and $D_\pi = ((\Delta\pi)^2 - (\Delta\pi)^2_{\rm vac})/(\Delta\pi)^2_{\rm vac}$ for displacement and momentum, respectively. If one of these quantities is smaller than zero, i.e., $D_u < 0$ or $D_\pi < 0$, the phonons are squeezed.

3. Results

For our calculations we consider a 5 nm large GaAs QD which is excited by a CW laser pulse with $\Omega_{\rm R} = 6\ {\rm ps}^{-1}$ at temperature $T = 0\,{\rm K}$. The switch on takes place on the time scale of 1 ps. Our results are summarized in Fig. 1. In the three columns we consider three different detunings, an excitation above the transition energy with $\Delta = +0.5$ meV (left column), resonant excitation with $\Delta = 0$ (middle column) and excitation below the transition energy with $\Delta = -0.5\,{\rm meV}$ (right column).

The first row shows the dynamics of the population f of the exciton state (solid red line) as well as the dynamics of the laser amplitude (blue dashed line). As soon as the laser sets in, the population exhibits Rabi oscillations. For $\Delta = 0$ the period of the Rabi oscillation is $T_{\rm R} = 2\pi/\Omega_{\rm R} \approx 1\,{\rm ps}$, which is determined by the strength of the laser excitation. For QDs, the coupling to the phonons is non-monotonic in energy [15] and the maximal coupling for our parameters is found at $\Omega_{\rm R} = 3\,{\rm ps}^{-1}$. An efficient coupling is reflected in the electronic system by strongly damped Rabi oscillations. Accordingly, for $\Omega_{\rm R} = 6\,{\rm ps}^{-1}$ the Rabi oscillations are almost undamped within the time window considered here, because the phonon system cannot react quickly enough to the fast change in the electronic system. Since in all cases studied here $\Delta^2/\hbar^2 \ll \Omega_{\rm R}^2$, the detuning plays a minor role for the oscillation period and only slightly affects the long time values of the occupation. Nevertheless it turns out that it strongly influences the phonon properties, as will be discussed below.

The second row depicts the lattice displacement $\tilde{u}$ (top) and the lattice momentum $\tilde{\pi}$ (bottom), which have been scaled to compensate for the radial decay. Each change in the electronic system is accompanied by the emission of a phonon wave packet, because the change in the electronic system leads to a rapid shift of the equilibrium position for the phonons. The creation of the exciton leads to the emission of a wave packet with negative amplitude in the displacement and one with alternating amplitude in the momentum. The switch on of the laser pulse leads to a rather intense wave packet, which is emitted from the QD. The amplitude of the following wave packet is lower and remains almost the same for all following wave packets, which is in agreement with the weak dephasing of the Rabi oscillations. The repetition rate of the wave packet emission matches the Rabi frequency. The shape of the wave packets is mostly independent of the detuning. In addition to the emitted wave packets, a polaron is created in the QD region which causes a localized lattice distortion which is persistent as long as the exciton occupation

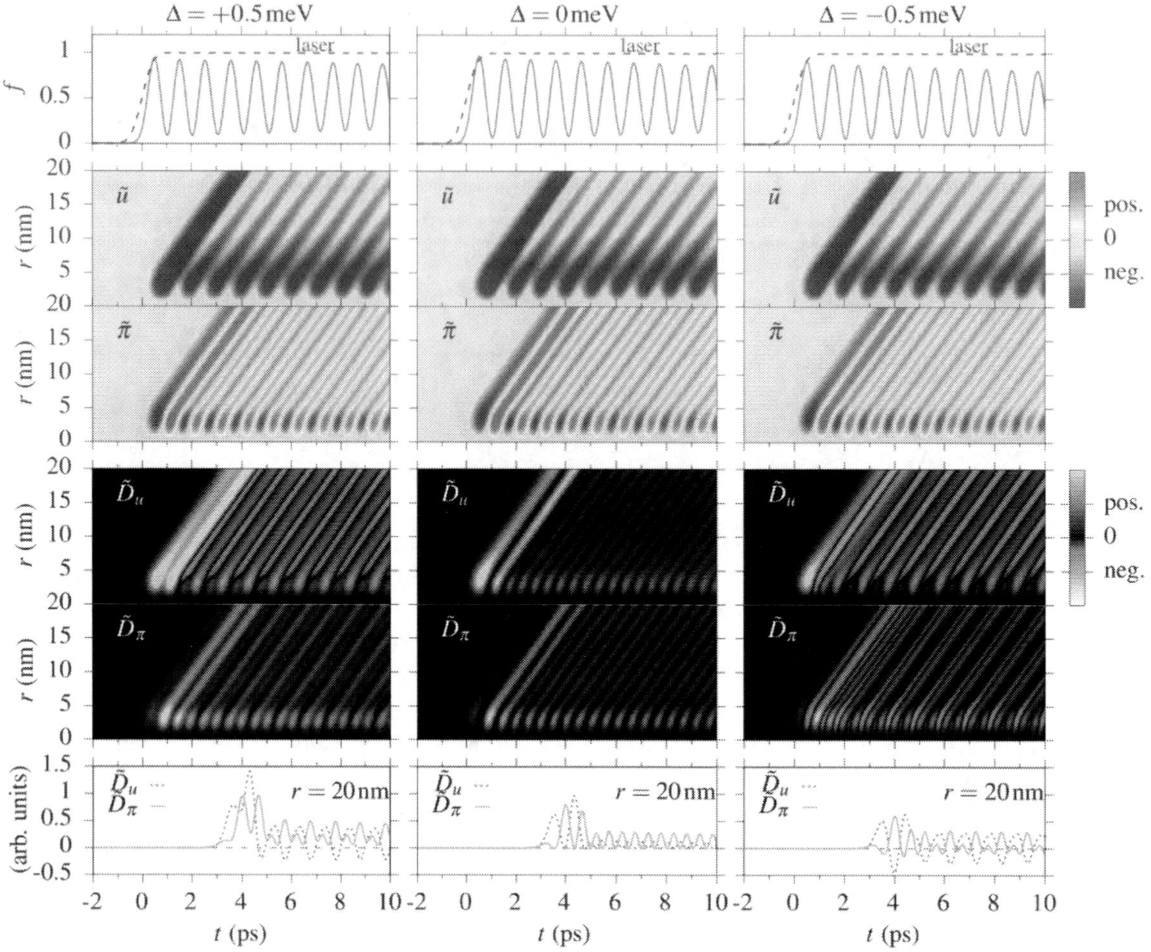

Figure 1. First row: occupation of the exciton state (solid red) and laser amplitude (dashed blue); second row: lattice displacement $\tilde{u} = u/r$ and momentum $\tilde{\pi} = \pi/r$; third row: scaled relative fluctuations of the displacement $\tilde{D}_u = D_u/r^2$ and of the momentum $\tilde{D}_\pi = D_\pi/r^2$; fourth row: cut of the fluctuations at $r = 20\,\text{nm}$. All displayed for three different detunings: positive detuning $\Delta = +0.5\,\text{meV}$, resonant excitation $\Delta = 0\,\text{meV}$ and negative detuning $\Delta = -0.5\,\text{meV}$. The scaling of the plotted quantities by r^{-1} and r^{-2}, respectively, compensates for the radial decay.

is non-vanishing.

The third row displays the fluctuations of the lattice displacement $\tilde{D}_u$ (top) and the lattice momentum $\tilde{D}_\pi$ (bottom), which have again been scaled to compensate for the radial decay. The fourth and last row shows the fluctuations of the lattice displacement $\tilde{D}_u$ (solid orange) and the lattice momentum $\tilde{D}_\pi$ (dotted green) at $r = 20\,\text{nm}$. Let us first consider the zero detuning case. Apart from the fluctuations of the initial wave packet, both fluctuations exhibit regular oscillations with twice the Rabi frequency. The fluctuations of the displacement $\tilde{D}_u$ clearly oscillate out of phase with the fluctuations of the momentum $\tilde{D}_\pi$. Note that oscillations with double frequency are not a clear proof for squeezing [10], i.e., it is not directly related to the fact that $\tilde{D}_{u/\pi}$ become negative. For zero detuning, the both fluctuations $\tilde{D}_u$ and $\tilde{D}_\pi$ are always positive, thus, no squeezing occurs.

For detuned excitation, the oscillation is less regular, because in addition to the double Rabi frequency single Rabi frequency components appear [14]. The amplitude of the fluctuations for displacement and momentum are still out of phase, i.e., when $\tilde{D}_u$ has a maximum $\tilde{D}_\pi$ has a minimum and vice versa. For negative detuning, we find that both fluctuations $\tilde{D}_u$ and $\tilde{D}_\pi$ fall below zero, i.e., the phonons exhibit displacement and momentum sqeezing. This is in agreement with findings from other systems, where squeezed states can be generated by negative detuning [8, 16, 17]. Previous studies have shown that for negative detuning squeezing is found for all excitation strengths [14]. It is interesting to note that for positive detunings, the appearance of squeezing depends on the excitation strength, i.e., on the Rabi frequency. If the Rabi frequency is too low, phonon emission processes dominate and $\tilde{D}_u$ becomes entirely positive [14]. For the high Rabi frequency taken here, we find that the fluctuations of the displacement $\tilde{D}_u$ fall below zero also for $\Delta > 0$. In contrast, the fluctuations of the momentum $\tilde{D}_\pi$ stay positive for all excitation strengths in this regime. Because for squeezing it is sufficient, when the fluctuations of one variable fall below their vacuum value, also here squeezed phonon wave packets emerge.

4. Conclusions

In summary, we have presented a detailed analysis of the properties of the phonons generated by the optical excitation of a QD. In particular we have compared the lattice momentum and its fluctuations with their respective displacement counterparts. Similar to the fluctuations of the displacement, the fluctuations of the momentum fall below their vacuum value for negative detuned excitation, confirming that in this case squeezed phonons are created. For positive detuning, squeezing has been found, because the lattice fluctuations fall below their vacuum value, however, the fluctuations of the momentum stay above their vacuum value and thus there is no momentum squeezing.

5. Acknowledgements

DER acknowledges financial support within the DAAD P.R.I.M.E. program.

References

[1] Ramsay A J 2010 *Semicond. Sci. Technol.* **25** 103001
[2] Reiter D E, Kuhn T, Glässl M and Axt V M 2014 *J. Phys. Cond. Matter* **26** 423203
[3] Reiter D E *et al.* 2012 *Acta Phys. Pol. A* **122** 1065
[4] Glässl M, Barth A M and Axt V M 2013 *Phys. Rev. Lett.* **110** 147401
[5] Quilter J H *et al.* 2015 *Phys. Rev. Lett.* **114** 137401
[6] Bounouar S *et al.* 2015 *Phys. Rev.* B **91** 161302
[7] Brüggemann C *et al.* 2011 *Nature Photon.* **6** 30
[8] Garrett G A *et al.* 1997 *Science* **275** 1638
[9] Hu X and Nori F 1997 *Phys. Rev. Lett.* **79** 4605
[10] Sauer S *et al.* 2010 *Phys. Rev. Lett.* **105** 157401
[11] Hussain A and Andrews S R 2010 *Phys. Rev.* B **81** 224304
[12] Misochko O V, Hu J and Nakamura K G 2011 *Phys. Lett. A* **375** 4141
[13] Wigger D, Reiter D E, Axt V M and Kuhn T 2013 *Phys. Rev.* B **87** 085301
[14] Wigger D *et al.* 2015 *Photonics* **2** 214
[15] Wigger D *et al.* 2014 *J. Phys. Cond. Matter* **26** 355802
[16] Papenkort T, Axt V M and Kuhn T 2012 *Phys. Rev.* B **85** 235317
[17] Gerry C C and Knight P L 2005 *Introductory Quantum Optics* (Cambridge: Cambridge University Press)

Coupled electron and thermal transport simulation of self-heating effects in junctionless MOSFET

M Mohamed[1], Z Aksamija[2], and U Ravaioli[1,3]

[1] Beckman Institute, University of Illinois, Urbana,IL, 61801, USA
[2] Electrical and Computer Engineering Department, University of Massachusetts, Amherst, MA 01003, USA
[3] Department of Electrical and Computer Engineering, University of Illinois, Urbana,IL, 61801, USA USA

E-mail: mohamed@illinois.edu, ravaioli@illinois.edu)

Abstract.
This paper explores self-heating effects on junctionless gate-all-around nanowire MOSFET using self-consistently coupled 3D full band electro-thermal transport. The self-consistent algorithm begins by supplying the heat generation data from the 3D electron Monte Carlo with 2D quantum correction to the phonon Monte Carlo. Subsequently, the phonon Monte Carlo transports the phonons introduced from the electron simulation and considers their scattering through the anharmonic three-phonon processes. The anharmonic three-phonon decay and the use of full dispersion facilitate a detailed description of heat transfer and the determination of the temperature map. We compare the performance of gate-all-around junctionless against the conventional inversion mode gate-all-around MOSFET. Our results indicate that junctionless MOSFET has less self-heating effects than the conventional inversion mode device, particularly at the limits of high currents.

1. Introduction

Thermal management and power dissipation issues in microprocessors are quickly emerging as the ultimate bottleneck to improving the performance of consumer/commercial electronics. Controlling device temperature and power dissipation is a key to sustaining electronic devices with longer battery life and improves the overall device reliability and life expectancy. As such, thermally conscious blueprints has become in recent years the focal concern of semiconductor roadmaps, at all design phases, as the temperature at the chip level and within a single transistor rises for future electronic devices [1, 2]. Without proper thermal management, inordinate power dissipation can potentially bring to end integrated circuit functionality. In this regard, coupled electro-thermal simulation model that self-consistently couple the electronic and phonon transport is essential in creating a cycle that will push designs to have lower carbon footprints and in creating environmentally conscious electronics that minimize waste.

Recently, junctionless (JL) transistor has been introduced to circumvent the adversity encountered in forming ultra-shallow junction [3, 4]. The idea behind junctionless transistors is similar to that of a gated resistor: the channel and the S/D regions are doped with the same polarity (n- or p-type) and concentration. As such, there are no junctions formed between the S/D regions and the channel; hence, considerably simplifying the fabrication process. In addition, junctionless transistors operate as depletion-mode (DM) devices and rely heavily on the use high workfunction gate and the ultrathin body. In this paper we employ a phonon Monte Carlo Transport (pMC) coupled to a 3D full band electrical Monte Carlo simulator (MoCa) with 2D quantum corrections [5], to examine the electro-thermal transport of junctionless Multigate MOSFET.

2. Thermal MoCa: Self-consistently Coupled electron and phonon Monte Carlo

In order to capture the details of self-heating in nanoscale devices, we need to carefully model phonon generation and transport and couple it to the electronic transport in a self-consistent manner. The self-consistency between the electronic and thermal simulations is critical to obtaining meaningful results in ultrascaled multi-gate MOS devices because of the strong self-heating effect which arises out of the small physical dimensions of the device. Thermal MoCa [5] self consistently couples 3D electron Monte Carlo with full phonon dispersion and 2D quantum correction (electrical MoCa) with a phonon Monte Carlo (pMC) which includes the influence of anharmonic phonon decay. The full self-consistent two-way coupling of electron and phonon populations is achieved in two steps added to the standard electron and phonon Monte Carlo simulations. The phonon data is transferred from the electrical MoCa to the pMC simulator. The pMC computes all the phonon interactions and then allows the phonons to travel using the velocities obtained from the full phonon dispersion. The energy of all the phonons in the phonon Monte Carlo step is calculated from their momenta using the full phonon dispersion relationship. Once all the phonon interaction has been computed by the pMC, the next step is to use that data to compute the temperature distribution across the device. The simulation domain in the thermal transport is divided up into a mesh of small cubes with 1 nm sides and the energy of all the phonons residing in each of the cubes is tallied. Subsequently, the total energy in each cube is used as a source term in the heat diffusion equation which is then solved iteratively to obtain the steady-state temperature distribution in the entire simulation domain. The temperature map obtained from the thermal transport step (pMC + heat equation) is fed back to the particle Monte Carlo with temperature dependent scattering table, and the process is repeated in a self-consistent manner until convergence criterion is met. The inclusion of anharmonic phonon decay, temperature dependent scattering rate, and full electron and phonon dispersion helps create a clear picture of heat transfer and provides a comprehensive way to access self-heating in nanoscale devices. The reader is referred to reference [5] for a more complete description of the details of the electrical and thermal transport used in this work.

3. Discussion and results

A key consideration in junctionless MOSFET design is to employ a large metal workfunction capable of fully depleting the channel at $V_G = 0V$. Our previous work [6] on n-channel junctionless MOSFET, using a metal workfunction of 5.15 eV, has demonstrated that even for cross-sections as thin as $t_{si} = 5\,\text{nm}$, the junctionless doping concentration must be less than $5 \times 10^{19}\ \text{cm}^{-3}$ in order to regulate the off performance (leakage current) and achieve acceptable $\text{I}_{on}/\text{I}_{off}$ ratio [6]. Based on this, we employ a square cross-sectional length $t_{si} = h_{si} = 5\,\text{nm}$ and a doping density of $4 \times 10^{19}\ \text{cm}^{-3}$. Additionally, the gate length (L_G) and gate oxide thickness (t_{ox}) for the simulations presented here are 20 nm and 1 nm, respectively.

We plot the average velocity profile and temperature distribution of a fully-depleted gate-all-around junctionless device (doping density = $4 \times 10^{19}\ \text{cm}^{-3}$ and $t_{si} = h_{si} = 5\,\text{nm}$) in Fig. 1 a) and b), respectively. The temperature distributions were obtained after solving the heat diffusion equation with Dirichlet (fixed) boundary conditions at the boundaries of the simulation region. This can be seen in the temperature profiles of Fig. 1b) where the temperature at the boundaries of the simulation region always stays at 300K. The results indicate inconsiderable self-heating effect, due to the low current drive, as demonstrated by the negligible change in

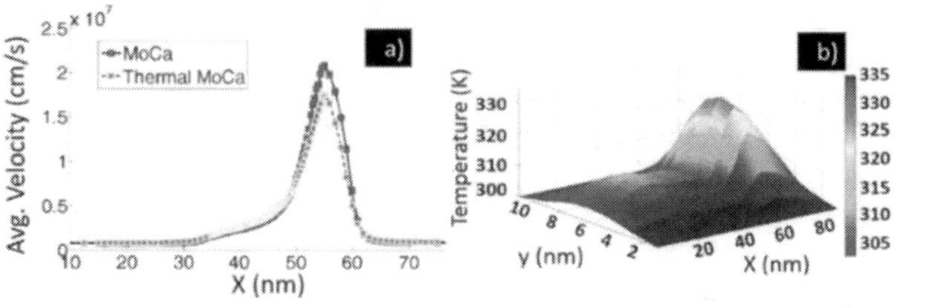

Figure 1: Comparing the (a) average velocity profile of thermal MoCa and the electrical (uncoupled) MoCa for a junctionless ; (b) corresponding temperature distribution

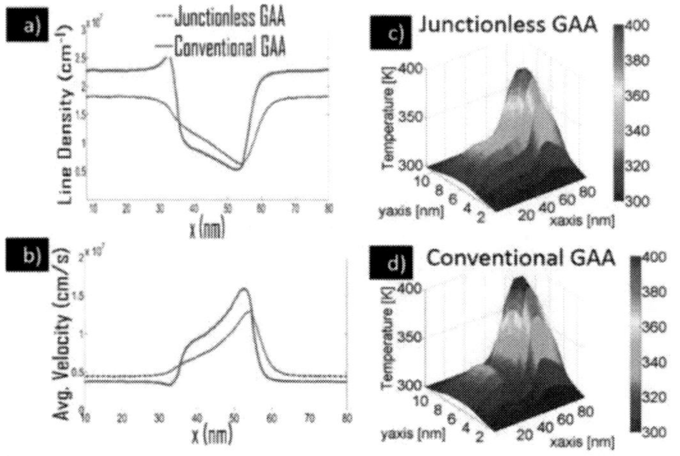

Figure 2: Comparison of JL and IM GAA MOSFET MOSFET with comparable $V_D \cdot I_D$. (a) line charge density (b) average velocity, temperature map of (c) JL and (d) IM GAA MOSFET.

average velocity between the electrical MoCa (uncoupled) and thermal MoCa, as well as the relatively low temperature distribution across the device. In order to explore this further, we benchmark the performance of JL GAA MOSFET with the conventional IM GAA. To ensure fair comparison, we simulate a JL and conventional IM GAA MOSFET with comparable power ($V_D \cdot I_D$) as shown in Fig. 2. We observe that the velocity degradation due to self-heating as well as the peak (maximum) temperature is higher for the conventional IM device. It is interesting to note that the average velocity of the inversion mode device (Fig. 2b) is higher than the saturation velocity for much of the channel (x = 35-55 nm), while the junctionless device shows that hot electrons (avg. velocity $\geq 1 \times 10^7$ cm/s) are mostly localized at the vicinity of the channel-drain region in both Figs. 1 a) and 2b). Furthermore, looking at the peak temperature as a function of the overall power dissipation as well as the variation of drain current as a function of doping (Fig. 3) reveals that the constraints set by the requirement to fully deplete the junctionless MOSFET limits that junctionless MOSFET to mainly low power applications. At such limits, despite the JL outperforming IM MOSFET, the differences in temperature is not very appreciable, as was observed in Fig. 2. Additionally, Fig. 3 highlights that the JL transistor will notably outperform the conventional MOSFET at the limit of high currents. Higher currents require doping density in excess of 5×10^{19} cm^{-3}; however, this leads to devices that are hard to turn off, as was discussed earlier even for cross-sections as small as

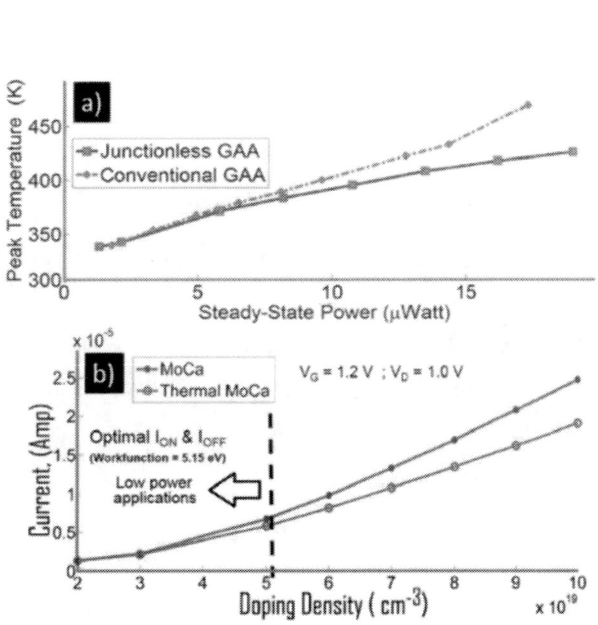

Figure 3: (a) Peak Temperature as a function of $V_D \cdot I_D$ and (b) Current degradation as function of doping density

$5\,\mathrm{nm} \times 5\,\mathrm{nm}$.

4. Conclusions

We have studied the self-heating behavior of junctionless gate-all-around MOSFET and benchmarked it against conventional inversion mode gate-all-around MOSFET using our in-house 3D full band coupled electro-thermal Monte Carlo simulator which explicitly takes into account the anharmonic three-phonon decay. Our results indicate that junctionless gate-all-around MOSFET has less self-heating effects as compared to conventional multi-gate MOSFET, particularly for applications that require high power. Suitable high workfunction metal gate ($> 5.15eV$) as well as novel JL design schemes which will constrain the off-current to acceptable limits at $V_G = 0V$ needs to explored in order to faciliate high power operation of JL MOSFET.

References

[1] D. G. Cahill, W. Ford, K. Goodson, G. Mahan, A. Majumdar, H. Maris, R. Merlin, and S. Phillpot, "Nanoscale thermal transport," *Journal of Applied Physics*, vol. 93, pp. 793–818, 2003.

[2] E. Pop, S. Sinha, and K. Goodson, "Heat generation and transport in nanometer-scale transistors," *Proceedings of the IEEE*, vol. 94, pp. 1587–1601, 2006.

[3] C.-W. Lee, A. Afzalian, N. D. Akhavan, R. Yan, I. Ferain, and J.-P. Colinge, "Junctionless multigate field-effect transistor," *Applied Physics Letters*, vol. 94, pp. 053 511–1–053 511–2, 2009.

[4] J. Colinge, C. Lee, A. Afzalian, N. Akhavan, R. Yan, I. Ferain, P. Razavi, B. O'Neil, A. Blake, M. White, A. Kelleher, B. McCarthy, and R. Murphy, "Nanowire transistors without junctions," *Nature Nanotechnology*, vol. 5, pp. 225–229, 2010.

[5] M. Mohamed, Z. Aksamija, W. Vitale, F. Hassan, K.-H. Park, and U. Ravaioli, "A conjoined electron and thermal transport study of thermal degradation induced during normal operation of multigate transistors," *Electron Devices, IEEE Transactions on*, vol. 61, no. 4, pp. 976–983, 2014.

[6] W. Vitale, M. Mohamed, and U. Ravaioli, "Monte carlo study of transport properties in junctionless transistors," in *In Proc. 14th Int. Workshop on Computational Electronics*, Pisa, Italy, 2010, pp. 1–3.

Journal of Physics: Conference Series **647** (2015) 012027 doi:10.1088/1742-6596/647/1/012027

Microscopic treatment of energy dissipation and decoherence via many-body Lindblad superoperators

R Rosati[1], R C Iotti[1], F Dolcini[1,2] and F Rossi[1]

[1] Department of Applied Science and Technology, Politecnico di Torino, Corso Duca degli Abruzzi 24, I-10129 Torino, Italy

[2] CNR-SPIN, Monte S. Angelo - via Cinthia, I-80126 Napoli, Italy

E-mail: roberto.rosati@polito.it

Abstract. Starting from a recent reformulation of the Markov limit, we apply the mean-field approximation to the resulting Lindblad-type many-electron dynamics, and derive a closed equation of motion for the electronic single-particle density matrix in the presence of one- and two-body scattering mechanisms. The proposed formulation preserves the positive-definite character of the single-particle density matrix. This result is in striking contrast with conventional Markov approaches, where the single-particle mean-field equations can lead to positivity violations and therefore to unphysical results.

One of the most challenging problems in the microscopic modeling of state-of-the-art electronic quantum systems and devices is the derivation of suitable scattering superoperators (see, e.g., [1]) to account for decoherence and dissipation phenomena (see, e.g., [2, 3, 4]). When the system-environment coupling becomes strong and/or the excitation timescale is extremely short, Markovian approaches are known to be unreliable, and memory effects have to be taken into account via quantum-kinetic approaches (see, e.g., [5, 6]); however, even in regimes where the Markov limit is applicable, conventional Markov approaches may lead to unphysical results, such as negative eigenvalues of the density matrix. To overcome this serious limitation, a few years ago an alternative and more general Markov procedure has been proposed [7]; however, in spite of its conceptual importance, the practical implementation of the latter is limited by the fact that the many-body evolution is in general not exactly solvable. So far, all relevant applications of such Markov treatment to semiconductor nanosystems are limited to the low-density limit, where a Lindblad-type scattering-induced evolution is demonstrated also for the single-particle density matrix (see, e.g., [8, 9]).

Aim of this contribution is to discuss a recently proposed kinetic treatment based on a nonlinear density-matrix equation [10]. More specifically, by applying the conventional mean-field approximation to the many-electron dynamics obtained via the alternative Markov limit recalled above, it is possible to derive a closed equation of motion for the electronic single-particle density matrix, in the presence of carrier-carrier as well as carrier-phonon scattering mechanisms. While in the low-density limit such nonlinear equation exhibits a Lindblad form –like for the many-body density matrix– at finite or high carrier concentrations the equation turns out to be non-Lindblad and highly nonlinear. Nevertheless, we can prove that the mean-field approximation does preserve the positive-definite character of the single-particle density matrix, an essential prerequisite of any reliable kinetic treatment of semiconductor quantum

Journal of Physics: Conference Series **647** (2015) 012027 doi:10.1088/1742-6596/647/1/012027

devices. This result is in striking contrast with the case of mean-field approximations applied to conventional (non-Lindblad) Markov approaches, where the corresponding single-particle equations may lead to positivity violations. The power and flexibility of the proposed density-matrix formalism has been confirmed by a recent investigation of phonon-induced quantum diffusion in state-of-the-art carbon nanotubes [11], where non-linear effects are found to play a relevant role.

The study of electro-optical processes in semiconductors mainly relies on physical quantities that depend on the electronic-subsystem coordinates only. This suggests the introduction of a many-electron density-matrix operator $\hat{\rho}$, defined as the trace of the global (i.e., electron plus various crystal excitations) density-matrix operator over non-relevant degrees of freedom (e.g., phonons, plasmons, etc.). Starting from the alternative Markov procedure proposed in Ref. [7], which allows one to derive a global Lindblad scattering superoperator, it is possible to show that also such a reduced many-electron dynamics is still of Lindblad type, i.e.,

$$\left.\frac{d\hat{\rho}}{dt}\right|_{\rm scat} = \sum_s \left(\hat{\mathbf{A}}^s \hat{\rho} \hat{\mathbf{A}}^{s\dagger} - \frac{1}{2}\left\{\hat{\mathbf{A}}^{s\dagger}\hat{\mathbf{A}}^s, \hat{\rho}\right\}\right) , \qquad (1)$$

where the explicit form of the reduced or electronic operators $\hat{\mathbf{A}}^s$ can be derived starting from the global Lindblad ones.

Within the above description, one has already performed a statistical average over the phonon bath, but the electronic subsystem is still treated via a many-body picture. However, in the investigation of semiconductor-based quantum materials and devices, many of the physical observables of interest (like, e.g., total carrier density, mean kinetic energy, charge current, and so on) may be well described via single-particle electronic operators. The study of the time evolution of such single-particle quantities requires the derivation of a closed equation of motion for the single-particle density matrix $\rho_{\alpha_1\alpha_2} = {\rm tr}\{\hat{c}^\dagger_{\alpha_2}\hat{c}_{\alpha_1}\hat{\rho}\}$. Combining the definition of $\rho_{\alpha_1\alpha_2}$ with the many-electron Lindblad dynamics in Eq. (1), and employing the cyclic property of the trace, one gets:

$$\left.\frac{d\rho_{\alpha_1\alpha_2}}{dt}\right|_{\rm scat} = \frac{1}{2}\sum_s {\rm tr}\left\{\left[\hat{\mathbf{A}}^{s\dagger}, \hat{c}^\dagger_{\alpha_2}\hat{c}_{\alpha_1}\right]\hat{\mathbf{A}}^s\hat{\rho}\right\} + {\rm H.c.} \qquad (2)$$

In order to get a closed equation of motion for the single-particle density matrix, it is now central to specify the form of the many-electron Lindblad operators $\hat{\mathbf{A}}^s$ which, in turn, depends on the particular interaction mechanism considered.

For the case of a generic carrier-phonon (cp) interaction mechanism the corresponding (one-body) Lindblad operator is always of the form

$$\hat{\mathbf{A}}^s = \sum_{\alpha\alpha'} A^{\rm cp}_{\alpha\alpha'}\hat{c}^\dagger_\alpha\hat{c}_{\alpha'} \ ; \qquad (3)$$

the latter describes the phonon-induced carrier transition (i.e., destruction plus creation) from the initial state α' to the final state α. In this case, the label $s \equiv \mathbf{q}, \pm$ shall correspond to the emission (+) or absorption (−) of a phonon with wavevector $\mathbf{q}$.

The Lindblad operator corresponding to carrier-carrier (cc) interaction is of the general (two-body) form

$$\hat{\mathbf{A}}^s = \frac{1}{2}\sum_{\alpha\overline{\alpha},\alpha'\overline{\alpha}'} A^{\rm cc}_{\alpha\overline{\alpha},\alpha'\overline{\alpha}'}\hat{c}^\dagger_\alpha\hat{c}^\dagger_{\overline{\alpha}}\hat{c}_{\overline{\alpha}'}\hat{c}_{\alpha'} \ ; \qquad (4)$$

the latter describes the transition (i.e., destruction plus creation) of our electronic pair from the initial (two-body) state $\alpha'\overline{\alpha}'$ to the final state $\alpha\overline{\alpha}$.

By inserting into Eq. (2) the above written carrier-phonon and carrier-carrier Lindblad operators, and employing the usual fermionic anticommutation relations, one ends up with terms involving average values of four and eight fermionic operators, respectively. At this point, the fundamental step to get a closed equation of motion for the single-particle density matrix consists of performing the well-known mean-field (or correlation-expansion) approximation (see, e.g., [4]). Employing this approximation scheme and omitting renormalization terms [10], the resulting single-particle equation is given by

$$\left.\frac{d\rho_{\alpha_1\alpha_2}}{dt}\right|_{\mathrm{scat}} = \frac{1}{2}\sum_s \sum_{\alpha'\alpha'_1\alpha'_2} \left[\left(\delta_{\alpha_1\alpha'} - \rho_{\alpha_1\alpha'}\right)\mathcal{P}^s_{\alpha'\alpha_2,\alpha'_1\alpha'_2}\rho_{\alpha'_1\alpha'_2} - \left(\delta_{\alpha'\alpha'_1} - \rho_{\alpha'\alpha'_1}\right)\mathcal{P}^{s*}_{\alpha'\alpha'_1,\alpha_1\alpha'_2}\rho_{\alpha'_2\alpha_2}\right] + \mathrm{H.c.} \tag{5}$$

with generalized carrier-phonon scattering rates

$$\mathcal{P}^{s=\mathrm{cp}}_{\alpha_1\alpha_2,\alpha'_1\alpha'_2} = A^{\mathrm{cp}}_{\alpha_1\alpha'_1} A^{\mathrm{cp}*}_{\alpha_2\alpha'_2} \tag{6}$$

and generalized carrier-carrier scattering rates

$$\mathcal{P}^{s=\mathrm{cc}}_{\alpha_1\alpha_2,\alpha'_1\alpha'_2} = 2\sum_{\overline{\alpha}_1\overline{\alpha}_2,\overline{\alpha}'_1\overline{\alpha}'_2} \left(\delta_{\overline{\alpha}_2\overline{\alpha}_1} - \rho_{\overline{\alpha}_2\overline{\alpha}_1}\right) \mathcal{A}^{\mathrm{cc}}_{\alpha_1\overline{\alpha}_1,\alpha'_1\overline{\alpha}'_1} \mathcal{A}^{\mathrm{cc}*}_{\alpha_2\overline{\alpha}_2,\alpha'_2\overline{\alpha}'_2} \rho_{\overline{\alpha}'_1\overline{\alpha}'_2}\,, \tag{7}$$

where $\mathcal{A}^{\mathrm{cc}}_{\alpha\overline{\alpha},\alpha'\overline{\alpha}'}$ denotes the totally antisymmetric parts of the two-body coefficients $A^{\mathrm{cc}}_{\alpha\overline{\alpha},\alpha'\overline{\alpha}'}$ in Eq. (4). It is worth noting that, opposite to the generalized carrier-phonon rates in Eq. (6), the generalized carrier-carrier rates in Eq. (7) are themselves a function of the single-particle density matrix; this is a clear fingerprint of the two-body nature of the carrier-carrier interaction.

The crucial issue related to the proposed kinetic treatment is the positivity analysis of the nonlinear density-matrix equation (5). Indeed, if the single-particle density matrix describes a physical state, its eigenvalues are necessarily positive-definite and smaller than one (Pauli exclusion principle); this feature must be preserved during the scattering-induced time evolution. As discussed in detail in Ref. [10], in order to prove that, it is crucial to move from the original single-particle basis $\{|\alpha\rangle\}$ to the time-dependent basis $\{|\lambda\rangle\}$ that instantaneously diagonalizes the single-particle density matrix; more specifically, by denoting with $\Lambda_\lambda = \rho_{\lambda\lambda}$ the generic density-matrix eigenvalue (corresponding to the eigenstate $|\lambda\rangle$), it is possible to show that its time evolution induced by the proposed density-matrix equation (5) is simply given by

$$\frac{d\Lambda_\lambda}{dt} = \sum_s \sum_{\lambda'} \left[(1-\Lambda_\lambda)P^s_{\lambda\lambda'}\Lambda_{\lambda'} - (1-\Lambda_{\lambda'})P^s_{\lambda'\lambda}\Lambda_\lambda\right]\,, \tag{8}$$

where $P^s_{\lambda\lambda'} = \mathcal{P}^s_{\lambda\lambda,\lambda'\lambda'}$ are positive-definite quantities given by the diagonal elements of the generalized scattering rates [see Eqs. (6) and (7)] written in the instantaneous density-matrix eigenbasis. This result is highly non-trivial: it states that, in spite of the partially coherent nature of the carrier dynamics in (5), the time evolution of the eigenvalues Λ_λ is governed by a non-linear Boltzmann-type equation formally identical to the semiclassical-case one.

We are now in the position to state that the physical interval [0, 1] is the only possible variation range of our eigenvalues Λ_λ; to this end, it is crucial to show that when the latter approach the extremal values, 0 or 1, their time derivatives do not allow them to exit the interval. Indeed, a closer inspection of the Boltzmann-like equation (8) shows that:

(i) if one of the eigenvalues is equal to zero, the corresponding time derivative is always non-negative;

(ii) if one of the eigenvalues is equal to one, its time derivative is always non-positive.

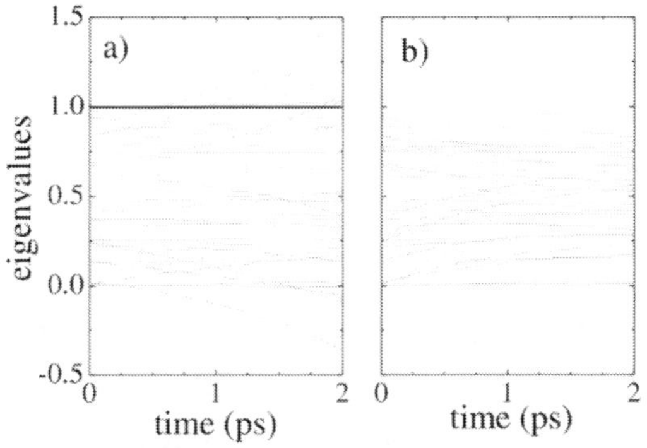

Figure 1. Time evolution of the density matrix eigenvalues, corresponding to a simple two-level system in the presence of carrier-phonon interaction. Panel a): conventional Markov approach; Panel b): proposed single-particle model.

This leads us to the important conclusion that, for both carrier-phonon and carrier-carrier scattering, the proposed nonlinear single-particle equation (5) will preserve the positive-definite character of the single-particle density matrix. We stress that the above positivity analysis is based on the fact that the matrix elements $P^s_{\lambda\lambda'}$ in (8) are positive-definite quantities. This property, which applies to the proposed single-particle equation (obtained starting from the global Lindblad-type scattering superoperator), is generally not fulfilled by conventional Markov models. In this case, starting from a non-Lindblad many-body scattering model, the system dynamics may exit the physical eigenvalue region, giving rise to positivity violations also in the low-density limit [7].

The outcome of the positivity analysis presented so far is fully confirmed by the subset of simulated experiments displayed in Fig. 1, where we report the time evolution of the density-matrix eigenvalues for the case of a simple two-level system, in the presence of carrier-phonon scattering. As expected, for the proposed nonlinear equation all the eigenvalue trajectories fall within the physical interval [0, 1], while for the nonlinear equation obtained within the conventional Markov limit a significant number of simulated eigenvalue trajectories exit the physical interval.

We finally stress that the proposed single-particle formulation has been also extended to the case of quantum systems with spatial open boundaries [10]; this provides a formal derivation of a recently proposed density-matrix treatment based on a Lindblad-like system-reservoir scattering superoperator [8].

We are extremely grateful to Salvatore Savasta and David Taj for stimulating and fruitful discussions. F.D. also acknowledges financial support from Italian FIRB 2012 project HybridNanoDev (Grant No. RBFR1236VV).

References

[1] Davies E 1976 *Quantum Theory of Open Systems* (Academic Press)
[2] Haug H and Koch S 2004 *Quantum Theory of the Optical and Electronic Properties of Semiconductors* (World Scientific)
[3] Datta S 2005 *Quantum Transport: Atom to Transistor* (Cambridge University Press)
[4] Rossi F 2011 *Theory of Semiconductor Quantum Devices: Microscopic Modeling and Simulation Strategies* (Springer)
[5] Axt V M and Mukamel S 1998 *Rev. Mod. Phys.* **70** 145–74
[6] Axt V M and Kuhn T 2004 *Rep. Prog. Phys.* **67** 433–512
[7] Taj D, Iotti R C and Rossi F 2009 *Eur. Phys. J. B* **72** 305–22
[8] Dolcini F, Iotti R C and Rossi F 2013 *Phys. Rev. B* **88** 115421
[9] Rosati R and Rossi F 2013 *Appl. Phys. Lett.* **103** 113105
[10] Rosati R, Iotti R C, Dolcini F and Rossi F 2014 *Phys. Rev. B* **90** 125140
[11] Rosati R, Dolcini F and Rossi F 2015 *Appl. Phys. Lett.* **106** 243101

Journal of Physics: Conference Series **647** (2015) 012028 doi:10.1088/1742-6596/647/1/012028

Correlation Effects of Localized Impurities on Electron Transport under 1-D Nano-Structures

N. Sano, M. R. Zulhidza, Y. Kaneno, S. Honda, A. Ueda, and K. Yoshida

Institute of Applied Physics, University of Tsukuba, Tsukuba 305-8573, Japan

E-mail: `sano@esys.tsukuba.ac.jp`

Abstract. We investigate the spatial correlation effects of ionized impurities on electron transport properties under quasi-1D nanowire structures. The impurity-limited resistance is evaluated with the Landauer formula under various spatial distributions of impurities and the simulation results are theoretically analyzed. We show that phase interference is significant even at room temperature when the separation between impurities along the axis direction is small, whereas strong phase randomization takes place as the separation increases.

1. Introduction

As the device dimension shrinks into nano-scale, the impurity density in the active channel region is inevitably high to attain proper gate-controllability in advanced Si FETs. Nevertheless, the number of impurities in the active region is rather small because of small volume. As a result, transport properties such as mobility could fluctuate, depending on the location and the number of impurities in the channel [1-3]. Such fluctuations in transport characteristics result from both incoherent and coherent effects and quantum mechanical simulations such as the nonequilibrium Green functions (NEGF) method indeed take into account of both effects. However, their physical mechanism is not well explored. In this report, we study numerically and analytically the physics behind the fluctuations of impurity-limited resistances in quasi-1D nanowires.

2. Methodology

2.1. Impurity-limited resistance

According to the Landauer picture, the conductance in quasi-1D nanowires could be calculated from the transmission coefficient through the doped channel region. Since the resistance R_{tot} given by the inverse of the conductance thus obtained includes both the quantum (contact) resistance R_0 and the impurity-limited resistance R_s, R_s is obtained from $R_s = R_{tot} - R_0$ and expressed by

$$R_s = \frac{1}{g_{vl}} \frac{\pi\hbar}{e^2} \frac{\sum\limits_{l,n} \int_{-\infty}^{\infty} dE \, \{1 - T_{ln}(E)\} \left(-\frac{\partial f_{FD}(E)}{\partial E}\right) \theta\left(E - \varepsilon_{ln}\right)}{\left\{\sum\limits_{l,n} \int_{-\infty}^{\infty} dE \left(-\frac{\partial f_{FD}(E)}{\partial E}\right) \theta\left(E - \varepsilon_{ln}\right)\right\} \left\{\sum\limits_{l,n} \int_{-\infty}^{\infty} dE \, T_{ln}(E) \left(-\frac{\partial f_{FD}(E)}{\partial E}\right) \theta\left(E - \varepsilon_{ln}\right)\right\}}, \quad (1)$$

where g_{vl} is the valley degeneracy, $T_{ln}(E)$ is the transmission coefficient for the in-coming electron in the subband (l, n) with total electron energy E and ε_{ln} is the subband energy. $f_{FD}(E)$ is the Fermi-Dirac distribution and $\theta(E)$ is the Heaviside step function. Here, we have assumed that there is no "contact" resistance associated with the difference in the local density of states at the junction between the lead and the channel. In other words, we consider the transport properties under the flat potential condition along the axis direction throughout this study. The transmission coefficient $T_{ln}(E)$ is obtained from the Fisher-Lee formula with the retarded Green function calculated by the recursive method. We employ Eq. (1) to calculate R_s in the present numerical simulations.

Under the extreme quantum limit where only the lowest subband A is involved, R_s reduces to

$$R_s = \frac{1}{g_{vl}}\frac{\pi\hbar}{e^2}\frac{1-\langle T_A\rangle}{\langle T_A\rangle} = \frac{1}{g_{vl}}\frac{\pi\hbar}{e^2}\frac{\langle R_A\rangle}{1-\langle R_A\rangle} \approx \frac{1}{g_{vl}}\frac{\pi\hbar}{e^2}\langle R_A\rangle, \quad (2)$$

where $\langle T_A\rangle$ and $\langle R_A\rangle$ are, respectively, the transmission and the reflection coefficients weighted with the derivative of the Fermi-Dirac distribution for the in-coming electrons in the lowest subband A. The last expression is valid under the weak scattering limit in which $\langle T_A\rangle$ is close to unity. Notice that R_s under the weak scattering limit is bounded above and physically inappropriate. In fact, we can show that the last expression, roughly equivalent to the Born approximation, is identical to the semi-classical expression derived from the Boltzmann transport equation (BTE) under the relaxation time approximation and, thus, the BTE indeed breaks down if the reflection coefficient $\langle R_A\rangle$ is large [4].

2.2. Device structure

For numerical simulations, we employ quasi-1D nanowires with rectangular cross-section area of 3.5×3.5 nm^2 and ionized donor or acceptor impurities are distributed at random in Si channel. The average impurity density in the channel is fixed at 10^{19} cm^{-3} so that the channel length L varies according to the number of doped impurities, namely, $L = 8$ nm for single-impurity, 16 nm for two impurities, etc. Since many simulations under various patterns of impurity distributions are required to perform, the scattering potential is assumed to be a simple screened potential (screening length: $\lambda_{sc} = 1.3$ nm) with image charges associated with the Si/Gate-Oxide interfaces.

3. Results and discussion

3.1. Numerical simulations

We first show numerical simulation results of the impurity-limited resistance R_s for 300 patterns of impurity distributions as a function of the number of (donor or acceptor) impurities in Fig. 1. R_s greatly varies depending on the position of impurities, but the average value of R_s's follows the linear relationship with the channel length and, thus, obeys Ohm's law. In the case of single-impurity (one impurity is doped in the channel), the fluctuations trivially result from the variations in the subband wave function and the scattering potential of ionized impurity. However, the physics behind such fluctuations for the cases of two and three impurities is rather involved. Figure 2 shows R_s for two donor impurities as a function of the axial separation Δ between the two impurities. There are three distinct regions, in reference with the averaged resistance indicated by the horizontal dashed line. In region I, fluctuations in R_s are small and most R_s's are close to the average value, which is nearly equal to twice of the resistance for the single-impurity, $2R_{single}$. As shown in the theoretical analyses below, this is because electron's phase is rapidly randomized as Δ becomes longer and the phase among the impurities becomes uncorrelated. Thus, two impurities behave as a series resistance and the total resistance is given by the simple sum of R_{single} due to each single-impurity. In regions II, R_s's greatly fluctuate and,

Journal of Physics: Conference Series **647** (2015) 012028 doi:10.1088/1742-6596/647/1/012028

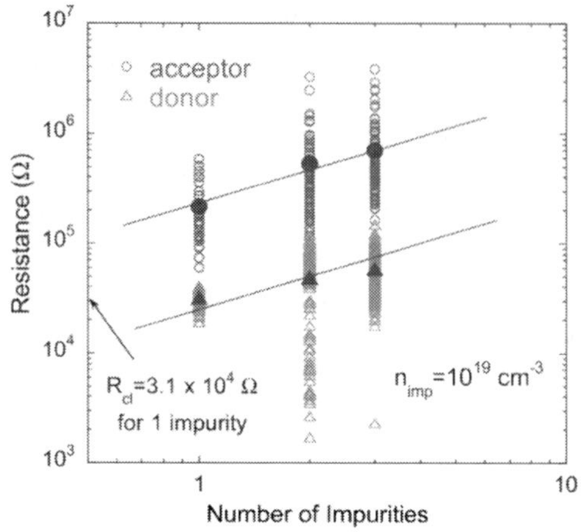

Figure 1. Impurity-limited resistance as a function of the number of donor or acceptor impurities. 300 patterns of impurity distributions are shown for each case.

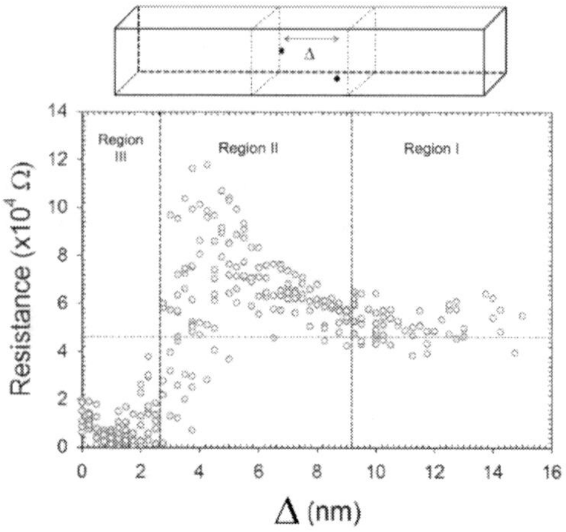

Figure 2. Impurity-limited resistance for two donor impurities as a function of the separation Δ between the two impurities along the wire axis, as shown on top.

in most cases, they are larger than the average value. This implies that the phase interference along the wire axis could play a crucial role even at room temperature. In region III, on the other hand, R_s's become smaller than the average value and this is because the two impurities closely located along the axis direction could be regarded as being on the same transversal plane and, thus, behave as a parallel resistance. It should be noticed that this scenario is possible only if the scattering potential has the long-range part, i.e., the finite screening length.

3.2. Theoretical analyses

In order to clarify the physics involved in R_s, let us consider a simple model in which two impurities are doped on the axis of the cylindrical wire and each impurity has the δ-scattering potential in the coordinate space.[1]

$$V(\mathbf{R}) = v_s(aS)\left\{\delta^{(3)}(\mathbf{R}-\mathbf{R}_{01}) + \delta^{(3)}(\mathbf{R}-\mathbf{R}_{02})\right\}, \tag{3}$$

where a represents the characteristic length along the axis direction over which the scattering potential is effective and S is the cross-sectional area of the wire. $\mathbf{R}_{01}$ and $\mathbf{R}_{02}$ are the position of impurities. For simplicity, we assume that each impurity has the same scattering potential energy v_s. The exact T_A and R_A are calculated from the Lippmann-Schwinger equation and the impurity-limited resistance R_s under the extreme quantum limit is found to be

$$R_s(\Delta) = \frac{1}{g_{vl}}\frac{\pi\hbar}{e^2}\frac{\left\langle \frac{2\gamma^2\{(1+\gamma^2)+(1-\gamma^2)\cos(2k\Delta)+2\gamma\sin(2k\Delta)\}}{1+2\gamma^2\{(1+\gamma^2)+(1-\gamma^2)\cos(2k\Delta)+2\gamma\sin(2k\Delta)\}}\right\rangle}{\left\langle \frac{1}{1+2\gamma^2\{(1+\gamma^2)+(1-\gamma^2)\cos(2k\Delta)+2\gamma\sin(2k\Delta)\}}\right\rangle}, \tag{4}$$

where k is the wave number of the in-coming electron and the scattering parameter γ is defined by $\gamma = v_s\frac{am}{\hbar^2 k}S|\xi_A(0)|^2$ where $\xi_A(r)$ is the subband wave function at radial position r.

Figure 3 shows R_s obtained from Eq. (4) as a function of the impurity separation Δ for $T = 300$ and 30 K. The scattering potential energy is assumed to be $v_s = 183$ meV, corresponding

[1] We notice that this simple analytical model is unable to explain the behavior in region III in Fig. 2.

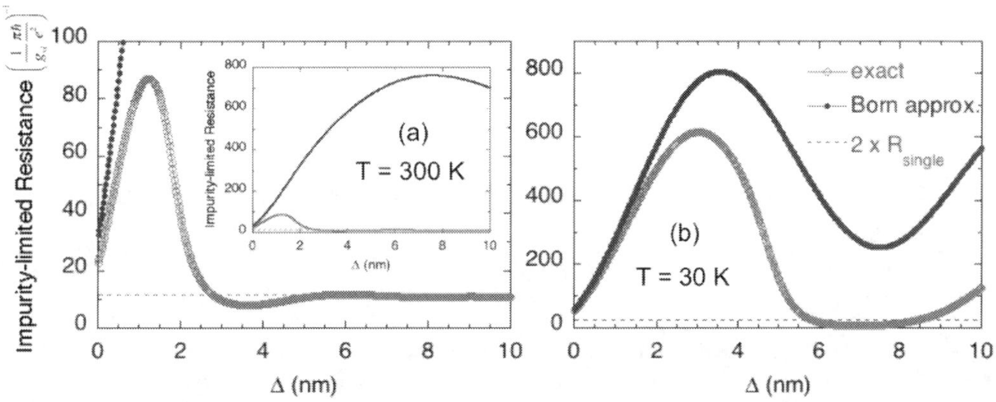

Figure 3. Impurity-limited resistance of two impurities from the analytical model as a function of the separation Δ between the two impurities along the wire axis. The radius of the cross-section is 2 nm and the resistances at (a) $T = 300$ K and (b) 30 K are shown.

to the screening length of $\lambda_{sc} = 2$ nm. The results from the exact expression and the Born approximation are plotted along with $2R_{single}$ represented by horizontal dashed lines. At low temperature ($T = 30$ K), the oscillatory behavior is clearly seen and this is due to the trigonometric dependence in Eq. (4). This, of course, results from electron's phase interference between the two impurities. As temperature increases ($T = 300$ K), this oscillation rapidly damps except in the first few nm where R_s is still greatly enhanced by the constructive phase interference. The asymptotic value at large Δ approaches $2R_{single}$, which is the uncorrelated limit. Since neither spatial averaging with respect to the position of impurities or energy dissipating scattering such as phonon interaction is included, this phase randomization is caused purely by the broadness in energy spectrum of the in-coming electrons from the reservoir (source). This is confirmed from the temperature dependence of the oscillatory behavior in Fig. 3. As a result, the average of R_s for various Δ approaches the uncorrelated limit $2R_{single}$. These findings coincide with those found from the numerical simulations and explain the fact that the constructive interference comes into play in R_s even at room temperature if the impurity separation is small and that R_s rapidly approaches the average value as the separation becomes larger. In addition, we should point out that the Born approximation completely breaks down at room temperature.

4. Conclusions

We have investigated numerically and theoretically the spatial correlation effects of ionized impurities on electron transport under the nanowire structures. We have demonstrated explicitly how the phase interference and phase randomization take place simultaneously in the impurity-limited resistance R_s at room temperature. Namely, the constructive phase interference is dominant even at room temperature in R_s if two impurities are located close each other, whereas R_s rapidly approaches the uncorrelated limit as the separation becomes larger. As a result, the ensemble average resistance follows classical Ohm's law even under nanostructures.

References

[1] Persson M P, Mera H, Niquet Y M, Delerue C and Diarra M 2010 *Phys. Rev.* B **82** 115318
[2] Martinez A Aldegunde M, Seoane N, Brown A R, Barker J R and Asenov A 2011 *IEEE Trans.* ED **58** 2209
[3] Niquet Y M Mera H and Delerue C 2012 *Appl. Phys. Lett.* **100** 153119
[4] Sano N to be published

Journal of Physics: Conference Series **647** (2015) 012029 doi:10.1088/1742-6596/647/1/012029

Energy Relaxation and Non-linear Transport in InAs Nanowires

R Hathwar, M Saraniti and S M Goodnick
Center for Computational Nanoscience, Arizona State University, USA

E-mail: rhathwar@asu.edu

Abstract. Using a full band Monte Carlo simulation the electron energy relaxation times of InAs nanowires along the [100] direction are shown to be greater than the energy relaxation time of bulk InAs. The full band structure and scattering rates of the nanowires are calculated using the $sp^3d^5s^*$ tight binding method and Fermi's Golden rule respectively. At moderately high electric fields, a runaway effect is also observed in the nanowires and is attributed to the 1D nature of the scattering rates.

1. Introduction

In recent years nanowires have become important as both a novel material and as a channel for transistors. Various companies are researching the usefulness of nanowire field effect transistors (FETs) as a possibility for devices with very small gate lengths. It is therefore very important to understand the material properties of nanowires and create models that can simulate transport in nanowires. In particular, the energy relaxation rates are of interest in terms of advanced concept photovoltaic devices based on hot carrier extraction [1]. Materials with a larger relaxation time lead to higher efficiency in solar cells based on hot carrier extraction. Also non-equilibrium behavior of nanowires is of high importance in simulating FETs and is therefore important to correctly model the behavior of nanowires in non-equilibrium conditions.

In section 2 the method used to simulate nanowires under photoexcitation as well as under high fields is discussed. In section 3 the results are discussed.

2. Simulation Method

The band structure of InAs nanowires are calculated using the $sp^3d^5s^*$ empirical tight binding method including spin. This method has accurately modelled the band gap and effective masses of nanowires and is used widely in the literature [2]. In this work a k-space Monte Carlo simulation is performed to simulate the high field and non-equilibrium properties of InAs nanowires. The 1D scattering rates needed for such a simulation are calculated using the Fermi's golden rule. Polar optical phonon scattering rates and deformation potential scattering rates are included in the simulation. The scattering rate for acoustic phonons using the equipartition approximation is [3],

$$W_{ac,\mu,\nu}(k,k') = \frac{E_{ac}^2 kT}{4\pi\rho v_s^2 \hbar} I_{\mu,\nu}(\mathbf{q})\,\mathrm{DOS(k')} \tag{1}$$

Journal of Physics: Conference Series **647** (2015) 012029 doi:10.1088/1742-6596/647/1/012029

and the scattering rate for optical phonons assuming dispersionless phonons is

$$W_{\mathrm{op},\mu,\nu}(k,k')=\frac{E_{op}^2}{4\pi\rho\omega_{op}}\left(N_{q,op}+\frac{1}{2}\pm\frac{1}{2}\right)I_{\mu,\nu}(\mathbf{q})\,\mathrm{DOS}(\mathrm{k}') \tag{2}$$

where

$$I_{\mu,\nu}(\mathbf{q})=\sum_{m}\sum_{m'}c_{\nu,m}^{*}(k')c_{\mu,m}(k)c_{\mu,m'}^{*}(k)c_{\nu,m'}(k')q_c\frac{J_1(q_c a_{diff})}{a_{diff}}e^{iq[\tau_{m,x}-\tau_{m',x}]} \tag{3}$$

where $a_{diff}=\sqrt{\left(\tau_{m,y}-\tau_{m',y}\right)^2+\left(\tau_{m,z}-\tau_{m',z}\right)^2}$ and the sum is over all orbitals in the supercell. $q_c\approx 1.9\pi/a$ is used as an effective cut-off radius to emulate the projection of the BZ of bulk material onto the confinement plane [4]. E_{ac} and E_{op} are the deformation potential constants for the acoustic and optical modes respectively, v_s is the sound velocity in InAs, μ and ν are the initial and final band indices and ω_{op} is the wavenumber for the optical phonons. The scattering rate for the polar optical phonons is given by [5]

$$W_{pop,\mu,\nu}(k,k')=\frac{\omega_0 e^2}{8\pi}\left[\frac{1}{\varepsilon_\infty}-\frac{1}{\varepsilon_0}\right]\left(N_{|\mathbf{q}|,\mathrm{op}}+\frac{1}{2}\pm\frac{1}{2}\right)I_{pop,\mu,\nu}(\mathbf{q}) \tag{4}$$

where

$$I_{pop,\mu,\nu}(\mathbf{q})=\sum_{m}C_{\mu,m}(k)C_{\nu,m}^{*}(k')\sum_{m'}C_{\mu,m'}^{*}(k)C_{\nu,m'}(k')\int_0^{q_c}\frac{J_0(q_t a_{diff})}{q^2+q_t^2}e^{iq[\tau_{m,x}-\tau_{m',x}]}q_t\,dq_t \tag{5}$$

where q_t is the phonon vector along the confinement plane. The integral in equation (5) can be pre-calculated for a fixed number of q_t values. This speeds up the computational time during the scattering rate calculation. It is important to note that when the scattering rates for bulk InAs and InAs nanowires are compared, they show an agreement at high energies for the same deformation potential and tight binding parameters as shown in figure (2). This agreement provides credibility to the tight binding method when scaling from bulk to nano-systems.

The agreement in the scattering rates result in an agreement of the high field velocity curves between bulk and nanowire InAs as shown in figure (3). To simulate hot carrier relaxation during photoexcitation the electrons are initialized with a mean energy corresponding to the excitation energy of the photons. A Gaussian half-width of 100meV is used. The momentum of the carrier is uniformly distributed across the Brillouin Zone. The Monte Carlo simulation is then ran until the average energy of the carriers remains a constant with respect to time. To calculate the energy relaxation time no electric field is applied to the carriers, and the carrier relax according to the scattering rates of the system.

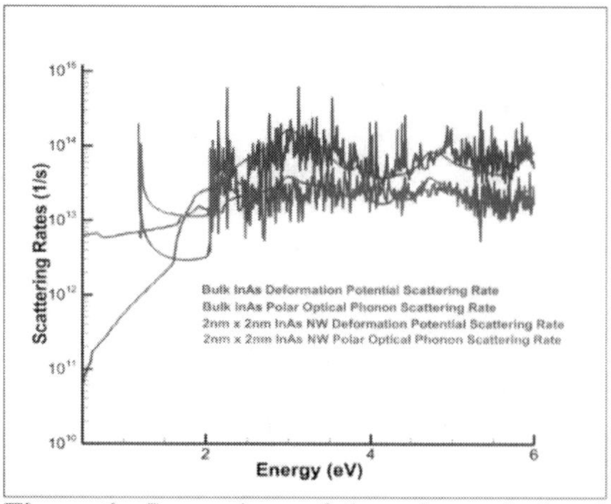

Figure 1. Comparison of the polar optical and deformation potential scattering rates between a 2nmx2nm InAs NW along [100] and bulk InAs using the same parameters.

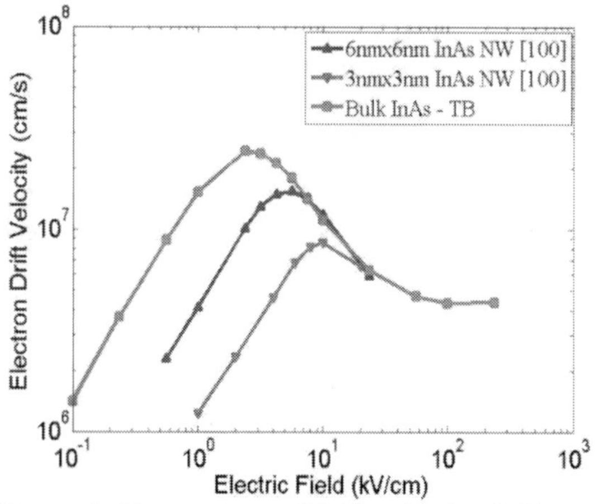

Figure 2. Comparison of drift velocity field curves between Bulk InAs and 3nmx3nm and 6nmx6nm InAs nanowires along [100].

3. Results

The electrons are initialized with a mean energy corresponding to the excitation energy shown in figure (3). After a few picoseconds the initial distribution thermalizes and the average energy decay is exponential with time. An interesting observation is that for a given nanowire size, the relaxation time does not vary much with the excitation energy.

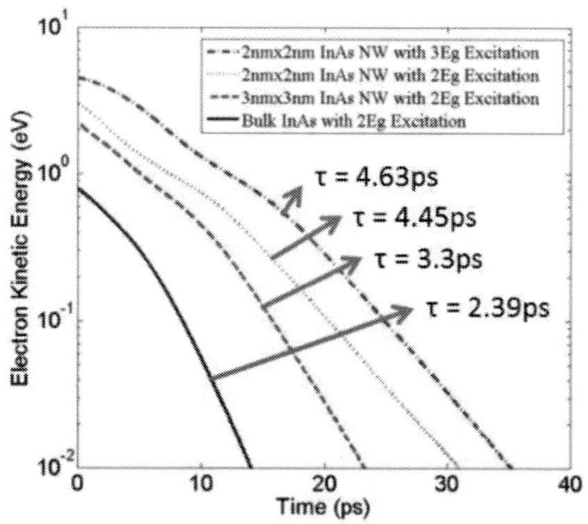

Figure 3. Energy relaxation of hot electrons in Bulk InAs, 2nmx2nm InAs NW [100] and 3nmx3nm InAs NW [100].

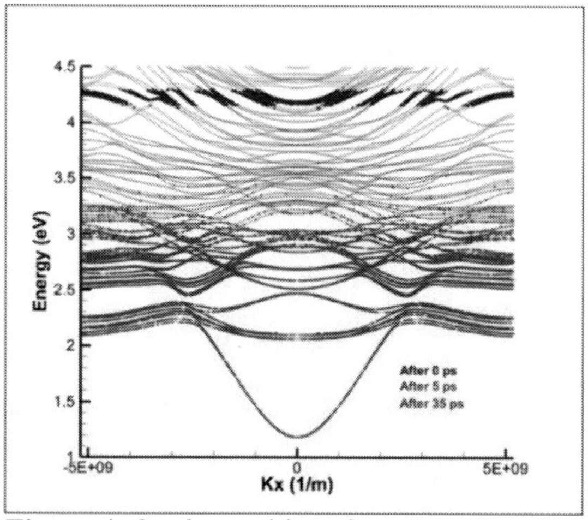

Figure 4. Carrier positions for a 2nmx2nm InAs NW [100] in k-space after photoexcitation. Each dot represents a carrier at a particular time. The initialization occurs at 0 ps.

In both cases it is noted that the relaxation time is greater for the case of nanowires when compared to the relaxation time of bulk InAs. Also smaller dimension nanowires lead to higher energy relaxation times. The initial non-exponential decay of the electrons is assumed to be due to the Gaussian nature of the initialization of carriers and the uniform distribution of carriers across the Brillouin Zone. Once the carriers undergo sufficient scattering, the distribution both in energy and momentum reaches a physically correct distribution for excited carriers and the decay then becomes exponential as expected. The change in the distribution in carriers as it cools down can be observed in figure (4). As the carriers cool, the energy width broadens and the carriers accumulate near the band minima.

At high fields an electron "runaway" effect is also observed. At low fields the distribution is Maxwellian with a temperature of 315K, but at high fields the distribution becomes more uniform due to the runaway effect [6] as shown in figure (5). This is attributed to the nature of the 1D scattering rates which decreases in value with increase in energy within a band as seen in the inset of figure (5). This highlights the importance of full Monte Carlo simulations to simulate such devices as an assumption of the distribution function cannot be made even at moderately high electric fields.

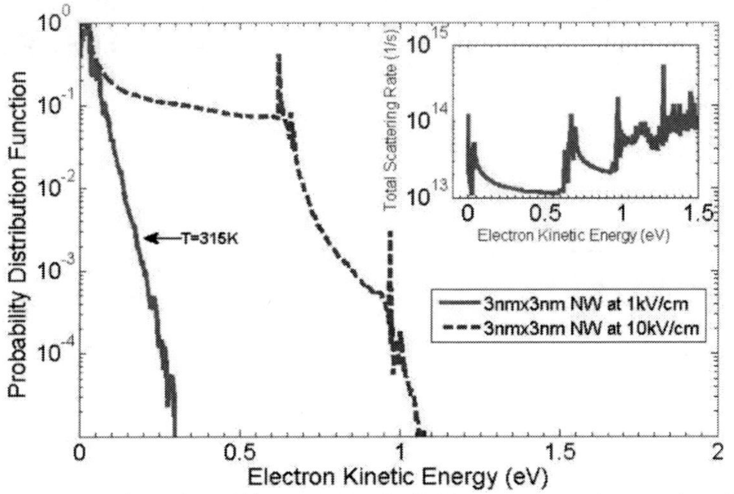

Figure 5. Distribution functions for different electric fields on a 3nmx3nm InAs NW [100].

4. Acknowledgements
The authors would like to thank the Numerical Device Modelling group at Intel, OR for providing the funding necessary for this research and for the support of the Technische Universität München - Institute for Advanced Study, funded by the German Excellence Initiative.

References
[1] Konig D, Casalenuovo K, Takeda Y, et al. 2009 *14th Intl. Conf. on Modulated Semicond. Structures*, Vol 42, Issue 10.
[2] Kotlyar R, Linton. T D, Rios R, Giles M D, et al. 2012 *J. Appl. Phys.* **111**, 12, pp. 123718-123718-11.
[3] Buin A K, Verma A, Anantram M P 2008 , *J. Appl. Phys.* **104**, 053716.
[4] Rössler U, *Solid State Theory – An Introduction,* 2004 Springer, New York.
[5] Hathwar R, Saraniti M, Goodnick S M 2014, *IEEE Nano,* Pg. 645.
[6] Dimitrev A P, Kachorovskii V Yu, Shur M S, Stroscio M 2000, *Solid State Communications* **113** pp. 565-568.

Journal of Physics: Conference Series **647** (2015) 012030 doi:10.1088/1742-6596/647/1/012030

Multi-scale Simulations of Metal-Semiconductor Nanoscale Contacts

M. Aldegunde[1], **S. P. Hepplestone**[2], **P. V. Sushko**[3] **and K. Kalna**[4]

[1]Warwick Centre for Predictive Modelling, School of Engineering, University of Warwick, Coventry CV4 7AL, UK.

[2]Deregallera Ltd., Caerphilly CF83 2HU, Wales, UK.

[3] Pacific Northwest National Laboratory, Richland, WA 99352, USA.

[4]Electronic Systems Design Centre, Swansea University, Swansea SA2 8PP, Wales, UK.

E-mail: k.kalna@swansea.ac.uk

Abstract. An electron transport simulations via a metal-semiconductor interface is carried out using multi-scale approach by coupling ab-initio calculations with 3D finite element ensemble Monte Carlo technique. The density functional theory calculations of the Mo/GaAs (001) interface show electronic properties of semiconductor dramatically change close to the interface having a strong impact on the transport. Tunnelling barrier lowers and widens due to a band gap narrowing near the interface reducing resistivity by more than one order of magnitude: from $2.1 \times 10^{-8}\Omega.\text{cm}^2$ to $4.7 \times 10^{-10}\Omega.\text{cm}^2$. The dependence of electron effective mass from the distance to the interface also plays a role bringing resistivity to $7.9 \times 10^{-10}\Omega.\text{cm}^2$.

1. Introduction

Metal-semiconductor contacts are integral components of any semiconductor device. When these devices are scaled to nanometer dimensions, transport through the contacts becomes affected by their atomic structure and atomic-scale variations of their electronic properties. Here, we present a multi-scale approach, in which the results of ab-initio calculations of a metal-semiconductor interface are mapped onto semi-classical transport simulations performed using a three-dimensional (3D) self-consistent finite element ensemble Monte Carlo method with atomic resolution. We demonstrate this approach on a Mo/GaAs(100) interface [1], a candidate for the source/drain contacts in III-V MOSFETs for future sub-16 nm CMOS technologies [2].

2. The Multiscale Computational Method

We use a two-level hierarchical method for predicting electrical current through a metal-semiconductor contact. The first level is based on the density functional theory (DFT) calculations to obtain the geometrical structure and electronic properties. The results allow to define characteristic electronic structure parameters, such as effective electron mass, as functions of the distance from the interface, which then transferred into semiclassical ensemble Monte Carlo transport simulations [3, 4].

2.1. Density Functional Theory Calculations

The Mo/GaAs (100) interface is modelled as a periodic heterostructure of an eight-bilayer thick slab of GaAs and a five monolayer thick slab of Mo, both terminated with (100) planes. The

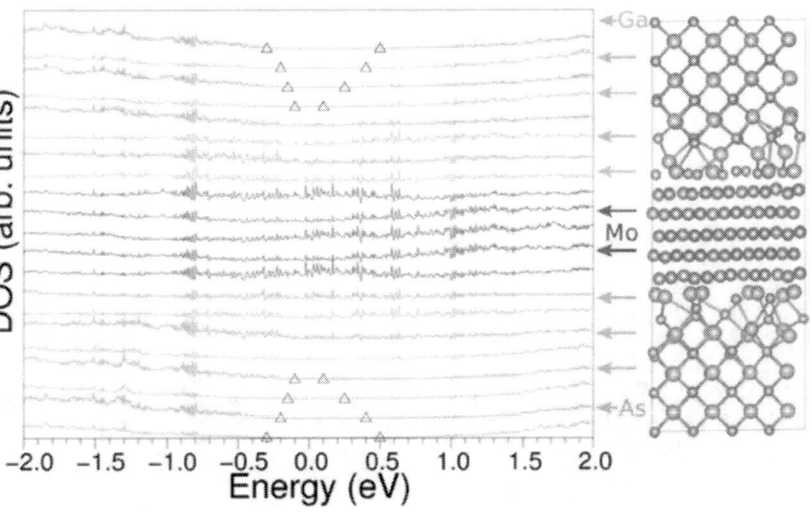

Figure 1. Layer averaged projected density of states (LPDOS)

Mo part of it was rotated by 33.7° around the axis perpendicular to the plane of the interface, resulting in lattice mismatch of less than 3%. We use the density functional by Perdew, Burke and Ernzerhof (PBE) [5] and the projected augmented waves method [6], as implemented in the Vienna ab-initio package (VASP) [7]. The total energy of the system was minimized with respect to coordinates of all atoms and the cell parameters. A combination of the conjugate gradient method and ab initio molecular dynamics, based on the Verlet algorithm, the cutoff of 450 eV and a $3 \times 3 \times 1$ Monkhorst-Pack k-point set were used for these calculations. The atom-projected density of states (PDOS) was calculated for the relaxed super-cell using the cutoff of 500 eV and a $5 \times 5 \times 2$ k-point set. Fig. 1 shows the geometric structure and density of states projected on atoms of each atomic plane in the Mo/GaAs (100) heterostructure. Analysis of the PDOS demonstrates the band gap narrowing and the presence of the metal induced gap states on both Ga- and As-terminated interfaces of GaAs with the Mo film, which we attribute to the charge density redistribution at the interface.

To find the dependence of the electron effective mass (m^*) in GaAs on the distance from the interface, we assume that the local value of m^* is determined by the local band gap, as defined using PDOS in Fig. 1. In turn, the value of the band gap correlates well with the GaAs lattice constant. Therefore, we first calculated the one-electron band gaps and the corresponding effective masses for several values of the lattice constants of the bulk GaAs. The CRYSTAL09 package [8], the Gaussian-type DEF2 basis set [9] and the PBE functional [5] were used in these calculations. The obtained values were recalculated to the distance from the interface and referenced to the bulk band gap and the electron mass. The band gap decreases linearly and disappears two bilayers away from the interface. Similarly, the electron effective mass decreases and becomes undefined when the gap closes.

2.2. Mapping of Material Parameters in Multiscale Approach

The variations of the conduction and valence band edges with respect to the Fermi level (Fig. 1) and the electron effective mass [10] were mapped into a 3D ensemble Monte Carlo transport simulation. It has been enhanced with a tunnelling transport model [11, 12] to treat the transport through the metal-semiconductor interface [13]. The reduction of the band gap (Fig. 2) changes the Schottky barrier, both its height and shape (Fig. 4). This affects dramatically tunnelling

Journal of Physics: Conference Series **647** (2015) 012030 doi:10.1088/1742-6596/647/1/012030

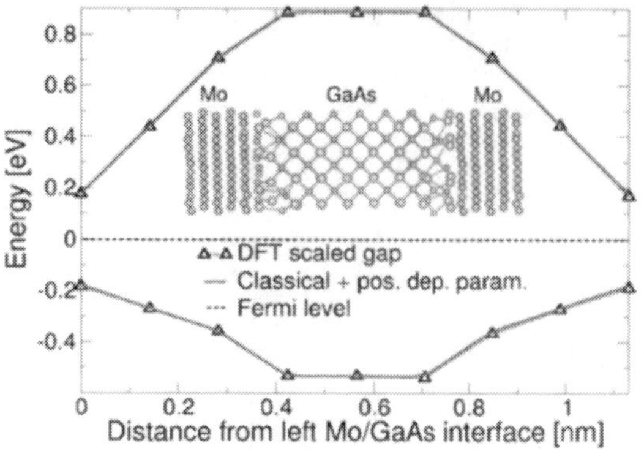

Figure 2. Conduction and valence band edges extracted from the DFT simulation (black diamond dotted line) for the position dependent parameters (continuous red line).

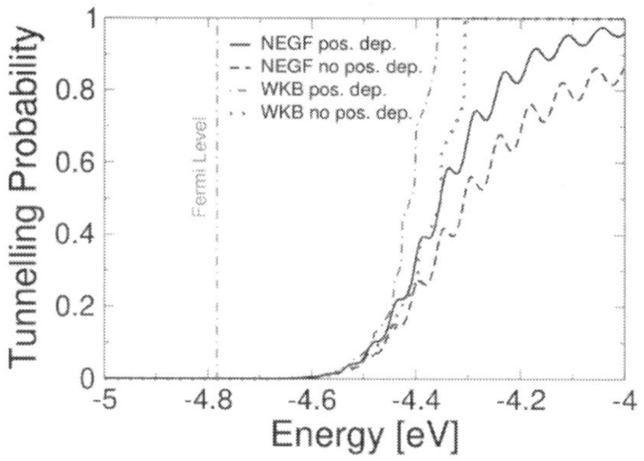

Figure 3. Tunnelling probabilities extracted from a 3D simulation of a Mo/n^+-GaAs interface for NEGF and WKB approximations.

probability at the interface as illustrated in Fig. 3. In addition, a small reduction in the mass of the electrons due to the local strain will further affects the probability. These changes in material characteristics of the interface impact the electrostatics through the self-consistent solution of 3D Poisson equation [4] which also takes into account the barrier lowering induced by the static image charge [14] as shown in Fig. 4.

2.3. Monte Carlo Simulated Transport Through Metal-Semiconductor Interface

To test our methodology, we have chosen a Mo/n^+-GaAs interface with a very high GaAs n-type doping of $5 \times 10^{19}\text{cm}^{-3}$. The high doping provides a very thin barrier which leads to the Ohmic behaviour of the contact. Such high doping can be fabricated using laser assisted annealing or rapid thermal annealing (RTA) [15]. The doping density of $4.5 \times 10^{19}\text{cm}^{-3}$ has been also reported in 30 nm gate length enhancement mode $In_{0.53}Ga_{0.47}As$ MOSFETs [16].

We simulate the contact resistivity using four different approaches to evaluate the impact of the two sources of the position dependent parameters: (i) no position dependent parameters, (ii) position dependence only for the electron effective mass, (iii) position dependence only for the conduction band and (iv) position dependences in both the electron effective mass and the conduction band. Note that the 3D ensemble Monte Carlo simulations were verified to provide the correct electron mobility in bulk materials. A contact resistivity of $2.1 \times 10^{-8}\Omega.\text{cm}^2$ was obtained using no position dependent parameters. The simulations with position dependent bands and masses has given a contact resistivity of $7.9 \times 10^{-10}\Omega.\text{cm}^2$, of $2.0 \times 10^{-8}\Omega.\text{cm}^2$ using only position dependent electron effective masses and, finally, of $4.7 \times 10^{-10}\Omega.\text{cm}^2$ using only position dependent bands.

3. Conclusions

In this work, we presented a hierarchical simulation model for predicting transport characteristics of metal-semiconductor interfaces. The electronic structure near the interface is calculated using DFT and the obtained variations of the conduction and valence band edges and the electron effective mass are then mapped into a semiclassical transport module within the 3D finite element self-consistent ensemble Monte Carlo technique enhanced with a tunnelling transport model. The resistivity of the Mo/n^+-GaAs contact is then investigated. We have found that the impact of the narrowing of the semiconductor band gap in the close vicinity of the interface

Journal of Physics: Conference Series **647** (2015) 012030 doi:10.1088/1742-6596/647/1/012030

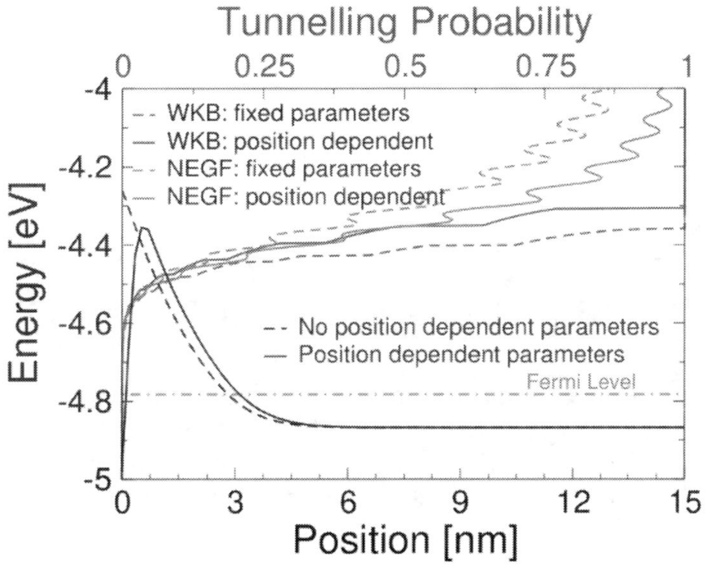

Figure 4. Conduction band and tunnelling probability obtained using position dependent (continuous line) and constant bulk material parameters (dashed line).

is a major reason for the modification of the contact resistivity. The band gap narrowing in semiconductor side of the interface reduces the resistivity by more than one order of magnitude: from $2.1\times10^{-8}\Omega.\text{cm}^2$ to $4.7\times10^{-10}\Omega.\text{cm}^2$. The position dependence of the electron effective mass is also playing an important role leading to a total resistivity of $7.9\times10^{-10}\Omega.\text{cm}^2$. Therefore, the fabrication process of a metal contact, which aims for minimal contact resistance should force the semiconductor band gap to narrow by inserting metal induced states not far below the conduction band or not far above the valence band.

This work was supported by the EPSRC grants EP/I010084/1, EP/I009973/1, and HECToR facility computer resource EPSRC grant EP/F067496. PVS was supported by the Royal Society.

References

[1] Truman J K and Holloway P H 1985 *J. Vac. Sci. Technol.* A **3** 992
[2] Singisetti U, Wistey M A, Zimmerman J D, Thibeault B J, Rodwell M J, Gossard A C, and Bank S R 2008 *Appl. Phy. Lett.* **93** 183502
[3] Jacoboni C and Lugli P 1989 The Monte Carlo Method for Semiconductor Device Simulation *Wien-New York: Springer-Verlag*
[4] Aldegunde M, Seoane N, García-Loureiro A J and Kalna K 2010 *Comput. Phys. Commun.* **181** 24
[5] Perdew J P, Burke K, and Ernzerhof M 1996 *Phys. Rev. Lett.* **77** 3865
[6] Blöchl P E 1994*Phys. Rev.* B **50** 17953
[7] Kresse G and Furthmiller 1996 *Phys. Rev.* B **54** 11169
[8] Dovesi R, Saunders V R, Roetti C, Orlando R, Zicovich-Wilson C M, Pascale F, Civalleri B, Doll K, Harrison N M, Bush I J, DArco P and Llunell M 2009 CRYSTAL09 User's Manual *University of Torino, Torino*
[9] Weigend F and Ahlrichs R 2005 *Phys. Chem. Chem. Phys.* **7** 3927
[10] Aldegunde M, Hepplestone S, Sushko P, and Kalna K 2014*Semicond. Sci. Technol.* **29** 054003 (7pp).
[11] Sun L, Liu X Y, Liu M, Du G and Han R Q 2003 *Semicond. Sci. Technol.* **18** 576
[12] Razavy M 2003 Quantum Theory of Tunneling *World Scientific*
[13] Rhoderick E H and Williams R H 1988 Metal-Semiconductor Contacts *Oxford University Press*
[14] Sze S M and Ng K K 2006 Physics of Semiconductor Devices *Wiley-Interscience*
[15] Kim S et al. 2013 *IEEE Trans. Nanotechnol.* **60** 2512
[16] Zhou X, Li Q, Tang C W, and Lau K M 2012 *IEDM Tech. Dig.* 773

Energy transfers in telecommunication-band region of (Sc,Er)$_2$ $_3$ thin films grown on Si(111)

S. Adachi[1], **Y. Kawakami**[1], **R. Kaji**[1], **T. Tawara**[2,3], **H. Omi**[2,3]

[1] Department of Applied Physics, Hokkaido University, Sapporo 060-8628, Japan
[2] NTT Basic Research Laboratories, NTT Corporation, Atsugi, Kanagawa 243-0198, Japan
[3] NTT Nanophotonics Center, NTT Corporation, Atsugi, Kanagawa 243-0198, Japan

E-mail: adachi-s@eng.hokudai.ac.jp

Abstract. The energy transfer dynamics in 1.5-μm region of $(Er_xSc_{1-x})_2O_3$ epitaxial thin films grown on Si(111) were investigated via temperature-dependence of the time-resolved photoluminescence (TR-PL). In order to exclude the dissipation via the cooperative up-conversion, the low power and resonant excitation of the third crystal-field level ($^4I_{13/2}$: Y'_3) of the Er^{3+} site with C_{3i} symmetry was employed. The TR-PL measurements of the Y'_1-Z'_1 transition have proved the existence of two decay components having fast (10–100 μs) and slow (0.1–1 ms) relaxation times in the range of 4–60 K. The model calculation including the temperature-sensitive and -insensitive non-radiative relaxations could explain well all the observed results. Moreover, the long averaged inter-Er-ion distance by decreasing Er density was found to be very effective to reduce both non-radiative relaxation rates.

1. Introduction

In the subject of the manipulations of the quantum states, the exciting activities are currently grown in different solid-state systems such as the semiconductor quantum dots [1] and the nitrogen-vacancy center in diamond [2] based on the interplay of the solid-state nature of the material engineering and the atomic-like character of the sharp discrete energy levels. The rare-earth-doped solids are alternative attractive candidates to exploit the coherent quantum phenomena because the intra-4f transitions in rare-earth ions are weakly perturbed by the crystalline environments, and they exhibit the resonance with a very narrow linewidth. In the various rare-earth-doped solids, we focus on Sc-doped erbium oxide ($(Er_xSc_{1-x})_2O_3$) [3, 4] since the Er ions (Er^{3+}) can interact with a telecomm-band photon ($\sim$1.54 μm) and can be grown on the Si(111) substrates. Therefore, the materials have a high possibility to be a platform for the coherent manipulations of the quantum information network using the existing fiber-optic network. However, for this purpose, the details of the energy transfers (ETs) have to be revealed.

2. Samples and Energy transfers in Er^{3+} levels

Er_2O_3 crystals have the lattice constant (a=10.54 Å) that is almost twice that of Si(111) (a=5.43 Å). $(Er_xSc_{1-x})_2O_3$ has the similar properties. Thus, the samples could be epitaxially grown on a Si(111) surface by MBE at 715°C, and were successful to be single-crystalline with very low density of defects [3]. Three $(Er_xSc_{1-x})_2O_3$ samples with different Er density were used in this work, and hereafter the samples with x=1.000, 0.054, and 0.012 (thickness$\sim$50 nm) are called

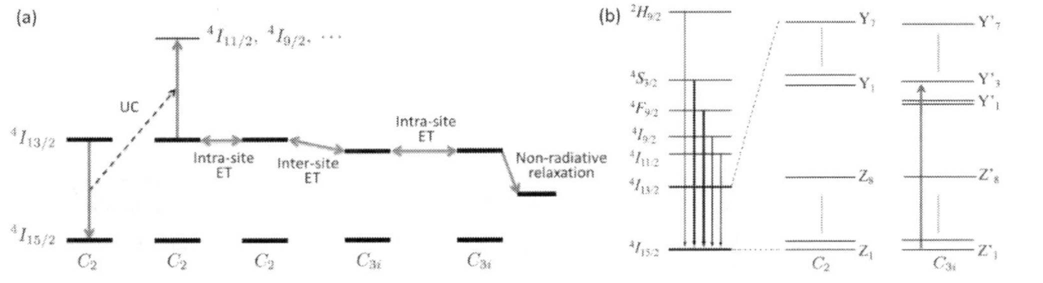

Figure 1. (a) Various ETs that are observed in the transition between the Stark levels of $^4I_{15/2}$ and $^4I_{13/2}$ in $(Er_xSc_{1-x})_2O_3$ crystal. (b) Energy level diagram of Er^{3+} in C_2 and C_{3i} symmetry sites (right) and UC-PL transitions (left).

as A, B, and C, respectively. The composition of the grown films was determined by Rutherford backscattering. Note that a large number of Er^{3+} ($\sim 3\times10^{20}$ cm^{-3}) are contained even in a sample C compared to a commercial Er-doped fiber amplifier material (typically $<10^{19}$ cm^{-3}). In this study, various types of ET shown in Fig. 1(a) are observed from one of the Stark levels in $^4I_{13/2}$ manifold of Er^{3+}, where the ETs between Er ions with same symmetry sites (intra-site ET) and with different symmetry sites (inter-site ET) are separated clearly. Er^{3+} has many optical transitions between $^4I_{13/2}$ manifold (Stark levels $Y_i^{(')}$, i =1-7, Fig. 1(b)) and $^4I_{15/2}$ (Stark levels $Z_j^{(')}$, j =1-8), which exhibits the photon emission and absorption at around 1.54 μm. The (forced) electric-dipole transitions between the Stark levels are allowed for Er^{3+} in C_2 symmetry sites ($Y_i \leftrightarrow Z_j$) and only magnetic dipole transitions are possible for Er^{3+} in C_{3i} sites ($Y'_i \leftrightarrow Z'_j$). Since Sc_2O_3 is transparent in visible to 1.5-μm region, the emission and absorption at $\sim$1.54 μm in $(Er_xSc_{1-x})_2O_3$ originate from Er^{3+}. In the experiments, the micro-PL setup with a streak camera (Hamamatsu Photonics C11293S) was used.

3. Results and Discussion

One of the important ETs is known as up-conversion process (UC in Fig. 1(a)) in the Er-doped materials. The UC-PL means the photon emission with shorter wavelength than the excitation one, and comes from the excited-state absorption and/or the cooperative UC (Auger process). [5] The UC process makes the ET dynamics complicated, and therefore avoiding the UC effect is preferable in this work. As an example of UC-PL, Fig. 2(a) summarizes the UC-PL spectra in the sample A at 4 K and Y'_3 resonant excitation with $P_{\rm exc}$=1 mW (upper panel) and $P_{\rm exc}$=30 mW (lower panel). While the observed UC-PLs were negligible at $P_{\rm exc}$=1 mW, the UC-PLs were significant in the high power excitation. Especially, the UC-PLs from $^4F_{9/2}$ and $^4S_{3/2}$ manifolds were much stronger than other UC-PLs and these UC-PLs become a severe population dissipation from a target state. Figure 2(b) shows the excitation power dependence of the integrated UC-PLs from $^4F_{9/2}$, $^4S_{3/2}$ and $^4I_{11/2}$ to the $^4I_{15/2}$, which indicates clearly that the appearance of the UC-PL needs a given critical excitation power and the UC-PLs become remarkable in high power excitation. The similar power dependence of the UC-PLs was observed in the samples B and C. Consequently, the influence of the UC-PL is considered to be negligible in the low power excitation region of 0.1-a few mW, and therefore, in the following experiments, all the measurements were carried out under $P_{\rm exc}$=1 mW. Considering that the observed UC-PL peaks correspond to the predicted transitions between Stark levels only in the C_2 site, the UC-PLs are found to emit only from Er^{3+} in C_2 site. This is supported also by the fact that the optical transitions between the higher-lying states and the ground states in C_{3i}

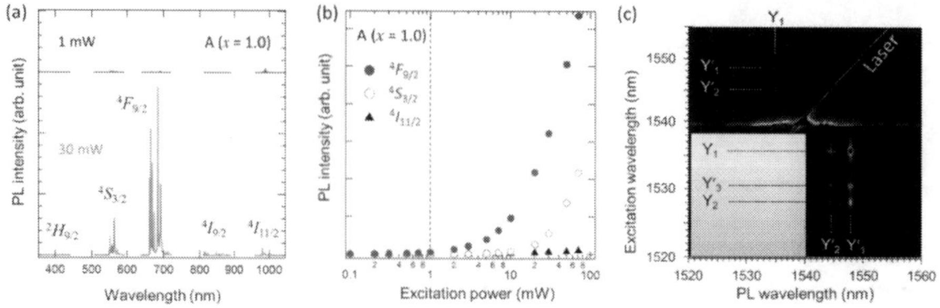

Figure 2. (a) The UC-PLs from various manifolds in the sample A with $P_{\rm exc}$=30 mW (lower panel, red) and $P_{\rm exc}$=1 mW (upper panel, black) at the Y'_3 resonant excitation ($\lambda_{\rm exc}$=1530.5 nm) and 4 K. (b) Excitation power dependence of the UC-PL intensities of three manifolds. (c) 2D plot of the PLE spectra of $^4I_{13/2}$ manifold in the sample A at 4 K with $P_{\rm exc}$=1 mW. The PLs below 1540 nm shown by a gray square are blocked by a low-pass filter. The central saturated region is due to the excitation laser stray.

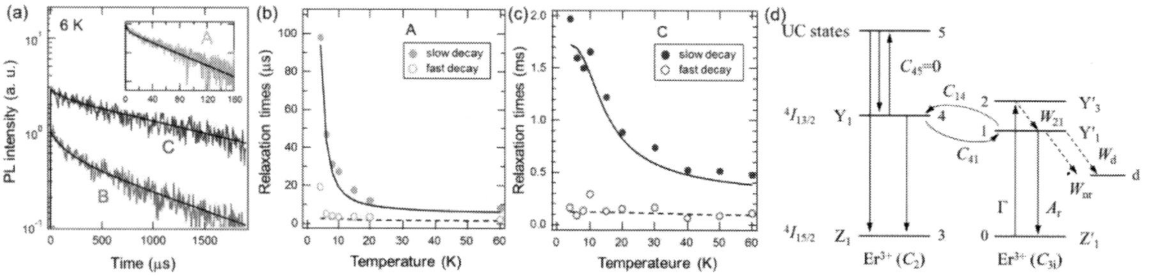

Figure 3. (a) TR-PL signals of the $Y'_1 - Z'_1$ transition at 6 K. The signal for the sample B is shifted down for easy view. (b), (c) Temperature dependence of the fast and slow relaxation components, $\tau_{\rm fast}$ and $\tau_{\rm slow}$ for the samples A and C. The solid curves in (a)-(c) are the calculated results based on the model shown in (d), but not the simple fitting curves.

site are forbidden. [6] Therefore, in the case of Y'_3 resonant excitation, the UC process occurs after the inter-site ET. Since the internal energy relaxation from Y'_3 to Y'_1 is considered to be very fast, the inter-site ET is supposed to occur between Y'_1 and Y_1. Even under the condition avoiding UC ($P_{\rm exc}$=1 mW) as shown in Fig. 2(c), other ETs were found the PL of the $Y_1 \rightarrow Z_1$ appears at the Y'_1 and Y'_2 resonances, and the strong PL of the Y'_1-Z'_1 transition are clearly seen at the Y_1, Y'_3, and Y_2 resonances. All the results suggest that the mutual inter-site ET can occur mainly between Y'_1 and Y_1 rather than between Y'_3 and Y_1. These findings were utilized to constract the model (Fig. 3(d)).

The inter-site and intra-site ETs avoiding the UC process can be reflected more directly to the TR-PL signals under the weak excitation condition ($P_{\rm exc}$=1 mW). The TR-PL measurements were carried out varying the temperature in the range of 4-60 K under the resonant excitation of Y'_3 level. The resonantly excited electrons to Y'_3 level relax quickly to the lowest energy level Y'_1 as mentioned above. Figure 3(a) shows the TR-PL signals of the $Y'_1 \rightarrow Z'_1$ transition in the three samples at 6 K in one instance. Though the signal decay in the sample A occurs in a different time scale, all the TR-PL signals for each sample and each temperature could be dissolved to two components by the double exponential fitting: fast decay and slow decay components with the relaxation times of τ_{fast} and τ_{slow}. Note that the solid lines in the figures are the calculated results based on the model (Fig. 3(d)) but not the simple double exponential fitting curves. Temperature dependences of τ_{fast} and τ_{slow} for the samples A and C are shown in Fig. 3(b) and (c). Both τ_{fast} and τ_{slow} are found to have the dependence on Er density. In addition, the τ_{fast} in all samples is insensitive to the temperature and is almost constant. Then the fast decay component is considered to be the temperature-insensitive non-radiative relaxation from Y_1' level due to the multi-phonon (MP) emissions (W_{d} in Fig. 3(d)). The temperature dependence of the MP relaxation time $\tau_{\text{MP}}(T)$ is expressed as $\tau_{\text{MP}}(T)/\tau_{\text{MP}}(0) = [1 - \exp(-E_{\text{p}}/k_{\text{B}}T)]^m$ in the single frequency model, where E_{p} is the phonon energy and m is the number of phonons. Since E_{p} is $\sim$74 meV in Sc_2O_3 [3], the MP relaxation for m=1$\sim$15 has no temperature dependence till 100 K according to the above equation. On the other hand, τ_{slow} shows the rapid reduction with increasing the sample temperature, particularly in the sample A. Since the radiative relaxation rate A_{r} is determined quantum-mechanically and is considered to be independent of the temperature, these results suggest strongly that the slow decay component includes the temperature-sensitive process such as a thermally-activation-type non-radiative relaxation $1/\tau_{\text{slow}} = A_{\text{r}} + W_{\text{nr}}^0 \exp(-E_{\text{A}}/k_{\text{B}}T)$, where W_{nr}^0 are the non-radiative relaxation rate at 0 K and E_{A} is the activation energy. From the results, A_{r} and W_{nr}^0 were estimated to be 3.92 and 122 ms^{-1}, 0.88 and 8.08 ms^{-1}, and 0.56 and 2.70 ms^{-1} for the sample A, B, and C, respectively. The large A_{r} for the sample A indicates the increase of the *effective* transition dipole moment due to the increase of the coherence volume. The smaller inter-Er^{3+} distance (3.3, 8.8, 14.5 Å for the sample A, B, and C) enhances the Er-Er interaction and induces the larger coherence volume, and therefore, it may lead to the larger effective dipole moment and the larger A_{r}.

To reproduce the TR-PL signals and their temperature dependences by the model calculation, the ET model shown in Fig. 3(d) is considered and the rate equations based on the model were solved. The calculated results indicated by the solid lines in Fig. 3(a), (b), and (c) agree well with the experimental results. The coincidence supports the validity of our model and assumption. From these analyzes, the temperature-insensitive non-radiative relaxation rate can be estimated to be 100, 3.3, and 1.4 ms^{-1} for the sample A, B, and C. It is found that the reduction of Er density is effective to suppress both non-radiative rates. This is because the longer inter-Er^{3+} distance suppresses the ETs in the same sites, and consequently the capturing events of the excited electrons by quenching centers are reduced.

4. Summary

We have investigated the ETs via the temperature dependence of the PL intensity and lifetime in $(Er,Sc)_2O_3$ epitaxial thin films grown on Si(111). The TR-PL measurements have proved the existence of two kinds of the decay component having fast and slow relaxation times in the range of 4–60 K. The model calculation introducing the inter-site ET, the temperature-sensitive and -insensitive non-radiative relaxations were found to explain all the experimental results. Thermal quenching could be suppressed greatly with decreasing Er density. The obtained results can contribute to realize the highly efficient optical devices based on Er-related materials for Si photonics and quantum-state manipulation devices in quantum information networks.

This work was financially supported by JSPS KAKENHI Grant Number 24360033.

References

[1] Greilich A, Carter S G, Kim D, Bracker A S, and Gammon D 2011 *Nature Photo.* **5** 702
[2] Fuchs G D, Burkard G, Klimov P V, and Awschalom D D 2011 *Nature Phys.* **7** 789
[3] Omi H and Tawara T 2012 *Jpn. J. Appl. Phys.* **51** 02BG07
[4] Tawara T, Omi H, Hozumi T, Kaji R, Adachi S, Goto H, and Sogawa T 2013 *Appl. Phys. Lett.* **102** 241918
[5] Pollnau M, Gamelin D R, Lüthi S R, Güdel H U, and Hehlen M P 2000 *Phys. Rev.* B **61** 3337
[6] Gruber J B, Nash K L, Sardar D K, Valiev U V, Ter-Gabrielyan N and Merkle L D 2008 *J. Appl. Phys.* **104** 023101

Nonequilibrium carrier dynamics in ultrathin Si-on-glass films

J Serafini[1], Y Akbas[1], L Crandall[2], R Bellman[3], C Kosik Williams[3] and Roman Sobolewski[1,2]

[1]Department of Physics and Astronomy and Laboratory for Laser Energetics University of Rochester, Rochester, NY, 14627-0231, USA

[2]Department of Electrical and Computer Engineering, University of Rochester, Rochester, NY, 14627-0231, USA

[3]Corning Incorporated, Corning, NY, 14831, USA

E-mail: roman.sobolewski@rochester.edu

Abstract. We present a femtosecond pump–probe spectroscopy approach for characterization of amorphous and microcrystalline silicon films grown on glass substrates. Such films are presently being considered as absorbers in tandem-type, Si-based photovoltaic cells. Our experiments consisted of time-resolved, two-colour femtosecond optical measurements, performed in the transmission mode in a wide range of delay times. Depending on the sample growth process, collected normalized transmissivity change ($\Delta T/T$) waveforms exhibited a bi-exponential relaxation dynamics with the characteristic times varying from picoseconds to nanoseconds. Experimental data were interpreted using a three-rate-equation models, and the relaxation was identified as, depending on the Si film type, being dominated by either carrier trapping or electron–phonon cooling and followed by electron-hole recombination. An excellent fit between the model and the $\Delta T/T$ transients was obtained and a correlation between the Si film growth process, its hydrogen content, and the associated trap concentration was demonstrated.

1. Introduction

The relaxation of optically excited carriers in amorphous microcrystalline silicon [(a-Si) and (mc-Si)] thin films has been part of many investigations that involved time-resolved optical measurements [1–4], driven by a better understanding of the carrier dynamics. It is also essential for determining the material's properties, which in this case are strongly influenced by a given deposition technique.

In this work, we demonstrate that a time-resolved, optical pump–probe technique provides an effective means for determining the details of the nonequilibrium carrier dynamics in ultrathin Si-on-glass films. We have collected a large family of normalized reflectivity change ($\Delta T/T$) waveforms of various a-Si and mc-Si films deposited under different conditions and have performed numerical fittings based on a general trap-dominated carrier-relaxation model to determine the optimum Si-deposition parameter space.

2. Sample fabrication and femtosecond optical pump–probe measurements

Two general sets of Si-on-glass samples were grown for this study. The first set ("**S**" samples) was deposited using a plasma-enhanced chemical-vapour-deposition (CVD) method and the fabricated films were primarily *p*-type SiCH and *n*-type a-Si films. Of notable interest, the H032811**S**1 sample was deposited with aCH_4/SiH_4 ratio of 0.85 and a BF_3/SiH_4 ratio of 1. The "**P**" samples were also deposited by the CVD method but with no dopants and with the increased hydrogen dilution, as compared to the growth of **S**-type samples, and were predominantly mc-Si films.

Femtosecond optical pump–probe spectroscopy is a unique investigation tool, extensively implemented to characterize nonequilibrium carrier dynamics in many condensed-matter systems, ranging from metals to various semiconducting materials [5]. In our case, for time-resolved characterization of carrier lifetime, we used a two-colour, pump–probe setup similar to that used earlier in our laboratory by Zhang *et al.* [6].

3. Experimental results and discussion

A very large family of $\Delta T/T$ waveforms was measured at room temperature. Each sample was tested at least three times at three different spots to average any film inhomogeneities. All collected transients had the same general shape and consisted of a pump-pulse–limited rising edge, followed by relaxation with a phenomenological, double-exponential decay. The initial fast relaxation time, varied in our samples from 2.4 ps (**S**-type) to ~80 ps (**P**-type) and could be ascribed to either direct trapping of hot, optically excited electrons or electron-optic–phonon cooling. The subsequent relatively slow relaxation could be interpreted as electron-hole recombination, and for the B032811**P**5 sample, the corresponding time was as long as ~1 ns. Although the above bi-exponential fitting analysis of the $\Delta T/T$ waveforms provides a direct indication of the desired growth conditions for maximizing the electron lifetime in Si photovoltaic absorbers, it does not give a physical insight into the actual carrier-relaxation dynamics.

Carrier-relaxation dynamics observed in our experiments are relatively simple to understand, keeping in mind that the 400-nm–wavelength (ultraviolet) pump excites electrons highly into the conduction band, creating a hot-electron population, while 800-nm–wavelength (infrared) probe photons, having an energy well below both the a-Si and mc-Si band gaps,[1] couple to photoexcited electrons through free-carrier absorption. For pump–probe measurements in the transmission mode, the experimental $\Delta T/T$ dependence in the thin-film limit can be expressed as:

$$\Delta T/T = -\Delta\alpha d, \tag{1}$$

where $\Delta\alpha$ is the photo induced absorption change and d is the film thickness. For the probe photons with energies below the band gap $\Delta\alpha$ does not depend on the details of the electronic excitation spectrum and is described by the Drude model, which relates it to the total free-carrier concentration N [7,8]:

$$\Delta\alpha = \frac{Ne^2}{\varepsilon_0 \varepsilon_b m^* \omega^2 \tau c}, \tag{2}$$

where ε_b is the material dielectric constant, m^* is the effective carrier mass, ω is the angular optical frequency, and τ is the momentum relaxation time. Combining equations (1) and (2), we see that in our experimental case, $\Delta T/T(t)$ is simply proportional to $N(t)$.

Recently, Jepsen *et al.* [9] have implemented a dynamic model that involves shallow trap sites in the band gap to reproduce the transient photoresponse of carriers in optically excited and THz-probed mc-Si wafers. Conceptually, a very similar model was used earlier to analyse femtosecond optical pump–probe studies of the transient photoresponse dynamics in low-temperature–grown GaAs films [5,10] that are known for their very large concentration of traps related to a distorted lattice and As precipitates.

Mathematically, the above scenario corresponds to the set of linear rate equations for time evolutions of concentrations of photoexcited electrons $N_{hot}(t)$, electrons at the bottom of the conduction band $N_{con}(t)$, and trapped electrons $N_{trap}(t)$, listed below:

---

[1]For our tested Si samples, the energy gap values varied from 1.65 to 1.82 eV.

Journal of Physics: Conference Series **647** (2015) 012032 doi:10.1088/1742-6596/647/1/012032

$$\frac{dN_{\text{hot}}}{dt} = I(t) - \frac{N_{\text{hot}} - N_{\text{con}}}{\tau_{\text{cool}}} - \frac{N_{\text{hot}}}{\tau_{\text{trap}}}, \qquad \frac{dN_{\text{con}}}{dt} = -\frac{N_{\text{con}} - N_{\text{hot}}}{\tau_{\text{cool}}} - \left(\frac{1}{\tau_{\text{rec}}} + \frac{1}{\tau_{\text{trap}}}\right) N_{\text{con}},$$

$$\frac{dN_{\text{trap}}}{dt} = -\frac{N_{\text{trap}} - N_{\text{trap,e}}}{\tau_{\text{rec}}} + \frac{N_{\text{con}} + N_{\text{hot}}}{\tau_{\text{trap}}}, \quad \text{and } \tau_{\text{trap}} = \frac{\tau_{\text{trap,max}}}{1 - \dfrac{N_{\text{trap}}}{N_{\text{trap,max}}}}, \tag{3}$$

where $I(t)$ is the optical pump-pulse intensity (which is assumed to have a Gaussian shape with the width of 100 fs) and $N_{\text{trap,e}}$ is the final concentration of electrons in the trap sites. We stress that τ_{trap} is, as in [9], a function of the number of available trap sites and their individual lifetimes, with the $N_{\text{trap,max}}$ and $\tau_{\text{trap,max}}$ representing the maximum concentration of trap sites in the sample and the longest trapping time, respectively.

We have solved equations (3) numerically and, as a result, obtained the total concentration of excited electrons in our system, given as

$$N(t) = N_{\text{hot}}(t) + N_{\text{con}}(t) - N_{\text{trap}}(t), \tag{4}$$

where the minus sign in front of N_{trap} reflects the fact that traps reduce concentration of free carriers.

Figure 1 [main panels] shows our experimental $\Delta T/T$ transients (red dots, left axis) and the corresponding simulated temporal evolutions of $N(t)$ (solid black line, right axis) and $N_{\text{trap}}(t)$ (solid red line, right axis) for **S**-type [figure 1(a)] and **P**-type [figure 1(b)] samples, representing the a-Si and mc-Si cases, respectively. In addition, we also present in figure 1 insets that show early stages of the time evolution of the components of $N(t)$, i.e., $N_{\text{hot}}(t)$ (solid blue line), $N_{\text{con}}(t)$ (solid green line), and $N_{\text{trap}}(t)$ (solid red line).

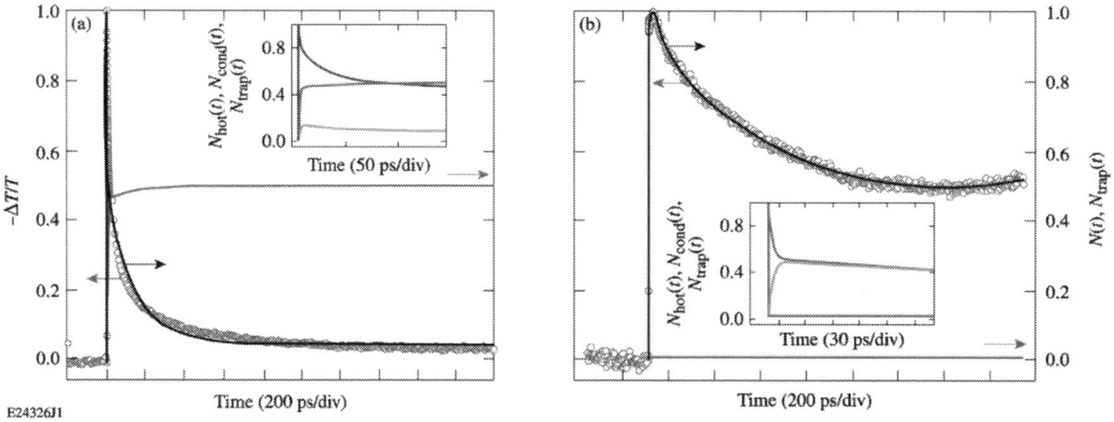

Figure 1. Experimental $\Delta T/T$ transients (red dots, left axis) of samples (a) H032811**S**1 and (b) B032811**P**5 and the corresponding $N(t)$ (solid black line, right axis) and $N_{\text{trap}}(t)$ (solid red line, right axis) simulation results. $N_{\text{trap}}(t)$ is normalized to $N(t)$. The insets present $N_{\text{hot}}(t)$ (solid blue line), $N_{\text{con}}(t)$ (solid green line), and $N_{\text{trap}}(t)$ (solid red line) components, corresponding in each case to the $N(t)$ presented in the main panel.

For an amorphous **S**-type sample [figure 1(a)], the concentration of trapped electrons is large and trapping at the very early stages of relaxation is very effective, resulting in rapid decay of the $\Delta T/T$

waveform or, equivalently, $N(t)$. The $N_{trap}(t)$ level reaches ~50% within the first 2 ps and remains approximately constant within our experimental window. Essentially a complete transfer of hot electrons to traps happens within ~60 ps [see inset in figure 1(a)] and the relaxation channel through the bottom of the conducting band represented by $N_{con}(t)$ is merely the background ($\tau_{cool} >> \tau_{trap}$). Traps remain occupied (N_{trap} almost constant) at the end of our ~2-ns-wide observation window; therefore, trap lifetimes for these a-Si samples are very long and might even exceed the pulse repetition time of our laser (~13 ns). In fact, in our experiments, traps are likely to be emptied by the next incoming ultraviolet (400-nm) pump pulse, contributing to the hot-electron population.

In contrast to **S**-type samples, for the B032811**P**5 sample [figure 1(b)], which represents the best mc-Si case, the density of trapped electrons can be, actually, neglected. As seen in the inset in figure 1(b), the hot-carrier relaxation is now completely dominated by the electron-optic–phonon cooling and the initial transfer to the bottom of the conduction time. With $N_{trap}(t)$ negligible, the final relaxation step is through the electron-hole recombination process and, indeed, in this film we observe the longest value of the experimental recombination time ($\tau_{trap} \approx 200$ ps).

4. Conclusions

Our femtosecond spectroscopy pump–probe experimental studies and subsequent simulations based on a general carrier relaxation in a photoexcited semiconducting materials model have proved that shallow trap sites in the band-tail states play a very important role in the relaxation of excited carriers in Si-on-glass samples intended for solar-cell absorbers. The observed relaxation dynamics in samples grown under different conditions reflects the trap-site densities and trapping lifetime, providing much-needed feedback to the control of wafer growth conditions. The latter is a crucial element for designing and optimizing new generations of Si-absorber–based photovoltaic solar cells with higher efficiencies.

Acknowledgment
This work is supported in part by the Corning, Inc., grant to the University of Rochester and by the New York State NYSTAR grant to the University of Rochester Center for Emerging and Innovative Sciences. Additional support came from the ARO Grant No. W911NF-12-2-0076. J. S. acknowledges support from the Frank Horton Graduate Fellowship Program at the University of Rochester's Laboratory for Laser Energetics, funded by the U.S. Department of Energy Office of Inertial Confinement Fusion under Cooperative Agreement No. DE-NA0001944 and the New York State Energy Research and Development Authority. The support of NYSTAR, ARO, and DOE does not constitute their endorsement of the views expressed in this article.

References
[1] Esser A, Seibert K, Kurz H, Parsons G N, Wang C, Davidson B N, Lucovsky G and Nemanich R J 1990 *Phys. Rev. B* **41** 2879
[2] Wraback M and Tauc J 1992 *Phys. Rev. Lett.* **69** 3682
[3] Fekete L, Němec H, Kadlec F, Kužel P, Stuchlík J, Fejfar A and Kočka J 2006 *J. Non-Cryst. Solids* **352** 2846
[4] Fekete L, Kužel P, Němec H, Kadlec F, Dejneka A, Stuchlík J and Fejfar A 2009 *Phys. Rev. B* **79** 115306
[5] See, e.g., Othonos A 1998 *J. Appl. Phys.* **83** 1789
[6] Zhang J, Belousov A, Karpiński J, Batlogg B, Wicks G and Sobolewski R 2011 *J. Appl. Phys.* **110** 113112
[7] Kim J 2009 *J. Korean Phys. Soc.* **55** 512
[8] Ganikhanov F, Burr K C, Hilton D J and Tang C L 1999 *Phys. Rev. B* **60** 8890
[9] Jepsen P U, Schairer W, Libon I H, Lemmer U, Hecker N E, Birkholz M, Lips K and Schall M 2001 *Appl. Phys. Lett.* **79** 1291
[10] Benjamin S D, Loka H S, Othonos A and Smith P W E 1996 *Appl. Phys. Lett.* **68** 2544

Steady-state and high-frequency electron transport in GaN nanowires

V. V. Korotyeyev[1], V.A. Kochelap[1], S.Vitusevich[2], V. Sydoruk[2] and L. Varani[3]

[1]Department of Theoretical Physics, Institute of Semiconductor Physics, Kyiv 03028, Ukraine
[2] Peter Grunberg Institute, Forschungszentrum Julich , Germany
[3]Institut d'Electronique du Sud, CNRS UMR 5214, University Montpellier 2, France

E-mail: `koroteev@ukr.net`

Abstract. The results of experimental and theoretical researches of steady-state and high-frequency electric characteristics of planar AlGaN/GaN heterostructure nanowires are presented. A strong depletion effect associated with electron trapping on the edge states is observed for narrow nanowires with widths $\sim 200 - 400$ nm. The spatial distributions of the electron concentration and electrostatic potential are calculated. At small applied voltages $(2 - 10$ V) for nanowires of length equal to 600 μm we found non-linear current-voltage characteristics related to a space-charge limited transport regime. We found that for narrow wires transversal transport is essentially ballistic: this dramatically modifies the high-frequency properties of the electron gas, in particularly, leading to a resonant absorption of THz radiation.

1. Introduction

Currently, different types of semiconductor nanowires (NW), for example, Si, $GaAs$, $InAs$, GaN, ZnO are intensively studied with the aims of nanoelectronic, optoelectronic, and sensor applications. In particular, they show promises as single photon detectors [1], low-power consumption nanoscale light-emitting diodes [2] and transistors [3]. The latter are widely used as sensors for various chemical and bio objects [4]. However, many fundamental physical questions about the internal electronic structure, effects of large surface, and size dependent transport phenomena remain unanswered up to now. Recent investigations of the electrical characteristics of different kinds of vertically-grown nanowires (whiskers) reveal common specific properties, in particular, the formation of depletion [5] or accumulation layers [6] at the NW interfaces and super-linear current voltage characteristics associated with the space-charge limited (SCL) transport regime [7]. Similar effects were found experimentally for planar heterostructure AlGaN/GaN NWs and will be discussed in this paper. Also, we will present an electrostatic model of the depletion layers and a kinetic theory of transversal high-frequency transport in NW.

2. Samples and experimental results

Our samples consist of long, spatially ordered ribbons (Fig.1(a)) with a 2D electron gas (2DEG). They were fabricated from AlGaN/GaN HEMT-heterostructures by Ar^+ ion beam etching (see details of fabrication in ref. [8]). Each set consists of 160 wires in parallel with length,

Journal of Physics: Conference Series **647** (2015) 012033 doi:10.1088/1742-6596/647/1/012033

L =620 μm with different widths, W, from 185 nm to 1110 nm for different NW sets. The initial characterization of the AlGaN/GaN quantum heterostructure (before NWs fabrication) gives a 2DEG concentration $n_0 = 2.2 \times 10^{12}$ cm^{-2}, which remains almost constant in the temperature range $T = 77-300$ K. The electron mobility, μ, varies from 4500 (77 K) to 1500 cm^2/Vs (300 K).

The dependencies of the sample conductance vs the geometrical width of the NWs, extracted from the slope of the low-field current-voltage (I-V) characteristics, are shown in Fig. 1(b). These

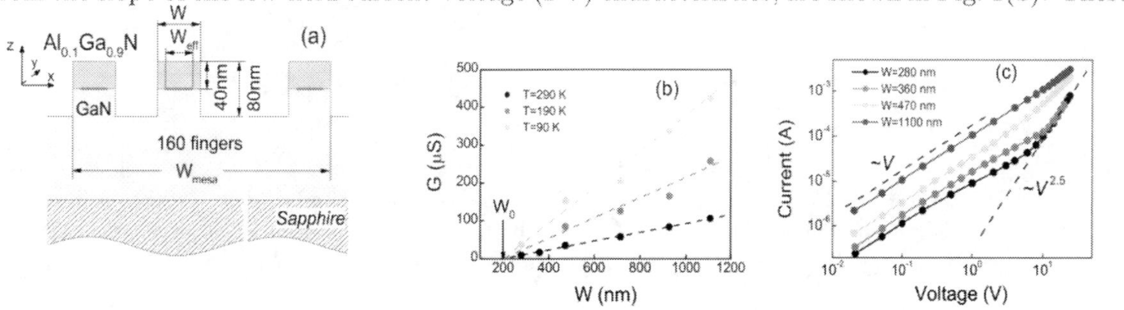

Figure 1. (a) Sketch of the vertical profile of the NWs structure. (b) Dependencies of the sample conductance vs the width of the NWs. (c) I-V characteristics of different NWs at T = 290 K.

dependencies are almost linear and simultaneously show a zero conductance at a critical width, $W_0 \sim$ 200 nm which does not depend on the temperature. The slopes of the fitting lines in Fig.1(b) agree with the measured mobility $\mu(T)$. Assuming that for the wide NWs the depletion effect is negligibly small, we can estimate the resistance of the sample as $R_0 = L/160en_0\mu W$, where e is the elementary charge. For example, for a sample with W = 1100 nm the estimation gives $R_0 \sim 2.2$ kΩ that almost coincides with the experimental value of 2.3 kΩ. This means that the contact resistance is much smaller than the resistance of the NWs.

The depletion effect, which is well-pronounced for the narrower NWs, is the result of the electron trapping on the edge states. Apparently, these states are quite deep because they are not activated by the temperature. The measured I-V characteristics of the different NWs (Fig.1(c)) show a linear behavior up to voltages near 1 V and a superlinear behavior at higher voltages, approaching a dependence of the type $I \sim U^{5/2}$. According to the theory of the SCL transport in the presence of charge traps [9], $I \sim V^{2+l}$ where the parameter $l > 0$. In our case $l \approx 0.5$, thus additionally proving the existence of sufficiently deep traps.

3. Electrostatic model of the depletion layers in a planar single nanowire

The effect of the electron trapping on the edge states induces a strong redistribution of the electron concentration across the NWs and gives rise to electrostatic fields both inside and around the NW. At equilibrium, the electrostatic model should include 3 equations: the 2D Poisson equation for the electrostatic potential φ, the relation between electron concentration and electrostatic potential and the charge conservation law:

$$\begin{array}{l} \frac{\partial^2\varphi}{\partial^2 x} + \frac{\partial^2\varphi}{\partial^2 z} = -\frac{4\pi e}{\kappa_0}\delta(z)\left[N_D - n(x) - N_s[\delta(x - W/2) + \delta(x + W/2)]\right] \\ n(x) = \frac{m^* k_B T}{\pi\hbar^2}\log\left[1 + \exp\left(\frac{\epsilon_F + e\varphi(x,0)}{k_B T}\right)\right] \quad N_D = 1/W \times \left(2N_s + \int_{-W/2}^{W/2} n(x)dx\right), \end{array} \tag{1}$$

where N_D and $n(x)$ are the 2D concentrations of the positive charges (ionized donors) and mobile electrons, respectively, N_s is the 1D concentration of the negative charge accumulated on an edge of the NW, κ_0, ϵ_F and m^* are the effective dielectric constant, Fermi level and electron effective mass, respectively. Using the Green function method [10], the system (1) is reduced to a non-linear integral equation that was solved numerically. The results of the calculations for different NWs are shown in Fig. 2 where, for the parameter N_s, we have used the values extracted from the conductance measurements of Fig. 1(b). Having the values of the estimated conductance $G_0 = 1/R_0$ and measured conductance, G, we can restore the percentage of trapped electrons, $\delta = (1 - G/G_0)/2$ and calculate the parameter $N_s = n_0 W \delta$. For the three considered samples, δ are equal to 0.35, 0.28 and 0.21 (for one edge).

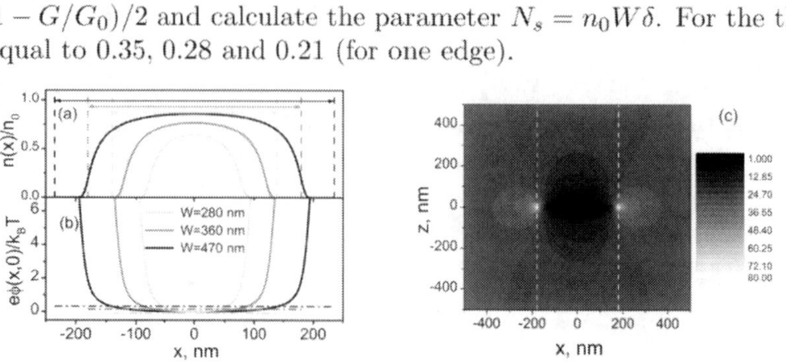

Figure 2. (a) and (b) Distributions of $n(x)$ and $\varphi(x, 0)$, respectively, at $T = 290$ K. The dash-dotted lines show the Fermi levels. (c) Distribution of the electrostatic energy around a NW with $W = 360$ nm. The vertical lines indicate the geometrical sizes of the NW.

As can be seen, wide depletion regions occur near the edges and the free electrons are mainly concentrated in the middle of the NW (see Fig. 2(a)). In turn, the self-consistent electrostatic potential (see Fig. 2(b)) sharply increases at the edges due to the repulsive action of the trapped electrons while it is almost flat in the middle due to the screening effect of free electrons. Such redistribution of the charges inside the NW forms a strongly inhomogeneous electrostatic field that spreads to distances of the order of 1 μm around the NW (see Fig. 2(c)).

4. High-frequency transport in the nanowires: geometrical resonance

The submicron transversal size of the considered NWs (comparable with the electron mean free path) makes them a good candidate for the observation of geometrical resonances (GR) that manifests themselves in the resonant absorption of electromagnetic waves polarized across NW. The physical reason of this resonance is a regular, oscillating dynamics of the electrons between the walls created by the confinement potentials, $\varphi(x, 0)$. The resonance frequencies and the shape of the resonance lines depend on the form of the built-in potential, temperature, electron mobility, etc. The GR was studied in the framework of the small-signal response, by solving the Boltzmann transport equation:

$$-i\omega \tilde{f}(\vec{p}, x) + \frac{p_x}{m^*}\frac{\partial \tilde{f}(\vec{p}, x)}{\partial x} - \frac{de\varphi(x, 0)}{dx}\frac{\partial \tilde{f}(\vec{p}, x)}{\partial p_x} - e\tilde{E}\frac{\partial f_0(\vec{p}, x)}{\partial p_x} = \hat{I}\{\tilde{f}\}, \tag{2}$$

where f_0 and $\tilde{f}$ are the steady-state (dc) and the high-frequency (ac) parts of the distribution functions, $\tilde{E}$ and ω are the amplitude and frequency of the ac signal, respectively. Analytical solutions of Eq.(2) were found in the limit of ballistic electron motion between the walls of confinement potentials ($\hat{I}\{\tilde{f}\} = 0$) by applying the method of characteristics and appropriate boundary conditions in the turning points (details of the solutions will be published elsewhere). Having the distribution function $\tilde{f}(\vec{p}, x)$ we calculated the ac current $\tilde{J}_\omega(x)$ in each point x and the specific ac mobility, defined as follows: $\langle \mu_\omega \rangle = 1/en_0\tilde{E}W \times \int_{-W/2}^{W/2} dx \tilde{J}_\omega(x)$. Here we consider the case of a rectangular potential that is approximately realized for planar NWs with charged edges. For a Boltzmann distribution f_0, the spectra of the real and imaginary parts of $\langle \mu_\omega \rangle$ are:

$$\begin{aligned} Re[\langle \mu_\omega \rangle] &= \mu_0 \frac{16}{\pi^{9/2}} \sum_{k=1}^{\infty} \frac{\Omega^2}{(2k-1)^5} \exp\left(-\frac{\Omega^2}{\pi^2 (2k-1)^2}\right) \\ Im[\langle \mu_\omega \rangle] &= \mu_0 \frac{1}{\Omega}\left[1 - \frac{4}{\sqrt{\pi}} \mathcal{P} \int_0^{\infty} \rho^2 \exp(-\rho^2) \frac{\tan(\Omega/2\rho)}{\Omega/2\rho} d\rho\right], \end{aligned} \tag{3}$$

where $\mu_0 = eW/m^* u_T$, $\Omega = \omega W/u_T$, u_T is the thermal velocity, the symbol $\mathcal{P}$ denotes Cauchy's principal values of the integral. The real part of $\langle \mu_\omega \rangle$ that is responsible for the absorption of an *ac* signal, has a maximum at frequency $\omega/2\pi = u_T/2W$ which corresponds to the electron oscillation period with thermal velocity between the walls of the potential well. The imaginary part of $\langle\mu_\omega\rangle$ changes its sign when real part reaches the maximum. These behaviors are illustrated in Fig. 3. For GaN NWs with $W = 50 - 200$ nm, effects of the GR can be observed

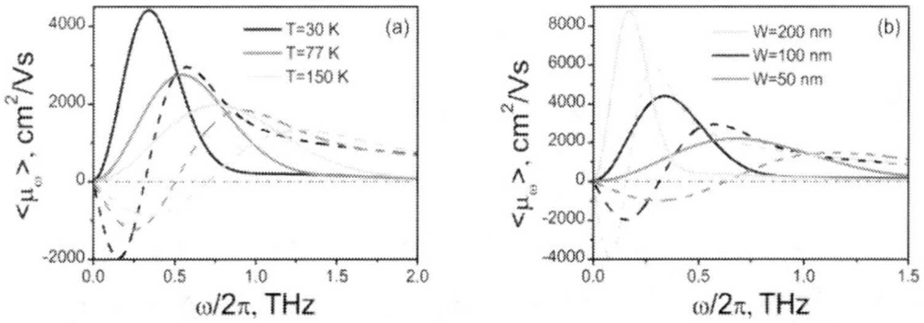

Figure 3. Spectra of $\langle\mu_\omega\rangle$. Solid and dashed lines correspond to the real and imaginary parts of $\langle\mu_\omega\rangle$, respectively. (a) Spectra for $W = 100$ nm and different temperatures, (b) spectra at $T = 30$ K for different widths.

in the sub-THz frequency range. With increasing temperature (see Fig. 3(a)) or decreasing NW width (see Fig. 3(b)), the resonance frequency is shifted to higher frequencies together with a broadening of the absorption line. It is important to stress that the resonance absorption arises for ballistic (collisionless) electrons which execute a regular, oscillating motion in the potential well. Obviously, the scattering processes will destroy the regular electron motion that will lead to the additional broadening of GR resonance. When the characteristic sizes of confinement potential will be much greater than electron mean free path, the GR resonance vanishes and electron response on *ac* perturbation will be described by standard Drude-Lorentz model. It should be noted, that GR can be experimentally studied using the propagation loss spectroscopy of the single NW embedded to the THz-waveguide or THz- transmission lines [11].

In summary, we have found that the existence of edge traps in planar AlGaN/GaN NWs leads to the formation of wide depletion regions, large electrostatic fields outside the samples and strongly non-linear current-voltage characteristics. We have also found that the transversal ballistic transport that may take place in narrow wires dramatically modifies the high-frequency properties of the wire structures, leading to new resonance effects. These phenomena could also be important for devices based on NWs: indeed, similar effects have been reported for self-switching GaN diodes [12]. We suggest that the large electrostatic fields outside the NWs can be useful for bio-sensing applications, particularly, for mobilization and separation of ions, polar molecules and cells.

Acknowledgments

Authors thanks to Federal Ministry of Education and Research (Germany), BMBF Project 01DK13016 and CNRS (France)- NASU (Ukraine) Project 26206 for support this work.

References

[1] Fujiwara A, Yamazaki K and Takahashi Y 2002 *Appl. Phys. Lett.* **80** 4567-4569
[2] Tchernycheva M et all 2014 *Nano Lett.* **14** 3515-3520
[3] Lu W, Xie P and Lieber Ch M 2013 *IEEE Trans. Electron Dev.* **55**(11) 2859-2876
[4] Ramgir N S, Yang Y and Zacharias M 2010 *small* **6** (16) 1705-1722.
[5] Simpkins B S, Mastro M A, Eddy Jr C R and Pehrsson P E 2008 *J. Appl. Phys.* **103** 104313
[6] Zhou X, Dayeh S A, Aplin D, Wang D and Yu E T 2006 *Appl. Phys. Lett.* **89** 053113
[7] Gu Y and Lauhon L J 2006 *Appl. Phys. Lett.* **89** 143102
[8] Lehnen P, Schapers T, Kaluza N, Thillosen N and Hardtdegen H 2007 *Phys. Rev. B* **76** 205307
[9] Chia A and LaPierre R R 2012 *J. Appl. Phys.* **112**, 063705
[10] Grinberg A A, Luryi S, Pinto M R, Schryer N L 1989 *IEEE Trans. Electron Dev.* **36**(6) 1162-1170
[11] Kadoya Y et al. 2008 *Radioengineering* **17**(2) 48-55
[12] Jeremie Torres et al. 2013 *Semicond. Sci. Technol.* **28** 125024

Journal of Physics: Conference Series **647** (2015) 012034 doi:10.1088/1742-6596/647/1/012034

Stepped Current-Voltage Relation and THz Oscillations in GaN MOSFET Due to Optical Phonon Emission: Monte Carlo simulation

V. Gružinskis[1], P. Shiktorov[1], and E. Starikov[1], H. Marinchio[2], C. Palermo[2], J. Torres[2], L. Varani[2]*

[1]Semiconductor Physics Institute, Center for Sciences and Technology,A. Goštauto 11, LT-01108 Vilnius, Lithuania

[2]Institut d'Electronique et des Systèmes (CC 05005), Université de Montpellier - Campus St Priest, 860 rue de St Priest, 34095 Montpellier cedex 5, France

*Email: luca.varani@umontpellier.fr

Abstract. Electron transport and drain current noise in wurtzite GaN MOSFET have been studied by Monte Carlo particle simulation that simultaneously solves the Boltzmann transport and pseudo-2D Poisson equations. It is shown that at positive gate voltages giving excess electron concentration in the *n*-region of the channel, drain current self-oscillations in the THz frequency range up to 6 THz are possible. These self-oscillations are driven by an electron plasma instability. Additionally a step-like drain current dependence on drain bias is demonstrated.

1. Introduction

In recent years terahertz (THz) technology gained great interest due to its potential applications in radio astronomy, remote sensing, commercial imaging, biomedicine and other. One of the most critical elements in modern THz technology is a generator of THz radiation. Semiconductor emitters are of great interest due to their compact sizes and possibilities to be integrated in a single chip. At the moment, one of the most perspective trends of THz sources development is related to plasma phenomena in field effect transistors. The aim of this report is to investigate the possibility of realization of plasma instabilities due to optical phonon emission already proposed some years ago in diodes, where the generation frequency has been found to be less than 1 THz [1,2]. To achieve higher frequencies it is necessary to increase the carrier concentration because the plasma oscillation frequency is proportional to the square root of the concentration. However, at higher doping levels the plasma instability is suppressed by impurity scattering. The advantage of MOSFET as compared to modern Schottky-barrier transistors (e.g. HEMT) is that the electron concentration under the gate in the channel can be considerably increased by a positive (not blocking) gate bias without changing the channel doping.

2. Model

We shall present the results of calculations of electron transport in MOSFETs based on a 20-200-20 nm n^+nn^+ GaN channel with a width of 50 nm. A 100 nm gate is centered in the n-region. The doping levels are n = 10^{15} cm^{-3} and n$^+$ = 4×10^{18} cm^{-3}. Calculations are performed by a simultaneous solution of Boltzmann and pseudo-2D Poisson equations [3] by means of the Monte-Carlo particle technique. A spherically symmetric nonparabolic wurtzite GaN conduction band [4] is assumed. All the results correspond to a lattice temperature T = 80 K if not indicated. The number of simulated particles, depending on the case, is varied from 10^5 to 5×10^5. The time step in all cases is 1 fs.

3. Numerical results

The current-voltage (I-V) relations calculated at liquid nitrogen temperature for different gate biases in GaN MOSFET are presented in Fig. 1. At gate bias U_g = 0 V the I-V relation exhibits the typical FET behavior. At increasing gate biases the electron concentration under the gate is growing and the electric field under the gate approaches the zero value. At sufficiently high values of U_g the full drain voltage U_d is applied to the un-gated regions.

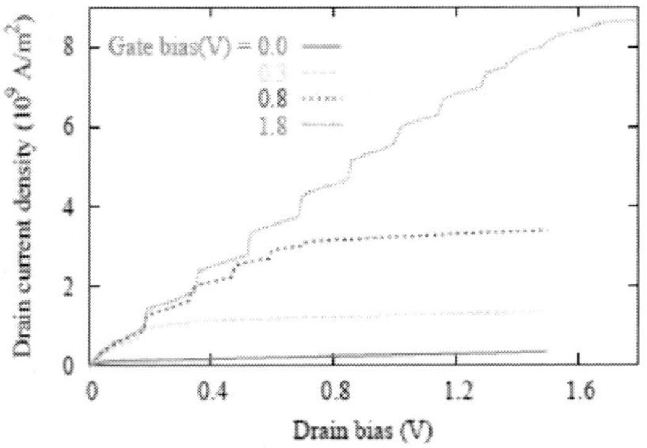

Figure 1. GaN MOSFET I-V relations.

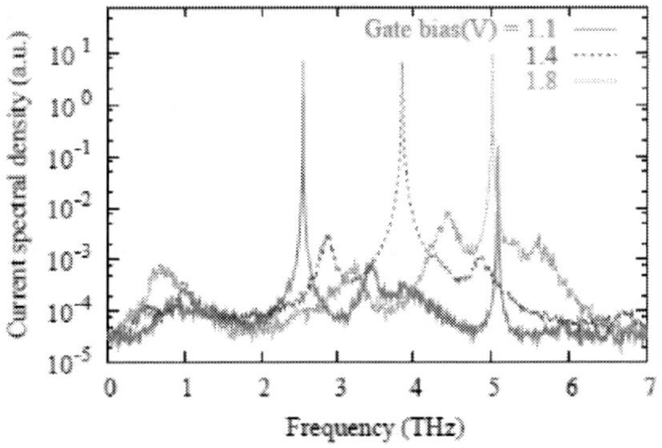

Figure 2. Drain current oscillation spectra at a drain bias 0.33 V and gate biases 1.1, 1.4, and 1.8 V.

The optical phonon emission in the un-gated regions leads to a free electron grating formation in these regions [1,2]. The number of grating spatial periods in the un-gated regions increases with U_d. The formation of such spatial periods produces evident current steps in the I-V relation.

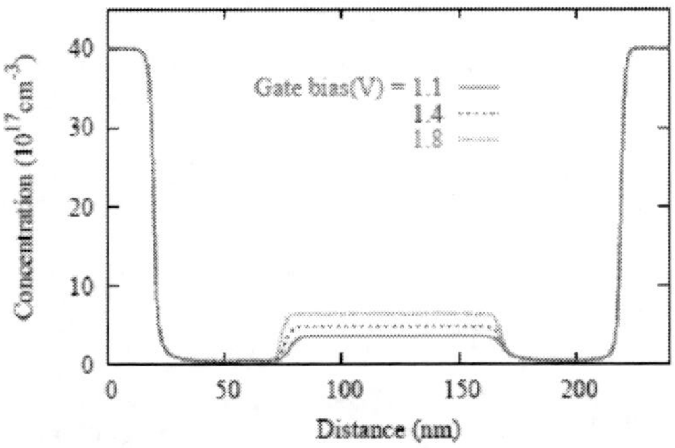

Figure 3. Concentration profiles in the channel at different gate biases: 1.1, 1.4, and 1.8 V.

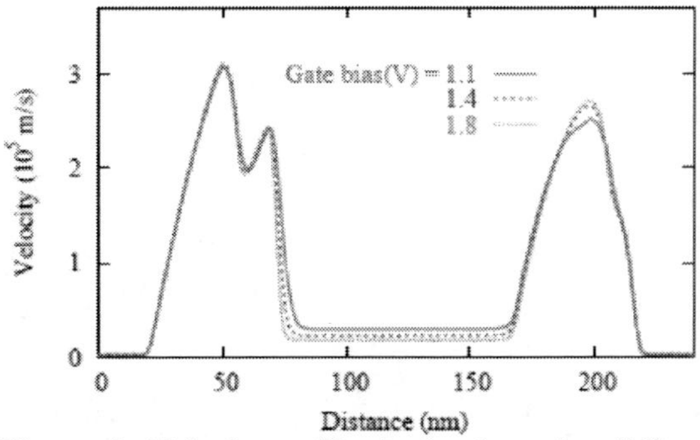

Figure 4. Velocity profiles in the channel at different gate biases: 1.1, 1.4, and 1.8 V

The formation of these current steps stops when the drain current flow at increased U_d washes out the excess electron concentration under the gate. Moreover, the simulation shows that the drain current oscillates at the drain contact just before the current step in the I-V relation: the corresponding oscillation frequency can reach a value of about 5 THz. This is demonstrated in Fig. 2 where the drain current self-oscillation spectra are calculated for the gate biases 1.1, 1.4, and 1.8 V and U_d = 0.33 V. These bias points are situated just before the second step in all three I-V relations (see Fig. 1). The drain current self-oscillation frequency strongly depends on gate bias: indeed, the frequencies are 2.5,

3.9, and 5.0 THz at U_g = 1.1, 1.4, and 1.8 V, respectively. The electron concentration and velocity profiles in the channel are shown in Fig. 3 and Fig. 4, respectively (here the reported values are averaged over more than a hundred drain current oscillation periods). One can observe an increase of the electron concentration (Fig. 3) and a decrease of the electron velocity (Fig. 4) under the gate at increasing U_g. This concentration and velocity behaviours fully exclude the classical time-of-flight mechanism under the gate as a possible explanation of the drain current self-oscillations: rather they are due to a streaming plasma instability where the self-oscillation frequency depends mainly on the electron concentration under the gate. Below we show the evolutions of various quantities in the GaN MOSFET channel in the presence of drain current self-oscillations. Fig. 5 shows one drain current oscillation period at U_d = 0.33 V and U_g = 1.8 V. The sampling points show the time moments at which the different profiles are calculated.

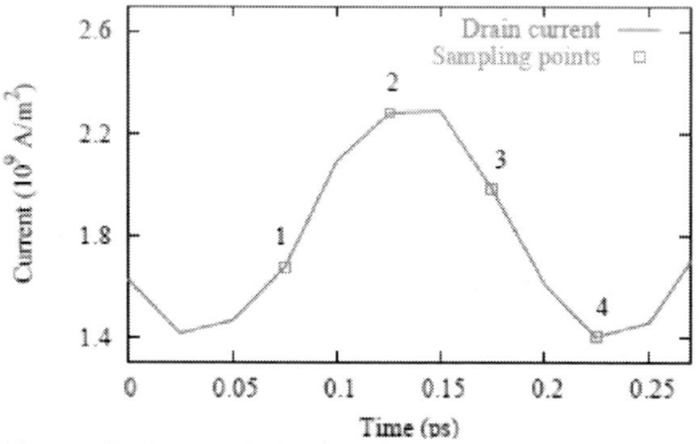

Figure 5. One period of drain current oscillations at U_d = 0.33 V and U_g = 1.8 V. Sampling points indicate the time moments for various quantity calculation. The time interval between the points is of 50 fs.

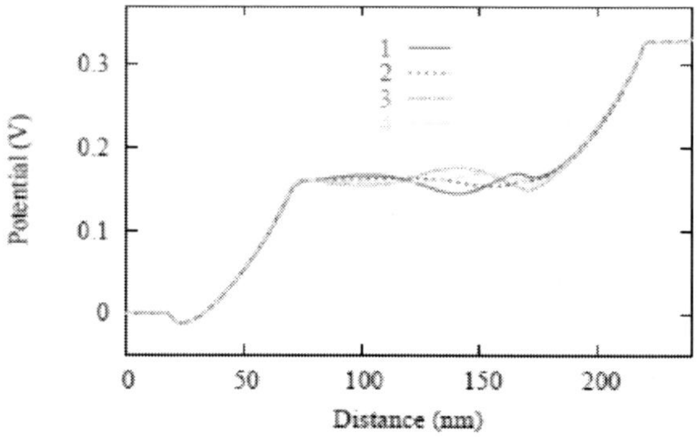

Figure 6. Potential profiles in the channel at time moments which are represented by sampling points from 1 to 4 and biases shown in Fig. 5.

The potential profile evolution in the channel of GaN MOSFET at U_d = 0.33V and U_g =1.8 V is presented in Fig. 6: such behaviour is typical for a standing space charge wave under the gate.

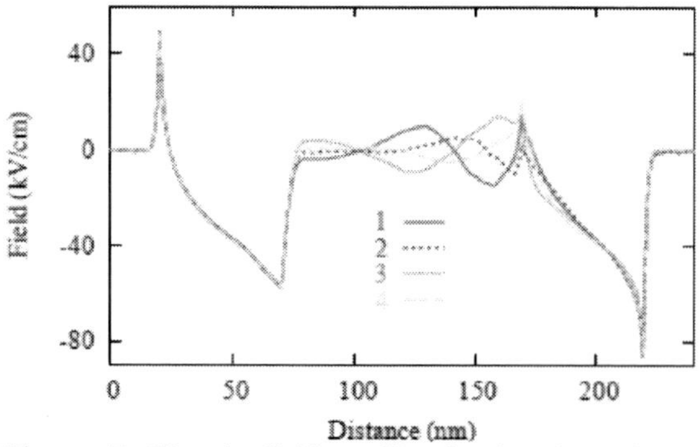

Figure 7. Electric field profiles in the channel at time moments represented by the sampling points from 1 to 4 and biases shown in Fig. 5.

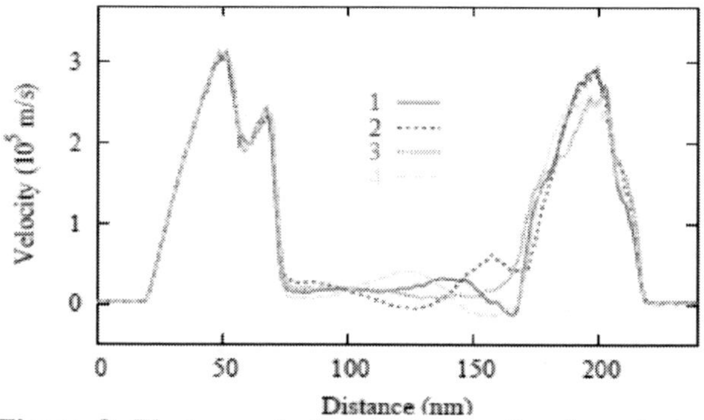

Figure 8. Electron velocity profiles in the channel at time moments represented by the sampling points from 1 to 4 and biases shown in Fig. 5

The electric field and electron velocity profiles are shown in Figs. 7 and 8, respectively: their evolutions confirm the presence of a standing space charge wave under the gate while the electron velocity evolution (Fig. 8) in the un-gated region near the drain is typical for a traveling space charge wave (e.g. accumulation layer or Gunn domain). Also Figs. 6, 7, and 8 show that no space charge wave is formed in the un-gated region near the source. Since most of the electrons are concentrated under the gate (see Fig. 3), the current oscillation amplitude and frequency are governed by the standing space charge wave while the flight time effects play a minor role. To evidence the electron stream in the channel under the gate, we report in Fig. 9 the electron distribution in the velocity-coordinate plane corresponding to the oscillating drain current maximum (sampling point 2 in Fig. 5).

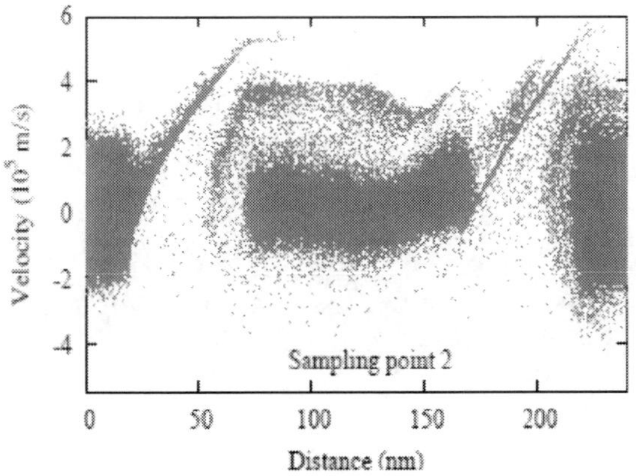

Figure 9. Electron distribution in the velocity-coordinate plane at time moment 2 and biases shown in Fig. 5. Each point represents a simulated electron.

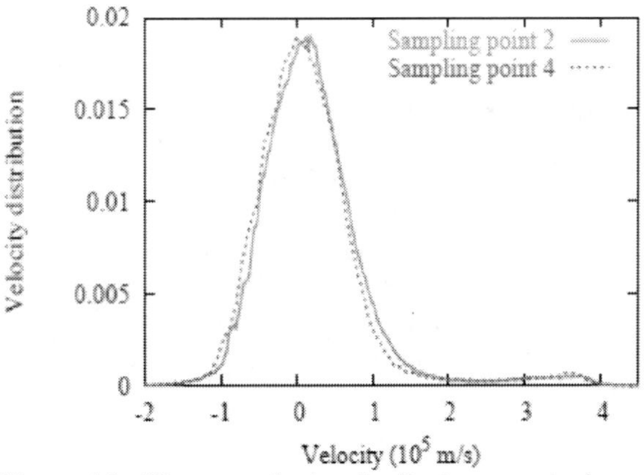

Figure 10. Electron velocity distribution vs velocity at time moments 2 and 4 for the biases shown in Fig. 5.

The electron dynamics in Fig. 9 can be explained as follows: most of the electrons entering from the source are simultaneously accelerated by the electric field in the un-gated region until they reach the optical phonon energy where they loose their velocity and energy after an optical phonon emission. Then these electrons are accelerated again until they reach the gated region of the channel where the time-averaged electric field is zero and they try to proceed ballistically in this region. However the ballistic motion of this fast electrons stream in the channel under the gate is impossible because of the presence of a high concentration of thermal electrons. In this case the well-known streaming plasma instability may arise, i.e. when a fast electron stream in the presence of vanishing external electric field penetrates a region with high concentration of thermal electrons. In Fig. 9 both the fast electron stream and the thermal electrons cloud are clearly visible. The velocity distributions of electrons in the channel under the gate are shown in Fig. 10. One can see the high concentration of thermal electrons around a zero velocity value and the low concentration of fast electrons with velocity around 3.5×10^5 m/s. During the instability the thermal electron cloud distribution corresponding to the drain current minimum (sampling point 4 in Fig. 10) is drifted to higher velocities at the drain current maximum (sampling point 2 in Fig. 10). This behaviour gives the main contribution to the drain current oscillations shown in Fig. 5. We have verified that these oscillations persist up to 120 K. Also it should be noted that the repeated simultaneous optical phonon emissions in the un-gated region lead to the formation of a free electrons grating that is at the origin of the stepped drain current I-V relation. We remark that slight steps in the I-V relation can be observed even at room temperature.

Acknowledgment
This work is supported by grant No. MIP-058/2013 of the Research Council of Lithuania.

References

[1] Gružinskis V, Shiktorov P, Starikov E, Reggiani L, Varani L and Vaissiere J C *Semicond. Sci. Technol.* 2004 **19** S173.
[2] Gružinskis V, Starikov E and Shiktorov P *J. Comput. Electron.* 2007 **6** 11.
[3] Marinchio H, Palermo C, Sabatini G, Varani L, Shiktorov P, Starikov E and Gružinskis V *J. Comput. Electron.* 2010 **9** 141
[4] Foutz B, O'Leary S, Shur M and Eastman L *J. Appl. Phys.* 1999 **85** 7727.

Journal of Physics: Conference Series **647** (2015) 012035 doi:10.1088/1742-6596/647/1/012035

Hybrid plasma modes in transistors: linear and non-linear responses

H. Marinchio[1], **V. Korotyeyev**[2], **C. Palermo**[1], **and L. Varani**[1]

[1] Institute of Electronics and Systems, UMR CNRS 5214, University of Montpellier, France
[2] Institute of Semiconductor Physics, Kiev 03650, Ukraine

E-mail: hugues.marinchio@ies.univ-montp2.fr

Abstract. We present an analytical model based on hydrodynamic equations and a pseudo-two-dimensional Poisson equation to study the response of a nanometric field-effect transistor channel in the THz domain. This model allows to study different kinds of external excitations of plasma modes and different geometries. We calculate the two first-order responses of the drain voltage or current, which are of peculiar interest in the perspective of THz wave generation and detection and THz electronics. Even at room temperature, each responses present resonances at the eigenfrequencies of the hybrid plasma modes sustained in the channel.

1. Introduction

The electron collective oscillations phenomena in solid-state devices received a close attention since the 90's and the theoretical works of Dyakonov and Shur [1, 2] which predict the instability, or at least the resonant behavior, of a gated two-dimensional (2D) electron gas constituting the channel of a field-effect transistor (FET). Such instabilities/resonances, occurring in the terahertz (THz) range for nanometric devices, can be used in view of the development of solid-state emitters or detectors of THz radiation. These possibilities have been confirmed experimentally by measurements of plasma instabilities at low temperature [3, 4] and plasma resonances to external excitations (THz wave or THz optical beating) at room-temperature [5].

Since then, analytical models have been developed to describe the stimulation of plasma modes in FETs and these are usually limited to strictly 2D plasma because they are based on the gradual channel approximation [2, 6, 7] which neglects the effect of the longitudinal electric field in the Poisson equation. We recently proposed an original analytical model based on hydrodynamic equations and a pseudo-2D (P2D) Poisson equation to overtake these limitations [8]. This model allows to study the different kinds of external excitations of plasma modes (THz, optical or both) and different geometries. In particular, we demonstrate the dispersivity of the plasma in the FET channel, the electron gas shifting progressively from a 2D plasma dynamic behavior at low frequency to a 3D plasma at high-frequency, that is why we introduced the concept of *hybrid* plasma modes [8].

Here we present the calculation of current and voltage responses for different boundary conditions and geometries. We will focus on the first-order harmonic response, interesting in the perspective of THz emission or amplification, the (second-order) average and second-harmonic responses, of interest in terms of THz detection or rectification and THz frequency multiplication, respectively.

2. Theoretical model

Our analytical approach consists in coupling the 3D carrier density n and mean velocity v conservation equations with the P2D Poisson equation [9]:

$$\begin{cases} \dfrac{\partial n}{\partial t} = -\dfrac{\partial (nv)}{\partial x} + G \\ \dfrac{\partial v}{\partial t} = -v\dfrac{\partial v}{\partial x} + \dfrac{q}{m}\dfrac{\partial V}{\partial x} - v\nu \\ \varepsilon_c \dfrac{\partial^2 V}{\partial x^2} + \varepsilon_s \dfrac{V_g - V_{th} - V}{d\delta} = qn \end{cases} \tag{1}$$

where n is the electron volumic density. d is the effective gate-to-channel distance and δ the channel thickness. d is related to the gate-to channel capacitance $C_{gc} = \varepsilon_s L/d$ with L the gate length and ε_s the permittivity of the insulator. ε_c is the dielectric constant, ν the velocity relaxation rate and m the electron effective mass which will be taken equal to their typic values in InGaAs at room-temperature [8]. V_g and V_{th} are the gate and threshold voltages, respectively. A sketch of the device is proposed in Fig. 1.

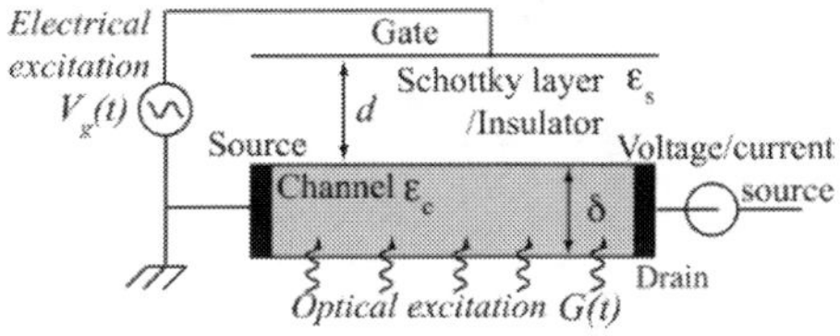

Figure 1. Pseudo-two dimensional representation of a FET.

Two kinds of THz excitations can be treated within this framework: (i) an electrical stimulation on the gate described by the harmonic part of the gate voltage $V_g(t) = V_{g0} + V_{g1}\cos(\omega t)$ and (ii) an optical beating excitation through the generation coefficient $G(t) = G_0 + G_1 \cos(\omega t)$.

As in Ref. [2], we assume that the excitations are sufficiently weak to only have to look for solutions of the form: $u(x,t) = u_0 + \delta u_1(x,t) + \delta\overline{u}(x) + \delta u_2(x,t)$, with $u = n, v, V$ or j (volumic current density defined below), where u_0 is the value in the absence of excitation, $\delta u_1(x,t)$ the first-order solution at the same pulsation ω than the stimulation, $\delta\overline{u}$ and $\delta u_2(x,t)$ are the average and second harmonic components. We introduce the complex quantities $u_1(x)$ and $u_2(x)$ verifying $\delta u_1(x,t) = \Re[u_1(x)\exp(\mathrm{i}\omega t)]$ and $\delta u_2(x,t) = \Re[u_2(x)\exp(\mathrm{i}2\omega t)]$. We additionally assume that the device is not or weakly biased implying that $v_0 = 0$, $V_0 = V_s = 0$ and other u_0 quantities are uniform.

The volumic current density can be expressed summing its conduction and displacement contributions: $j(x,t) = -qnv - \varepsilon_c \frac{\partial^2 V}{\partial t \partial x}$.

3. Linear response

Rewriting the system (1) at first-order, we obtain:

$$\mathrm{i}\omega n_1 = -n_0 v_1' + G_1 \; ; \; \mathrm{i}\omega v_1 = \frac{q}{m} V_1' - \nu v_1 \; ; \; \varepsilon_c V_1'' + \varepsilon_s k_t^2 (V_{g1} - V_1) = q n_1 \tag{2}$$

where $k_t^2 = 1/d\delta$ and the prime notation stands for the derivative with respect to x.

Then, combing Eqs. (2) we get:

$$\delta V_1'' + \beta(\omega)^2 \delta V_1 = \beta(\omega)^2 e_1 \text{ where } \beta(\omega)^2 = \frac{\epsilon_s {k_t}^2 \left(\omega^2 - i\nu\,\omega\right)}{\epsilon_c \left(\omega_{3D}^2 - \omega^2 + i\nu\,\omega\right)} \text{ and } e_1 = V_{g1} - \frac{q}{\mathrm{i}\omega\varepsilon_s k_t^2} G_1. \tag{3}$$

$\omega_{3D} = \sqrt{\frac{q^2 n_0}{m\varepsilon_s}}$ is the bulk plasma frequency of the material constituting the channel.

The solution has the form: $V_1 = A\cos(\beta x) + B\sin(\beta x) + e_1$ where A and B depend on the chosen boundaries conditions (BC). We consider two cases: i) asymmetrical BC, i.e. the drain current is fixed ($j(L) = 0$); ii) symmetrical BC, i.e. $V(L) = 0$. In the former case, the quantity of interest is the amplitude of the drain-to-source voltage potential $V_{ds1} = |V_1(L)|$ while in the latter case, it is the drain current amplitude $j_{d1} = |j_1(L)|$. Both quantities are plotted as functions of the stimulation frequency f on Fig. 2.

The sharp resonances can be associated with the excitation of *hybrid plasma modes*. Asymmetrical BC allow the stimulation of odd modes and symmetrical ones the sustainment of even modes. In both cases, the resonances frequencies f_p cannot overreach the 3D plasma frequency [see Ref.[8] for further details].

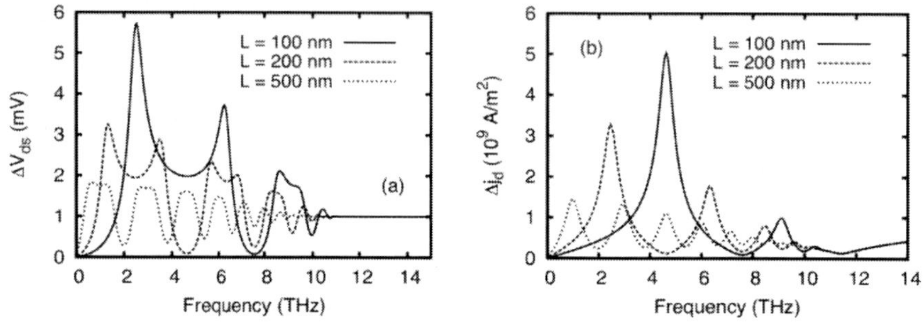

Figure 2. (a) Amplitude of the drain-to-source voltage oscillation and (b) amplitude of the drain current density oscillation as functions of the excitation frequency for the reported gate lengths. Each structure is stimulated by a harmonic oscillation of the gate voltage ($V_{g1} = 1$ mV) and presents $d_t = 1/k_t = 14$ nm.

4. Non-linear response

At second order, the system (1) becomes:

$$\begin{cases} \dfrac{\partial \delta n_2}{\partial t} = -\dfrac{\partial(\delta n_1 \delta v_1 + n_0 \delta v_2 + n_0 \delta\overline{v})}{\partial x} \\ \dfrac{\partial v_2}{\partial t} = -\dfrac{\partial(\delta v_1^2/2)}{\partial x} + \dfrac{q}{m}\dfrac{\partial(\delta V_2 + \delta\overline{V})}{\partial x} - \nu(v_2 + \delta\overline{v}) \\ \varepsilon_c \dfrac{\partial^2(\delta V_2 + \delta\overline{V})}{\partial x^2} - \varepsilon_s k_t^2(\delta V_2 + \delta\overline{V}) = q(n_2 + \delta\overline{n}) \end{cases} \tag{4}$$

4.1. Average response

The carrier conservation implies that $\delta\overline{j}$ is uniform: i.e. $\delta\overline{j} = 0$, obviously for an open drain but also for a shortened device because of its symmetry and bias conditions. As a consequence, no DC response to an external stimulation can be seen under symmetrical BC (at least at zero applied drain-to-source voltage).

We obtain the stationary responses under asymmetrical BC by time-averaging (4) and using the property $\langle \delta x_1 \delta y_1 \rangle = \frac{1}{2}\Re[x_1 y_1^*]$ with y_1^* the complex conjugate of y_1:

$$\delta\overline{v} = -\frac{1}{2n_0}\Re[n_1 v_1^*] \text{ and } \delta\overline{V} = \frac{m}{q}\left(\int_0^x \nu\delta\overline{v}\mathrm{d}x + \frac{1}{2}(|v_1^2| - |v_1(0)^2|)\right) \tag{5}$$

The average drain-to-source response $\Delta V_{ds} = \delta\overline{V}(L)$ is calculated using the expression obtained in the previous section and presented on Fig. 3(a). Peaks at the different plasma frequencies f_p are seen. Let us remark that this average bias is here slightly negative; nevertheless, Ref. [9] shows that, when a moderate-to-high drain current is imposed, ΔV_{ds} becomes positive and of much greater amplitude since the overall non-linearity of the system is increased.

4.2. Second harmonic response

By only considering the time-dependent part of the quantities appearing in Eq. (4), one can obtain the differential equation verified by V_2:

$$V_2'' + \beta(2\omega)^2 V_2 = \beta(2\omega)^2 e_2 \text{ with } e_2 = \frac{q}{2\mathrm{i}\varepsilon_c k_t^2}\left(\frac{1}{2}(n_1 v_1^*)' + \frac{1}{2\mathrm{i}\omega + \nu}(|v_1|^2/2)'\right) \quad (6)$$

V_2 has consequently the same form than V_1, substituting ω with 2ω and e_1 with e_2. Consequently, resonances at frequencies equal to $f_p/2$ appear in the frequency response of $V_{ds2} = |V_2(L)|$ presented on Fig. 3(b).

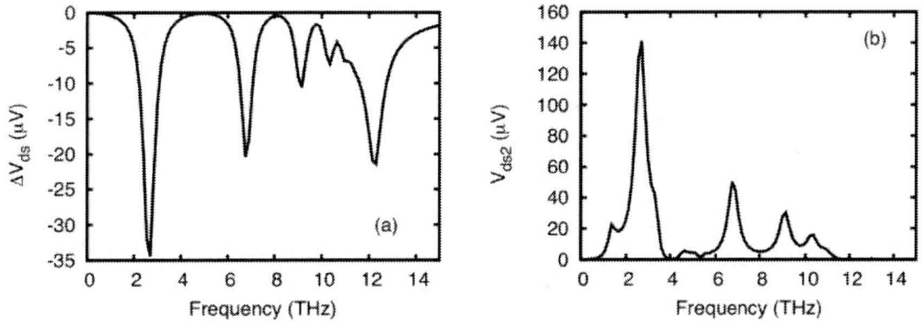

Figure 3. (a) Average value and (b) amplitude of the doubled-frequency part of the drain-to-source voltage. $V_g = 1$ mV, $d_t = 14$ nm and $L = 100$ nm.

5. Conclusion

A model allowing the calculation of the first and second order parts of the voltage and current responses of a nanometric FET channel to a THz stimulation has been presented. Such a stimulation not only induced significant drain voltage or current oscillations at the frequency of the excitation f but also a variation of the DC quantities (only for symmetrical BC) and also oscillations at $2f$ when a hybrid stationary plasma mode is stimulated. The amplitudes and frequencies of the corresponding resonances highly depend on the geometry.

References

[1] Dyakonov M and Shur M 1993 *Phys. Rev. Lett.* **71** 2465–2468
[2] Dyakonov M and Shur M S 1996 *IEEE Trans. Electron Devices* **43** 380–387
[3] Knap W, Deng Y, Rumyantsev S, Lu J Q, Shur M S, Saylor C A and Brunel L C 2002 *Appl. Phys. Lett.* **80** 3433–3435
[4] Nouvel P, Torres J, Blin S, Marinchio H, Laurent T, Palermo C, Varani L, Shiktorov P, Starikov E, Gruzinskis V, Teppe F, Roelens Y, Shchepetov A and Bollaert S 2012 *J. of Appl. Phys.* **111** 103707
[5] Nouvel P, Marinchio H, Torres J, Palermo C, Chusseau L, Gasquet D, Varani L, Shiktorov P, Starikov E and Gružinskis V 2009 *J. Appl. Phys.* **106** 013717
[6] Veksler D, Teppe F, Dmitriev A P, Kachorovskii V Y, Knap W and Shur M S 2006 *Phys. Rev. B* **73** 125328
[7] Ryzhii V, Khmyrova I, Satou A, Vaccaro P O, Aida T and Shur M 2002 *J. Appl. Phys.* **92** 5756–5760
[8] Marinchio H, Palermo C, Mahi A, Varani L and Korotyeyev V 2014 *J. Appl. Phys.* **116** 013707
[9] Marinchio H, Sabatini G, Pousset J, Palermo C, Chusseau L, Varani L, Starikov E and Shiktorov P 2009 *Appl. Phys. Lett.* **94** 192109

InP Double Heterojunction Bipolar Transistor for broadband terahertz detection and imaging systems

D Coquillat[1,4], V Nodjiadjim[2], A Konczykowska[2], N Dyakonova[1], C Consejo[1], S Ruffenach[1], F Teppe[1], M Riet[2], A Muraviev[3], A Gutin[3], M Shur[3], J Godin[2], W Knap[1]
[1] Laboratoire Charles Coulomb (L2C), UMR 5221 CNRS-Université de Montpellier, Montpellier, France
[2] III-V Lab (Bell Labs, TRT and CEA/LETI joint Lab), Route de Nozay, 91460 Marcoussis, France
[3] Rensselaer Polytechnic Institute, Troy, New York 12180, USA
E-mail: Dominique.Coquillat@univ-montp.fr

Abstract. This paper presents terahertz detectors based on high performance 0.7-μm InP double heterojunction bipolar transistor (DHBT) technology and reports on the analysis of their voltage responsivity over a wide frequency range of the incoming terahertz radiation. The detectors operated without any spatial antennas to couple terahertz radiation to the device and have been characterized in the 0.25 – 3.1 THz range with the responsivities (normalized to 1 W radiant power) of 5 V/W and 200 μV/W measured at 0.35 THz and 3.11 THz, respectively. The InP DHBTs also performed as the imaging single-pixels at room temperature in the raster scanned transmission mode. A set of the sub-terahertz images of plant leaves suggest potential utility of InP DHBT detectors for terahertz imaging dedicated to non-invasive testing of plants.

1. Introduction

Terahertz (THz) imaging has a large potential in the field of contact-free testing, for example in the field of non-destructive diagnosis of plants [1] and non-destructive testing of VLSI. Such applications require compact devices that operate at room temperature over a wide frequency range (0.1 to 10 THz). The recent developments of coded-aperture imaging with the single pixel detectors have led to new approaches for low cost real-time THz imaging [3,4]. This technique uses the dynamic spatial light modulators to multiplex and acquire the high-frame-rate images and requires the fast single-pixel detectors.

The existing THz single-pixel detectors that work at room temperature are bulky, slow, and prohibitively expensive. The development of low cost real-time THz imaging requires sensitive, fast, and robust room temperature detectors with a high dynamic range. Recently, InP DHBTs have attracted interest as sub-THz detectors [5,6] and compared favourably with the existing detectors based on the Schottky diodes [7], field effect transistors [8], or Si heterojunction bipolar transistors [9]. In this work, we evaluate InP DHBTs designed for 100 Gbit/s applications as single-pixel elements for the THz detection in the wide frequency range of 0.25 - 3.11 THz at room temperature. We also present two-dimensional images obtained at 0.67 THz in the raster scanned transmission mode and show that the resulting absorption directly reflects the quantity of water content in plant leaves.

[4] To whom any correspondence should be addressed.

Journal of Physics: Conference Series **647** (2015) 012036 doi:10.1088/1742-6596/647/1/012036

2. Technology and experimental set up

To address high bit-rate optical communications applications, in particular 100 Gbit/s transmission, development of a submicron process was undertaken. To achieve such high bit-rate, the transistor should have f_t/f_{max} in the 250-300 GHz range. The development of such a submicron InP DHBT technology was presented in Ref. 10, where in-depth study of device geometry and structure has allowed to get the needed performances and yield. The InP DHBTs used in this work have a 0.7-µm wide emitter contact. They demonstrated cut-off frequency and the maximum frequency of operation, f_T and f_{max}, above 320 and 280 GHz, respectively [10,11]. The static current gain is around 40 and the common-emitter breakdown voltage is above 5V. In this work, they operate as the THz detectors without any spatial antenna to couple the THz radiation to the device. The metallization of the contact pads and bonding wires act as effective antennas. For sub-THz photoresponse measurements, the DHBTs were exposed to the focused beam from one of the two tunable electronic sources covering the ranges from 0.26 to 0.37 THz and from 0.64 to 0.69 THz with output power ~ 1 mW (Fig. 1a). For frequencies above 1 THz, the photoresponse $\boldsymbol{\Delta u}$ was measured with a continuous-wave CO_2 –pumped molecular gas laser emitting at 1.40, 2.52, and 3.11 THz with the maximum output powers of ~ 100 mW, 150 mW, and 90 mW, respectively. The radiation was focused on the transistor surface using parabolic mirrors and the radiation intensity was modulated by a chopper. The emitter terminal of the device was grounded and the base-emitter bias V_{BE} was controlled by a Keithley Source Meter. The base-emitter bias and radiation frequency dependence was measured using a lock-in amplifier. For THz imaging, parabolic mirrors were used to focus the beam to the focal point, and the object (a fresh or dry plant leaf) was mechanically scanned using computer controlled translation stages.

3. Experimental results and discussion

In Fig. 1b, we have plotted the induced *dc* collector-emitter voltage Δu (photoresponse) of the 0.7-µm DHBT as a function of base-emitter bias V_{BE} applied in order to increase the sensibility of the THz detector. In this case, zero collector-emitter bias was applied, $V_{CE} = 0$. The optimal THz detection bias conditions were achieved when V_{BE} was close to 0.6 V. Above this value, the photoresponse was a decreasing function of V_{BE}.

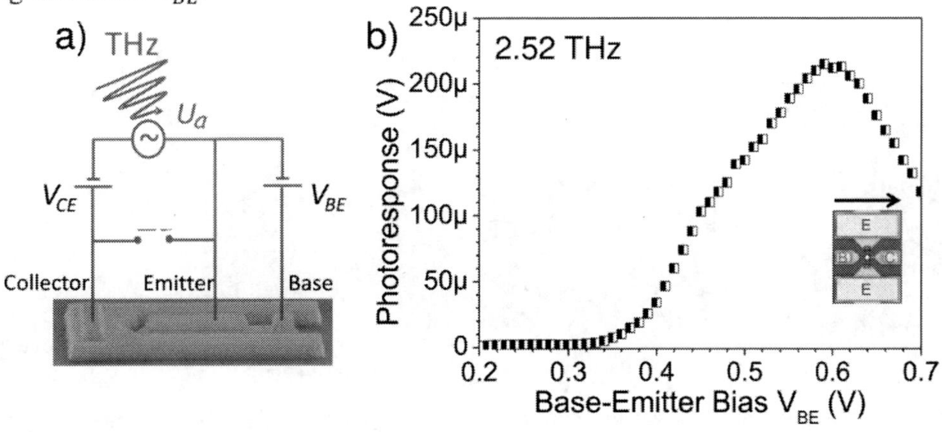

Figure 1. a) InP DHBT operating in detector mode for induced *ac* photovoltage. b) Photoresponse for 2.52 THz radiation as a function of the base-emitter bias V_{BE}. Inset: The polarization of the electric field of the incident THz radiation is parallel to the base-to-collector direction.

The value of the optimal photoresponse at 2.52 THz was ~ 200 µV for the beam focused by a *f*/1 off-axis parabolic mirror. For a power at the focal plane of approximately 150 mW, the responsivity, R_{1W}, normalized to 1 W radiant power was ~1.33 mV/W. It should be noted that the corresponding area-normalized responsivity calculated with regard to the area of the device taken equal to diffraction limited area [12] is 15.9 times larger than this value. The log-log plot of the responsivity measurements is depicted in Fig. 2. The device responsivity rolls off non-linearly with increasing frequency and follows a ω^{-4} law as shown by the guide for the eye included as a dashed dotted line. This dependence can be explained by the combined effects of the capacitance, *C*, shunting the active nonlinear device resistance, *R*, and parasitic series inductance, *L*. At high frequencies the voltage drop V_R across *R* becomes independent of *R* and proportional to $1/(\omega^2 LC)$. The detector response is proportional to V_R^2, and, therefore inversely proportional to $1/(\omega^4 L^2 C^2)$.

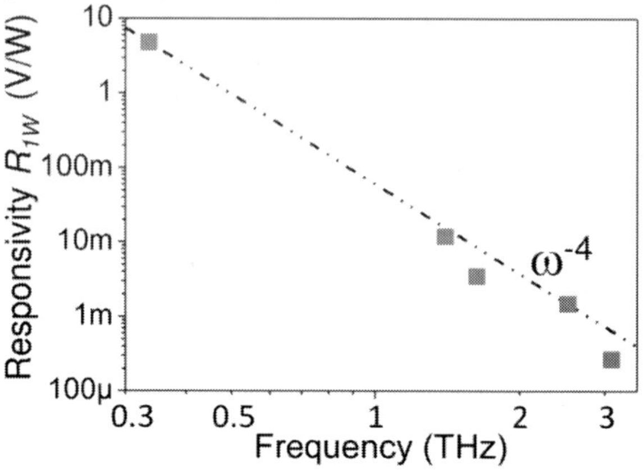

Figure 2. Log-log plot of responsivity (normalized to 1 W radiant power) as a function of frequency of the incident radiation (for $V_{BE} \sim 0.7$ V). The black dashed dotted line is a guide for the eye.

Using the same InP DHBT, we have performed raster scan imaging in transmission mode as shown in Fig. 3. The objects to image (Fig. 3a) are a fresh and a dry leaves of a viburnum tinus plant (71 mm × 80 mm). By focussing the source beam with a parabolic mirror, we achieved an almost Gaussian profile exhibiting full width at half maximum of 1.2 mm at 0.67 THz. The THz image (Fig 3b and 3c) is a result of the volume of water at the measured spot. It illustrates that small changes in water content of a leaf can be detected very sensitively using the InP DHBT detectors. Fig. 3d shows the measured transmission in the 0.64 – 0.69 THz range through a fresh leaf and a dry leaf of a viburnum tinus plant.

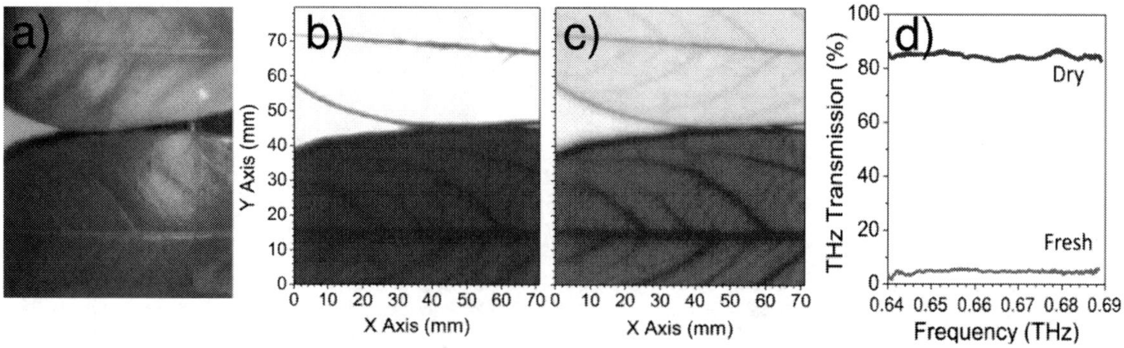

Figure 3. a) Visible image of a fresh leaf (bottom) and a dry leaf (top) of viburnum tinus. b) The transmitted 0.67 THz intensity measured using an unbiased InP DHBT detector reveals the detailed structure of the leaves in linear scale and c) in logarithmic scale. Applying a logarithm operator to the image can be useful when the dynamic range is too large to be displayed. d) Transmission of the leaves as a function the frequency in the 0.64 – 0.69 THz frequency range.

Journal of Physics: Conference Series **647** (2015) 012036 doi:10.1088/1742-6596/647/1/012036

Conclusion

In conclusion, we demonstrated the detection in wide frequency range of 0.25 - 3.1 THz electromagnetic radiation by InP DHBTs at room temperature. They can operate far above the frequencies at which they have gain and can still rectify THz current and voltage. These cost-efficient detectors can be used in a wide frequency range to monitor the water content of leaves and plants.

Acknowledgments

This work was partly supported by the ANR P2N NADIA "Integrated NAno-Detectors for terahertz Applications" (ANR-13-NANO-0008), by the "Contrat de Plan Etat Région du Languedoc Roussillon 2015-2020: BioNanoImaging Foundry", and by the Region of Languedoc-Roussillon through the "Terahertz Platform". The work at RPI is supported by Army Research Laboratory (ARL) Multiscale Multidisciplinary Modeling of Electronic Materials (MSME) Collaborative Research Alliance (CRA) (Grant No. W911NF-12-2-0023, Program Manager: Dr. Meredith L. Reed).

References

[1] Born N, Behringer D, Liepelt S, Beyer S, Schwerdtfeger M, Ziegenhagen B and Koch M 2014 *Plant Physiol.* **164** 1571

[2] Stillman W, Veksler D, Elkhatib T A, Salama K, Guarin F and Shur M S 2008 *Electronics Letter* **44** 22

[3] Chan W L, et al. 2009 *Appl. Phys. Lett.* **94** 213511

[4] Watts C M, et al. 2014 *Nature Photonics* **8** 605

[5] Coquillat D, Nodjiadjim V, Konczykowska A, Riet M, Dyakonova N, Consejo C, Teppe F, Godin J, Knap W 2014 *Proc. 39th IRMMW-THz: Int. Conf. on Infrared, Millimeter, and Terahertz Waves* (Tucson, AZ, USA)

[6] Suszek J, Siemion A, Coquillat D, Nodjiadjim V, Konczykowska A, Riet M, Sobczyk A, Zagrajek P, Palka N, Czerwińska E, Błocki N, Kolodziejczyk A, Dyakonova N, Teppe F, Consejo C, Knap W, Sypek M 2015 *Proc. 40th IRMMW-THz: Int. Conf. on Infrared, Millimeter, and Terahertz Waves* (Hong Kong, China)

[7] Han R , Zhang Y, Coquillat D, Videlier H, Knap W, Brown E, Kenneth K O 2011 *IEEE J. of Solid-State Circuits* **46** 2602

[8] Knap W, Nadar S, Videlier H, Boubanga-Tombet S, Coquillat D, Dyakonova N, Teppe F, Karpierz K, Lusakowski J, Sakowicz M et al. 2011 *J. of Infrared Millimeter and Terahertz Waves* **32** 618

[9] Al Hadi R, Grzyb J, Heinemann B, Pfeiffer U 2012, *Proc. IEEE Bipolar/BiCMOS Circuits and Tech. Meeting*

[10] Godin J, Nodjiadjim V, Riet, M, Berdaguer P, Drisse O, Derouin E, Konczykowska A, Moulu J, Dupuy JY, Jorge F, Gentner JL, Scavennec A, Johansen T, Krozer V 2008 *Proc. Compound Semiconductor Integrated Circuits Symposium, CSIC'08, IEEE* 1 (Monterey, CA, USA)

[11] Dupuy JY, et al. 2013 *IEEE Transactions on Microwave Theory and Techniques* **61** 517.

[12] Tauk R, Teppe F, Boubanga S, Coquillat D, Knap W, Meziani Y, Gallon C, Boeuf F, Skotnicki T, Fenouillet-Beranger C, Maude D K, Rumyantsev S, and Shur M S 2006 *Applied Physics Letters* **89** 253511

Journal of Physics: Conference Series **647** (2015) 012037 doi:10.1088/1742-6596/647/1/012037

Terahertz studies of 2D and 3D topological transitions

M Marcinkiewicz[1], F Teppe[1], C Consejo[1], N Dyakonova[1], W Desrat[1], D Coquillat[1], S Ruffenach[1], W Knap[1], N N Mikhailov[2], S A Dvoretskii[2], F Gonzalez-Posada[3], J-B Rodriguez[3] and E Tournié[3]

[1]L2C, UMR 5221 CNRS, Université Montpellier, GIS Teralab, 34095 Montpellier, France
[2]A.V. Rzhanov Institute of Semiconductor Physics, Siberian Branch, Russian Academy of Science, 630090 Novosibirsk, Russia
[3]IES, UMR 5214, Université Montpellier, 34000 Montpellier, France

E-mail: Michal.Marcinkiewicz@univ-montp2.fr

Abstract. We report terahertz measurements on bulk HgCdTe crystals at the semiconductor-to-semimetal topological transition and InAs/GaSb inverted band structure quantum well. We show that physical parameters of these narrow gap systems driven in specific topological phases can be efficiently extracted by means of terahertz photoconductivity studies.

1. Introduction

In the past years, the discovery of novel topological states of matter has attracted significant attention [1]. The most salient feature of such topological insulators (TIs) is that they have a full insulating gap in the bulk but gapless edge or surface states at boundaries. For two-dimensional (2D) TIs, the 1D edge hosts gapless spin-helical edge channels linearly crossing at a Dirac point [2], as was demonstrated experimentally in HgTe quantum wells (QWs) and InAs/GaSb broken gap QWs [3], whereas a 3D TI has protected spin-helical surface states usually with a Dirac cone-like dispersion [4]. Recently, such Dirac-type excitation has been extended to 3D, leading to the concept of Dirac semimetals [5]. These remarkable states have nontrivial topological properties without having a bulk band gap. Experiments on Na_3Bi and Cd_3As_2 have indeed successfully confirmed their Dirac semimetal character. Three-dimensional ultra-relativistic massless electrons have also been observed in bulk $Hg_{1-x}Cd_xTe$ (MCT) at the point of semiconductor-to-semimetal topological transition [6].

2. Band properties

2.1. InAs/GaSb

InAs, GaSb, AlSb and related semiconductor alloys form a family with a sufficient lattice match (6.1 Angstrom family [7]) for epitaxial growth without significant strains in grown structures. The structure considered in this work was a AlSb/GaSb/InAs/AlSb double layer QW with AlSb barriers. The characteristic feature of such systems is the alignment of bands of GaSb and InAs – the conduction band of InAs overlaps with the GaSb valence band. That system is referred to as Broken-Gap QW (BGQW). In InAs/GaSb BGQW, the energy of the valence band edge of

GaSb is higher (tens of meV) than the conduction band edge of InAs – the structure exhibits semimetallic behaviour. Hole and electron states at the interface are mixed. This hybridisation of the bands opens a small energy gap with inverted band structure.

InAs/GaSb QWs allow us to study many interesting physical phenomena, such as the Quantum Spin Hall Effect (QSHE) [8] or magnetic field-induced semimetal-to-semiconductor transition [9].

2.2. MCT

The $Hg_{1-x}Cd_xTe$ compounds are widely used in infrared optics because they provide an unprecedented degree of freedom in band gap tunability with x [10]. As the Cd composition increases, the energy gap increases from a negative value for pure HgTe to a positive one with CdTe. For $x > x_c$ MCT is a conventional semiconductor, where for $x < x_c$ the band gap is negative and the whole structure becomes a semimetal. The two distinct phases are not topologically equivalent, as characterized by a Z_2 topological invariant [11]. It means that for a critical Cd composition $x_c = 0.17$ the energy gap vanishes. At this point the system exhibits a Dirac cone-like dispersion relation.

3. Results

3.1. InAs/GaSb BGQW

The sample designed for this study consists of a 3 nm GaSb cap layer, two 50 nm AlSb barriers with 8 nm GaSb and 15 nm InAs layers between them. The structure was grown on an AlSb/GaSb superlattice buffer layer and a n-doped GaAs substrate. The sample was processed and multiple Hall bars were prepared for measurements. A picture of one of the Hall bars is shown in the inset of Fig.2.

To probe the carrier density, magnetotransport and photoconductivity measurements were performed. The results shown on Fig.1 were obtained at $T = 1.65$ K during illumination with a 292 GHz Schottky source. The photovoltage curve was obtained using lock-in technique – the beam was modulated mechanically with a chopper and signal was measured by lock-in. The THz radiation and lock-in technique revealed the Shubnikov - de Haas (SdH) oscillations at much weaker magnetic fields, compared to standard magnetotransport measurements. A cyclotron resonance peak was observed at $B = 0.37$ T, which points to an effective electron mass of $m^* = 0.035\ m_e$. Those results are in good agreement with previous results obtained on similar samples by transmission measurements, not presented in this work. The carrier concentration, estimated by SdH measurements, was $n = 5.7 \cdot 10^{11}$ cm^{-2}.

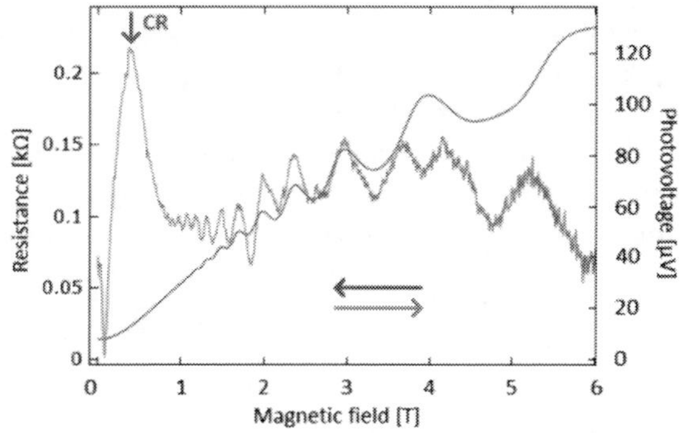

Figure 1. The photoconductivity of InAs/GaSb QW at 1.65 K and under illumination by 292 GHz radiation as a function of magnetic field. The red marker indicates the position of the cyclotron resonance. Purple line (left y axis) represents the standard magnetoresistance curve, measured for $I = 10\ \mu$A. Teal line (right y axis) represents the photoconductivity curve.

Our sample does not have a gate – the only way to change the Fermi level position was to use the negative and positive persistent photoconductivity (NPPC and PPPC respectively) [12]. To probe the persistent photoconductivity effect, the sample was placed with an LED in liquid nitrogen (T = 77 K) and the conductivity of the sample was measured as a response to an illumination with red (blue) LED with peak wavelength λ = 660 (470) nm. The results, presented on Fig.2, are a sign of the PPPC effect. It was possible to increase the conductivity a few times before saturation was reached. The effect disappeared after warming up the sample to the room temperature.

The measurements were repeated in a liquid helium cryostat (T = 4.2 K). In contrast to measurements at T = 77K consecutive illuminations decreased the conductivity – the sample exhibited the NPPC effect. Those results are consistent with those obtained in [12], except that the change between PPPC and NPPC occurred at different temperature. PPPC and NPPC effects arise mainly from two competing mechanisms – PPPC occurs due to the creation of electron-hole pairs in InAs well and the capture of photoexcited holes by deep defects. On the other hand NPPC occurs due to the pumping of electrons into the local minima at the well-barrier interface, existing because of compositional fluctuations, which leads to decrease of electron concentration. The influence of those two processes changes with the temperature – in high temperatures PPPC effect dominates, in lower – NPPC. We claim that our system is different structurally and qualitatively, that is why the temperature range of PPPC and NPPC domination differs. More data are required to explain this dependence in a quantitative way.

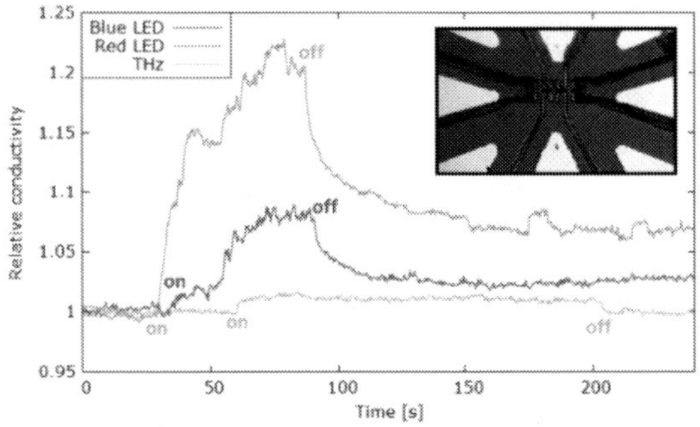

Figure 2. The results of illumination of the InAs/GaSb QW by blue and red LED light at T = 77 K. We observed an increased conductivity. This effect lasted after illumination was terminated. Light-blue curve presents change of conductivity while the sample was illuminated by THz light. In inset a photo of processed InAs/GaSb Hall bar is shown.

3.2. MCT bulk

A 3.2 μm thick MCT sample was grown using standard molecular-beam epitaxy on a 400 μm (013)-oriented semi-insulating GaAs substrate with a 50 nm thick ZnTe buffer layer and a 5 μm thick CdTe buffer layer. The cadmium concentration in the sample was x = 0.165. Therefore, an inverted band structure and a small negative gap are expected. The carrier density was estimated by magnetotransport and photoconductivity measurements.

THz induced SdH-like oscillations allow us to extract a carrier concentration of about $1 \cdot 10^{15}$ cm^{-3}, comparable to the electron concentration measured by Hall effect. Also, the amplitude of THz induced SdH oscillations changed with the temperature (Fig.3). Using Lifshitz-Kosevich formula [13] we were able to estimate the effective carrier mass in this sample (Fig.4).

The carrier effective mass found in sample A ($m^* = 2 \cdot 10^{-3}$ m_e) cannot be used to fit our results, but much higher mass values are necessary. The best fitting value is definitely $m^* = 2.1$ m_e, thus, three times higher than the larger heavy hole mass ever reported in this compound [10]. However, our experimental results are not accurate enough to validate this precise mass value and further experiments at lower temperatures have to be performed.

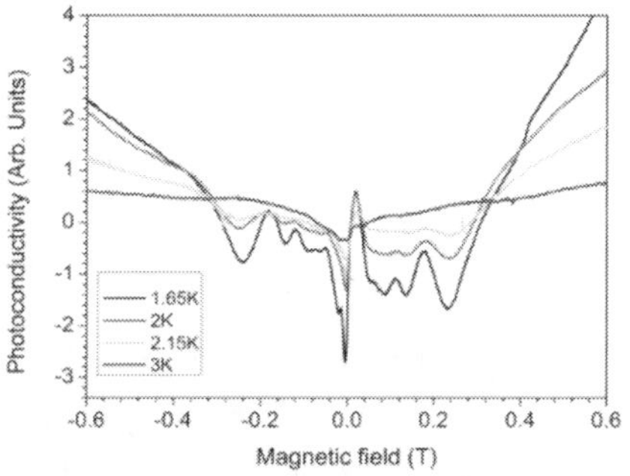

Figure 3. The close view of photoconductivity of bulk MCT sample B at 1.65 K under 292 GHz incident radiation as a function of magnetic field. Each line corresponds to a different temperature in range 1.65 K – 3 K. The rapid vanishing of the SdH oscillations with increasing temperature indicates that carriers have a relatively large effective mass.

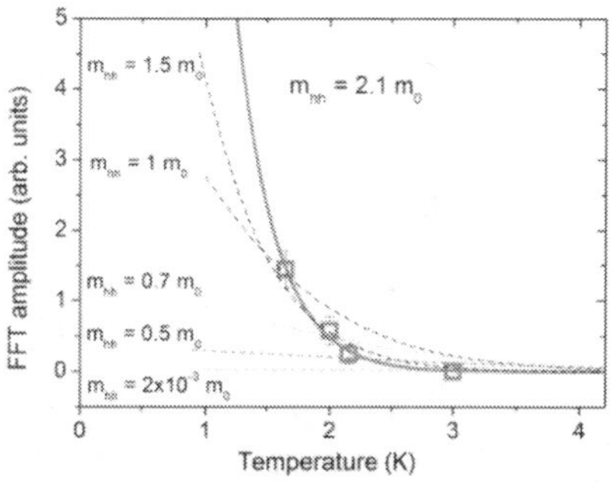

Figure 4. The amplitude of Fourier transform peaks corresponding to low field oscillations in MCT, as a function of temperature (empty squares). Dashed and straight lines are the fits for different values of effective mass.

4. Conclusion

We have studied by THz photoconductivity MCT bulk crystals and GaSb/InAs QW, which are known to exhibit semimetal-to-semiconductor topological phase transitions and Quantum Spin Hall Effect. Our first results allowed us to determine the heavy holes effective mass in MCT sample. In the case of GaSb/InAs QW , our THz photoconductivity results gave us information about the effective mass of electrons. We were also able, using PPPC and NPPC effect, to change the Fermi level in the structure, which is essential for future studies of topological phase transitions.

References

[1] Hasan G M Z and Kane C L 2010 *Rev. Mod. Phys.* **82** 3045-67
[2] König M et al 2007 *Science* **318** 766-70
[3] Halvorsen E, Galperin Y and Chao K A 2000 *Phys. Rev.* B **61** 16743
[4] Ando Y 2013 *J. Phys. Soc. Japn.* **82** 102001
[5] Young S M, Zaheer S, Teo J C Y, Kane C L, Mele E J and Rappe A M 2012 *Phys. Rev. Lett.* **108** 140405
[6] Bernevig B A, Hughes T L and Zhang S-C 2006 *Science* **314** 175761
[7] Kroemer H 2004 *Physica* E **20** 196-203
[8] Du L, Knez I, Sullivan G and Du R-R 2015 *Phys. Rev. Lett.* **114** 096802
[9] Lo Ikai and Mitchel W C 1993 *Phys. Rev.* B **48** 9118
[10] Rogalski A 2005 *Rep. Prog. Phys.* **68** 22672336
[11] Orlita M et al 2014 *Nature* **10** 233-38
[12] Wang W C, Tsai L C, Fan J C, Chen Y F and Lo Ikai 1999 *J. Appl. Phys.* **86** 3152
[13] Lifshitz I M and Kosevich A M 1955 *Zh. Eksp. Teor. Fiz.* **29** 730

Journal of Physics: Conference Series **647** (2015) 012038 doi:10.1088/1742-6596/647/1/012038

Nonequilibrium transport in GaAs Schottky mixers at 2.5 THz

D Pardo[1], **J Grajal**[1], **and S Pérez**[2]

[1] Department of Signal, Systems and Radiocommunications, Technical University of Madrid, Av. Complutense 30, Madrid 28040, Spain

[2] Department of Applied Physics, University of Salamanca, Plaza de la Merced s/n, Salamanca 37008, Spain

E-mail: dpardo@gmr.ssr.upm.es

Abstract. This work presents an analysis of the electrical and the noise performances of a 2.5 THz mixer. Reliable and self-consistent simulations of the circuit are carried out by means of a Monte Carlo model of the diode coupled to a harmonic balance technique. Simulations with this tool have shown that hot electrons and intervalley transitions can highly degrade the performance of this mixer. Additionally, these phenomena can be mitigated by decreasing the epilayer length of the diode.

1. Introduction

Schottky diodes are one of the most used devices for applications at terahertz frequencies such as Earth observation, planetary science, medicine, security, etc. These devices have the advantage over other sensor technologies to work at room temperature with good sensitivities [1, 2]. Actually, planar GaAs Schottky-based mixers operating at frequencies up to 2.5 THz have been designed and fabricated [3].

Physics-based modeling of the carrier transport in semiconductor devices is essential to understand and optimize the performance of nonlinear circuits at millimeter and submillimeter wavelengths. At these high frequencies, the electrical response of the diodes is limited by different physical phenomena such as velocity saturation, nonlocal or plasma effects among others. In addition, the noise performance of the diode is mainly degraded due to the presence of hot electron noise and intervalley scattering events. An accurate and reliable description of these phenomena requires high level physics-based models like the Monte Carlo model which provides a unified and self-consistent description of the electrical and noise performances of the diode.

This work presents an analysis of the carrier transport in the diode of the 2.5 THz receiver [3] and its impact on its two main figures of merit, i.e. the conversion losses and the equivalent input noise temperatures. To carry out this task, we have developed a compact simulator based on a multi-tone harmonic balance (HB) technique [4] coupled to a Monte Carlo (MC) model of the diode [5] (denoted by MCHB). This tool avoids the need of any additional analytical or empirical model to carry out physics and circuits optimizations of the mixer.

2. Description of the simulation tool

This section presents a brief description of the structure of a Schottky receiver and the simulation tool used in this work to calculate the conversion losses and noise temperatures of this circuit.

A Schottky receiver is composed of different elements which contribute to the conversion losses and noise temperature of the circuit. The total conversion losses of the receiver L_{rec} are:

$$L_{rec} = L_{qo} L_{rf} L_d L_{if}, \tag{1}$$

where L_{qo} represents the losses in the quasi-optical parts of the systems (lens, diplexers, etc.), L_{rf} represents the losses in the feed-horn, filters and mixer waveguides, L_d are the diode conversion losses, and L_{if} are the resistive and mismatch losses in the IF matching circuit [1, 6]. By using the Friis' formula, the equivalent input noise temperature of the receiver is:

$$\begin{aligned} T_{rec,DSB} &= (L_{qo}-1)T_0 + L_{qo}(L_{rf}-1)T_0 + L_{qo}L_{rf}T_{d,DSB} + L_{qo}L_{rf}L_{d,DSB}(L_{if}-1)T_0 \\ &+ L_{qo}L_{rf}L_{d,DSB}L_{if}T_{if} = (L_{qo}-1)T_0 + L_{qo}T_{mix,DSB} + L_{qo}L_{mix,DSB}T_{if} \end{aligned} \tag{2}$$

where DSB and SSB mean double and single sideband, respectively, ($L_{d,SSB}(dB) = L_{d,DSB}(dB) + 3\ dB$ and $T_{d,SSB}(K) = 2T_{d,DSB}(K)$ have been assumed [6]), T_0 is the physical temperature, T_d is the diode noise temperature, and T_{if} is the noise temperature of the IF chain. L_d and T_d are calculated with our MCHB simulator, [7]. Then, L_{rec} and T_{rec} are evaluated using (1) and (2), and the data provided in the literature for L_{qo}, L_{rf}, L_d, L_{if} and T_{if} [7].

Our circuit simulator is based on a HB technique properly adapted for multi-tone excitations, coupled to a self-consistent MC model of the diode [7]. The MC model [5] is based on the microscopic simulation of the electron transport in the semiconductor, providing a reliable description of the electrical and noise performances of the devices. Shot noise, thermal noise, hot electron noise and intervalley scattering noise as well as the effects due to the modulation of the noise sources by the voltage waveforms are inherently taken into account with this method. The multi-tone HB takes into account the different intermodulation products of the LO and the RF signals, ($mf_{LO} + nf_{RF}$, where m and n are integers, and f_{LO} and f_{RF} are the LO and RF frequencies, respectively). It uses the fast Fourier transforms when f_{LO} and f_{RF} are commensurate and the almost periodic Fourier transforms when they are not [6].

3. Analysis of a 2.5 THz mixer

This section presents an analysis of the performance of the single-ended fundamental 2.5 THz mixer presented in [3], see table 1. This mixer has been selected as a representative example to analyze the impact of the high frequency physical phenomena in the performance of the circuit. Some parameters of the mixer [3] like the ideal barrier height of the diode ϕ_b, the bias voltage V_{bias}, and circuit impedances are not provided in the literature, see table 1. In the simulations, ϕ_b and V_{bias} have been selected according to typical values for GaAs mixers and the circuit impedance at the LO, RF and IF frequencies have been optimized with the MCHB tool to minimize the conversion losses at the available LO power P_{LO}= 5 mW indicated in [3]. According to [3], L_{qo}= 2 dB, L_{if}= 0.4 dB and T_{if}= 150 K (L_{rf} is not provided in [3] and L_{rf}= 0 dB has been assumed in the simulations).

Figure 1 shows L_{rec} and T_{rec} of the 2.5 THz mixer simulated with the MCHB tool. In order to optimize the performance of the mixer, a diode with epilayer length L_e= 30 nm has been also simulated (L_e= 100 nm in [3]). Some discrepancies between measured and simulated results are observed, due to the indicated uncertainties in some parameters of the mixer.

L_{rec} for the mixer with L_e= 100 nm in figure 1(a) shows an abrupt increase at LO power around 6 mW (flat band conditions are achieved at LO power around 4 mW). According to MCHB simulations, the occupation of the L valley is around 5 % at P_{LO}= 2.5 mW, and 20 % at 5 mW, rising quickly up to 60 % at 6.3 mW, see figure 2(a). In addition, when the LO power

Table 1. Parameters used in the simulation of the 2.5 THz fundamental mixer. Data in bold type are provided in [3]. $f_{IF} = f_{RF} - f_{LO} = 25$ GHz has been selected to speed up MCHB.

Circuit parameters		Diode parameters	
RF frequency, f_{RF}, (GHz)	**2500**	Epilayer doping, N_e, (cm^{-3})	**1x10^{18}**
LO frequency, f_{LO}, (GHz)	2475	Epilayer length, L_e, (nm)	**100**
Bias voltage, V_{bias}, (V)	0.6	Area, (μm^2)	**0.35**
Circuit impedance at f_{RF}, Z_{RF}, (Ω)	70+j113	Ideal barrier height, ϕ_b, (V)	0.99
Circuit impedance at f_{LO}, Z_{LO}, (Ω)	86+j36	Series resistance, R_s, (Ω)	20.3
Circuit impedance at f_{IF}, Z_{IF}, (Ω)	250	Junction capacitance, $C_j(0)$, (fF)	1.1

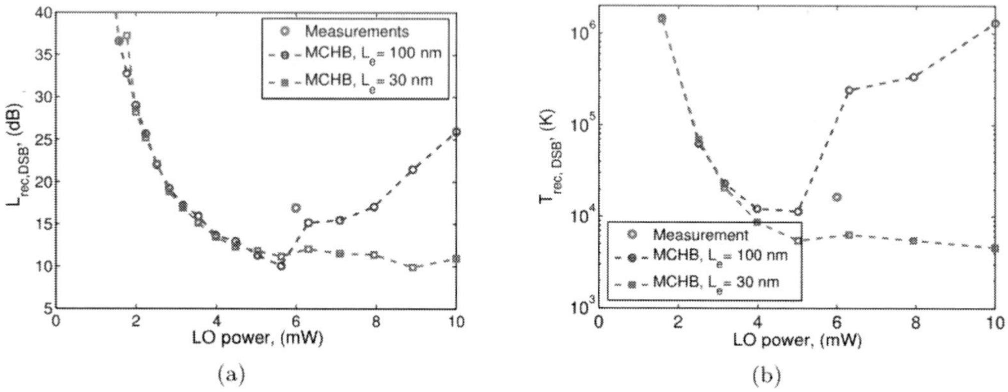

Figure 1. (a) $L_{rec,DSB}$ and (b) $T_{rec,DSB}$ of the 2.5 THz fundamental receiver in [3] with epilayer length L_e= 100 nm [3] and 30 nm. Simulated results obtained with the MCHB tool.

is increased from 5 mW to 6.3 mW, the occupation of the X valley increases from values lower than 5 % up to 20 %. These data suggest that the abrupt increase of L_{rec} is due to velocity saturation effects originated by the occupation of the upper low mobility valleys (L and X in GaAs) in the undepleted region of the epilayer. Figure 1(b) shows a significant degradation of the noise temperature of the mixer with L_e= 100 nm at P_{LO} higher than 4 mW. As the LO power increases, the energy of the electrons in the undepleted epilayer of the diode increases with respect to the equilibrium value, leading to hot electron noise. When the energy of the electrons is high enough for intervalley transfer to take place, the stochastic transitions between the lowest and upper valleys of the semiconductor generate an additional contribution to the noise spectrum usually known as intervalley scattering noise. The abrupt increase of the noise temperature in figure 1(b) for the diode with L_e= 100 nm at P_{LO} around 6 mW is due to the quick increase of the intervalley transfer at these powers (as it has been depicted in figure 2 (a)).

In order to optimize the performance of the 2.5 THz mixer, the epilayer length of the diode has been set to 30 nm. Figure 1 shows that L_{rec} and T_{rec} for the mixer with L_e= 30 nm do not present the abrupt increases observed for the mixer with L_e= 100 nm. There are three important phenomena which cause this enhancement of L_{rec} and T_{rec} in the optimized mixer:

- The epilayer is totally depleted during a high temporal fraction of the LO period, see figure 2(b), mitigating velocity saturation effects.
- The electrons that come from the substrate (low field region) enter into the epilayer (high field region), where they move with a velocity higher than expected from the steady-state velocity-field curve (nonlocal transport).

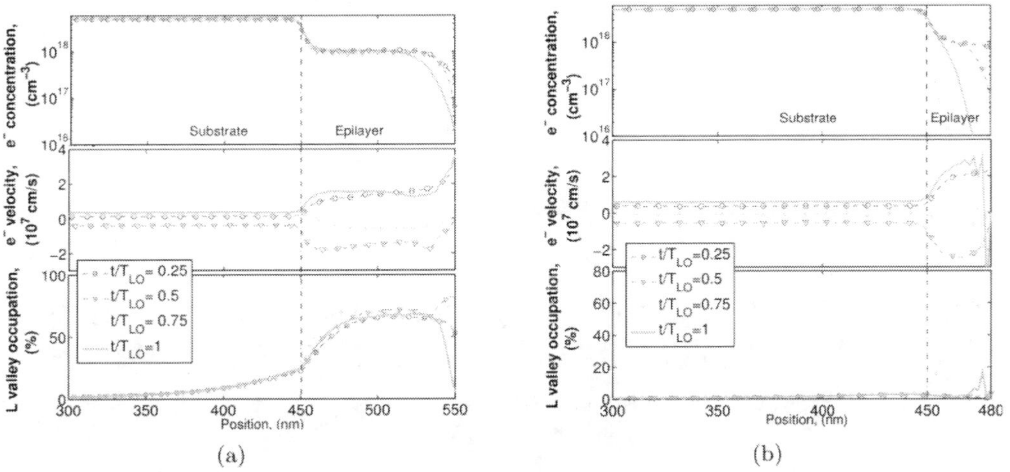

Figure 2. Profiles of the electron concentration, the electron velocity, and the occupation of the L valley for the 2.5 THz mixer with (a) L_e= 100 nm, and (b) L_e= 30 nm at P_{LO}= 6.31 mW. Results obtained with the MCHB tool at different time instants in a LO period (T_{LO}).

- The mean free path of the electrons in the epilayer is of order of the epilayer length, and hence, electron transport is strongly affected by ballistic carriers.

Due to these phenomena, for a long time of the period of the signal most of the electrons in the epilayer of the diode with L_e= 30 nm remain in the Γ valley, see figure 2(b), leading to L_{rec} and T_{rec} nearly constant at P_{LO} higher than 4 mW, see figure 1.

4. Conclusion

An accurate Monte Carlo harmonic balance circuit simulator has been used to analyse the electrical and the noise performances of a 2.5 THz mixer. It has been shown that velocity saturation limits the conversion losses of this mixer. In addition, the generation of hot electrons and intervalley transfer degrade its noise temperature. Nonlocal transport and ballistic transport in short epilayers have demonstrated to mitigate these limiting phenomena, leading to an important enhancement of the performance of the mixer.

Acknowledgments

This work was supported by the Spanish National Research and Development Program under projects TEC2011-28683-C02-01, TEC2013-41640-R, by the Madrid Regional Government under project S2013/ICE-3000 (SPADERADAR-CM), and by the Consejería de Educación de la Junta de Castilla y León under project SA052U13.

References

[1] Maestrini A, Thomas B, Wang H, Jung C, Treuttel J, Jin Y, Chattopadhyay G, Mehdi I and Beaudin G 2010 *C. R. Phys.* **11** 480–495
[2] Chattopadhyay G 2011 *IEEE Trans. THz Sci. Technol.* **1** 33–53
[3] Siegel P H, Smith R P, Graidis M C and Martin S C 1999 *IEEE Trans. Microw. Theory Techn.* **47** 596–604
[4] Grajal J, Krozer V, Gonzalez E, Maldonado F and Gismero J 2000 *IEEE Trans. Microw. Theory Techn.* **48**
[5] Shiktorov P, Starikov E, Gruzinskis V, Perez S, Gonzalez T, Reggiani L, Varani L and Vaissiere J 2004 *IEEE Electron Device Lett.* **25** 1–3
[6] Maas S 1993 *Microwave Mixers* 2nd ed (Boston, MA: Artech House)
[7] Pardo D, Grajal J and Pérez S *IEEE Trans. THz Sci. Technol.* Submitted

Temperature and Surface Traps Influence on the THz Emission from InGaAs Diodes

A. Rodriguez-Fernandez[1], I. Íñiguez-de-la-Torre[1], Ó. García-Pérez[1], S.García[1], A. Westlund[2], P- Å Nilsson[2], J. Grahn[2], T. González[1], J. Mateos[1] and S. Pérez[1]

[1] Applied Physics Department, Salamanca University, Salamanca 37008, Spain
[2] Department of Microtechnology and Nanoscience, Chalmers University of Technology, SE-412 96 Gothenburg, Sweden

E-mail: susana@usal.es

Abstract. Monte Carlo simulations forecast Gunn-like oscillations at ~0.75-1.25 THz in InGaAs planar recessed diodes (slot diodes); however, up to date no experimental evidence of this effect has been observed. The effects of temperature and surface charges on the emission parameters from InGaAs diodes are analyzed by means of an ensemble Monte Carlo simulator. Cooling the device down to 77 K strongly improves the amplitude of the oscillations and can increase their frequency. On the other hand, the ratio between cap and recess charges plays an important role for the onset of oscillations. A high level of traps in the recess region may completely attenuate the emission.

1. Introduction

The millimeter and sub-millimeter frequency range has found an increasing number of applications in several fields as defence, pharmacy, engineering or medicine among others [1]. However, compact sources of signals in this region of the electromagnetic spectrum are scarce and this fact is limiting the progress of terahertz (THz) technologies and applications [2,3]. Previous studies [4] of electronic transport in InGaAs slot diodes, based on Monte Carlo (MC) simulations, have shown the possibility to generate ultrafast Gunn-like oscillations. Nevertheless, up to date no experimental evidence of this effect has been observed in the performed experiments neither with a Schottky nor with a bolometer detectors [5]. The main objective of this work is trying to identify possible mechanisms that prevent the onset of oscillations in real devices. By means of MC simulations we analyse the influence of temperature and surface charges, located on the cap and recess region of the devices, on the THz emission.

2. Device structure and Monte Carlo simulation

The layer structure of the simulated slot diodes, similar to that used for the fabrication on InGaAs HEMTs [6], is shown in Figure 1. For the calculations we have used an ensemble MC tool, which is self-consistently coupled with a two dimensional Poisson solver. The material parameters based on a three valley system are reported in [6]. The devices are divided into 5 nm long and 1-10 nm wide

Journal of Physics: Conference Series **647** (2015) 012039 doi:10.1088/1742-6596/647/1/012039

meshes depending on the doping and the required resolution of the potential along the structure.

Ohmic boundary conditions are applied in the source and drain contacts, which are vertically placed adjacent to the different materials. Accordingly, no uniform potential and concentration profiles are considered along these contacts, but those that would be obtained if real top electrodes were simulated.

The effect of degeneracy has been introduced by using locally the classical rejection technique, with electron heating and non-equilibrium screening effects introduced by means of the local electron temperature. No other quantum effects are considered in the simulation in order to have reasonable computer simulation times. The validity of this approach (mainly under high field conditions) and that of the whole MC model has been confirmed in previous work [6]. In order to detect the presence of ultrafast Gunn oscillations, special attention is devoted to the calculation of the Fast Fourier Transform of the current.

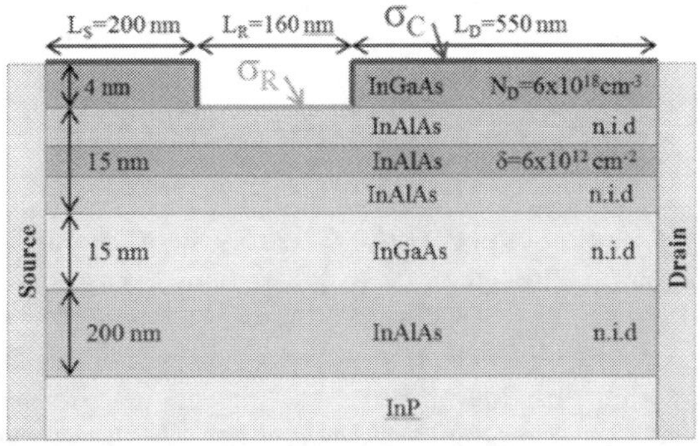

Figure 1. Geometry of the simulated InAlAs/InGaAs slot diode. n.i.d. stands for non-intentionally doped, and σ_C, σ_R for the surface charges density on the cap layer and recess region respectively.

3. Results and discussion

As shown in previous works [4,7], a planar InAlAs/InGaAs slot diode exhibits self-generated Gunn-like terahertz current oscillations operating under DC bias due to the modulation of the injection of electrons into the recess-to-drain region, which alternatively takes place in the Γ or L valley. The influence of bias, recess-to-drain length and δ-doping has been also reported in previous studies [8]. But unfortunately up to date no experimental evidence of this effect is obtained. In order to satisfactorily achieve emission in future attempts, the influence of lowering the operating temperature of the device and the impact of the surface charges density on the cap layer, σ_C, and recess region, σ_R, are analyzed.

3.1. Influence of the operating temperature

Figure 2(a) shows MC simulated time sequences of the current of a slot diode at different temperatures. In the inset the IV characteristics are plotted. We have used the following values for the surface charge density: σ_C=4×10^{12} cm^{-2} for the cap layer and

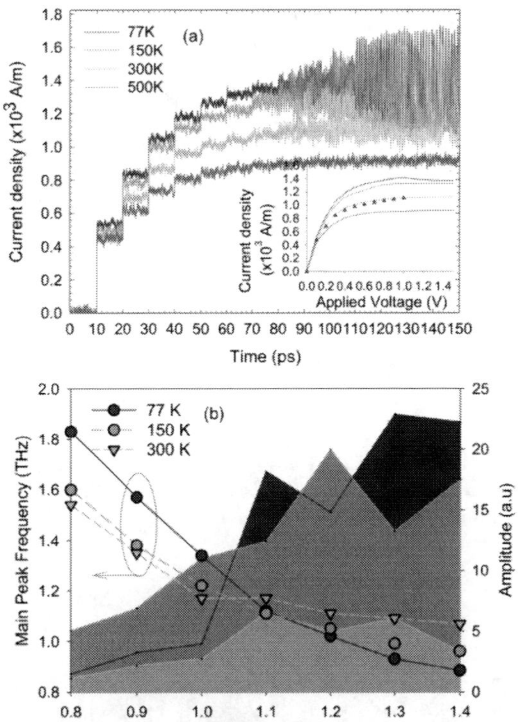

Figure 2. (a) Simulated time sequences of the current for a bias range from 0 to 1.5 V (steps of 0.1 V of 10 ps each), at different temperatures (500, 300, 150 and 77 K). Inset: I-V characteristics. For comparison, the experimental DC current measured in a fabricated device with the same geometry is also plotted (triangles). (b) Oscillation frequencies (left y-axis) and amplitudes (right y-axis) in a slot-diode obtained by means of Monte Carlo simulations at different temperatures. (V_{DC} > 0.8 V).

$\sigma_R=5\times10^{12}$ cm^{-2} for the recess. These values have been carefully chosen in order to reproduce the 300 K experimental DC I-V characteristic of the devices fabricated in Chalmers[5]. Lowering the temperature in the MC simulations clearly benefits the onset of the oscillation for $V_{DC}\sim1$ V. Conversely, excessive temperatures around 500 K inhibit the oscillatory behavior. Figure 2(b) shows, for stable oscillation conditions ($V_{DC}>0.8$ V), a ×3 factor of increase in the oscillation amplitude at 77 K besides a slight decrease in the operation frequency. The decrease in the amplitude around 1.1 V is attributed to the beginning of the population of the X-valley. However, the predicted MC oscillations around 1 THz at 300 K were not observed in several experimental attempts neither with a Schottky detector up to 900 GHz with a noise floor <50 nW nor with a bolometer detector up to 1 THz with a noise floor <10 nW.

3.2. Dependence on the surface charges of the cap and recess regions

Other possible origin of the absence of oscillations in the real device can be the existence of an excessive amount of carrier traps on the surface of the device. These surface charges are very sensitive, and not easily controllable in the fabrication process. Therefore, here we analyze with our MC tool the effect on the oscillating phenomena of several values of trap densities on the cap and recess regions. In particular, the presence of detectable oscillations is strongly determined by the value of the surface charge in the recess region, which can almost attenuate the emission as can be seen below. Simulated time sequences of the current are shown in Figure 3. Comparing the results shown in Figure 4(a) and Figure 4(b), it is observed that the increase of σ_R degrades the

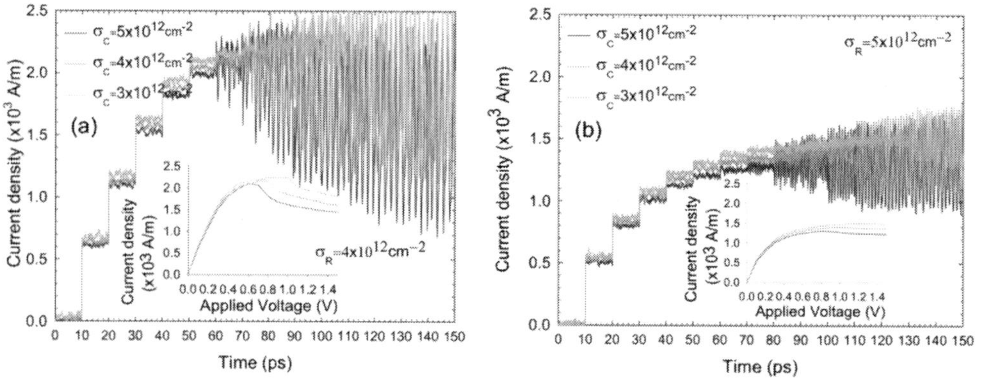

Figure 3. Simulated time sequences of the current for a bias range from 0 to 1.5 V (steps of 0.1 V of 10 ps each) at values of σ_R: (a) $\sigma_R=4\times10^{12}$ cm^{-2} and (b) $\sigma_R=5\times10^{12}$ cm^{-2}. Values of σ_C=(5,4 and 3)×10^{12} cm^{-2} are considered. Insets: IV characteristics.

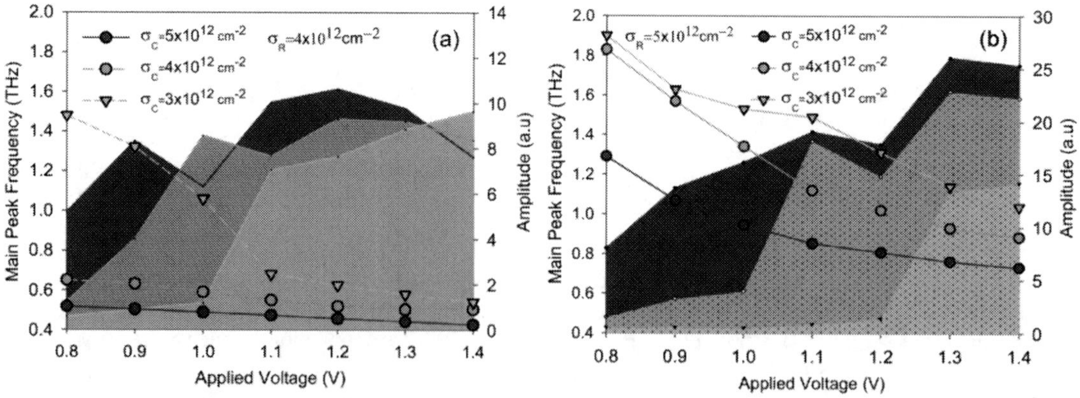

Figure 4. Oscillation frequencies (symbols and left y-axis) and amplitudes (right y-axis) in a slot-diode obtained with Monte Carlo simulations at values of σ_R: (a) $\sigma_R=4\times10^{12}$ cm^{-2} and (b) $\sigma_R=5\times10^{12}$ cm^{-2}. Values of σ_C=(5,4 and 3)×10^{12} cm^{-2} are considered.

performance of the emitter by reducing very significantly the amplitude of the oscillations, although the fundamental oscillation frequency slightly increases for higher σ_R. Note that the scale for the amplitude in the y-axis in (b) is 2 times smaller than in (a). With regard to the surface charge on the cap, Figure 4 shows that an increase of σ_C causes a reduction of the fundamental frequency. The amplitude of the oscillation increases with the cap surface charge density in the case of Figure 4(b), whereas it remains similar in the case of Figure 4(a). In addition, the lower the value of σ_C, the higher the DC threshold voltage for the onset of oscillation. Thus the presence of detectable oscillations is strongly determined by σ_R and σ_C. Note that an increase of σ_R from 4 to 5×10^{12} cm^{-2} strongly attenuates the emission and for σ_R=6×10^{12} cm^{-2}, the emission disappears (result not shown here). Consequently the surface charges can be considered as a key parameter in order to obtain the onset of oscillations.

4. Conclusions

We have shown by means of MC simulations, how the operation of the slot diodes as THz emitters benefits from cooling the devices down to 77 K with a ×3 factor in the oscillation amplitude. This result opens the possibility that the fabricated slot diodes could show some evidence of THz emission at low temperature. In addition, we have studied the amplitude and frequency of the oscillations for different values of the surface charges densities in cap and recess finding that they are key parameters for the onset of the oscillating phenomena.

Acknowledgements

This work has been partially supported by the Dirección General de Investigación Científica y Técnica (MINECO) through the Project TEC2013-41640-R and by the Consejería de Educación de la Junta de Castilla y León through the Project SA052U13.

5. References

[1] Miles R E, Zhang X-C, Eisele H and Krotkus A, 2007 Terahertz Frequency Detection and Identification of Materials and Objects (Dordrecht: Springer Netherlands)
[2] Carr G. L, Martin M. C, McKinney W. R, Jordan K, Neil G. R and Williams G. P, 2002 High-power terahertz radiation from relativistic electrons. Nature 420 153–6
[3] Köhler R, Tredicucci A, Beltram F, Beere H. E, Linfield E. H, Davies A. G, Ritchie D. A, Iotti R C and Rossi F, 2002Terahertz semiconductor-heterostructure laser. Nature 417 156–9
[4] Pérez S, González T, Pardo D and Mateos J, 2008 Terahertz Gunn-like oscillations in InGaAs/InAlAs planar diodes J. Appl. Phys. 103 1–5
[5] Westlund A, Nilsson P-Å and Grahn J, 2011 Fabrication and Characterization of InGaAs/InAlAs Slot diodes 37th WOCSDICE Works 2–3
[6] Mateos J, Gonzalez T, Pardo D, Hoël V and Cappy A, 2000 Monte Carlo simulator for the design optimization of low-noise HEMTs IEEE Trans. Electron Devices 47 1950–6
[7] Pérez S, Mateos J, Pardo D and González T, 2008 Excitation of millimeter-wave oscillations in InAlAs/InGaAs heterostructures Phys. Status Solidi Curr. Top. Solid State Phys. 5 146–9
[8] Pérez S, Mateos J, Pardo D and González T, 2009 On the geometrical tunabililty of THz Gunn-like oscillations in InGaAs/InAlAs slot diodes J. Phys. Conf. Ser. 193 012090

Terahertz spectroscopy of a multilayers flake of graphene

V Clericò[1], J A Delgado Notario[2], N Campos[3], D Gómez[3], E Diez[1], J E Velazquez[2] and Y M Meziani[2]

[1] Dept. Física Fundamental, Universidad de Salamanca, 37008 Salamanca, Spain
[2] Dept. Física Aplicada, Universidad de Salamanca, 37008 Salamanca, Spain
[3] Energy Area. Fundación ITMA. C/Calafates 11. 33417 Avilés, Spain

E-mail: vito_clerico@usal.es

Abstract. Terahertz Time Domain Spectroscopy (THz-TDS) technique has been used to characterize two graphene based samples produced by exfoliation and CVD methods. In both cases, we find two interesting bands, in both CVD and exfoliated graphene, in the 0.2-2 THz range. Quality information can be extracted with this non-destructive technique. The obtained values of the conductivity are in the same range of those obtained in the literature.

1. Introduction

Graphene based devices are attracting a large interest in nowadays researches due to its unique properties. For example, graphene flakes exhibit record electronic mobility at room temperature that would permit the development of transparent and flexible devices. Many approaches are used to produce 'graphene' [1-4] like Hammer, epitaxial and CVD methods along with the basic one: exfoliation. However, optoelectronic properties of the obtained samples are far from the incredible ones of exfoliated graphene [5]. As graphene fabrication is under intense research, it's very important to find new techniques for identifying and characterizing graphene flakes. Terahertz Time Domain (THz-TDS) spectroscopy technique is non-destructive and permits the extraction of different material parameters (conductivity, dielectric constant, refractive index…) [6-8]. The possibility to obtain a THz-imaging of a big flake and mapping local conductivity has been recently demonstrated [9] which could extend the use of TDS-THz technique for quality control of large graphene flakes made with different techniques. Here we present THz-TDS transmission measurements on different samples of graphene fabricated using two different methods: exfoliated multilayer graphene and monolayer CVD graphene. In literature different experiments on CVD graphene [9], epitaxial graphene [10] and terahertz spectroscopy can be found but results on the terahertz spectroscopy measurements of exfoliated graphene are uncommon [11]. In our measurements, we obtained similar behaviours in CVD and exfoliated samples. The optical conductivity was extrapolated from the exfoliated graphene measurements. The comparison in terahertz region of graphene produced with different techniques with exfoliated graphene could be used to optimize the recipes to obtain better large area graphene flakes using CVD and epitaxial techniques.

2. Preliminary results and discussion

The first sample was a single large multilayers graphene obtained after a different number of exfoliations (from #1 to #10) on Magic tape (see Figure 1(a)). They had an irregular shape and typical

Journal of Physics: Conference Series **647** (2015) 012040 doi:10.1088/1742-6596/647/1/012040

size of 7 mm^2. The second monolayer graphene sample was prepared by the CVD technique [12] deposited on polyimide (PI). Two silver contacts were defined at the extremity of the flake, so that the final size of the nude graphene was 0.35*15 mm^2 (Figure 1(b)).

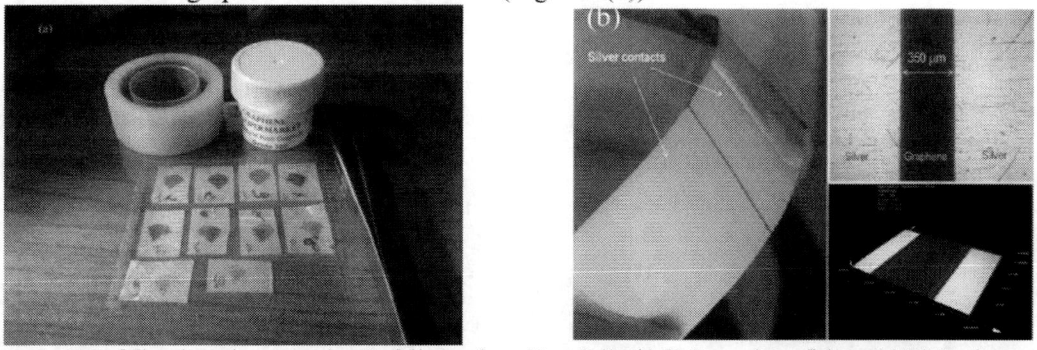

Figure 1: (a) Exfoliated graphene samples (Exfoliation from #1 to #10) (b) CVD graphene flake and 2D/3D microscope images of the gap between contacts.

Figure 2 shows the terahertz pulse intensity versus time delay for the exfoliated graphene samples. The reference signal was obtained on magic tape. It can be seen clearly the increase of the transmitted intensity as a function of the numbers of exfoliations. Since the signal with graphene samples is in the same temporal location, we did not take into account the thickness of graphene in our calculations. Figure 3 shows the amplitude spectra of the reference and graphene samples (from #1 to #10) where a window from 0.2 to 2.5 THz and water absorption lines (~0.51THz, ~1.1THz,...) are observed. The noise floor cuts the signal around 2.3 THz. For this, we focus our study within the range 0.2-2 THz.

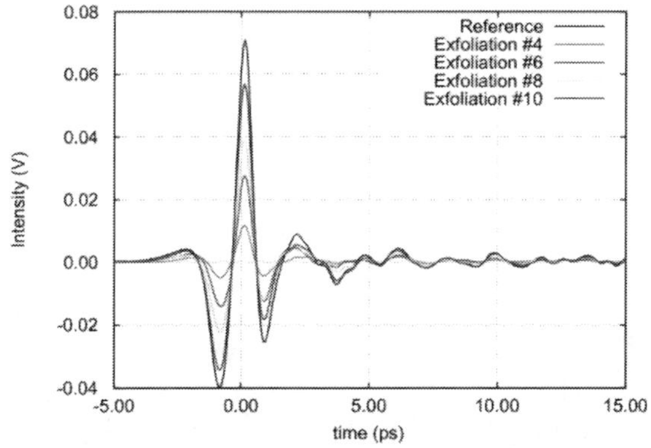

Figure 2: Terahertz pulse intensity signal vs. delay time for different samples (exfoliation #4, #6, #8, #10) and reference.

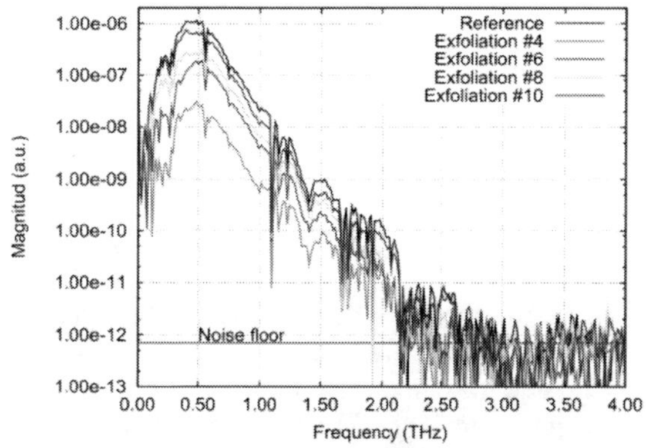

Figure 3: Amplitude spectra of the reference (magic tape) and the transmitted pulse of graphene samples.

The transmission is obtained as a ratio between the spectra of graphene samples and their respective references. Figure 4 shows the transmission signal at different numbers of exfoliations. The reference one is obtained as a ratio between the reference signals obtained at the beginning and at the end of the experiment to demonstrate the stability of the system. The obtained signal has a constant value around 1 except for some lines related to water absorptions (Fig. 3). The transmission increased with the number of exfoliations, as expected. Two very interesting absorption bands at 0.5 and 1.1 THz are observed after a sufficient number of exfoliations, revealing the presence of graphene in the sample. These two bands are not related to possible multiple reflection into the magic tape for two reasons. First of all no second echo was observed in the temporal response (Fig.2). Second because these two bands are not visible neither in the transmission spectra of the thicker graphene (4-6 exfoliations), nor in the ratio between the two references. Instead the sharp line at 1.1 THz is related to water absorption

Journal of Physics: Conference Series **647** (2015) 012040 doi:10.1088/1742-6596/647/1/012040

lines and not to the graphene (Fig.3). Figure 5 shows the transmission signal obtained in the case of CVD graphene. The overall shape is similar to the one of exfoliated graphene. In particular, comparing with the exfoliated #6 sample, we can notice a shift of the two bands to 0.3 and 0.8 THz.

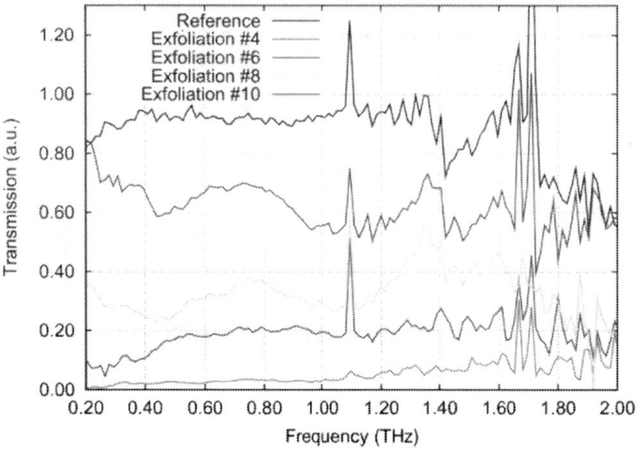

Figure 4: Transmission signal vs. frequency of for different samples (exfoliation #4, #6, #8, #10) and reference

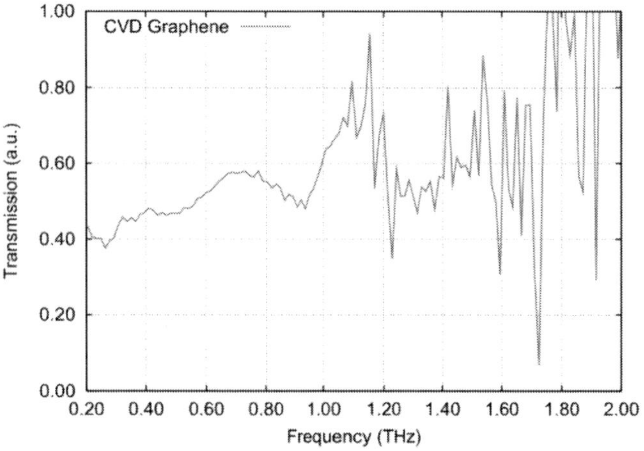

Figure 5: Transmission signal vs. frequency of for CVD graphene.

From exfoliated graphene, in which the transmission had an absolute value, it is possible to extrapolate quantitative information. In particular we calculated the average optical conductivity σ_a of multilayer exfoliated graphene without the use of metal contacts. The same model (thin film Fresnel coefficients and the Drude model) was used in the far infrared region [9], [13]. We treated graphene as zero-thickness conductive film (the thickness of graphene, *d*, is much smaller than the wavelength) as it was explained before. Because of the Fresnel coefficient for the reflection term since possible interference terms are negligible, the transmission was given by:

$$\mathrm{T}(\sigma_a) = \frac{1}{|1 + \frac{Z_0}{1+n_{tape}}\sigma_a|^2} \tag{1}$$

where Z_0= 376.7 Ω was the vacuum impedance, n_{tape} is the index of refraction of the tape (n_{tape}=1.48). This model works well for a high number of exfoliations for which zero thickness is a good approximation ($d<<\lambda$). Taking into consideration the inhomogeneity of the sample, we estimated the average conductivity from 0.2 to 2 THz for the samples with #8 and #10 exfoliations. In the case #10 these values were between 0.001-0.002 Ω^{-1}, and between 0.004-0.006 Ω^{-1} for #8. These results are comparable with those of Tomaino et al. [9], and two order of magnitude higher than the optical conductivity due to interband transitions [14]. So, as for CVD graphene, the intraband transitions are predominantly in terahertz region also for multilayer exfoliated graphene.

Conclusions

Even if these results are preliminary and it will be necessary a deeper understanding of the meaning of the shift and the intensities of bands in the terahertz regions, the obtained results give us both qualitative and quantitative information on graphene samples and confirm the THz-TDS technique as a method suitable for quality control of production of large area flakes.

Acknowledgments: The authors greatly acknowledge the support from Junta de Castilla y Leon, the DGICYT under project FIS2009-07880 and JCYL SA049A10-2, MEC through grant AP2009-2619 and MINECO through Project No. TEC2012-32777. We thankfully acknowledge Centro de Láseres Pulsados Ultraintensos (CLPU) for his material help to set up the THz-TDS system under the 2012 Collaboration Agreement.

References

[1] Zhang Y, Zhang L and Zhou C 2013 *Acc Chem. Res.* **46**(10) 2329
[2] Dreyer DR, Park S, Bielawski CW and Ruoff RS 2010 *Chem. Soc. Rev.* **39** 228
[3] Pei S and Cheng H-M 2011 *Carbon* **50** 3210
[4] Norimatsu W and Kusunoki M 2014 *Phys. Chem.Chem. Phys.* **16** 3501
[5] Novoselov KS, Fal'ko VI, Colombo L, Gellert PR, Schwab MG and Kim K 2013 *Nature* **490** 192
[6] Ulbricht R, Hendry E, Shan J, Heinz TF, Bonn M 2011 *Rev. Mod. Phys.* **83**, 543
[7] Docherty CJ and Johnston MB 2012 *J. Infrared Milli Terahz Waves* **33 (8)** 797
[8] Yasuda H and Hosako I 2008 *Jpn J. Appl. Phys.* **47** 1632
[9] Tomaino JL, Jameson AD, Kevek JW, Paul MJ, van der Zande AM, Barton RA, McEuen PL, Minot ED and Lee Y-S 2011 *Optics Express* **19**(1) 141
[10] Dawlaty JM, Shivaraman S, Strait J, George P, Chandrashekhar M, Rana F, Spencer MG, Veksler D and Chen Y 2008 *Appl. Phys. Lett.* **93**, 131905
[11] Boubanga-Tombet S, Chan S, Watanabe T, Satou A, Ryzhii V and Otsuji T 2012 *Phys. Rev.B* **85** 035443
[12] Campos N and Gómez D 2014 *Conference paper, Nanotechnology Materials and Devices Conference (NMDC), 2014 IEEE 9th*, Aci Castello, Catania (Italy)
[13] Lee C, Kim JY, Bae S, Kim KS, Hong BH and Choi EJ 2011 *Appl. Phys. Lett* **98**, 071905
[14] Mak K F at al. 2008 *Phys. Rev. Lett* **101**, 196405

Journal of Physics: Conference Series **647** (2015) 012041 doi:10.1088/1742-6596/647/1/012041

TCAD study of sub-THz photovoltaic response of strained-Si MODFET

J.A. Delgado Notario, Y.M. Meziani, and J.E. Velázquez-Pérez

Departamento de Física Aplicada, Universidad de Salamanca
Pza. de la Merced s/n – Edificio Trilingüe
E-37008 Salamanca, Spain

Email. js@usal.es

Abstract. This paper reports on a bi-dimensional TCAD study of the sub-THz response of strained-Si MODFETs. A charge boundary condition for the floating drain contact was implemented to obtain the photovoltaic (PV) response of the device. A non-resonant THz PV response was obtained in agreement with theoretical and experimental works. A main result of this study is that the PV response is strongly influenced by both the gate length and the gate topology. In particular, it was found that a dual-finger gate gives the maximum response at 300GHz.

1. Introduction

During the past two decades a considerable effort was devoted to the development of Field-Effect Transistors (FETs) using the Si/SiGe material system. N-channel strained-Si FETs with buried [1] (s-Si MODFET, modulation doped FETs) and surface channels [2] (s-Si MOSFET) were developed. Minimum noise figures as low as 0.4dB at 2.5GHz and cut-off frequencies (f_T) in excess of 70 GHz at 300K were achieved in s-Si MODFET [3]. A number of analog building blocks were demonstrated [4], nevertheless, the final break through the market of these technologies was hindered by low-yields of global-strain wafers. Recently, it has been shown the potential of s-Si MODFETs as room temperature detectors of sub-THz radiation [5, 6], i.e. in frequency ranges far beyond f_T. As Schottky-gated s-Si MODFETs exhibit large values of channel electron mobility ($\approx$2,000 $cm^2V^{-1}s^{-1}$, [7]) they can be used as detectors of THz electromagnetic radiation [8]. This has been experimentally demonstrated and the mechanism of detection identified as the excitation of resonant or overdamped plasma waves in the two-dimensional electron plasma in the device channel [6].

2. Device description and models used in simulation

In this work, we simulated n-channel s-Si MODFETs. Devices had a thick SiGe virtual substrate over a p-Silicon wafer. The final Ge molar concentration in the virtual substrate is 0.3. The device has a 10 nm tensile strained (in terms of biaxial deformation) Si channel, sandwiched between two heavily doped SiGe electron supply layers (doped at ~10^{19} the top supply layer and ~10^{18} the bottom one) to generate an electron dual-channel in the strained-Si quantum well. The structure of the simulated device is shown in the left plot of Fig 1 along with the energy band diagram calculated using Taurus-Medici [9]. The transistor was studied with different gate lengths (L_G) in the range 50 nm to 500 nm while keeping constant the Source-to-Drain distance (L_{SD}) at 2 μm. The source-to-gate length (L_{SG}) was modified to study the effect of asymmetry in the device by changing the position of the gate.

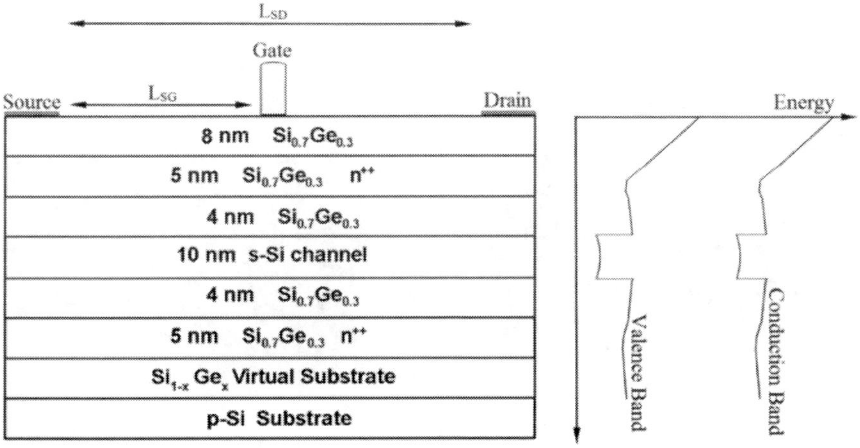

Figure 1. Vertical layout of the (left) and energy band diagram (right) of the MODFET under study.

A 2D numerical study was performed using Synopsys TCAD to understand the response found in THz measurements. Modeling the electrical behavior of transistors with deep-submicron gates needs an accurate description of the complex nature of the carrier transport in the channel. To this aim we used a two-dimensional hydrodynamic (2DHD) self-consistently coupled to a two-dimensional solution of the Poisson equation. In the 2DHD simulations impurity de-ionization, Fermi-Dirac statistics and mobility degradation due to both longitudinal and transverse electric field were taken into account. The source and drain regions were simulated as non-self-aligned implanted contacts.

The study of the THz photovoltaic (PV) response of the transistor was implemented by biasing source and gate and floating the drain contact while, as in measurements, a sub-THz sinusoidal signal was superimposed to the gate voltage as described in [8]. The amplitude of the gate signal was fixed to 5mV; the induced drain voltage exhibit both the same shape (sinusoidal) and frequency of the gate AC voltage ensuring that no frequency conversion takes place but its amplitude is considerably smaller than the one of the gate's signal as in the sub-THz range the device is unable to amplify signals, additionally the mean value of the induced drain voltage were negative in good agreement with the theoretical model [8]. The charge boundary condition for the floating is specified as:

$$\oint \vec{D} \cdot \overrightarrow{dS} = Q$$

where $\vec{D}$ is the displacement vector, Q is the total charge and the integral is evaluated over the drain contact surface.

Table 1. Electrical parameters of the s-Si MODFETs.

	Simulated structures			
L_G (nm)	50	100	250	500
V_{th} (V)	-1.2	-0.9	-0.75	-0.675
g_m (mS/mm)	72	98	112	105
$g_{m,exp}$ (mS/mm)	-	66	76	74
f_T (GHz)	36	31	22	11

3. Results and discussion

Table 1 summarizes the most relevant DC and AC parameters obtained in simulations. The second row gives the threshold voltage (V_{th}).The third and fourth rows contain the maximum values of the transconductance obtained in simulations and in measurements, respectively. The differences found in transconductance are attributed, essentially, to differences in contact resistances and implantations between the simulated and the actual transistors.

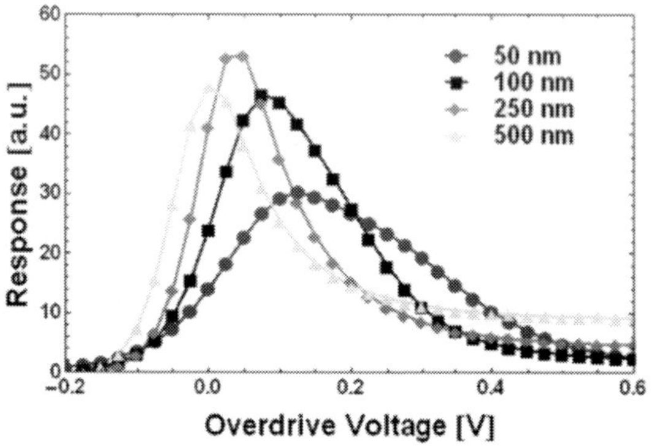

Figure 2. PV response of s-Si MODFET with different gate-lengths under THz excitation of 300GHz vs. gate overdrive.

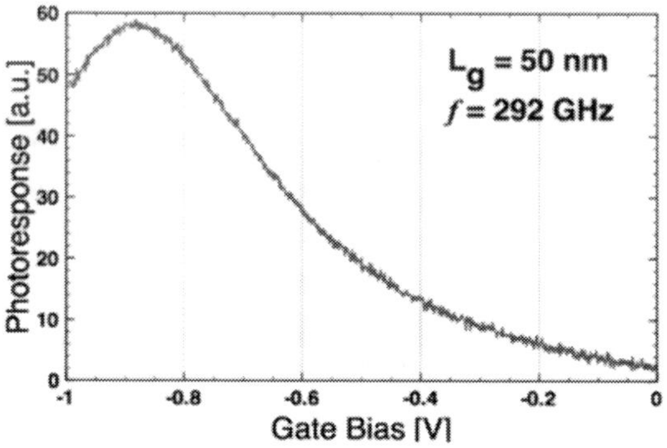

Figure 3. Experimental PV response of a s-Si MODFET with L_G=50nm under THz excitation of 292GHz [10]. The value of the threshold voltage was $V_{th} \approx$ -0.84V.

The gate structure is a key design parameter to enhance the responsivity of FETs as THZ sensors. Figure 2 gives the simulated PV response of the four simulated transistors when excited at 0.3THz. Results show a non-resonant response in agreement with theoretical [8] and experimental results (see Fig. 3 [10]).

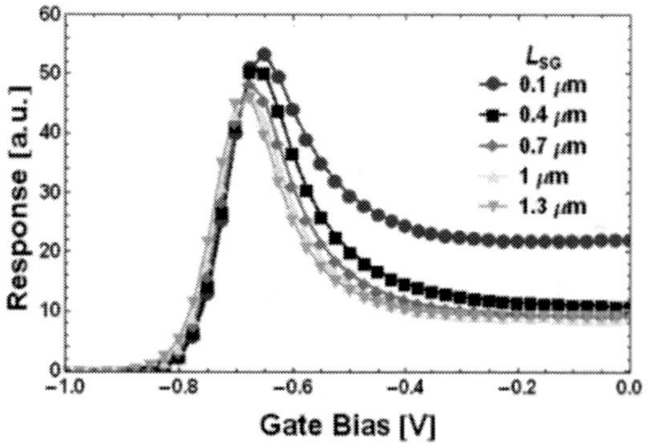

Figure 4. PV response of a 500-nm gate transistor placed for four different values of L_{SG} (under a 300GHz excitation).

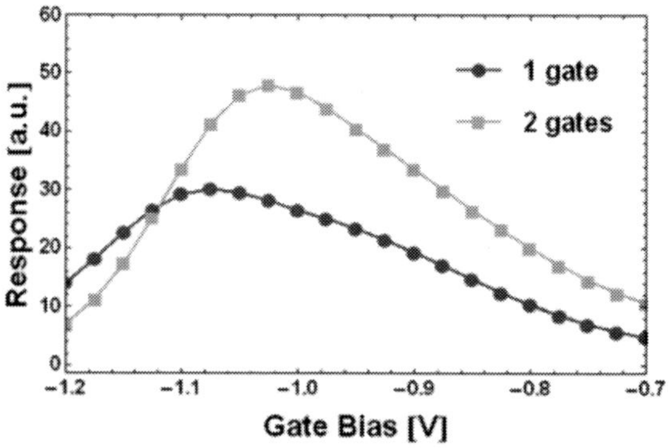

Figure 5. PV response of a 50-nm gate transistor for a gate with 1 and 2 fingers equally distributed between source and drain (under a 300GHz excitation).

Following the same trend as transconductance, the THz PV response exhibit lower values as the gate is shortened owing to a weakened channel control by the gate. The main conclusion to be drawn from Fig. 2 is that neither self-aligned structures, nor very short gate lengths will lead to optimum structures in terms of the THz photoresponse. We also found a significant dependence of the PV maximum with the excitation frequency.
Figure 4 gives the PV response of a MODFET when its 500-nm gate is placed at different distances (L_{SG}) from the source contact: only marginal changes in the PV response are obtained; in strong contrast to some commonly accepted ideas, an asymmetric gating of the channel will not necessarily lead to significant enhancements of the THz response. On the contrary, changes of the gate topology, such as a grating of the gate, has a significant impact on the THz PV response in agreement with FDTD (Finite Differences Time Domain) calculations using a simpler model for the electrons transport in the channel [11]. In Fig. 5 the PV response of s-Si MODFETs with two configurations of the gate is given. A large enhancement, close to 60%, of the PV response is found using a two-finger gate as compared to a single-finger gate transistor.

Conclusions
A 2D TCAD study of strained-Si MODFETs is performed to investigate their sub-THz photovoltaic (PV) response. To this aim a charge boundary condition for the floating drain contact was implemented. In agreement with measurements, a non-resonant sub-THz PV response is found for the simulated devices. A main result of this study is that the PV response is strongly influenced by both the gate length and gate topology. As the maximum PV response changes with both gate length and excitation frequency a careful design of the device is needed to optimize the device parameters (such as its gate length) to obtain a maximum of PV response In particular, it was found that a dual-finger gate is suitable to obtain maximum response at 300GHz.

Acknowledgement
This work was supported by the Spanish MINECO and FEDER under grant # TEC2012-32777.

References
[1] Zeuner M, Hackbarth T, Enciso-Aguilar M, Aniel F and von Känel H 2003 Jpn. J. Appl. Phys. **42** 2363
[2] Olsen SH, O'Neill AG, Driscoll L S, Kwa KSK, Chattopadhyay S, Waite AM, Tang YT, Evans AGR, Norris DJ, Cullis AG, Paul D J, Robbins D J 2003 IEEE Trans. Elec. Dev. **50** 1961
[3] Aniel F, Enciso-Aguilar M, Giguerre L, Crozat P, Adde R, Mack T, Seiler U, Hackbarth T, Herzog H J, König U, Raynor B 2003 Solid-State Electron. **47** 283
[4] Vilches A, Fobelets K, Michelakis K, Despotopoulos S, Papavassiliou C, Hackbarth T, and König U 2003 IEE Electron. Lett. **39** 884
[5] Meziani Y M, García-García E, Velázquez-Pérez J E, Coquillat D, Dyakonova N, Knap W, Grigelionis I, Fobelets K 2013 Solid-State Electron **83** 113
[6] Meziani Y M, García E, Velazquez E, Diez E, El Moutaouakil A, Otsuji T and Fobelets K 2011 Semicond. Sci. Technol. **26** 105006
[7] Fobelets K, Jeamsaksiri W, Papavasilliou C, Vilches T, Gaspari V, Velazquez-Perez J E, Michelakis K, Hackbarth T, König U 2004 Solid-State Electron. **48** 1401
[8] Dyakonov M and Shur M S Phys. 1993 Rev. Lett. **71** 2465
Dyakonov M and Shur M S 1996 IEEE Trans. Electron Dev. **43** 380
[9] Synopsys, Taurus User Guide, Version Z-2007.3 Synopsys Inc., Mountain View, CA (2007).
[10] Meziani Y M, García E, Calvo J, Diez E, Velazquez E, Fobelets K and Knap W 2012 *Bolometers*, Ed U Perera (Rijeka: InTech) Chapter 6
[11] Otsuji T, Popov V, Knap W, Meziani Y M, Dyakonova N, Coquillat N, Teppe F, Fateev D, and Velazquez Perez J E, Japanese Patent PCT/JP2010/007074 (3 December 2010).

Non-Markovian Effects in the Spin Transfer Dynamics in Diluted Magnetic Semiconductors due to Excitation in Proximity to the Band Edge

M. Cygorek, V. M. Axt

Theoretische Physik III, Universität Bayreuth, 95440 Bayreuth, Germany

E-mail: `Moritz.Cygorek@uni-bayreuth.de`

Abstract. The non-Markovian effects in the spin dynamics in diluted magnetic semiconductors found in quantum kinetic calculations can be reproduced very well by a much simpler effective single electron theory, if a finite memory is accounted for. The resulting integro-differential equation can be solved by a differential transform method, yielding the Taylor series of the solution. From the comparison of both theories it can be concluded that the non-Markovian effects are due to the spectral proximity of the excited electrons to the band edge.

1. Introduction

Diluted magnetic semiconductors (DMS) are a class of workhorse materials in the field of semiconductor spintronics, since they combine the magnetic degree of freedom with the versatility and highly developed fabrication schemes of the semiconductor technology. Usually, Mn doped II-VI or III-V semiconductors are studied and a localized *s*-*d* interaction between the carrier and Mn spins modelled by a Kondo-like Hamiltonian has been found to describe the magnetic properties and the spin dynamics of DMS very well.

A numerical calculation based on a quantum kinetic theory (QKT) for the spin dynamics in DMS governed by the *s*-*d* interaction[1] showed that, among other phenomena, non-Markovian effects, such as overshoots or oscillations of the total spin polarization, can be found in one- and two-dimensional systems[2, 3]. The quantum kinetic theory can be presented in a more easy-to-use and intuitive way, by eliminating the correlations at the cost of a memory integral. Because it was found that, in doing so, it is crucial to account for a precession-like dynamics of the carrier-impurity correlations, the equations are referred to as *precession of electron spins and correlations* (PESC) equations[4]. In the present article, we show that the non-Markovian spin dynamics in DMS, found in the quantum kinetic theory, can be well described by an approximation of the PESC equations. The resulting integro-differential equation can be solved by a differential transform method (DTM)[5]. An analysis based on this simplified approach reveals that the non-Markovian effects are due to the proximity of the electronic excitations to the band edge.

2. Equation of motion

In Ref. [4] effective equations of motion for the correlation-induced spin dynamics in DMS were derived. For initially vanishing magnetization of the magnetic impurities, the time evolution of the conduction band electron spin polarization in a DMS quantum well structure can be found from Eq. (7a) of Ref. [4]:

$$\frac{\partial}{\partial t}s_{\omega_1}(t) = -\frac{\eta}{\pi}\int_0^t dt' \int_0^{\omega_{BZ}} d\omega \cos[(\omega_1-\omega)(t-t')]\left[s_{\omega_1}(t') + \frac{1}{4}(s_\omega(t') - s_{\omega_1}(t'))\right], \tag{1}$$

where s_{ω_1} is the mean electron spin of electrons with energy $\hbar\omega_1$ (relative to the band minimum), η is the spin transfer rate in the Markov limit and $\hbar\omega_{BZ}$ is the energy at the end of the first Brillouin zone. If we assume a parabolic band structure, we find $\omega_1 = \frac{\hbar k_1^2}{2m^*}$ with effective mass m^* for an electron with wave vector $\mathbf{k}_1$ and $\eta = \frac{35}{12}\frac{J_{sd}^2 m^* n_{Mn}}{\hbar^3 d}$ with coupling constant J_{sd}, magnetic ion density n_{Mn} and quantum well width d.

It is noteworthy that in the time derivative for the total spin, where Eq. (1) is integrated over ω_1, the term $(s_\omega(t') - s_{\omega_1}(t'))$ cancels. Since this term can be expected to lead to an insignificant contribution to the total spin, we henceforth neglect this term which simplifies the analysis of the spin dynamics drastically. Despite this argument being valid only for the total spin, we shall show by numerical calculation that also the individual spin dynamics for an electron at the energy ω_1 is reasonably well described by this approximation (cf. Fig 1(c) and (d)). Thus, Eq. (1) can be reduced to

$$\frac{\partial}{\partial t}s_{\omega_1}(t) = -\frac{\eta}{\pi}\int_0^t dt' \left[\frac{\sin((\omega_{BZ}-\omega_1)(t'-t))}{t'-t} + \frac{\sin(\omega_1(t'-t))}{t'-t}\right] s_{\omega_1}(t') \tag{2}$$

The phyiscal meaning of Eq. (2) becomes most obvious when the Markov limit is regarded, which assumes that s_{ω_1} changes on a much slower timescale than the oscillations of the integral kernel. Then, on the r. h. s. of Eq. (2), $s_{\omega_1}(t')$ can be evaluated at $t' = t$ and drawn out of the integral. Keeping in mind that $\lim_{\omega\to\infty}\sin(\omega t)/t = \pi\delta(t)$ and that the integral ranges only over one half of the $\sin(\omega t)/t$ peak, one finds: $\frac{\partial}{\partial t}s_{\omega_1} = -\eta s_{\omega_1}$, which shows a simple exponential decay of s_{ω_1} with the rate η. This corresponds to a golden rule-type transfer of the electron spin to the impurity system.

However, the condition for the applicability of the Markov limit was $\eta \ll \omega_1$ and $\eta \ll \omega_{BZ}-\omega_1$. For realistic parameters (e. g., the parameters used in Ref. [2, 3] yield $\hbar\eta \approx 0.45$ meV) and excitations far away from the end of the first Brillouin zone, only the latter condition is fulfilled, while for excitations close to the band edge, ω_1 can be of the same order of magnitude as η. Thus, we apply the Markov limit ($\omega_{BZ} \to \infty$) only on the first term of Eq. (2). The number of parameters can be reduced by substituting $\tau := \eta t$ and $\xi := \omega_1/\eta$. Then, the problem is transformed to:

$$\frac{\partial}{\partial \tau}\Phi_\xi(\tau) = -\frac{1}{2}\Phi_\xi(\tau) - \frac{1}{\pi}\int_0^\tau d\tau' \frac{\sin(\xi(\tau'-\tau))}{\tau'-\tau}\Phi_\xi(\tau'), \quad \Phi_\xi(0) := 1, \tag{3}$$

where $s_{\omega_1}(t) = s_{\omega_1}(0)\Phi_{\omega_1/\eta}(\eta t)$. Thus, the shape of the time evolution depends only on the ratio between ω_1 and η.

3. Numerical Evaluation of the non-Markovian Spin Dynamics

We solve the integro-differtial equation (3) by a technique similar to Zhou's differential transform method (DTM)[5], which consists of Taylor-expanding all terms in Eq. (3) at $\tau = 0$. This yields

Journal of Physics: Conference Series **647** (2015) 012042 doi:10.1088/1742-6596/647/1/012042

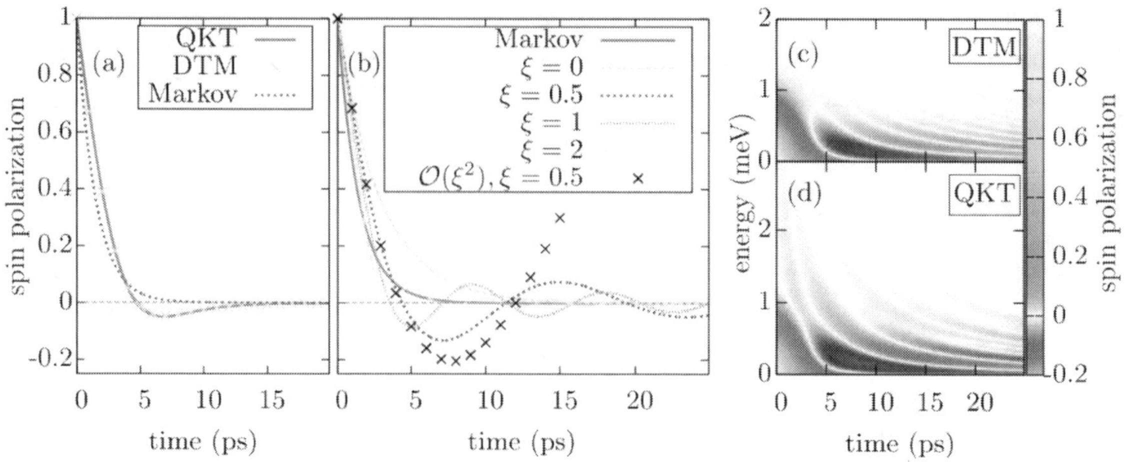

Figure 1. (a): Spin dynamics in a 4 nm wide $Zn_{0.93}Mn_{0.07}Se$ quantum well with $\eta \approx 0.67$ ps^{-1} with Gaussian excitaton at $\hbar\omega = 0$ and standard deviation $\Delta = 0.4$ meV (same as in Ref. [2]) according to the full quantum kinetic theory (QKT), the differential transform method (DTM, from Eq. (4)) and the Markov limit. (b): Time evloution of the spin of single electrons with fixed energies $\hbar\omega = \hbar\eta\xi$, compared with the Markov limit and the expression in Eq. (5) for the low-ξ approximation. The spectrally resolved time evolution of the spin polarization is shown in (c) for the DTM calculation, and in (d) for the QKT (cf. Ref. [2]).

a recursion relation between the derivatives of Φ_ξ:

$$\Phi_\xi^{(i)} = -\tfrac{1}{2}\Phi_\xi^{(i-1)} - \tfrac{1}{\pi} \sum_{0 \le 2m \le i-2} \tfrac{(-1)^m}{2m+1} \xi^{(2m+1)} \Phi_\xi^{(i-2-2m)}, \tag{4}$$

where $\Phi_\xi^{(i)}$ is the i-th derivative of Φ_ξ evaluated at $\tau = 0$. The numerical evaluation of Eq. (4) is very efficient and $\Phi_\xi(\tau)$ can be calculated to high orders by substituting the derivatives into the Taylor expansion. We refer to this algorithm as the DTM calculation.

It is noteworthy that from the recursion relation (4) closed expressions can be derived for $\Phi_\xi(\tau)$ to a certain order in the ratio ξ by combinatoric analysis of the paths from $\Phi_\xi^{(0)} = 1$ to $\Phi_\xi^{(n)}$ and comparing the Taylor series with that of known functions. E. g., to second order in ξ, we find:

$$\Phi_\xi(\tau) = e^{-\frac{\tau}{2}} + \tfrac{\xi}{\pi}[(2\tau+4)e^{-\frac{\tau}{2}} - 4] + \left(\tfrac{\xi}{\pi}\right)^2 \left[(2\tau^2 + 16\tau + 48)e^{-\frac{\tau}{2}} + 8\tau - 48\right] + \mathcal{O}(\xi^3) \tag{5}$$

which should be valid for excitations near the band edge where $\xi \ll 1$ can be fulfilled.

4. Results

To check the validity of the approximation of neglecting the last term of Eq. (1), we compare the DTM calculation with the results of a full quantum kinetic treatment. Fig. 1(a) shows that the non-Markovian dynamics of the total spin given in Fig. 1(b) of Ref.[2] can be reproduced almost perfectly with the DTM calculation. Also, the time evolution of an individual spin of an electron with energy $\hbar\omega_1$ is very similar in both calculations except for a high-energy tail appearing in the full quantum kinetic result, as can be seen from the spectrally resolved time evolution presented in Figs. 1(c) and (d) for DTM and QKT calculations, respectively. This finding confirms that Eq. (3) indeed captures the main non-Markovian features of the full quantum kinetic theory.

Fig. 1(b) shows the results of the DTM calculation for different values of ξ. For $\xi = 0$, the dynamics is given by an exponential decay with half the rate η, as can be seen also in the low-ξ approximation in Eq. (5). For larger values of ξ, the decay rate approaches η and oscillations start to appear whose amplitudes eventually decrease for even larger values of ξ, where the time evolution converges to the exponential decay of the Markov limit ($\omega_1 \to \infty$). Thus, the non-Markovian features are only present if the approximation $\xi \gg \eta$ breaks down, i. e., if the excited electrons are spectrally close to the band edge, where the characteristic energy scale is given by $\hbar\eta$.

This can easily be understood if another derivation of the Markov limit starting from Eq. (1) is considered. If the assumption of a vanishing memory is made and on the r. h. s. the functions $s_\omega(t')$ are evaluated at $t' = t$, we can first integrate over dt' and then over $d\omega$. Calculating the first interal gives

$$\frac{\partial}{\partial t} s_{\omega_1}(t) = -\frac{\eta}{\pi} \int\limits_0^{\omega_{BZ}} d\omega \frac{\sin[(\omega_1 - \omega)t]}{\omega_1 - \omega} \left[s_{\omega_1}(t) + \frac{1}{4}(s_\omega(t) - s_{\omega_1}(t)) \right]. \tag{6}$$

Using again the fact that $\lim_{t\to\infty} \frac{\sin[\Delta\omega t]}{\Delta\omega} \to \pi\delta(\Delta\omega)$, one again ends up with the Markov limit. For finite time t, however, the integral kernel is not yet contracted to a δ-distribution and the finite integral limits cut off tails of the $\frac{\sin[\Delta\omega t]}{\Delta\omega}$ function. This cut-off is particularly significant, if the peak of the integral kernel, which is given by ω_1 is close to one of the integral limits.

Furthermore, it can be seen in Fig. 1(b) that the low-ξ approximation in Eq. (5) yields reasonable results for $\xi = 0.5$ for the initial exponential decay while it fails to reproduce the long-term oscillations.

5. Conclusion

The non-Markovian overshoots and oscillations in the time evolution of the carrier spins in DMS found in a quantum kinetic theory can be reproduced by integro-differential equation of a much simpler form that also simplifies the interpretation considerably. A differential transform method (DTM) is employed to solve the resulting equation and allows to find closed-form expressions for low excitation energies of electrons.

It is found that a non-exponential behaviour of the time evolution of the electron spin is only present for electrons excited close to the band edge, where the decay predicted by the rate and the oscillations with frequency corresponding to the electron energies take place on the same time scale. Technically, this is due to the fact that a sinc-function that converges to a δ-distribution in the Markov limit is cut off by the band edge. It is noteworthy that similar time evolutions have also been found in different setups, e. g., for the hole spin dynamics due to phonon scattering in a GaAs quantum well when the scattering rate is close to the phonon frequency[6].

Acknowledgments

We acknowledge the support by the Deutsche Forschungsgemeinschaft through the Grant No. AX 17/9-1.

[1] Thurn C and Axt V M 2012 Phys. Rev. B **85** 16520
[2] Thurn C, Cygorek M, Axt V M, and Kuhn T 2013 Phys. Rev. B **87**, 205301.
[3] Thurn C, Cygorek M, Axt V M, and Kuhn T 2013 Phys. Rev. B **88**, 161302
[4] Cygorek M and Axt V M 2014 Semicond. Sci. Technol. **30**, 085011.
[5] Zhou J K 1986 *Differential Transformation and its Applications for Electrical Circuits* (Wuuhahn: Huarjung University Press)
[6] Zhang P and Wu M W 2007 Phys. Rev. B **76**, 193312

Journal of Physics: Conference Series **647** (2015) 012043 doi:10.1088/1742-6596/647/1/012043

Confined acoustic phonon-mediated spin relaxation in a two-dimensional quantum dot in the presence of perpendicular magnetic field

K A Vardanyan, A L Vartanian, A G Stepanyan and A A Kirakosyan
Department of Solid State Physics, Yerevan State University,
1, A.Manoogian, 0025 Yerevan, Armenia

E-mail: vardan@ysu.am

Abstract. The spin-relaxation time due to the electron-acoustic phonon scattering in GaAs quantum dots is studied after the exact diagonalization of the electron Hamiltonian with the spin-orbit coupling. It has been shown that in comparison with flexural phonons, the electron coupling with the dilatational phonons causes 3 orders faster spin relaxation. We have found that the relaxation rate of the spin-flip is an order of magnitude smaller than that of the spin-conserving.

1. Introduction

Manipulation and readout of spins in nanostructures is a promising way not only for developing spin based quantum logic gates, spin memory devices for various quantum information processing applications, but also because they have greatly improved our understanding of the role of different spin-dependent interaction mechanisms [1,2]. The spin-orbit coupling, which is one of the main mechanisms of spin relaxation, is a relevant intrinsic interaction in nonmagnetic semiconductor nanostructures. In this paper we have considered spin-relaxation in a two dimensional parabolically confined quantum dot embedded inside a free-standing semiconductor slab. The relaxation occurs via acoustic phonon-mediated scattering processes in the presence of both perpendicular magnetic field and spin-orbit coupling, represented by the Rashba and Dresselhaus terms. Our study has been motivated by a recent experiment, where the long spin relaxation times (30ns) have been measured in GaAs quantum dots [3].

2. Theory

We consider a QD system, where the QD is confined by a parabolic potential with frequency ω_0 in the quantum well of width a. A magnetic field **B** is applied perpendicularly to the surface of the quantum well. The total Hamiltonian is given by

$$H = H_0 + \alpha(\sigma_x P_y - \sigma_y P_x) + \beta(-\sigma_x P_x + \sigma_y P_y) + H_{ph} + H_{e-ph} \quad (1)$$

where $H_0 = \mathbf{P}^2/2m^* + m^*\omega_0^2(x^2+y^2)/2 + g^*\mu_B B\sigma_z/2$ is electron Hamiltonian without the spin-orbit coupling, second and third terms represent the Rashba and Dresselhaus spin-orbit couplings, respectively, $\boldsymbol{P} = -i\hbar\boldsymbol{\nabla} + (e/c)\boldsymbol{A}$ with $\boldsymbol{A} = (B/2)(-y, x.0)$, m^* is the electron effective mass, $g^*\mu_B B\sigma_z/2$ is the Zeeman energy, $\boldsymbol{\sigma} = (\sigma_x, \sigma_y, \sigma_z)$ represents the Pauli matrices.

Recently, the freestanding quantum well structures have been developed using advanced nanofabrication techniques, which have provided the possibility of observation of effect of confinement of acoustic phonon modes in a quantum plane. In elastic continuum model the confined acoustic phonon modes in freestanding quantum wells have been classified, according to their spatial symmetries, as shear, dilatational, and flexural modes. The acoustic phonon Hamiltonian in free standing quantum well (FSQW), (H_{ph} in Eq. (1)), has the standard form $H_{ph} = \sum_q \hbar\omega_q b_q^+ b_q$, where the quantum number $q = (\beta, n, \boldsymbol{q})$ includes the phonon symmetry β (β takes values from the set dilatational, flexural, shear), the mode number m, and the in plane wave vector $\boldsymbol{q}$. The most important mechanism of the interaction between electrons and acoustic vibrations in semiconductors is via the deformation potential coupling. In a FSQW structure, the electron-confined acoustic phonon interaction Hamiltonian is given by

$$H_{el-ph} = \sum_{\beta,n,q} e^{iqr} \Gamma_\beta(\boldsymbol{q}, n, z)\left[b_{\beta,n,q} + b^+_{\beta,n,-q}\right].$$

The functions $\Gamma_d(\boldsymbol{q}, n, z)$ and $\Gamma_f(\boldsymbol{q}, n, z)$, describing the intensity of the electron interactions with dilatational and flexural phonon modes (shear waves don't interact with electrons through the deformation potential) given in [4].

The wave functions, $\Psi_{n,l,\sigma}$, and the energy spectrum, $E_{n,l,\sigma}$, of the Hamiltonian H_0 can be written as:

$$\Psi_{v,l,\sigma} = \sqrt{v!/\pi R^2(v + |l|)!}\, \zeta^{|l|} e^{-\zeta^2/2} e^{il\theta} L_v^{|l|}(\zeta^2)\chi_\sigma,$$
$$E_{v,l,\sigma} = \hbar\Omega(2v + |l| + 1) - \hbar\omega_c l + g^*\mu_B B\sigma_z/2,$$

where χ_σ represents the eigenfunction of σ_z. $v = 0,1,2,\ldots$, $l = 0, \pm1, \pm2, \ldots$ are quantum numbers, $\Omega = \sqrt{\omega_0^2 + \omega_c^2/4}$, $\omega_c = |e|B/m^*c$, ζ denotes the scaled polar coordinate ($\zeta = \rho/R$, with $R = \sqrt{\hbar/m^*\Omega}$), $L_n^{|l|}(x)$ is the generalized Laguerre polynomial. The spin-orbit coupling can be treated as a small perturbation to the discrete orbital energy level spectrum in the QD. As a result, the electron spin states are admixtures of spin and orbital states. Thus, in first-order perturbation theory, we get $\varepsilon = E_{n,l,\sigma}$, and

$$\Phi_{0,0,\uparrow} = \Psi_{0,0,\uparrow} + \sum_{n,l,\sigma} \frac{\left\langle \Psi_{n,l,\downarrow} \middle| H_{SO}^{D(R)} \middle| \Psi_{0,0,\uparrow} \right\rangle}{E_{0,0,\uparrow} - E_{n,l,\sigma}} \Psi_{n,l,\downarrow}, \quad \Phi_{0,0\downarrow} = \Psi_{0,0,\downarrow} + \sum_{n,l,\sigma} \frac{\left\langle \Psi_{n,l,\uparrow} \middle| H_{SO}^{D(R)} \middle| \Psi_{0,0,\downarrow} \right\rangle}{E_{0,0,\downarrow} - E_{n,l,\sigma}} \Psi_{n,l,\uparrow},$$

$$\Phi_{0,1,\uparrow} = \Psi_{0,1,\uparrow} + \sum_{n,l,\sigma} \frac{\left\langle \Psi_{n,l,\downarrow} \middle| H_{SO}^{D(R)} \middle| \Psi_{0,1,\uparrow} \right\rangle}{E_{0,1,\uparrow} - E_{n,l,\sigma}} \Psi_{n,l,\downarrow}, \quad \Phi_{0,1,\downarrow} = \Psi_{0,1,\downarrow} + \sum_{n,l,\sigma} \frac{\left\langle \Psi_{n,l,\uparrow} \middle| H_{SO}^{D(R)} \middle| \Psi_{0,1,\downarrow} \right\rangle}{E_{0,1,\downarrow} - E_{n,l,\sigma}} \Psi_{n,l,\uparrow}.$$

As spin-orbit couplings mix the spin states with different orientations in the Zeeman sublevels, the spin relaxation is possible via electron-phonon interaction. Using Fermi's golden rule for the transition from the initial state i to the final state f,

$$W_{i\to f}^{\pm} = \frac{2\pi}{\hbar}\left|\langle f | H_{el-ph} | i \rangle\right|^2 \delta\left(E_f - E_i \pm \hbar\omega_n^\beta(\boldsymbol{q})\right),$$

and the expressions for the four lowest levels with the Dresselhaus (H_{SO}^D) and Rashba (H_{SO}^R) SO couplings, we get the relaxation rate assisted with the dilatational and flexural acoustic phonon scattering as follows:

Journal of Physics: Conference Series **647** (2015) 012043 doi:10.1088/1742-6596/647/1/012043

$$W_{i\to f}^{\pm}(\beta,n,\boldsymbol{q}) = \frac{\pi E_d^2 |F_{\beta,n}|^2}{A\rho\omega_n^{\beta}(\boldsymbol{q})} \left|\left\langle \Psi_{n,l,\downarrow}\left|H_{SO}^{D(R)}\right|\Psi_{m,p,\uparrow}\right\rangle\right|^2 \left(N_n^{\beta}(\boldsymbol{q}) + \frac{1}{2} \pm \frac{1}{2}\right)(q_{t,n}^2 - q^2)^2 (q_{l,n}^2 + q^2)^2 \left(tsc1_{\beta}\left(\frac{q_{t,n}a}{2}\right)\right)^2 \left|\int_0^a \varphi_{s'}(z)\, tsc2_{\beta}(q_{l,n}z)\, \varphi_s(z)dz\right|^2 \delta\left(E_f - E_i \pm \hbar\omega_n^{\beta}(\boldsymbol{q})\right)$$

where + (-) is used to denote emission (absorption), β distinguishes the dilatational (d) and flexural (f) confined acoustic modes, the label $n = 1,2,3,\dots$ is the index for different branches of confined modes, $N_n^{\beta}(\boldsymbol{q})$ is the Bose distribution of phonon with mode $(\beta, n, \boldsymbol{q})$, E_d is the acoustic deformation potential constant, ρ is the density of the material, and A is the area of the FSQW structure. The symbols representing trigonometric functions are $tsc1_{\beta} = sin$ and $tsc2_{\beta} = cos$ if $\beta = d$ (dilatational) and $tsc1_{\beta} = cos$ and $tsc2_{\beta} = sin$ if $\beta = f$ (flexural). The function $F_{\beta,n}$ is the normalization constant defined in terms of $q_{l,n}$ and $q_{t,n}$ [4], $\varphi_s(z)$ are eigenfunctions of the 1D Schrodinger equation (in calculations the Fang–Howard wavefunction [5] is used). The spin relaxation rate, Γ (time, τ), can be determined by

$$\Gamma = 1/\tau = \sum_i f_0(E_i) \sum_f \left(1 - f_0(E_f)\right) \sum_{\pm} \sum_{\beta,n,\boldsymbol{q}} W_{i\to f}^{\pm}(\beta,n,\boldsymbol{q}).$$

3. Numerical results and discussion

In figure 1 we present the spin-flip relaxation rate as a function of magnetic field due to Rashba spin-orbit coupling mediated by the electron interaction with both dilatational and flexural acoustic phonon modes. As we can see, the relaxation rate rises with the increase of the intensity of the magnetic field. This feature is quite opposite to the bulk, two- and one- dimensional cases where the spin relaxation rate always decreases with the magnetic field. This is because in the dot case there are only discrete energy levels and the magnetic field helps to increase the spin-flip scattering.

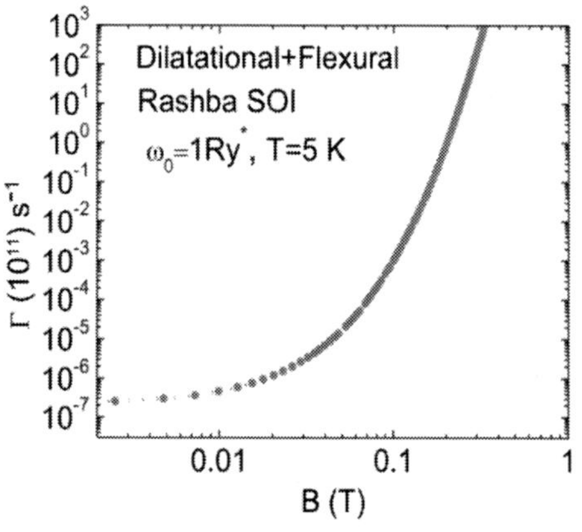

Figure 1. Spin relaxation rate as a function of a magnetic field.

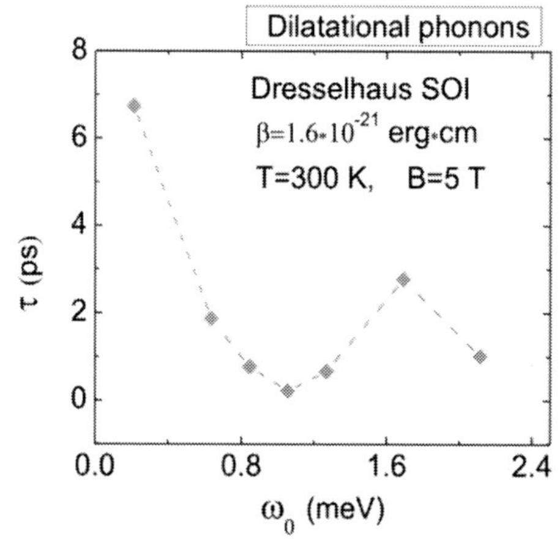

Figure 2. Spin relaxation time as a function of the parabolic confinement potential frequency.

The calculated spin-flip relaxation time from parabolic confinement potential frequency is plotted in figure 2. Dots represent calculated values. Only a few values of the relaxation time have been calculated, which don't allow us to determine the frequency behavior of the spin relaxation. The spin relaxation time as a function of temperature for the spin flip transitions caused by Dresselhaus spin-orbit interaction is presented in figure 3 by considering the electron-dilatational acoustic phonon (figure 3a) and the electron-flexural acoustic phonon (figure 3b) interactions. It is clearly seen that in comparison with flexural phonons, the electron coupling with the dilatational phonons causes 3 orders faster spin relaxation.

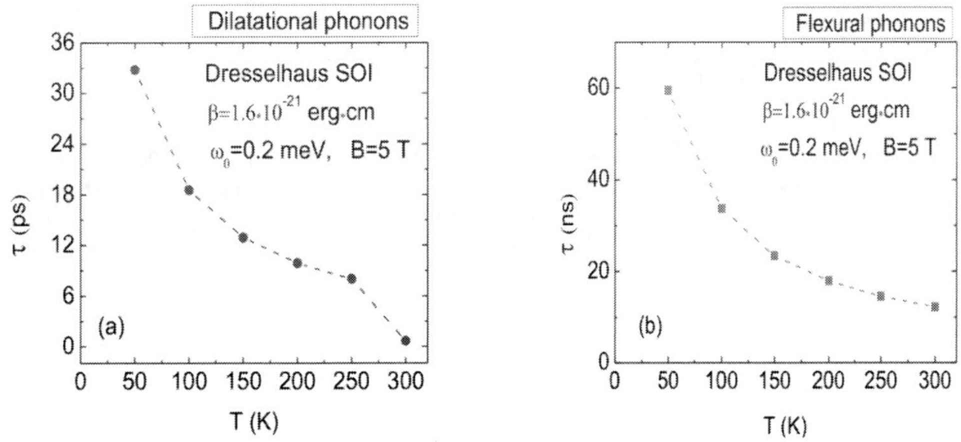

Figure 3. Spin relaxation time as a function of temperature for the spin flip transitions caused by Dresselhaus spin-orbit interaction: (a) via electron-dilatational acoustic phonon scattering, (b) via electron-flexural acoustic phonon scattering.

One can also notice from the figures 3a and 3b that with the increase of temperature, the spin relaxation time decreases. These features can be understood as follows: with the increase of the temperature, the average phonon number gets larger. This enhances the electron-phonon scattering and leads to the larger transition probability. We have found that the spin-flip relaxation rate is one order of magnitude smaller than that of the spin-conserving scattering.

Acknowledgements
This work was supported by the State Committee of Science MES RA (Projects No.13-1C196, 13IE-038, 13RF-093, 13YR-1C0044) and the ANSEF grant (condmatth-3821) from the Armenian National Science and Education Fund based in New York, USA.

References

[1] *Semiconductor Spintronics and Quantum Computation* 2002 ed. by D D Awschalom, D Loss and N Samarth (Berlin: Springer-Verlag)
[2] Zuti´c I, Fabian J and Das Sarma S 2004 *Rev. Mod. Phys.* **76** 323
[3] Balocchi A, Duong Q H, Renucci P, Liu B L, Fontaine C, Amand T, Lagarde D and Marie X 2011 *Phys. Rev. Lett.* **107** 136604
[4] Bannov N, Mitin V and Stroscio M 1994 *Phys. Stat. Sol.* B **183** 131
[5] Fang F F and Howard W E 1966 *Phys. Rev. Lett.* **16** 797

Canted antiferromagnetic to ferromagnetic phase transition in bilayer graphene

S.Pezzini[1,2,4], **C.Cobaleda**[2,5], **B.A.Piot** [3], **V.Bellani**[1] **and E.Diez**[2]

[1] Dipartimento di Fisica and CNISM, Universitá degli studi di Pavia, I-27100 Pavia, Italy
[2] Laboratorio de Bajas Temperaturas, Universidad de Salamanca, E-37008 Salamanca, Spain
[3] Laboratoire National des Champs Magnétiques Intenses, CNRS-UJF-UPS-INSA, F-38042 Grenoble, France

E-mail: `s.pezzini@science.ru.nl`

Abstract. We performed electrical transport measurements in strong tilted magnetic fields (up to 30 T) on a high quality bilayer graphene sample. The possibility of tuning the system into a topological state at zero filling factor, associated to a ferromagnetic phase, is demonstrated on the basis of an insulator-to-metal transition and simultaneous arising of strong non-local response. Based on recent theoretical predictions, relevant aspects of the phase transition are quantitatively extracted. New experimental developments on the topics are proposed in the final section.

1. Introduction

The quantum spin Hall (QSH) effect is a new phase of matter where a gapped bulk coexist with gapless helical edge states, which are protected from backscattering by a symmetry of the physical system [1, 2]. While time-reversal symmetry plays this role in two-dimensional topological insulators (2D TIs) such as inverted HgTe [3, 4] and InAs/GaSb quantum wells [5, 6], an analogous topological phase is expected in single and double layer graphene at zero filling factor, provided that an intense in-plane component of the magnetic field ($B_{\|}$) strongly enhances the Zeeman splitting with respect to competing interaction anisotropies [7, 8]. In particular, bilayer graphene is expected to be continuously tunable from a canted antiferromagnetic (CAF) to a ferromagnetic (F) state at the charge neutrality point (CNP), with associated closing of the energy gap for the one-dimensional edge channels and realization of two copies of the QSH effect [9, 10, 11].

2. Sample and experimental set-up

We studied this phenomenology in a van der Waals heterostructure [12] made of a stacking of hexagonal boron nitride (h-BN)[13], bilayer graphene and h-BN flakes obtained by micro-mechanical cleavage. The sample, produced with the dry pick-up technique described in Ref.[14], was previously characterized with experiments both at zero magnetic field [15] and at finite filling factor [16]. We performed electrical transport measurements with standard low-frequency (≈ 13

[4] Present address: High Field Magnet Laboratory (HFML-EMFL), Radboud University, 6525 ED Nijmegen, The Netherlands
[5] Present address: ASML, 5504 DR Veldhoven, The Netherlands

Journal of Physics: Conference Series **647** (2015) 012044 doi:10.1088/1742-6596/647/1/012044

Hz) ac lock-in technique, using an excitation current of 100 nA and varying the carrier density with a dc back-gate voltage (V_g). The angle between the magnetic field and the graphene plane was varied continuously from 0° to 180°; the correct orientation of the sample was tested by carefully checking the carrier density corresponding to the integer quantum Hall states at filling factor ±4, which depends on the perpendicular component of the magnetic field ($B_\perp$) only.

3. Four-probe and non-local measurements in tilted magnetic fields

Figure 1 (a) shows the evolution of the four-probe resistance R_{xx} as a function of V_g when an increasingly strong $B_\parallel$ is applied in addition to a constant quantizing $B_\perp$. This is achieved by progressively increasing the tilting angle of the sample's plane with respect to the field axis and consequently adjusting the total intensity of the field (B_{tot}). As $B_\parallel$ grows, the resistance at the CNP (R_{xx}^{CNP}) drops dramatically, eventually evolving into a local minimum. The suppression of R_{xx}^{CNP} is indicative of the creation of new conduction channels at charge-neutrality. When the highest values of $B_\parallel$ are considered, the local minimum at the CNP assumes values $\frac{1}{4}h/e^2 < R_{xx}^{CNP} < \frac{1}{3}h/e^2$. The lower limit $\frac{1}{4}h/e^2$ represents the theoretical value for the four-probe resistance in the F phase of bilayer graphene, which is calculated taking into account that six ohmic contacts to the graphene were employed in the present experiment and that each edge path between electrical contacts corresponds to a resistor of value $\frac{1}{2}h/e^2$. The effect of the strong $B_\parallel$ applied is therefore consistent with the formation of four helical edge states, i.e. with the realization of two copies of the QSH effect. The not exact quantization detected in our measurements (also observed in Ref.[10] for bilayer graphene, Ref.[8] for single-layer graphene and Ref.[4, 3] for 2D TIs) indicates residual backscattering, likely to be due to short-circuiting via bulk states and/or non-ideal contacts acting as uncontrolled de-coherence sites [4].

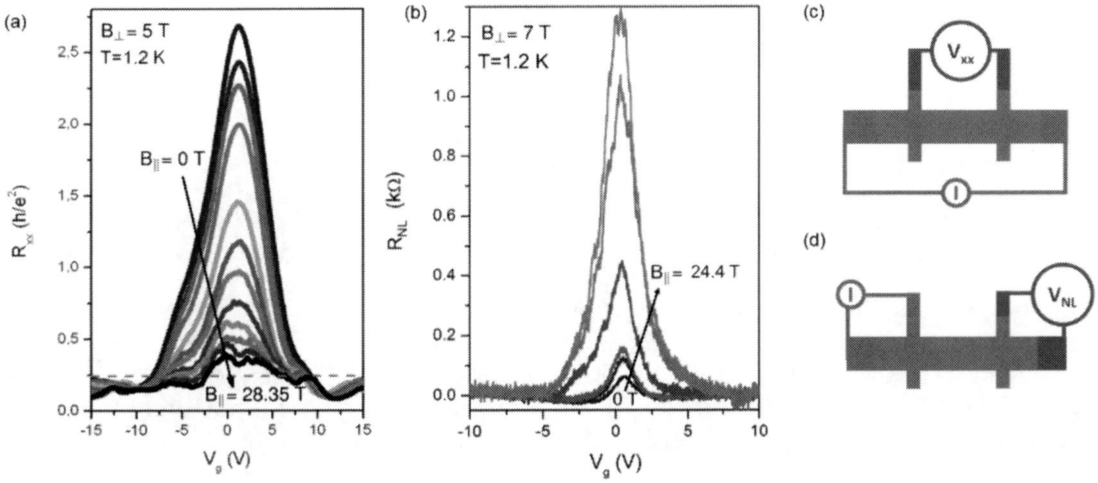

Figure 1. (a) R_{xx} as a function of V_g for increasing $B_\parallel$ with fixed $B_\perp = 5$ T, at $T \simeq 1.2$ K. (b) Non-local resistance as a function of V_g at $B_\perp = 7$ T, $T \simeq 1.2$ K, for increasing values of $B_\parallel$. The corresponding contacts' configurations are shown in (c) for R_{xx} and in (d) for R_{NL}. I indicates the current injection-collection, V the measurement of the voltage drop.

A further indication of transport via edge states at the CNP in the presence of strong $B_\parallel$, supporting the scenario of the F phase, is obtained with non-local measurements. The contacts' configuration employed is sketched in Figure 1 (d): the voltage probes are placed as far as possible from the source and drain current contacts. In the case of diffusive transport via bulk states, the non-local signal $R_{NL} = V/I$ is expected, according to the de van der Pauw formalism [17], to be exponentially small in the separation between the current contacts and the voltage probes. This is the case of the first trace in Figure 1 (b), which corresponds to $B_{\|} = 0$, $B_{\perp} = B_{tot} = 7$ T, at which $R_{NL}^{CNP}/R_{xx}^{CNP} < 10^{-3}$. The application of increasing $B_{\|}$ has again a tremendous effect, with an increase in R_{NL}^{CNP} of more than one order of magnitude. Note that, on the other hand, R_{xx}^{CNP} falls drastically: this rules out the possibility of a dominant bulk contribution to R_{NL}^{CNP}. The strong non-local signal at the CNP, together with the contrasting behavior of the four-probe resistance, can be explained only with transport via edge states. The theoretical value, calculated within the same assumption as in the previous paragraph, is $R_{NL}^{CNP} = \frac{1}{12}h/e^2$: also in this case the data show a non-perfect quantization.

4. Temperature-dependent measurements and identification of the critical point

We measured R_{xx}^{CNP} at different values of T and B_{tot}, while keeping $B_{\perp}$ constant. In Figure 2 (a) we show the data for $B_{\perp} = 5$ T; analogous ones were acquired also at $B_{\perp} = 4$, 6 and 7 T. A smooth insulator-to-metal transition is observed, as expected for the CAF-F [7, 9], while a T-independent crossing point for the isotherms of R_{xx}^{CNP} separates the two phases, providing a clear experimental identification of the critical point.

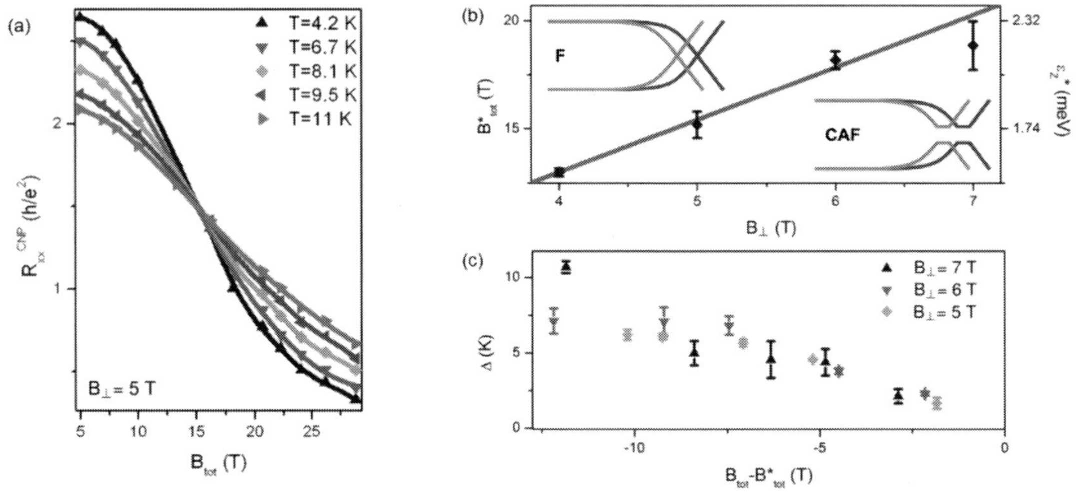

Figure 2. (a) Isotherms of R_{xx}^{CNP} as a function of B_{tot} for $B_{\perp} = 5$ T. The symbols are the experimental points, while the continuous lines are obtained by interpolation with grade-five polynomials. (b) Critical total field (B_{tot}^*, left axis) and critical Zeeman energy (ϵ_Z^*, right axis) for the CAF-F transition as a function of $B_{\perp}$. (c) Energy gap Δ as a function of $B_{tot} - B_{tot}^*$, estimated from the thermally-activated R_{xx}^{CNP} in the insulating phase for different values of $B_{\perp}$.

Figure 2 (b) shows the values of critical field B_{tot}^* corresponding to the T-independent crossing points as a function of $B_{\perp}$. B_{tot}^* clearly increases with $B_{\perp}$ and a linear dependence can be inferred. From a linear fit to the data (see green line in Figure 2 (b)), we obtain

$$B_{tot}^* = (2.42 \pm 0.21) \times B_{\perp} \tag{1}$$

Equation (1) provides a clear experimental determination of the $B_{\perp}$-dependence of this critical field. The Zeeman energies at the critical points ϵ_Z^* (see right axis in Figure 2 (b)) are consistent with the CAF-F boundary $\epsilon_Z \approx 0.8$ meV reported in Ref.[10] for $B_{\perp} = 1.75$ T, which was estimated on the basis of the saturation of R_{xx}^{CNP} rather than at the insulator-to-metal transition as done here. The reliability of our estimation of the critical point is confirmed by the evolution of the energy gap Δ of the insulating phase as the in-plane field is increased. In Figure 2 (c) we show that Δ (obtained by fitting the thermally-activated resistance $R_{xx}^{CNP} \propto \exp(\Delta/2k_BT)$) continuously decreases towards $B_{tot} = B_{tot}^*$, eventually pointing at vanishing values, perfectly matching the evolution of the energy gap for the edge excitations which is reported in Ref.[9].

5. Conclusions and future developments

The experimental work we performed, which combines non-local and four-probe measurements of the resistance of a high quality sample, together with the T-dependence of the latter, consistently confirm the possibility of inducing a topological phase in bilayer graphene with the use of a strong in-plane magnetic field. We provide a clear identification of the critical field for the CAF-F transition as corresponding to the T-independent crossing point for the isotherms of R_{xx}^{CNP}. In Ref.[11] we proceed further in the analysis of this transition and propose a scaling behavior of R_{xx}^{CNP} with $(B_{tot} - B_{tot}^*)/T$ in the vicinity of the critical point.

Future work will be mainly centred on detailed non-local measurements with systematic permutation of the contacts' configuration (see e.g. Ref.[18]): this will allow determining the exact contribution of each edge path and possibly identifying the origin of the non-exact quantization reported here. Top-gated samples will be employed in order to study transmission, reflection and interference effects of the helical edge channels in the presence of a potential barrier, which will be represented by a fully gapped layer polarized region, locally induced by a vertical electric field. An alternative way of probing the CAF-F transition is represented by resistively-detected NMR measurements [19], which can give a direct estimate of the evolution of the spin polarization at the CNP in bilayer graphene.

This work has been supported by the Spanish MINECO MAT2013-46308-C2-1-R and FPU-AP2009-2619 (C.C.), Junta de Castilla y León SA226U13 and European Union CTA-228043-EuroMagNET II Programme.

References

[1] C. L. Kane and E. J. Mele, Phys. Rev. Lett. **95**, 146802 (2005).
[2] M. Z. Hasan and E. J. Mele, Rev. Mod. Phys. **82**, 3045 (2010).
[3] M. König, S. Wiedmann, C. Brüne, A. Roth, H. Buhmann, L. W. Molenkamp, X.-L. Qi and S.-C. Zhang, Science **318**, 766 (2007).
[4] A. Roth, C. Brüne, H. Buhmann, L. W. Molenkamp, J. Maciejko, X.-L. Qi and S.-C. Zhang, Science **325**, 294 (2009).
[5] C. Liu, T. L. Hughes, X.-L. Qi, K. Wang, and S.-C. Zhang,Phys. Rev. Lett. **100**, 236601 (2008).
[6] I. Knez, R.-R. Du, and G. Sullivan, Phys. Rev. Lett. **107**, 136603 (2011).
[7] M. Kharitonov, Phys. Rev. Lett. **109**, 046803 (2012).
[8] A. F. Young, J. D. Sanchez-Yamagishi, B. Hunt, S. H. Choi, K. Watanabe, T. Taniguchi, R. C. Ashoori and P. Jarillo-Herrero, Nature **505**, 528 (2014).
[9] M. Kharitonov, Phys. Rev. B **86**, 075450 (2012).
[10] P. Maher, C. R. Dean, A. F. Young, T. Taniguchi, K. Watanabe, K. L. Shepard, J. Hone and P. Kim, Nat. Phys. **9**, 154-158 (2013).
[11] S. Pezzini, C. Cobaleda, B. A. Piot, V. Bellani and E. Diez, Phys. Rev. B **90**, 121404(R) (2014).
[12] A. K. Geim and I. V. Grigorieva, Nature **499**, 419 (2013).
[13] C. R. Dean, A. F. Young, I. Meric, C. Lee, L. Wang, S. Sorgenfrei, K. Watanabe, T. Taniguchi, P. Kim, K. L. Shepard and J. Hone, Nat. Nanotech. **5**, 722726 (2010).
[14] P. J. Zomer, M. H. D. Guimarães, J. C. Brant, N. Tombros and B. J. van Wees, Appl. Phys. Lett. **105**, 013101 (2014)..
[15] C. Cobaleda, S. Pezzini, E.Diez and V. Bellani, Phys. Rev. B **89**, 121404(R) (2014).
[16] C. Cobaleda, S. Pezzini, A. Rodriguez, E. Diez and V. Bellani, Phys. Rev. B **90**, 161408(R) (2014)
[17] L. J. van der Pauw, Philips Research Report **13**, 1-9 (1958).
[18] K. Suzuki, Y. Harada, K. Onomitsu, and K. Muraki, Phys. Rev. B **87**, 235311 (2013).
[19] W. Desrat, D.K. Maude, M. Potemski, J.C. Portal, Z.R. Wasilewski and G. Hill, Phyica E **12**, 149 (2002).

Journal of Physics: Conference Series **647** (2015) 012045 doi:10.1088/1742-6596/647/1/012045

Phonon-induced quantum diffusion in Carbon-based materials

R Rosati[1], **F Dolcini**[1,2] **and F Rossi**[1]

1 Department of Applied Science and Technology, Politecnico di Torino, C.so Duca degli Abruzzi 24, 10129 Torino, Italy

2 CNR-SPIN, Monte S.Angelo - via Cinthia, I-80126 Napoli, Italy

E-mail: roberto.rosati@polito.it

Abstract. Employing as prototypical systems the metallic single-walled carbon nanotubes (SWNTs), we investigate the electronic propagation in Carbon-based materials, showing that their linear electronic spectra protect the spatial shape of electronic wavepackets from phonon-induced diffusion, up to micrometric scale already at room temperature. To this end, we employ a recently proposed nonlinear Lindblad-based density-matrix approach, which allows us to account for the interplay between phase coherence and dissipation/decoherence, avoiding both huge computational costs of non-Markovian approaches or the limitations of oversimplified dephasing models.

Nowadays, a quite fascinating challenge in nanotechnology is to control the dynamics of electron waves as accurately as in optics.[1] To this purpose, Carbon-based materials such as graphene[2, 3] or metallic single-walled carbon nanotubes (SWNTs)[4] are optimal candidates for dispersionless electronic propagations, thanks to their linear electronic spectra where the Fermi velocity $v_F \sim 10^6$ m/s replaces the speed of light c for photons. In particular, metallic SWNTs can be regarded as the one-dimensional electron waveguides that can in principle provide an electronic alternative to photon-based quantum information processing, as they are synthesized with high accuracy [4, 5, 6]. Although scattering with vacancies or defects can be fairly neglected at first, any implementation in realistic devices requires to determine the impact of electron-phonon processes, which are indicated by experiments as the main source of scattering at intermediate and room temperature [7, 8]. Here, along the lines of recent results [9], we investigate their impact by employing a recently developed density-matrix approach, which enables us to account for both energy dissipation and decoherence effects.[10] We demonstrate that, while in semiconducting SWNTs an electronic wavepacket spreads already for a scattering-free propagation, in metallic SWNTs its shape is essentially unaltered, even in the presence of electron-phonon coupling. Differently from predictions based on the conventional relaxation-time approximation (RTA), our results thus indicate metallic SWNTs as a realistic electron-based platform for information transfer.

In the neighborhood of **K** and **K**' valleys, the free-electron states belonging to the lowest energy subband may be labelled by $\alpha = (k, b, v, \nu)$, where k denotes the continuous wavevector component along the SWNT axis, $b = \pm 1$ and $v = \pm 1$ stands respectively for conduction/valence band and for **K**/**K**' valley, while ν may take the value $\nu = 0, \pm 1$, depending on the geometry of

Journal of Physics: Conference Series **647** (2015) 012045 doi:10.1088/1742-6596/647/1/012045

the SWNT. The corresponding spectrum, which is given by

$$\epsilon_\alpha = b\hbar v_F\sqrt{k^2 + (\nu/3R)^2} \ , \qquad (1)$$

is gapped (semiconducting nanotube) and, near $k = 0$, parabolic-like (similarly to conventional semiconductors) for $\nu \neq 0$, while for $\nu = 0$ it is gapless (metallic case), with the typical massless Dirac-cone structure.[11] The SWNT phonon spectrum near the $\mathbf{K}$ and $\mathbf{K}'$ points includes[11, 12] longitudinal and transverse acoustic modes with sound velocities of about $1.9 \cdot 10^4$m/s and $1.5 \cdot 10^4$m/s respectively, the breathing modes orthogonal to the nanotube surface, with a roughly q-independent spectrum $\hbar\omega_{Br} \simeq 0.14\,\mathrm{eV\AA}/R$, and a couple of optical modes whose energies are about 0.2 eV.

In order to account for energy dissipation and decoherence induced by the nanotube phonon bath on the otherwise phase-preserving electron dynamics, we have applied to the carbon nanotube the formalism introduced in [10] via a numerical solution of the Lindblad-based non-linear density-matrix equation (LBE)

$$\frac{d\rho_{\alpha_1\alpha_2}}{dt} = \frac{\epsilon_{\alpha_1} - \epsilon_{\alpha_2}}{\imath\hbar}\rho_{\alpha_1\alpha_2} + \qquad (2)$$
$$+\frac{1}{2}\sum_{\alpha'\alpha'_1\alpha'_2,s}\left[(\delta_{\alpha_1\alpha'} - \rho_{\alpha_1\alpha'})\,\mathcal{P}^s_{\alpha'\alpha_2,\alpha'_1\alpha'_2}\rho_{\alpha'_1\alpha'_2} - \left(\delta_{\alpha'\alpha'_1} - \rho_{\alpha'\alpha'_1}\right)\mathcal{P}^{s*}_{\alpha'\alpha'_1,\alpha_1\alpha'_2}\rho_{\alpha'_2\alpha_2}\right] + \mathrm{H.c.}$$

Here, the first term describes the scattering-free propagation, whereas the second term is a non-linear scattering superoperator expressed via generalized scattering rates $\mathcal{P}^s_{\alpha_1\alpha_2,\alpha'_1\alpha'_2}$, whose explicit form is microscopically derived from the electron-phonon Hamiltonians, with s labelling the various phonon modes. The fully quantum-mechanical density-matrix equation (2) enables us to go beyond the conventional Boltzmann transport equation, which is recovered in the diagonal limit ($\rho_{\alpha_1\alpha_2} = f_{\alpha_1}\delta_{\alpha_1\alpha_2}$),[13] where the generalized scattering rates reduce to the semiclassical rates provided by the standard Fermi's golden rule, $P^s_{\alpha\alpha'} = \mathcal{P}^s_{\alpha\alpha,\alpha'\alpha'}$.

In order to show that carbon nanotubes can be utilised as quantum-mechanical channels for the non-dispersive transmission of electronic wavepackets, we have performed simulated experiments where the shape of an initially prepared wave packet is monitored while it evolves under the effect of the phonon bath. Any initial state can always be written as $\rho = \rho^\circ + \delta\rho$, where ρ° is the homogeneous equilibrium state and $\delta\rho$ describes a localised excitation, whose spatial shape (e.g. Gaussian) can in principle be generated experimentally via a properly tailored optical excitation. The description of the specific optical generation is beyond the aim of the present paper. However, the localisation of the initial wave packet is the crucial aspect in our analysis. A simple choice that captures this essential physical ingredient is an initial state described by an intra-valley conduction-band density matrix $\rho_{\alpha_1\alpha_2} = \delta_{v_1,v_2}\delta_{b_1,1}\delta_{b_2,1}\rho_{k_1k_2}$, where

$$\rho_{k_1k_2} = \sqrt{f^\circ_{k_1}f^\circ_{k_2}}\,e^{-\ell|k_1-k_2|} \ . \qquad (3)$$

Here f°_k is the Fermi-Dirac distribution of the conduction-band states k, and the parameter ℓ plays the role of a delocalization length. Indeed for $\ell \to \infty$ the spatially homogeneous equilibrium state $\rho^\circ_{k_1k_2} = f^\circ_{k_1}\delta_{k_1k_2}$ is recovered, whereas for finite ℓ the excitation $\delta\rho$ consists of interstate phase coherence (intraband polarization) determined by the parameter ℓ and whose presence is necessary for producing localized distributions. In particular, in the limit $\ell \to 0$ one obtains the maximally localized wavepacket. Moreover, a distinguished feature of the initial condition in (3) is the absence of nonequilibrium diagonal contributions ($\rho_{kk} = f^\circ_k$), which implies that energy dissipation and decoherence will affect the non-diagonal contributions only; this is the typical situation produced by a weak interband optical excitation.

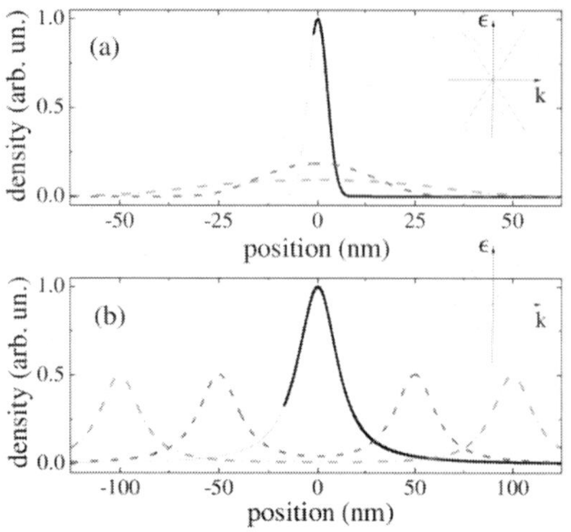

Figure 1. Room-temperature scattering-free evolution in a semiconducting (10,0) SWNT ($\nu = 1$) (a) and in a metallic (12,0) ($\nu = 0$) SWNT (b): Charge density of the maximally-localized ($\ell \to 0$) electronic wavepacket in Eq. (3) as a function of the position along the SWNT axis at three different times: $t = 0$ fs (solid), $t = 50$ fs (dashed) and $t = 100$ fs (dashed-dotted).

We start our discussion by analysing the scattering-free propagation of the initial state (3), i.e. switching off the electron-phonon interaction in Eq. (2). Figure 1 shows the spatial carrier density $n(r_\parallel)$ at different times for the initial state (3), taken at room temperature and in the maximally-localized limit ($\ell \to 0$). The cases of semiconducting and metallic nanotubes are shown in panels (a) and (b), respectively. For the semiconducting nanotube ($\nu \neq 0$) the dispersion relation (see Eq. (1)) at small k reduces to the parabolic spectrum of conventional semiconductor materials and gives rise to the typical classical-like diffusion process, preventing any information transfer via electronic wavepackets. In contrast, for the case of the metallic nanotube ($\nu = 0$), characterised by a linear dispersion, the initial charge peak splits into two identical and shape-preserving components which travel in opposite directions with velocity $\pm v_F$ (see Fig 1(b)), i.e., $n(r_\parallel, t) = n^+(r_\parallel - v_F t) + n^-(r_\parallel + v_F t)$. This non-dispersive propagation is the interesting starting point of our analysis.

We now focus on the metallic nanotube and switch on the electron-phonon coupling to address the crucial question of whether and how energy dissipation and decoherence modify such ideal dispersion-free scenario. To this end, we have performed a set of simulated experiments based on the LBE (2) including all the above introduced phonons. In order to expand the space scale with respect to the ideal scenario displayed in Fig. 1(b), here the delocalization length ℓ in the initial condition (3) is chosen such to get a FWHM value of the initial peak of 0.4μm.

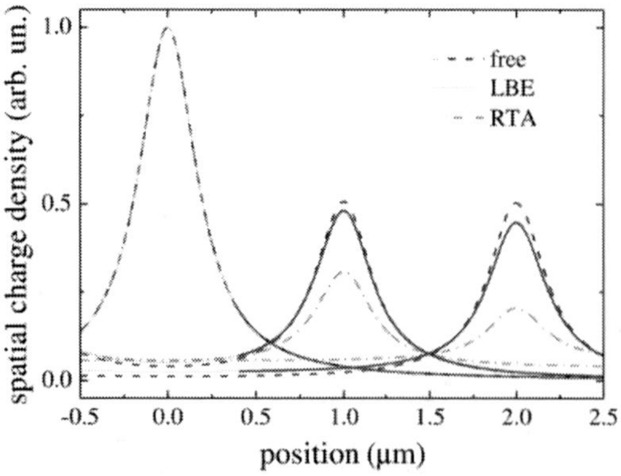

Figure 2. Room-temperature evolution of the electronic wavepacket (3), with ℓ such to provide an initial FWHM of 0.4 μm, in a metallic (12,0) SWNT as a function of the position $r_\parallel$ along the SWNT axis, at three different times: $t = 0$ ps (left peaks), $t = 1$ ps (central peaks) and $t = 2$ ps (right peaks). The solid lines show the effects of the carrier-phonon coupling, accounted for by the LBE (2), compared to the scattering-free (dashed lines) as well as RTA propagations (dot-dashed lines).

Figure 2(a) shows the wavepacket propagation in a (12,0) SWNT at room temperature. The electron-phonon coupling, accounted for by our LBE simulation (solid curve), does not significantly alter the shape with respect to the ideal scattering-free result (dashed curve), so that the transmission is essentially dispersionless up to the micrometric scale (the diffusion time would still increase at lower temperatures, e.g. at 77 K one has a signal attenuation of less than 1% after 1μm). Such a strong suppression of the phonon-induced diffusion predicted by our LBE is mainly due to the negligible effect produced by forward processes (i.e., those which preserve velocities), which do not lead to the scattering non-locality and quantum diffusion speed-up phenomena that occur in semiconducting materials [13]. Remarkably, such uneffectiveness of forward scattering processes occurs despite their electron-phonon couplings are a priori one order of magnitude bigger than the backward ones [11] for acoustic as well as breathing phonons (here the optical modes have a negligible impact due to their high energy[7]). In sharp contrast, the oversimplified RTA strongly overestimates the phonon-induced dissipation, due to its inability to distinguish between forward and backward processes, as shown by the dash-dotted line of Fig.2. Notice that, while semiclassical approaches show a somehow similar forward-scattering suppression, here the off-diagonal nature of the density matrix makes such phenomena highly nontrivial and strictly related to the genuine quantum phenomenon of scattering-induced nonlocality.

In conclusion, differently from what happens in semiconducting materials[10], in metallic SWNTs the impact of the phonon-induced diffusion on the wavepacket propagation is very limited. The scattering-free propagation is thus not just an ideal scenario, and the initial wavepacket propagates at the Fermi velocity with an almost unaltered shape, despite dissipation and decoherence processes are in principle present. Our results indicate that metallic SWNTs are a promising platform to realise quantum channels for the non-dispersive transmission of electronic wavepackets. We thus expect these results will pave the way for advanced optoelectronics with metallic SWNTs.

Acknowledgments
We gratefully acknowledge funding by the Graphene@PoliTo laboratory of the Politecnico di Torino, operating within the European FET-ICT Graphene Flagship project (www.graphene-flagship.eu). F.D. also acknowledges support from Italian FIRB 2012 project HybridNanoDev (Grant No.RBFR1236VV).

References
[1] Kok P, Munro W J, Nemoto K, Ralph T C, Dowling J P and Milburn G J 2007 *Rev. Mod. Phys.* **79**(1) 135–174
[2] Neto A C, Guinea F, Peres N, Novoselov K S and Geim A K 2009 *Rev. Mod. Phys.* **81** 109
[3] Basov D N, Fogler M M, Lanzara A, Wang F and Zhang Y 2014 *Rev. Mod. Phys.* **86**(3) 959–994
[4] Charlier J C, Blase X and Roche S 2007 *Rev. Mod. Phys.* **79**(2) 677–732
[5] Maruyama S, Kojima R, Miyauchi Y, Chiashi S and Kohno M 2002 *Chem. Phys. Lett* **360** 229–234
[6] Arnold M S, Green A A, Hulvat J F, Stupp S I and Hersam M C 2006 *Nat. Nanotechnol.* **1** 60–65
[7] Yao Z, Kane C L and Dekker C 2000 *Phys. Rev. Lett.* **84** 2941
[8] Park J Y, Rosenblatt S, Yaish Y, Sazonova V, Üstünel H, Braig S, Arias T, Brouwer P W and McEuen P L 2004 *Nano Lett.* **4** 517–520
[9] Rosati R, Dolcini F and Rossi F 2015 *Appl. Phys. Lett.* **106** 243101
[10] Rosati R, Iotti R C, Dolcini F and Rossi F 2014 *Phys. Rev. B* **90**(12) 125140
[11] Ando T 2005 *J. Phys. Soc. Jpn.* **74** 777–817
[12] Suzuura H and Ando T 2002 *Phys. Rev. B* **65** 235412
[13] Rosati R and Rossi F 2014 *Phys. Rev. B* **89** 205415

Effect of charged impurity scattering on the electron diffusivity and mobility in graphene

R Rengel, J M Iglesias, E Pascual and M J Martín
Department of Applied Physics, University of Salamanca, Salamanca 37008, Spain
E-mail: raulr@usal.es

Abstract. The influence of charged impurities on the transport properties of graphene, including the diffusion coefficient at low and high electric fields, is investigated by means of an ensemble Monte Carlo simulator. Three different possible substrates are considered, h-BN, SiC and SiO_2. The results show the importance of impurity scattering in degrading the diffusion coefficient, electronic mobility and drift velocity, particularly at low and intermediate electric fields. The influence of the substrate dielectric constant in relation to impurities and surface polar phonon scattering is also evidenced; this turns SiC into the most suitable substrate of choice from this point of view.

1. Introduction

Graphene is commonly recognized as one of the most promising materials for future electronics [1]. However, its excellent intrinsic electronic properties are significantly affected by several effects appearing in fabricated samples. In particular, Coulomb impurities acting as long-range scatterers are an important scattering source in graphene, limiting the conductivity in this material [2]. Therefore, it is of primary importance to consider this type of interaction in realistic models for the study of graphene. Moreover, it becomes particularly interesting to analyze the interplay of this kind of scattering with intrinsic phonon interactions and with remote phonons from the underlying substrate, and to study how the substrate type affects the Coulomb screening of the charged impurities.

In this work, we have analyzed the influence of charged impurities on the transport properties of graphene, including the low-field mobility and the diffusion coefficient at low and high electric fields, by means of an ensemble Monte Carlo simulator. For relatively high carrier concentration, slightly away from the Dirac point, the semiclassical Boltzmann transport theory provides a good description of charge transport physics in graphene [3]. Consequently, the Monte Carlo method becomes a powerful tool for the investigation of the electronic transport properties of this material [4]. Three different substrates have been considered, i.e., h-BN, SiC and SiO_2.

2. Monte Carlo simulator

The calculations have been carried out by means of an ensemble Monte Carlo simulator incorporating the main properties of graphene, such as a linear dispersion relation close to the Dirac point, which implies that electrons behave as massless particles. The most relevant scattering mechanisms are included in the simulation [5, 6].

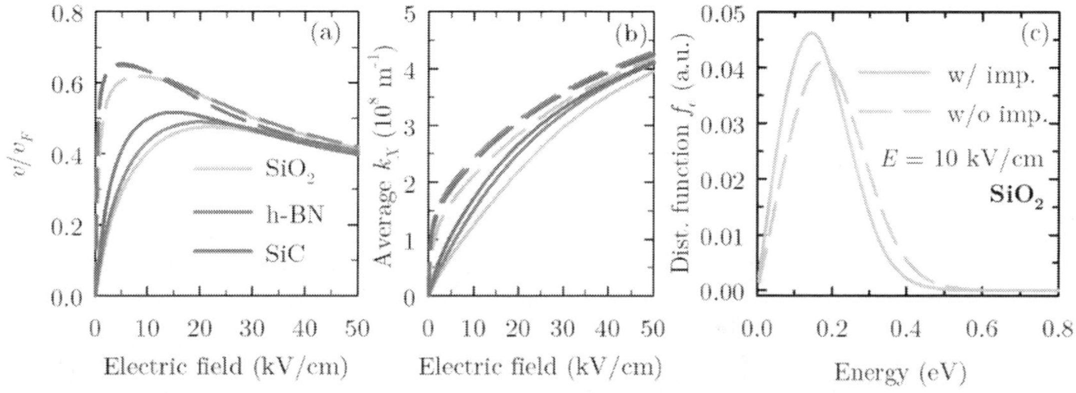

Figure 1. Average drift velocity (a), average wavevector in the direction of the applied field (b) and energy distribution function for the case of SiO_2 (c). Solid lines represent the results including impurity scattering and dashed lines without it. The electron density is 10^{12} cm^{-2}.

In the case of Coulomb interactions with charged impurities, the scattering matrix element is obtained from the 2D Fourier transform of the electrostatic potential [2, 3, 7]. Applying Fermi's Golden rule it is possible to obtain the transition probability as a function of the energy, which is proportional to the impurity density. The anisotropy of Coulomb scattering is rigorously treated for the selection of the final angle after a scattering event.The screening of 2D Coulomb interactions is included taking into account its anisotropic nature and the static polarizability to account for the dielectric function [8], considering the electronic temperature self-consistently determined at each timestep [9]. The calculation of the diffusion coefficient is carried out as explained in [5], considering an excess electron population coupled to the background population by means of a carrier-carrier exchange scattering mechanism.

3. Results and discussion

The effect of charged impurities on the electron drift velocity as a function of the applied field is shown in Figure 1(a). The electron density concentration considered in this work is equal to 10^{12} cm^{-2}. The impurity density is $5{\cdot}10^{11}$ cm^{-2} for all the substrates under study. As it can be observed, impurities noticeably reduce the drift velocity, particularly at low and moderate (below 20 kV/cm) electric fields, and they also moderate the negative differential conductance observed.

At very high fields the differences tend to reduce, not only between the case with and without impurities, but also in relation to the different substrate types. It is interesting to remark that, while in the absence of impurities graphene on h-BN and on SiC present similar velocity peak values, when impurities are considered graphene on SiC clearly outperforms graphene on h-BN. The average wavevector in the direction of the applied field is shown in Figure 1(b). Charged impurity scattering reduces the average wavevector value, with the largest influence in the range 0-10 kV/cm. This is directly related to a reduction in the average energy provoked by this kind of interaction; it also shifts the energy distribution function f_e, as evidenced in Figure 1(c), which shows f_e at 10 kV/cm for the case of considering SiO_2 as the supporting substrate. A similar behavior of f_e for h-BN and SiC was observed.

In order to discuss the influence of charged impurity scattering, it is relevant to check the quantitative influence of this type of interactions as compared to other scattering events in the graphene layer. Figure 2 shows the average number of scattering events per unit time for the three cases under consideration (graphene on SiO_2, on h-BN and on SiC). The maximum

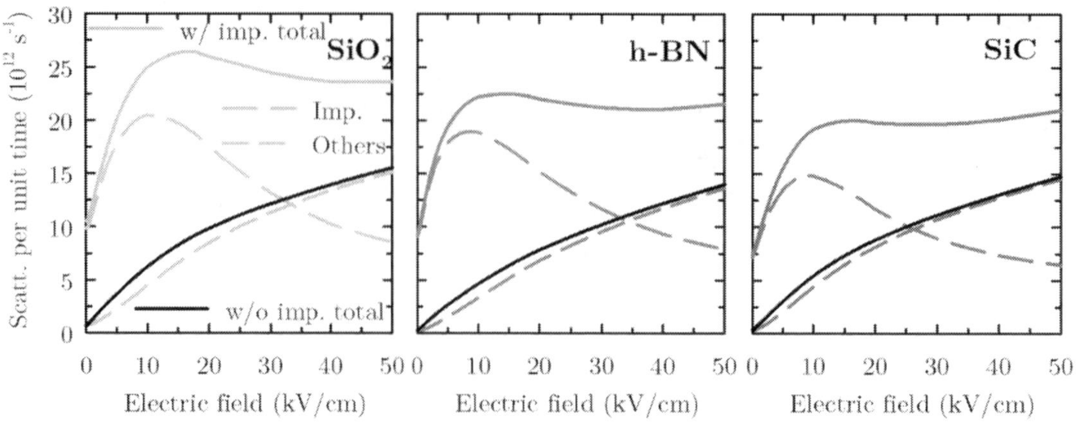

Figure 2. Average number of scatterings per unit time, for SiO_2 (a), h-BN(b) and SiC (c). 'w/imp total' refers to the total number of scatterings when impurities are included and 'w/o imp total' the total number when impurities are not considered.

intensity of impurity scattering is obtained for electric fields close to 10 kV/cm in all cases; it is also noticeable the reduced influence of impurity scattering in the case of SiC. This is related to the larger dielectric constant of this substrate type, which provides a more intense screening of the impurities and therefore reduces the strength of this interaction. Due to the reduction of the average energy in the presence of charged impurities, the other types of scattering events have a reduced rate as compared to the case without impurities. At very high electric fields, over 10 kV/cm, the influence of impurity scattering becomes progressively reduced: its Coulomb nature makes it less prone to affect electronic transport at high energies.

Charged impurity scattering also affects to how much power is transferred to the graphene layer and the underlying substrate via inelastic phonons (Figure 3(a)). Below 10 kV/cm, there is a significant reduction of the power exchange due to inelastic interactions, which would translate into less self-heating in the samples.

Let us now check the evolution of macroscopic parameters as the low-field mobility or the diffusion coefficient as a function of the impurity concentration (Figures 3(b) and 3(c)). In the absence of impurities, graphene on h-BN provides the best results (mobility equal to 260,000 cm^2/Vs, diffusion coefficient equal to 15,800 cm^2/s), with graphene on SiC closely following (221,000 cm^2/Vs and 13,200 cm^2/s) and graphene on SiO_2 clearly underperforming (96,500 cm^2/Vs and 5,800 cm^2/s). However, when impurities are present, this behaviour changes: both parameters are significantly reduced (one order of magnitude approximately) and graphene on SiC shows the best performance of the cases under study, with graphene on h-BN only slightly better than graphene on SiO_2. This is directly related to the lower dielectric constant of h-BN as compared to SiC, which affects the screening of impurities and surface polar phonon scattering.

At high fields (over 10 kV/cm) the diffusion coefficient presents similar values for all the substrates under consideration. In this case, even so if charged impurities are still the dominant scattering type, its strong anisotropic nature and the large energy values attained in this rage moderate their influence on the microscopic diffusion processes.

The high-field drift velocity at 10 kV/cm (Figure 3(d)) also shows important differences, with SiC also presenting the largest values as compared to the other substrates under consideration. This is clearly related to the reduced influence of impurities in this case, as previously discussed. However, at very intense fields (50 kV/cm) the differences tend to reduce, as inelastic scattering processes tend to minimize the differences between the different cases under consideration.

Journal of Physics: Conference Series **647** (2015) 012046 doi:10.1088/1742-6596/647/1/012046

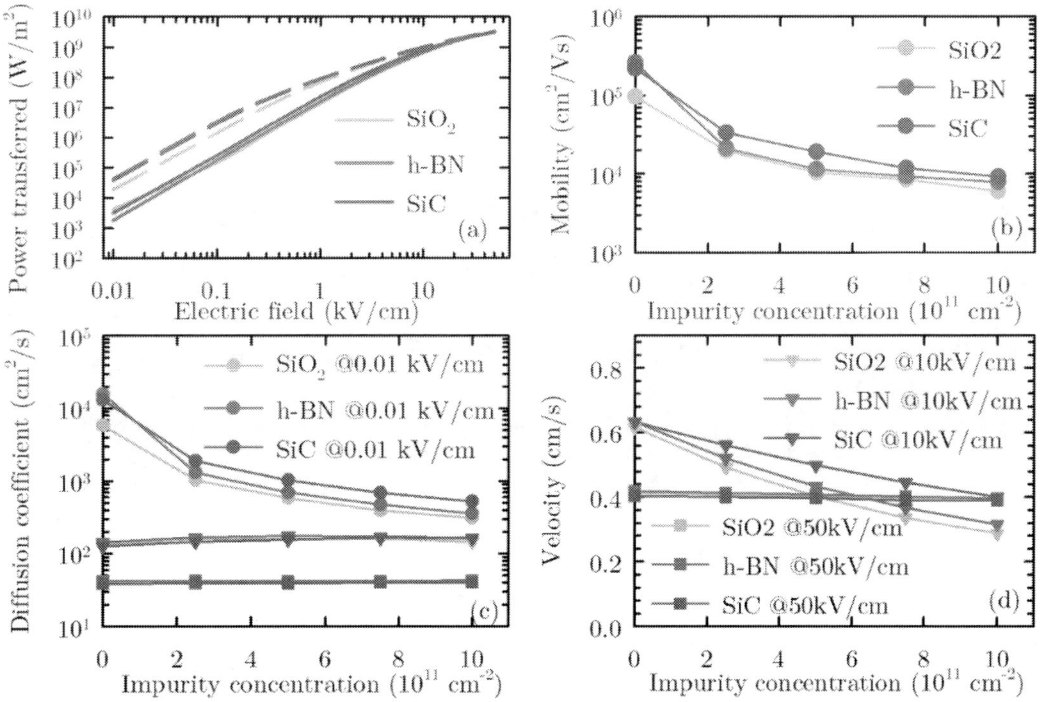

Figure 3. Power transferred as a function of the electric field (a), low-field mobility as a function of the impurity concentration (b), diffusion coefficient (c) and high-field drift velocity (d).

4. Conclusions

The influence of charged impurities on macroscopic parameters such as the low-field mobility, diffusion coefficient or drift velocity has been analyzed by means of a microscopic ensemble Monte Carlo approach. The results evidence that charged impurities significantly affect the transport properties of graphene on a substrate, degrading the overall performance. Moreover, the substrate type is of key importance: in the absence of impurities graphene on h-BN shows the best performance, while in the presence of impurities graphene on SiC gives the best results in terms of drift velocity, low-field mobility and diffusion coefficient, due to a more efficient screening of charged impurities.

Acknowledgments

This work was supported by research project TEC2013-42622-R from the Ministerio de Economía y Competitividad.

References

[1] Novoselov K S, Fal'ko V I, Colombo L, Gellert P R, Schwab M G and Kim K 2012 *Nature* **490** 192–200
[2] Ando T 2006 *Journal of the Physical Society of Japan* **75** 074716
[3] Adam S, Brouwer P W and Das Sarma S 2009 *Phys. Rev. B* **79**(20) 201404
[4] Ferry D 2013 *Journal of Computational Electronics* **12** 76–84 ISSN 1569-8025
[5] Rengel R and Martín M J 2013 *Journal of Applied Physics* **114** 143702
[6] Rengel R, Pascual E and Martín M J 2014 *Applied Physics Letters* **104** 233107
[7] Hamaguchi C 2003 *Journal of Computational Electronics* **2** 169–171 ISSN 1569-8025
[8] Hwang E H and Das Sarma S 2007 *Phys. Rev. B* **75**(20) 205418
[9] Fang T, Konar A, Xing H and Jena D 2011 *Phys. Rev. B* **84**(12) 125450

Journal of Physics: Conference Series **647** (2015) 012047 doi:10.1088/1742-6596/647/1/012047

Evaluation of the Collection Length and Optical Path Enhancement in a-Si:H Solar Cells

M. Fortes, E. Comesaña, J. A. Rodríguez, P. Otero, A. J. Garcia-Loureiro

Centro de Investigación en Tecnoloxías da Información da Universidade de Santiago de Compostela, CiTIUS. Santiago de Compostela, Spain

E-mail: montserrat.fortes@usc.es

Abstract. The characterization of the electrical parameters of solar cells is fundamental to improve their performance. In the case of a-Si:H thin film solar cells, the degrading effect of light has to be mitigated, among others, with thin absorber layers and light confinement techniques. The novelty of this paper is the development of method that evaluates both, the light trapping enhancement and the collection length of the carriers in these and other field-driven devices. In this research we compared experimental results of a-Si:H cells with our simulation model, obtaining good agreement.

1. Introduction

Thin film hydrogenated amorphous silicon (a-Si:H) solar cells have small ambipolar diffusion lengths (L_d). For undoped a-Si:H this value is 0.1-0.4 μm [1]. Therefore these cells use drift, rather than diffusion, as dominant transport mechanism. In the other hand, a-Si:H presents a light-induced degradation (LID) during the first hours of illumination. This phenomenon is known as Staebler-Wronski effect (SWE) and its impact is inversely proportional to the thickness of the i-layer [2]. Since thin i-layers are preferred, light confinement is imperative to increment the path of light and thus improve the absorbance of the cell. In the present article we disclose a method to evaluate the electric and optic performance of textured a-Si:H cells. We extract from the internal quantum efficiency parameters as important as the collection length and the optical enhancement factor.

2. Experimental sample

The 1 cm^2 thin film a-Si:H solar cell studied in this article follows a pin structure. Thin p and n dopped layers of around 10-20 nm induce an electric field on a 200 nm intrinsic layer thus, carriers can move a larger distance before recombination. This so-called collection length (L_c) is at least 10 times greater than the ambipolar L_d of the material [3]. The materials of the device were deposited over a glass substrate that has a Transparent Conductive Oxide (TCO) layer. The front electrode (TCO) is randomly textured in order to scatter the light and it is made up of tin oxide doped with fluorine (SnO_2:F). Following, plasma enhanced chemical vapor deposition (PECVD) technique is used to add the doped and intrinsic a-Si:H layers. Finally, the back contact consists of Al-doped Zinc Oxide (AZO) and an aluminum coating.

Figure 1. Schematic representation of the cross section of the simulated pin a-Si:H solar cell.

3. Simulation tool

Our simulation tool is Sentaurus TCAD. The implemented configurations and the employed parameters are described in reference [4]. The 2D simulated solar cell has the following structure: TCO/p-layer a-SiC:H/i-layer a-Si:H/n-layer a-Si:H/AZO/Al, as depicted in figure 1. The propagation of light was simulated using Monte Carlo Raytracing (RT) in Sentaurus, which needs the complex refractive index (CRI) of all the materials. These values were obtained from reflection, transmission and photothermal deflection spectroscopy measurements. The simulation was performed in the 300-800 nm range, to which a-Si:H is sensitive.

4. Proposed methodology

Based on Basore [5], in diffusive devices it is possible to get the average L_d of the minority carriers plotting the inverse internal quantum efficiency (IQE) in function of the experimental absorption length of the material ($L_{abs} = 1/\alpha$). A linear fitting must be done in the sub-bandgap energy range and the inverse slope would correspond to L_d. This method was extended to flat thin film a-Si:H solar cells, by Tobail et al. [6]. They considered that J_{sc} is proportional to the optical generation (G) and to the collection length [7]. The absorption coefficient ($\alpha(\lambda)$) was obtained experimentally. The photon flux (for each wavelength) at a given depth of the i-layer can be found assuming a Poisson statistical distribution:

$$\phi(d) = \phi_0 exp(-\alpha d), \tag{1}$$

being ϕ_0 the incident photon flux density, and d the depth along the propagation direction of light [8]. Considering that each photon produces one electron-hole pair, the optical generation contribution of each wavelength is:

$$G = \phi_0 \alpha exp(-\alpha d_i), \tag{2}$$

where d_i is the thickness of the i-layer. One may note that generation decreases monotonically, this is because the light intensity diminishes with depth, due to the absorption. However, eq. 2 is not fulfilled in textured cells. Due to light confinement, the average optic path (d_{opt}) in our device is greater than d_i. Additionally, we have to bear in mind the effective absorption of the whole device, which is $1 - R$. Thus we propose the following correction:

$$G = \phi_0(1 - R)\alpha exp(-\alpha d_{opt}). \tag{3}$$

Regarding eq. 3 and expressing IQE as a function of L_{abs}:

$$IQE = \frac{J_{sc}}{q\phi_0(1-R)} = \alpha exp(-\alpha d_{opt}) \times L_c. \quad (4)$$

We obtain a linear expression for 1/IQE in function of L_{abs} if we take into account the range where L_{abs} is greater than d_{opt}. This can be done because short wavelengths are absorbed before reaching the back reflector. The optical path enhancement can only be experienced by long wavelengths. Therefore:

$$IQE^{-1} = \frac{L_{abs}}{L_c}\left(1 + \frac{d_{opt}}{L_{abs}}\right) = \frac{L_{abs}}{L_c} + \frac{d_{opt}}{L_c}. \quad (5)$$

Here the inverse slope is the average L_c and the y-intercept equals d_{opt}/L_c. We use this last one to obtain the average optical path of light in the long wavelength range. Moreover, since d_i is known we can define the optical enhancement factor (OEF) as:

$$OEF = \frac{d_{opt}}{d_i}. \quad (6)$$

Although we do not know beforehand the value for d_{opt}, we do know that its minimum value at long wavelengths will be d_i. The rays at long wavelengths will rebound at the reflective back contact, therefore we can start the adjustment at $L_{abs} > d_i$ to approximate to the linear range.

5. Results

Since these are field-driven devices, in contrast with Tobail, we suggest that the result of any L_c should be reported as well together with the electric field at which it was obtained. We defined E as the the field at $d_i/2$, which is around 5.90×10^4 V/cm. From the simulation model we obtained a reflection profile for the whole device, and the external (EQE) and internal (IQE) quantum efficiencies at the initial and stabilized states of the cell. A low-pass filter was applied to these raw data, using the moving average of every 10 simulated points. The rms of the extracted noise was not larger than an absolute 1.6% in any of the six simulated curves. In general, the simulated and experimental data agree (see figures 2a) and 2b)). However, the overestimation

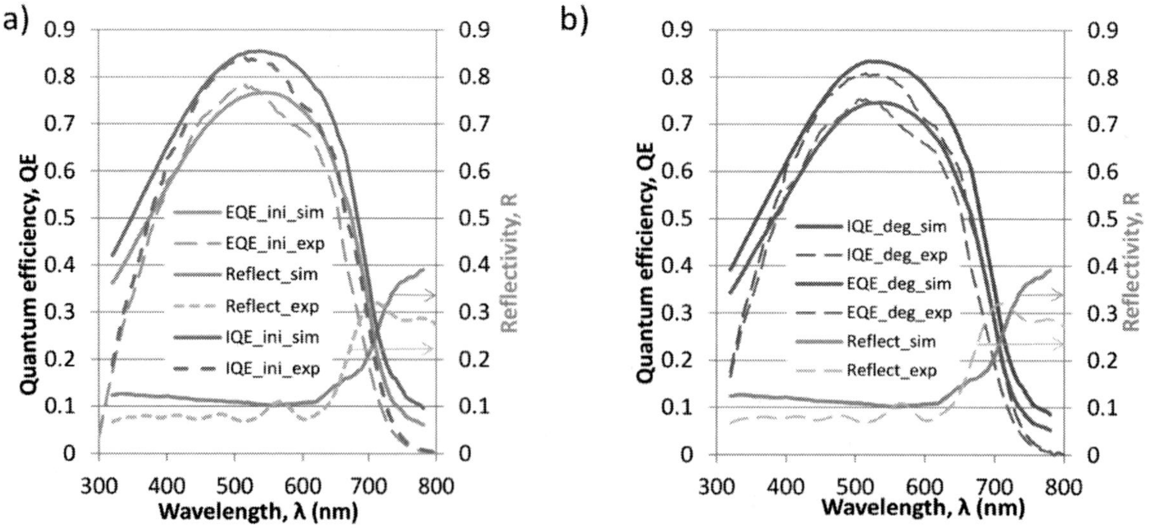

Figure 2. Simulated reflectivity, EQE and IQE in the a) initial and b) degraded state.

Journal of Physics: Conference Series **647** (2015) 012047 doi:10.1088/1742-6596/647/1/012047

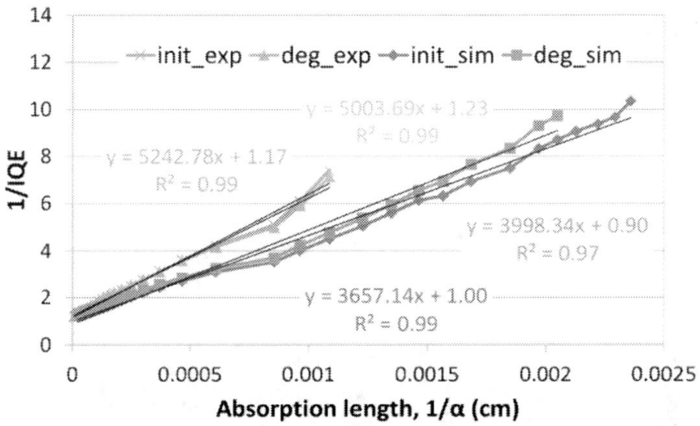

	Initial	
	Experimental	Simulated
Lc (um)	1.91	2.73
OEF	11.12	13.67

	Degraded	
	Experimental	Simulated
Lc (um)	2.00	2.50
OEF	12.31	11.25

Figure 3. Inverse IQE against absorption length for the experimental and simulated curves.

at wavelengths shorter than 350 nm can be explained due to the absence of the front glass in the simulation. It was not included due to lack of experimental CRI data. In figure 3 we show the $1/IQE$ versus L_{abs} plot, for absorption lengths greater than the thickness of the intrinsic layer (200 nm). The results for the obtained L_c and the OEF (calculated using equation 6) are summarized in the right part of the aforementioned figure. We found good agreement too. For the device under study, the carriers generated from low energy photons experience an average collection length of 2 μm and the light trapping techniques increase around 12 times the path of light in the long wavelength range (700-800 nm). This factor agrees as well with the expected, because it does not exceed the geometrical Yablonovitch limit ($4n^2$) [9], which averages 60 in the given wavelength range.

6. Conclusions

A noninvasive technique to evaluate the optical and electrical parameters of texturized a-Si:H solar cells is presented. The collection length of the photogenerated carriers under a given electric field can be extracted from the simulated or measured IQE. The optical enhancement factor of the light trapping techniques can also be determined. A collection length of 2 μm (under a 5.9×10^4 V/cm field) was found for the solar cell under analysis. This number is coherent with the expected since it is around ten times greater than the reported diffusion lengths of a-Si:H. For its part, knowing the estimated thickness of the i-layer we calculated an optical enhancement factor of 12, value in agreement with the theory.

References

[1] Pattnaik, Sambit., Xiao, Teng., Shinar, R., Shinar, J., and Dalal, V.L. 2013. IEEE Journal of Photovoltaics 3 (1), 295-299.
[2] Staebler, D. L., Wronski, C. R. 1980. Journal of Applied Physics 51 (6), 3262.
[3] Shah, A.V., Platz, R., Keppne, H. 1995. Solar Energy Materials and Solar Cells 38 (1-4), 501 - 520.
[4] Fortes, M., Comesaña, E., Rodriguez, J.A., Otero, P. and Garcia-Loureiro, A.J. 2014. Solar Energy 100, 114-123.
[5] Basore, P.A. 1993. IEEE Photovoltaic Specialists Conference 1993 Conference Record of the Twenty Third, 147-152.
[6] Tobail, O., Kim, J., and Sadana, D. 2011. Energy Procedia 100, 213 - 219.
[7] Schiff, E.A. 2003. Solar Energy Materials and Solar Cells 78 (14), 567 - 595.
[8] Verzellesi, G., Vecchi, M.C., Zen, M., and Rudan, M. 1991. Proc. 4th Int. Conf. Simulation of Semiconductor Devices and Processes (SISDEP91), 57 - 64.
[9] Yablonovitch, E. 1982. J. Opt. Soc. Am. 72 (7), 899-907

Journal of Physics: Conference Series **647** (2015) 012048 doi:10.1088/1742-6596/647/1/012048

The influence of electric field on the quasi-confined and interface polaron basic parameters in a wurtzite nitride quantum wire

A L Asatryan, K A Vardanyan, A L Vartanian and A A Kirakosyan
Department of Solid State Physics, Yerevan State University,
1, A.Manoogian, 0025 Yerevan, Armenia

E-mail: annaa@ysu.am

Abstract. The influence of an external electric field on the quasi-confined and interface Fröhlich polaron basic parameters in a wurtzite nitride cylindrical quantum wire embedded in a non-polar matrix is studied theoretically. By using the Lee-Low-Pines variational approach, the dependences of the polaron self-energy and effective mass are obtained as functions of the wire radius and the strength of the electric field. The contributions of different phonon modes to the polaron basic parameters are estimated.

1. Introduction

Quantum heterostructures based on the group III nitride semiconductors are of substantial current interest in view of their potential applications in optoelectronic devices operating in the green, blue, and ultraviolet spectral region [1,2]. Among the numerous studies of physical properties in nitride quantum structures with low-dimensionalities, the quasi-one dimensional nitride quantum wires (QWRs) have attracted special interest. This is mainly because the confinement of carriers in two directions and freedom in the third direction in QWRs promise more efficient lasers and optical gain as well as possible applications for optical waveguide and photovoltaic cells in comparison with quantum wells and quantum dots. For example, the self-organized wurtzite GaN nanowire ultraviolet lasers have revealed a narrow emission linewidth and relatively low threshold [3,4].

Compared with the well-studied zinc-blende QWR's (see e.g. [5–9]), the polaron effects in wurtzite nitride QWR's become much more complicated due to the complication of the optical-phonon modes. Polaron effects become more interesting in the presence of external fields and need to be investigated in depth. The properties of an interface polaron in a wurtzite nitride cylindrical nanowire in the presence of an external electric field have been studied in our previous work [10]. In the present paper the influence of electric field on the polaron self-energy and effective mass in the wurtzite nitride cylindrical QWR including quasi-confined (QC) and interface optical (IO) phonon modes is investigated by using the Lee-Low-Pines (LLP) variational approach.

2. Theory

Let us consider an electron moving in a wurtzite cylindrical QWR of radius R with infinite confining barriers. The wire is surrounded by a nonpolar matrix with dielectric constant ε_d. An electric field is applied perpendicular to the QWR axis (z axis). The z axis is taken to be along the direction of the crystallographic c axis of the wurtzite material. To obtain the polaron basic parameters, we perform the standard LLP variational method [10]. As a result, the following expressions for polaron self-energy and effective mass are obtained

$$E^0_{self}(F,R) = -\sum_{mq_z} \frac{\left|\Gamma^m_{QC}(q_z)\right|^2 \left|M_{QC}(m,q_z,F,R)\right|^2}{\hbar\omega + \frac{\hbar^2 q_z^2}{2m^*}} - \sum_{mq_z} \frac{\left|\Gamma^m_{IO}(q_z)\right|^2 \left|M_{IO}(m,q_z,F,R)\right|^2}{\hbar\omega + \frac{\hbar^2 q_z^2}{2m^*}}, \quad (1)$$

$$m_{pol} = m^* \left(1 + \sum_{mq_z} \frac{2\hbar^2 q_z^2}{m^*} \frac{\left|\Gamma^m_{QC}(q_z)\right|^2 \left|M_{QC}(m,q_z,F,R)\right|^2}{\left(\hbar\omega + \frac{\hbar^2 q_z^2}{2m^*}\right)^3} + \sum_{mq_z} \frac{2\hbar^2 q_z^2}{m^*} \frac{\left|\Gamma^m_{IO}(q_z)\right|^2 \left|M_{IO}(m,q_z,F,R)\right|^2}{\left(\hbar\omega + \frac{\hbar^2 q_z^2}{2m^*}\right)^3} \right), \quad (2)$$

where

$$\left|\Gamma^m_{QC,IO}(q_z)\right|^2 = \frac{4e^2\hbar\omega}{L(q_zR)^2 K_m^2(q_zR)} \left\{ \gamma^2 \bar{\varepsilon}_t \left[f^2_{m-1}(\gamma q_z R) + f^2_{m+1}(\gamma q_z R) - f_m(\gamma q_z R) \right.\right.$$
$$\left.\left. \times \left[f_{m-2}(\gamma q_z R) + f_{m+2}(\gamma q_z R) \right]\right] + 2\bar{\varepsilon}_z \left[f^2_m(\gamma q_z R) - f_{m-1}(\gamma q_z R) f_{m+1}(\gamma q_z R) \right] \right\}^{-1}, \quad (3)$$

$$M_{QC,IO}(m,q_z,F,R) = \langle \Phi | K_m(q_zR) f_m(\gamma q_z \rho) | \Phi \rangle, \quad (4)$$

$$f_m(\gamma q_z \rho) = \begin{cases} J_m(\gamma q_z \rho), & QC \\ I_m(\gamma q_z R), & IO \end{cases} \quad (5)$$

m^* is the electron's effective mass, L is the length of the wire, q_z is the phonon wave vector along wire direction, F is the electric field strength, $\Phi(\rho,\varphi,z)$ is the wave function of the electron moving along the wire axis with momentum $\hbar k_z$ [10], $J_m(x)$ is the m th order first kind Bessel function, $I_m(x)$ and $K_m(x)$ are the m th order first and second kind modified Bessel functions, respectively, $\gamma(\omega) = \sqrt{\left|\varepsilon_z(\omega)/\varepsilon_t(\omega)\right|}$, the frequency dependent dielectric functions in radial (t) and axial (z) directions are given by

$$\varepsilon_\nu(\omega) = \varepsilon_{\nu\infty} \frac{\omega^2 - \omega^2_{\nu L}}{\omega^2 - \omega^2_{\nu T}}, \qquad \nu = t, z. \quad (6)$$

In Eq. (3) $\bar{\varepsilon}_\nu$ $(\nu = t, z)$ are the effective dielectric functions of the wire material, which are defined as

$$\bar{\varepsilon}_\nu = \left(\frac{1}{\varepsilon_\nu - \varepsilon_{\nu 0}} - \frac{1}{\varepsilon_\nu - \varepsilon_{\nu\infty}} \right)^{-1}, \quad \nu = t, z. \quad (7)$$

Here, $\varepsilon_{\nu\infty}$ $(\varepsilon_{\nu 0})$ is the optical (static) dielectric constant, $\omega_{zL}, \omega_{zT}, \omega_{tL}$ and ω_{tT} are the zone center characteristic frequencies of A_1(LO), A_1(TO), E_1(LO) and E_1(TO) modes, respectively. According to the analysis in Ref. [11], QC optical phonon modes in a wurtzite cylindrical QWR exist in two frequency regions, $[\hbar\omega_{zT}, \hbar\omega_{tT}]$ and $[\hbar\omega_{zL}, \hbar\omega_{tL}]$, while IO phonon modes exist only in the frequency region $[\hbar\omega_{tT}, \hbar\omega_{zL}]$. The frequencies we obtain from the dispersion relations for the QC and IO modes in wurtzite QWRs [11].

QC and IO phonon subsystems become independent and can be considered separately, therefore (1) and (2) allow us to investigate electron-QC phonon and electron-IO phonon interactions and calculate the total contributions in polaron self-energy and effective mass by simply adding up the two parts.

3. Numerical results and discussion

To illustrate our numerical results, we consider a wurtzite GaN QWR with an electron-phonon coupling constant $\alpha = 0.49$. Energies are expressed in terms of effective Rydberg energy, $E_R \approx 32.4\,\text{meV}$ and lengths - in the effective Bohr radius, $a_B \approx 24.2\,\text{Å}$. Figures 1a and 1b are respectively the plots of the polaron self-energy and effective mass versus wire radius in an electric field $F = 50kV/cm$. It can be clearly seen that the polaron self-energy monotonically decreases with the increase of R due to electron coupling with the IO and QC phonon modes. As a result, the total polaron self-energy, which incorporates the contributions of electron interaction with both IO and QC phonons, also decreases. The same behavior is observed for the quasi-1D polaron effective mass (figure 1b). These results are expected, as the increase of the electron spatial confinement implies an enhancement of the effective electron-phonon coupling [12,13].

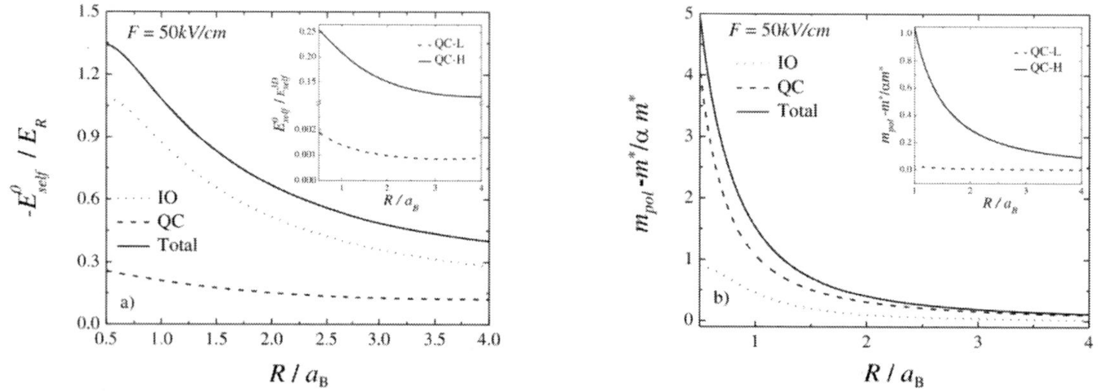

Figure 1. Polaron self-energy (a) and effective mass (b) as a function of GaN wire radius at $F = 50kV/cm$. Insets show the contributions of the higher and the lower frequency QC modes.

Our calculation shows that at $R = 3a_B$ and $F = 50kV/cm$ the IO phonon contribution to the total self-energy is about 74% and the QC phonon contribution to the total effective mass is about 81.7%. These results indicate that the importance of QC and IO phonon modes is variously displayed in different polaronic parameters. A similar result has been reported for wurtzite GaN QWR without an electric field in [11]. The comparison with the results obtained for the cubic GaAs-based cylindrical QWR [7] shows that the polaron self-energy due to QC (IO) phonon modes in GaN QWRs is about 2.2 (3) times larger than that in GaAs QWR structures, when $F = 50kV/cm$. It can be found from the inset figures that the contribution of QC-L modes on the polaron basic parameters can nearly be neglected relative to the QC-H modes.

Figure 2 shows the polaron self-energy and the effective mass as a function of the electric field applied perpendicular to the QWR axis in case of $R = 3a_B$. It can be seen that opposite to polaron effective mass the IO phonon contribution in the polaron self-energy dominates over contribution of QC phonons. A similar result has been reported for cubic GaAs-based cylindrical QWR in [7]. The increase of the IO polaron basic parameters with the increase of the electric field can be explained by the fact that the maximum of the electronic charge distribution with increasing electric field displaces from the GaN wire axis towards the interface. As a result, the role of IO phonons increases. The presence of the minimum value in the effective mass of QC polaron (figure 2b) can be understood as follows. With the increasing electric field the role of QC phonons decreases. Further increase of the electric field (Stark effect) implies an additional confinement of the particle and, therefore, the electron-phonon coupling enhances. The above-mentioned two factors take place in the case of self-

Journal of Physics: Conference Series **647** (2015) 012048 doi:10.1088/1742-6596/647/1/012048

energy too (figure 2a), with the difference that the energy increase due to Stark confinement is larger than its decrease due to the shift of the wave peak of an electron. Therefore the minimum existing in the polaron effective mass dependence does not appear in this case. Figure 2 shows that the change in polaron self-energy and effective mass is about 11% and 2.1%, respectively, when the electric field increases up to $120kV/cm$.

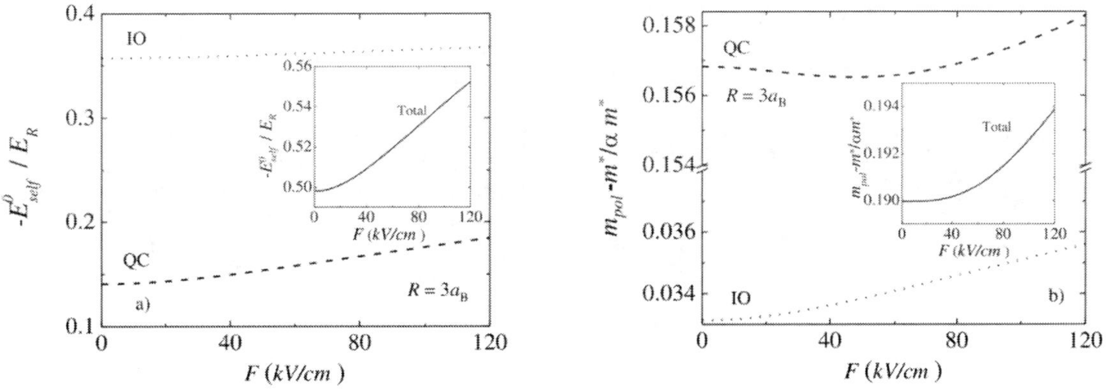

Figure 2. Polaron self-energy (a) and effective mass (b) as a function of electric field for GaN wire with radius $R = 3a_B$. Inset shows the total (QC and IO) contribution.

In conclusion, we have calculated the Fröhlich polaron self-energy and effective mass in a wurtzite nitride cylindrical QWR in the presence of an electric field. Calculations for GaN QWR reveal that i) the contribution of QC-L modes on the polaron self-energy and effective mass can nearly be neglected relative to QC-H modes, ii) the polaron self-energy due to QC and IO phonon modes in GaN QWRs is larger than that in GaAs QWR structures, iii) the interplay between the Stark and spatial confinements leads to the existence of a minimum in the electric field dependence of the polaron effective mass.

Acknowledgement
This work was supported by the State Committee of Science MES RA (Projects No.13-1C196, 13IE-038, 13RF-093, 13YR-1C0044) and the research grant (condmatth-3821) from the Armenian National Science and Education Fund (ANSEF) based in New York, USA.

References
[1] Nakamura S and Chichibu S F 2000 *Introduction to Nitride Semiconductor Blue Lasers and Light Emitting Diodes* (London: Taylor & Francis)
[2] Morkoc H 1999 *Nitride Semiconductors and Devices* (New York: Springer)
[3] Gradečak S, Qian F, Li Y, Park H -G and Lieber C M 2005 *Appl. Phys. Lett.* **87** 173111
[4] Johnson J C, Choi H J, Knutsen K P, Schaller R D, Yang P D and Saykally R J 2002 *Nature Mater.* **1** 106
[5] Buonocore F, Iadonisi G, Ninno D and Ventriglia F 2002 *Phys. Rev.* B **65** 205415
[6] Xie H -J 2004 *Physica* E **22** 906
[7] Vartanian A L 2011 *Eur. Phys. J.* B **83** 481
[8] Vartanian A L, Asatryan A L, Vardanyan L A and Arzumanyan V Y 2012 *J. Phys.: Conf. Ser.* **350** 012022
[9] Vartanian A L, Asatryan A L and Vardanyan L A 2013 *Physica* E **47** 134
[10] Asatryan A L, Vardanyan K A and Vartanian A L, 2015 *Superlattices Microstruct.* **77** 286
[11] Zhang L, Shi J-J and Gao S 2008 *Semicond. Sci. Technol.* **23** 045014
[12] Das Sarma S 1986 *Phys. Rev.* B **27** 2590
[13] Peeters F M, Wu X -G and Devreese J T 1986 *Phys. Rev.* B **33** 3926

Time-dependent current of interacting quantum capacitors subjected to large amplitude pulses

M. I. Alomar[1,2], **Jong Soo Lim**[3] **and David Sánchez**[1,2]

[1] Institut de Física Interdisciplinària i Sistemes Complexos IFISC (CSIC-UIB), E-07122 Palma de Mallorca, Spain

[2] Departament de Física, Universitat de les Illes Balears, E-07122 Palma de Mallorca, Spain

[3] School of Physics, Korea Institute for Advanced Study, Seoul 130-722, Korea

E-mail: misabel@ifisc.uib-csic.es

Abstract. We investigate the time-dependent response of an interacting mesoscopic capacitor using the Floquet-Green function formalism applied to a single-level Anderson model. We obtain closed expressions for the current and the occupation valid for arbitrary values of the applied ac potential in the limit of small frequencies. In the noninteracting case, we obtain nonsinusoidal responses when the ac amplitude allows crossings between the dot level and the Fermi energy of the attached reservoir. For interacting electrons treated within the Hartree approximation, we self-consistently calculate the capacitor current as a function of time and find a decrease of the peak amplitudes due to the on-site Coulomb repulsion.

1. Introduction

Electron dynamics in phase-coherent conductors yield a universal value of the charge relaxation resistance for single-channel capacitors and small ac amplitudes, independently of the transmission probability [1, 2]. When the ac amplitude applied to the capacitor is large such that the dot level crosses the reservoir Fermi energy, Fève et al. found [3] that during the first (second) half-cycle of the voltage pulse the capacitor emits (absorbs) one electron. The resulting quantized current in terms of the elementary charge e has immediate applications for metrology and quantum computation and is crucial, in general, for the electronic manipulation of quantum conductors in real time.

Since quantum dots are small cavities, adding extra electrons causes electron-electron repulsion to become a major issue. The relation between electronic interactions and time-dependent transport in quantum capacitance-resistance circuits has been investigated in a number of works [4, 5, 6, 7, 8, 9]. However, most of these works assume that the pulse amplitude is small. Interestingly, Ref. [10] has formulated a scattering theory for arbitrary pulse strengths valid for noninteracting electrons and fast electron sources. It is the aim of this work to consider *both* electron-electron interactions and large amplitude voltages in order to investigate the role of interaction effects in the time-dependent current of quantum capacitors coupled to ac fields.

We propose an Anderson model for a single level in the vicinity of a gate that generates the oscillating ac voltage. The current is expressed in terms of the dot Green's function. We first address the noninteracting case and then Coulomb interactions are treated within the Hartree approximation. We find that the dot occupation becomes a nonsinusoidal (but periodic) function of time for high ac amplitudes. Correspondingly, the current peaks increase as the ac amplitude

Journal of Physics: Conference Series **647** (2015) 012049 doi:10.1088/1742-6596/647/1/012049

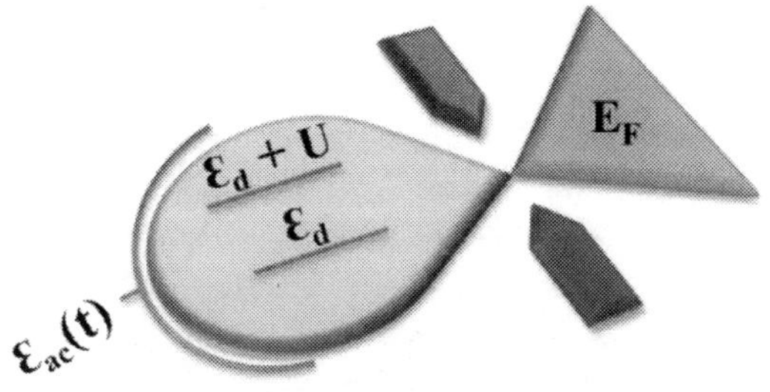

Figure 1. Sketch of a mesoscopic capacitor. A single-level quantum dot with energy ε_d and Coulomb interaction U is connected via one lead to an electron reservoir of Fermi energy E_F. In addition, the dot is capacitively coupled to a harmonic modulation $\varepsilon_{ac}(t)$.

is enhanced. In the presence of Coulomb interactions, the mean occupation decreases as the interaction strength increases due to repulsion effects and the current peaks thus shrink.

2. Theoretical Model

We consider a mesoscopic capacitor formed with an ac-gated quantum dot that can exchange carriers through a constriction with a coupled electron reservoir (see Fig. 1). In our case, the dot has a single energy level and we take into account the electron-electron interactions via the on-site charging energy U. Thus, our time-dependent Hamiltonian reads

$$\mathcal{H} = \sum_{k\sigma} \varepsilon_{k\sigma} C^\dagger_{k\sigma} C_{k\sigma} + \sum_{k\sigma} \left(V_k^* d^\dagger_\sigma C_{k\sigma} + V_k C^\dagger_{k\sigma} d_\sigma \right) + \sum_\sigma \widetilde{\varepsilon}_\sigma(t) d^\dagger_\sigma d_\sigma + U n_\uparrow n_\downarrow . \quad (1)$$

The first term in the right-hand side describes noninteracting electrons in the reservoir and the second term represents the coupling between localized and conduction electrons with tunnel strength V_k. Here, $\sigma = \{\uparrow, \downarrow\}$ labels the spin state. In the third term of Eq. (1), $\widetilde{\varepsilon}_\sigma(t) = \varepsilon_\sigma + \varepsilon_{ac}(t)$ is the energy level driven by the ac monochromatic potential $\varepsilon_{ac}(t) = \varepsilon_{ac} \cos \Omega t$ with ε_{ac} the ac amplitude and Ω the frequency. In the following, we assume spin degeneration and $\varepsilon_\uparrow = \varepsilon_\downarrow \equiv \varepsilon_d$. Finally, the fourth term takes into account the on-site Coulomb repulsion with $n_\sigma = d^\dagger_\sigma d_\sigma$ the number operator for dot electrons.

The time-dependent current is given by the rate of change in time for the occupation expected value, $I(t) = e\partial_t \sum_\sigma \langle n_\sigma \rangle (t) = e\partial_t \sum_\sigma \int d\varepsilon\, G^<_{\sigma,\sigma}(t,\varepsilon)/(2\pi i)$, where $G^<(t,\varepsilon)$ is the lesser Green's function in the mixed energy-time representation [11], which is obtained from the definition $G^<(t,t') = i\langle d^\dagger(t') d(t)\rangle$ and the relation $G^<(t,t') = \int d\varepsilon e^{-i\varepsilon(t-t')} G^<(t,\varepsilon)/(2\pi)$. Then, our problem is reduced to finding the dot occupation. We are interested in an expansion up to leading order in Ω:

$$I(t) = I^{(1)}(t) + \ldots = e\partial_t \sum_\sigma \left(< n^f_\sigma > (t) + \ldots \right) \quad (2)$$

where the indices f (frozen) and (1) indicate zero and first order in Ω, respectively. The frozen term corresponds to a stationary problem with time-dependent parameters [12].

2.1. Noninteracting case

Let us first consider noninteracting electrons ($U = 0$). After lengthy but straightforward algebra, we find

$$< n^f_\sigma > (t) = \int d\varepsilon f(\varepsilon) D_\sigma(t,\varepsilon) \quad (3)$$

$$I^{(1)}(t) = -e \sum_\sigma \int d\varepsilon \left(-\partial_\varepsilon f(\varepsilon)\right) \left(\partial_t \varepsilon_{ac}(t)\right) D_\sigma(t,\varepsilon) \quad (4)$$

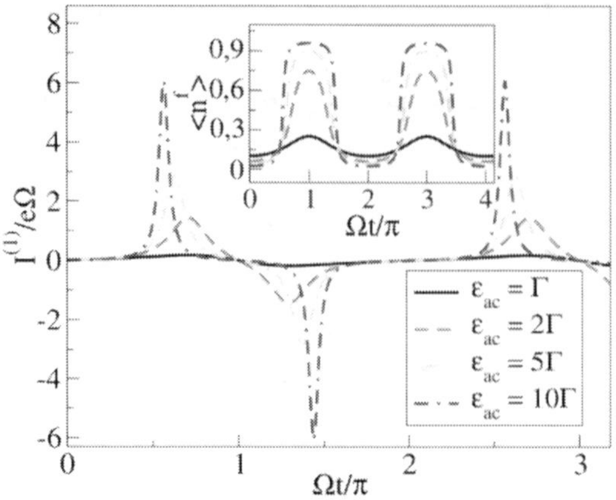

Figure 2. Time-dependent current up to leading order in the ac frequency as a function of Ωt for different values of the pulse amplitude ε_{ac}. The frozen dot occupation per spin is shown in the inset as a function of time for the same values of ε_{ac}. Parameters: $\varepsilon_d = 2\Gamma$ and $k_B T = 0$.

where $D_\sigma(t,\varepsilon) = \Gamma/\left[\pi\left((\varepsilon - \varepsilon_\sigma - \varepsilon_{ac}(t))^2 + \Gamma^2\right)\right]$ is the instantaneous density of states for the coupled dot with $\Gamma = 2\pi\sum_{k\sigma}|V_k|^2\delta(\varepsilon - \varepsilon_{k\sigma})$ the broadening due to hybridization. The Fermi distribution is $f(\varepsilon) = (1 + \exp\left[(\varepsilon - E_F)/k_BT\right])^{-1}$, where the Fermi energy $E_F = 0$ is set as the reference energy and T is the temperature. Time is measured in units of $1/\Omega$ such that current is given in terms of $e\Omega$.

We show in the inset of Fig. 2 the dot occupation per spin, $\langle n^f\rangle \equiv \langle n^f_\uparrow\rangle = \langle n^f_\downarrow\rangle$, as a function of Ωt for different values of ε_{ac} and a fixed position of the energy level. In the frozen regime, the electronic potential is instantaneously adjusted to the variation of the ac field. As a consequence, $\langle n^f\rangle$ attains its maximum value when $\Omega t = \pi + 2\pi m$ $(m = 0, 1, \ldots)$. For small ac amplitudes (black full line) the response is sinusoidal, as expected. For larger amplitudes, during the first half-cycle of the pulse the occupation increases (an electron is absorbed) whereas during the second half-cycle the occupation decreases (an electron is emitted). This effect is more visible for increasing values of ε_{ac}, which allow ε_d to cross over E_F. Interestingly, the modulation of $\langle n^f\rangle(t)$ becomes nonsinusoidal for high values ε_{ac} (blue dot-dashed line). The main panel of Fig. 2 presents the current to first order in the ac frequency. For small pulses the current is nearly zero. Larger amplitudes generate current peaks (dips) during the absorption (emission) part of the process. As ε_{ac} increases, the current pulses become narrower and more separated. Between pulses the occupation remains roughly constant and the current is close to zero.

2.2. Interacting electrons

Now we take into account the Coulomb interaction ($U \neq 0$). For metallic dots with good screening properties, it is a valid approximation to consider the Hartree approach, which amounts to decoupling higher-order correlators in the following way: $\ll d_\sigma, d^\dagger_\sigma n_{\bar\sigma} \gg (t,t') \approx G_{\sigma,\sigma}(t,t')\,\langle n_{\bar\sigma}\rangle\,(t')$. Hence, the equation of motion for the Green's function can be exactly solved. We find that both the frozen dot occupation and the current to first order in Ω take the same form as in Eqs. (3) and (4) but replacing $\varepsilon_{ac}(t)$ with $\varepsilon_{ac}(t) + U <n^f_{\bar\sigma}> (t)$. As a consequence, Eq. (3) becomes a self-consistency equation for $\langle n^f\rangle$ and the dot occupation has to be solved iteratively.

The inset of Fig. 3 shows the results of our numerical evaluations for the frozen occupation in the Hartree approximation at a fixed value of the dot level varying the interaction strength. We keep the ac amplitude with a large value ($\varepsilon_{ac} = 5\Gamma$). We observe that the peak maxima decrease while the minima stay constant when U increases. We attribute this behavior to a repulsion effect that hampers the process of emission and absorption of electrons since an extra energy is to be paid in order to populate the dot. This implies that the current peaks are significantly lower than in the noninteracting case (see main panel of Fig. 3). Our results suggest that

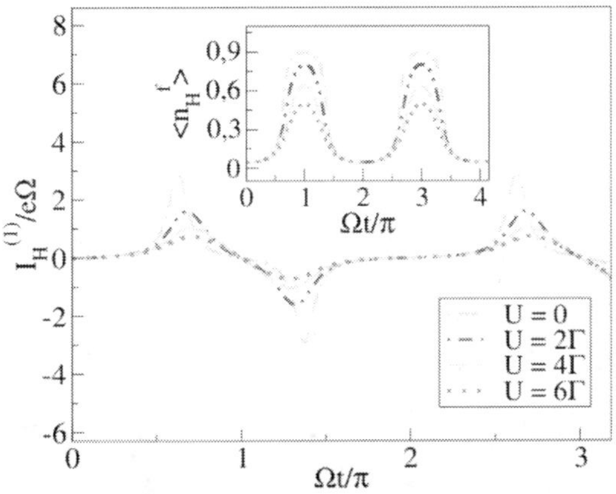

Figure 3. Time-dependent current in Hartree approximation versus time for different values of the Coulomb interaction strength U. In the inset, we plot the frozen dot occupation per spin (also in Hartree approximatioin) as a function of Ωt for the same values of U as in the main plot. Parameters: $\varepsilon_d = 2\Gamma$, $\varepsilon_{ac} = 5\Gamma$ and $k_B T = 0$.

mean-field interactions can be understood as an effective ac potential with a lowered amplitude. Elucidating this effect further will be subject of a future work.

3. Conclusions

We have investigated the dynamical response of a single-level quantum dot attached to a reservoir and in the presence of an external ac potential of arbitrary amplitude. We have described the current and occupation behavior in the adiabatic limit (small ac frequencies) using a mixed energy-time Green's function formalism applied to an Anderson Hamiltonian model. We have found strong effects of the electron-electron interactions that lower the amplitude of the current peaks. This result is important in view of recent advances that propose mesoscopic capacitors as useful resources for metrology due to the quantization of the ac current. Further work is needed to study higher-order currents beyond the adiabatic limit [10], stronger electronic correlations (Kondo limit) [5, 8] or the role of temperature driven potentials in thermoelectric setups [13].

Acknowledgments

This work has been supported by MINECO under Grant No. FIS2014-52564.

References

[1] M. Büttiker, H. Thomas, and A. Prêtre, Phys. Lett. A **180**, 364 (1993).
[2] J. Gabelli, G. Fève, J.-M. Berroir, B. Plaçais, A. Cavanna, B. Etienne, Y. Jin, and D. C. Glattli, Science **313**, 499 (2006).
[3] G. Fève, A. Mahé A, J. M. Berroirm, T. Kontos, B. Plaçais, D. C. Glattli, A. Cavanna, B. Etienne, and Y. Jin, Science **316**, 1169 (2007).
[4] S. E. Nigg, R. López, and M. Büttiker, Phys. Rev. Lett. **97**, 206804 (2006).
[5] C. Mora and K. Le Hur, Nature Phys. **6**, 697 (2010).
[6] Y. Hamamoto, T. Jonckheere, T. Kato, and T. Martin, Phys. Rev. B **81**, 153305 (2010).
[7] J. Splettstoesser, M. Governale, J. König, and M. Büttiker, Phys. Rev. B **81**, 165318 (2010).
[8] M. Lee, R. López, M.-S. Choi, T. Jonckheere, and T. Martin, Phys. Rev. B **83**, 201304(R) (2011).
[9] M. Filippone, K. Le Hur, and C. Mora, Phys. Rev. Lett. **107**, 176601 (2011).
[10] M. Moskalets, P. Samuelsson and M. Bütiker, Phys. Rev. Lett. **100**, 08660 (2008).
[11] M. F. Ludovico, J. S. Lim, M. Moskalets, L. Arrachea, and D. Sánchez, Phys. Rev. B **89**, 161306(R) (2014).
[12] M. V. Moskalets, *Scattering matrix approach to non-stationary quantum transport*, World Scientific (2011).
[13] J. S. Lim, R. López, and D. Sánchez, Phys. Rev. B **88**, 201304(R) (2013).

High-field electron transport in GaN under crossed electric and magnetic fields

V.A. Kochelap[1], V. V. Korotyeyev[1], G.I. Syngayivska[1] and L. Varani[2]
[1]Department of Theoretical Physics, Institute of Semiconductor Physics, Kyiv 03028, Ukraine
[2]Institut d'Electronique du Sud, CNRS UMR 5214, University Montpellier 2, France

E-mail: korotееv@ukr.net

Abstract. High-field electron transport studied in crossed electric and magnetic fields in bulk GaN with doping of 10^{16} cm^{-3}, compensation around 90% at the low lattice temperature (30 K). It was found the range of the magnetic and electric fields where the non-equilibrium electron distribution function has a complicated topological structure in the momentum space with a tendency to the formation of the inversion population. Field dependences of dissipative and Hall components of the drift velocity were calculated for the samples with short- and open-circuited Hall contacts in wide ranges of applied electric ($0 - 20$ kV/cm) and magnetic ($1 - 10$ T) fields. For former sample, field dependences of dissipative and Hall components of the drift velocity have a non-monotonic behavior. The dissipative component has the inflection point which corresponds to the maximum point of the Hall component. For latter sample, the drift velocity demonstrate a usual sub-linear growth without any critical points. We found that GaN samples with controlled resistance of the Hall circuit can be utilized as a electronic high-power switch.

1. Introduction

Semiconductor compounds of group-III nitrides, in particularly GaN, InN and AlN, are the base elements of the modern high-power and high-frequency micro- and optoelectronics. The strong electron-optical-phonon coupling, large optical phonon energy, sufficiently high electron mobility make them perspective for the development of the electrical pumping THz sources. The basic physical mechanism underlying of such devices is the quasiballistic electron motion associated with a emission of the optical phonon arising at sufficiently strong applied electric field. In literature this mechanism is widely discussed for nitride quantum heterostructures [1, 2, 3, 4] and short $n^+ - n - n^+$ diodes [5]. In the case of the unlimited sample this transport regime leads to the formation of the strongly-anisotropic distribution function in momentum space and the dynamic negative differential conductivity in the specific frequency ranges.

The application of an external magnetic field, $\vec{H}$, transverse to the direction of the electric field, $\vec{E}$, can essentially affect on the dynamic of the quasiballistic (*streaming*) electrons and their distribution function. A qualitative analysis of the possible transport regimes in crossed electric and magnetic fields have been done in the pioneer paper [6]. In particularly, it was predicted two effects: a formation of the inverted electron distribution and a suppression of dissipative transport with optical phonon emission. The first effect can be realized in the magnetic fields satisfying inequality $m^*cE/P_0 < H < 2m^*cE/P_0$ (I), where $P_0 = \sqrt{2m^*\hbar\omega_0}$ is the electron momentum corresponding to the optical phonon energy, $\hbar\omega_0$, m^* is the electron effective mass and c is the light velocity and the second effect can be realized at magnetic fields $H > 2m^*cE/P_0$ (II).

Transport phenomena in crossed electric and magnetic fields at low temperature and at dominant inelastic scattering have been intensively studied in the past both theoretically and experimentally. These investigations were aimed to reveal conditions of the formation of an inversion population necessary for a development of high-frequency generators. These ideas were realized in $p-$Ge where the inversion population between light- and heavy-hole bands and emission of the high-frequency radiation were observed under the conditions of the deep cooling [7]. However, in conventional polar semiconductors (n-GaAs, n-InSb, n-InP), in spite of many theoretical predictions [8, 9], the emergence of the electron inverted distribution associated with electron-optical phonon interaction was not found experimentally.

Here we present a theoretical analysis of steady-state electron magneto-transport in bulk GaN in which the strong electron-optical interaction and the large optical phonon energy promote favorable conditions for the observation of the above-mentioned effects. We focus on the study of the distribution function and current-voltage characteristics for different schemes of the external electrical circuits.

2. Transport model and analysis of the non-equilibrium distribution function

Analysis of the magneto transport was carried out for bulk-like sample of the compensated GaN. Compensation process allows to reduce free electron concentrations and avoid the undesirable influence of the non-elastic $e-e$ scattering [2] on the formation of the *streaming* transport regime. Calculations were performed in the frames of the numerical solution of Boltzmann transport equation by single-particle Monte-Carlo method [10]. In calculations we assumed the low ambient temperature of, 30 K, impurity and electron concentrations of $N_i = 10^{16}\text{cm}^{-3}$ and 10^{15}cm^{-3}, respectively. We took into account the main three scattering mechanisms: scattering by ionized impurities, acoustic phonons and polar optical phonons. For similar parameters it was found that electron streaming effect at zero magnetic field occurs in the range of electric fields $3-10$ kV/cm [11]. Under the study of the hot magneto transport we considered the range of the applied magnetic fields of a few tesla when the quantization effects can be neglected.

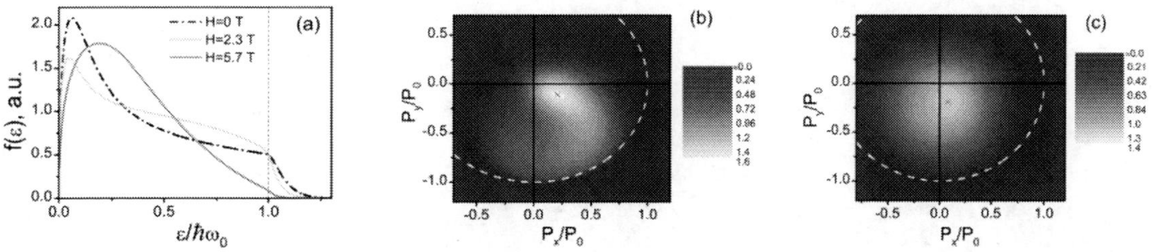

Figure 1. (a): Distribution function $f(\epsilon)$ for different magnetic fields. (b) and (c): The contour plots of the distribution function $f(P_x, P_y)$, at $H = 2.3$ T and $H = 5.7$ T, respectively. The white dashed circle separates the *passive* ($\epsilon < \hbar\omega_0$) and *active* ($\epsilon > \hbar\omega_0$) energy regions. The sign $\times$ indicates the maximum of the distribution. E = 5 kV/cm. E and H are applied along the $x-$ and $z-$axis, respectively.

Figs. 1 demonstrate behavior of the electron distribution function in the energy, $f(\epsilon)$, and in the momentum, $f(P_x, P_y)$, spaces for several magnetic fields. For magnetic field, $H = 2.3$ T satisfying inequality (I) the streaming distribution (dashed-doted curve) is strongly modified. The effect of accumulation of high-energy electrons in the energy interval $[0.5..0.9] \times \hbar\omega_0$ is clear seen. The corresponding distribution in the momentum space acquires the complicated topological structure (b) stipulated by the coexistence of two groups of carriers: electrons executing *streaming* motion (interacting with optical phonons) and electrons moving on cyclotron orbits closed in the *passive* regions (non-interacting with optical phonons). The center of $f(P_x, P_y)$ is essentially shifted along both dissipative (x) and Hall (y) directions. However, for assumed parameters of GaN the inverted distribution function is still not formed due to the intensive scattering on the ionized impurities. Such distribution can be obtained only in the limit of the small impurity concentration, $N_i \sim 10^{14}$ cm^{-3}. For $H > 5$ T satisfying inequality (II) most of electrons move along the cyclotron trajectories closed in the *passive* region and interaction with optical phonons is strongly suppressed. The distribution function in the energy space is localized in the *passive* region and have axial symmetry in the momentum space. The discussed evolution of the distribution function is reflected in the specific behavior of the transport characteristics.

3. Current-voltage characteristics

In generally, the results of measurements of the transport characteristics in the magnetic field depend on the configuration of the external electrical circuit. There are two typical schemes of measurements: scheme with short-circuited Hall contacts ($R_h = 0$) and scheme with open-circuited Hall contacts ($R_h = \infty$). In the case of $R_h = 0$, the effect of the charge accumulation at Hall contacts can be neglected and component of the electric field in the Hall direction, $E_h^{(0)}$, is equal to zero. In turn, both components of the drift velocity: dissipative, $V_d^{(0)}$, (describes the current between Ohmic contact) and Hall, $V_h^{(0)}$, (describes the current between Hall contacts) are nonzero. The developed Monte Carlo procedure allows to calculate both components of the drift velocity, simultaneously.

In the case of the sample with $R_h = \infty$, only dissipative component, $V_d^{(\infty)}$, is the nonzero. In turn, due to the charge accumulation on Hall contacts, electric field acting on electron has two nonzero components: dissipative , $E_d^{(\infty)}$, (applied field) and Hall, $E_h^{(\infty)}$. If the sample is so long in the Hall direction that spatial inhomogeneity of the electric field near the Hall contacts can be neglected, electrical characteristics $V_d^{(\infty)}(E_d^{(\infty)})$ and $E_h^{(\infty)}(E_d^{(\infty)})$ can be easily obtained using results for the case of the sample with $R_h = 0$. From simple geometrical considerations, it follows: $V_d^{(\infty)} = \sqrt{V_d^{(0)2} + V_h^{(0)2}}$, $E_d^{(\infty)} = E_d^{(0)} \cos\theta$ and $E_h^{(\infty)} = E_d^{(0)} \sin\theta$ where θ is the Hall angle and defined as $\tan\theta = V_h^{(0)}/V_d^{(0)}$.

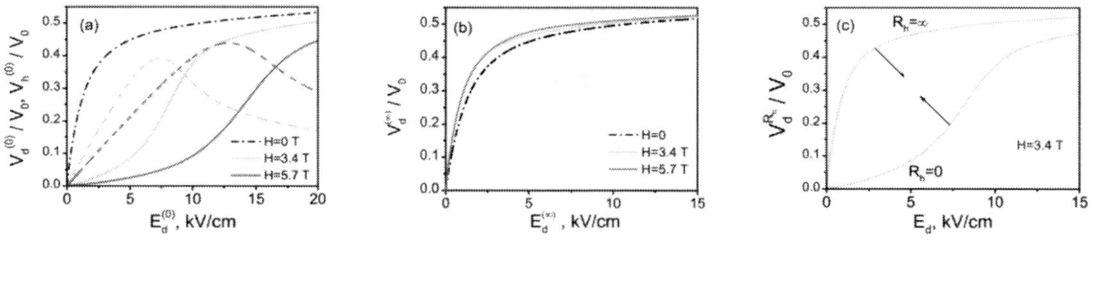

Figure 2. (a): Dependences of the dissipative (solid lines) and Hall (dashed lines) components of the drift velocity vs applied electric field for sample with $R_h = 0$. (b): dissipative components vs applied electric field for sample with $R_h = \infty$. (c): Comparison of the dissipative components of the drift velocity for the samples with short- and open-circuited Hall contacts. The characteristic velocity $V_0 = 4 \times 10^7$ cm/s.

Typical electric fields dependences of the nonzero components of the drift velocity for the electrical circuits with $R_h = 0$ and $R_h = \infty$ are shown in Figs. 2(a) and (b), respectively. As seen, $V_d^{(0)}(E_d^{(0)})$ and $V_h^{(0)}(E_d^{(0)})$ have three specific intervals. For example, for $H = 3.4$ T (green pair of curves in (a)) in the interval of applied electric fields of $0 - 5$ kV/cm the Hall current dominates over the dissipative one. The dependence $V_d^{(0)}(E_d^{(0)})$ shows superlinear growth, while $V_h^{(0)}(E_d^{(0)})$ remains almost linear. This situation corresponds to the electron distribution with axial symmetry (see Fig.1(c)) and amplitudes of the applied fields have to satisfy the inequality (II). With further increasing of the electric fields when inequality (I) is already satisfied, dissipative and Hall currents become of the same order. At this, the dependence $V_d^{(0)}(E_d^{(0)})$ contains the inflection point which corresponds to the maximum point in the dependence of $V_h^{(0)}(E_d^{(0)})$. Electron distribution have the form close to the one depicted in Fig.1(b). At electric fields > 15 kV/cm, Hall current is suppressed and dissipative current tends to the curve obtained at zero magnetic field. At stronger applied magnetic field (red pair of curves in (a)) the inflection and maximum points are shifted to the range of stronger electric fields and dissipative current is negligibly small up to 10 kV/cm (it is observed the collapse of the dissipative current).

For the electrical circuit with $R_h = \infty$, electric-field dependences of the drift velocity (b) have monotonic behavior and in contrast to the sample with $R_h = 0$ there is no inflection point. Note that the drift velocity under applied magnetic fields can exceed its value at zero magnetic field. The similar result was obtained and discussed in ref.[9] relating to n-GaAs and n-InP samples.

In summary, we compare dissipative currents calculated for the circuits with $R_h = 0$ and $R_h = \infty$ (see Fig.2(c)). Apparently, for the Hall circuit with the finite values of the R_h dissipative current-applied electric field characteristic is placed between the solid and dashed curves. Thus, changing the resistance of the Hall circuit one can control the dissipative current between Ohmic contacts (analysis of such circuit will be published elsewhere). We suggest, that peculiarities of the high-field electron magneto transport in GaN can be used for the development of the four-terminal devices, in particularly, high power electronic switches.

Acknowledgments

Authors thanks to CNRS (France)- NASU (Ukraine) Project 26206 for support this work.

References

[1] Starikov E, Shiktorov P, Gruzinskis V, Varani L, Palermo C, Millithaler J-F and L. Regiani 2008 *J. Phys: Condens. Matter* **20** 384209; 2007 *Phys. Rev. B* **76** 045333
[2] Lu J T and Cao J C 2005 *Semicond. Sci. Technol.* **20** 829-833
[3] Korotyeyev V V, Kochelap V A, Kim K W and Woolard D L 2003 *Appl. Phys. Lett.* **82** 2643.
[4] Kim K W, Korotyeyev V V, Kochelap V A, Klimov A A and Woolard D L 2004 *J. Appl. Phys.* **96** 6488.
[5] Iniguez-de-la-Torre A, Mateos J and Gonzalez T 2010 *J. Appl. Phys.* **107** 053707
[6] Vosilyus E and Levinson E, 1966 *Zh. Eksp. Teor. Fiz.* **50** 1660-1666 (in Russian)
[7] Vorob'ev L E , Osokin F I, Stafeev V I and Tulupenko V N 1982 *JETP Lett* **35**(9), 440-443.
[8] H. Maeda, T. Kurosawa 1972 *J. Phys. Soc. Japan* **33** 562
[9] S. Kachyulis, I. Parshyalyunas, G. Tamulaitis 1983 *Hot Electrons in Semiconductors: Streaming and Anisotropic Distributions in Crossed Fields* ed. Andronov A A and Pozela Yu K (Gorki) pp. 101-113.
[10] Jacoboni C and Reggiani L 1983 *Rev. Mod. Phys* **55**(3) 645-705.
[11] G.I. Syngayivska and V.V. Korotyeyev 2013 *Ukrainian Journal of Physics* **58**(1) 40-55.

Modeling of a Back-Illuminated HgCdTe MWIR Avalanche Photodiode with Alloy Gradients

A. K. Storebø and T. Brudevoll

Norwegian Defence Research Establishment, PO Box 25, 2007 Kjeller, Norway

E-mail: asta-katrine.storebo@ffi.no

Abstract. We present results from 2D Monte Carlo simulations of a mid-wave infrared (MWIR) back-illuminated planar n-on-p HgCdTe electron-initiated avalanche photodiode (e-APD). The main objective of this work has been to study the dependence of the multiplication gain, excess noise factor, and response time on the position where the incident photon is absorbed. We also quantify the effects of the naturally occurring vertical alloy gradient electric fields in APDs grown by liquid phase epitaxy. The simulated gain is relatively independent of the excitation position, but a small increase is observed near the edges of the APD. The excess noise factor is around 1.2 for all positions. It is found that the alloy gradient field helps to speed up the response of the device, especially for excitation positions far away from the multiplication region.

1. Introduction

HgCdTe (MCT) electron-initiated avalanche photodiodes (e-APDs) are characterized by high sensitivity and gain combined with low avalanche multiplication noise [1], [2]. These properties make them well suited for detection of very low radiation levels, such as in space imaging systems and laser-based detection systems. MCT APDs designed for the mid- and long-wavelength infrared regions, with cadmium alloy fractions less than 53%, operate mainly via single-carrier (electron) initiated multiplication. Therefore the gain becomes an exponential function of the reverse bias voltage. This simple gain mechanism results in an excess noise factor close to unity, largely independent of the gain. In addition, the response time can be very short, thus gain-bandwidth products in excess of 1 THz become possible. A great deal of modeling work on APDs have focused on obtaining phenomenological analytical relations describing the dependence of multiplication gain on reverse bias voltage or depletion layer geometry [3]. Other efforts have explored the various carrier interactions in more detail with Monte Carlo techniques [4]. In the present paper, we study the dependence of these properties on the position of the initial photo-excitation for a back-illuminated mid-wave infrared MCT APD, using 2D Monte Carlo simulation. We also study the effects of the naturally occurring vertical alloy gradient electric fields in the p-type absorption layer of APDs grown by liquid phase epitaxy [5], [6], [7].

2. APD structure

Details of the simulated APD are shown in figure 1. It is smaller, but geometrically similar to the planar structure discussed in [8], and larger than the other designs investigated in [4]. The large devices studied in [8] allowed the position-dependent multiplication gain to be probed experimentally by focusing radiation onto different spots on the APD surface. Then a two-gain model consisting of a constant gain in the central region and an enhanced gain at the edges of the APD was introduced as a means of explaining the observations. It is therefore of interest to see if some of these effects could be confirmed by the Monte Carlo simulations in the present work.

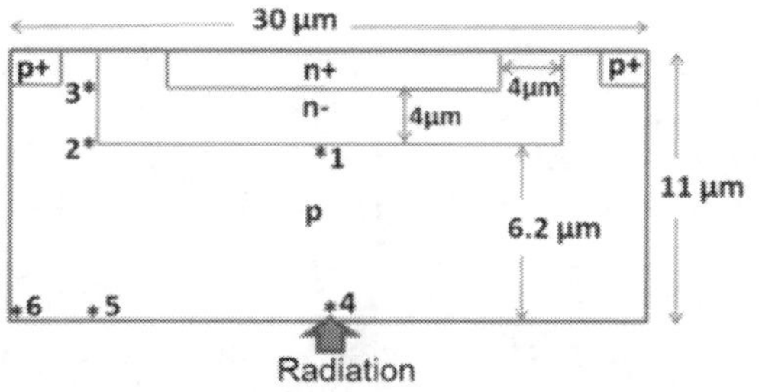

Figure 1. Layout of the planar electron-initiated APD. The doped-in carrier concentrations are $p = 2{\cdot}10^{16}$ cm^{-3}, $n- = 5{\cdot}10^{14}$ cm^{-3}, and $p+ = n+ = 2{\cdot}10^{17}$ cm^{-3} for the contacts. Numbers 1-6 indicate different initial photo-excitation positions used in the simulations.

It has been speculated whether the alloy gradient quasi-electric fields (AGFs) naturally arising in LPE-grown MCT materials may play a role in speeding up the transport of the initial photo-generated carriers, thus resulting in shorter response times [8]. Simulations were therefore carried out both with and without these AGFs. In simulations without alloy gradients, the Cd alloy fraction, x, was set to a constant value of 0.28. In simulations with alloy gradients, x was set to decrease exponentially from 0.33 at the entrance face (i.e. cut-off at 4 µm) to a value of 0.28 at the p-n junction, according to the expression $x = 0.27+0.06{\cdot}\exp(-z/3.7)$, where z is the distance from the irradiated back surface in µm [5]. The effect of the gradient can be described by two bare AGFs, one affecting holes and the other affecting electrons. Free, doped-in electric charges then respond and move according to one of the two bare AGFs, resulting in a self-consistent equilibrium bias electric field affecting both carrier types. The alloy gradient and associated electron- and hole bare AGFs are shown in figure 2.

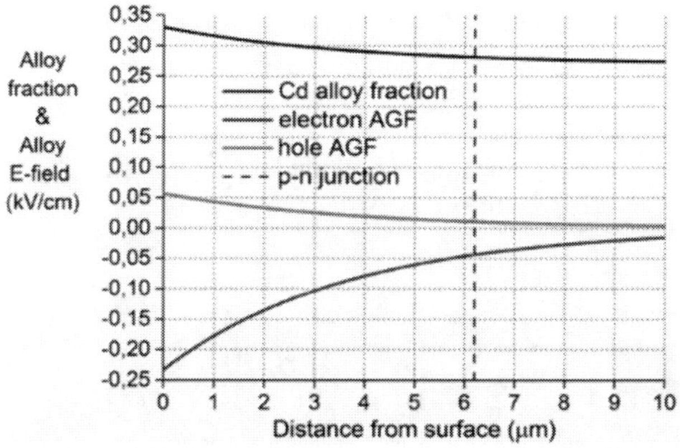

Figure 2. Cd alloy fraction and bare alloy gradient quasi-electric fields in undoped HgCdTe as a function of distance from the irradiated back surface at 77 K. The dashed line indicates the position of the p-n junction.

The importance of the alloy gradient field may depend on the thickness of the p-type absorption layer, since the gradients are located near the outer surface. A thick p-type absorption layer is important for the ability to operate near the cutoff wavelength of the MCT material. The p-layer is chosen to be 6.2 µm thick in the present work, and the AGF is fairly weak at the p-n junction.

3. Simulation model

In order to obtain a small photo-generated signal from a large background of free charge carriers, we performed a linearised, small signal analysis with single electron/hole resolution on top of a background bias electric field calculation. The bias calculation was carried out in the super-particle picture, and the resulting electric field was smoothed before it was used in the small signal analysis as a 'frozen field'. Each APD simulation consisted of 1000 different single-photon absorption events at the same position. Six excitation positions were considered, as indicated in figure 1. Impact ionization was modeled with a simple Keldysh type expression, also used in other work [9], and terminal currents were calculated according to [10], [11].

Carrier transport in the p-type absorption layer can be affected by ambipolar effects varying with irradiation and doping, and at high reverse bias carrier tunneling through the p-n junction may create additional electron-hole pairs. Such processes were not included here, only ionized impurity and phonon scattering mechanisms were considered.

4. Results and discussion

All simulation results presented below are for 7 V reverse bias and 77 K temperature. Figure 3 shows four examples of simulated electron tracks with/without an AGF, for photo-excitation at two different positions (4 and 5) on the irradiated surface. It is evident that the AGF helps to bring the electrons on a more direct path towards the p-n junction.

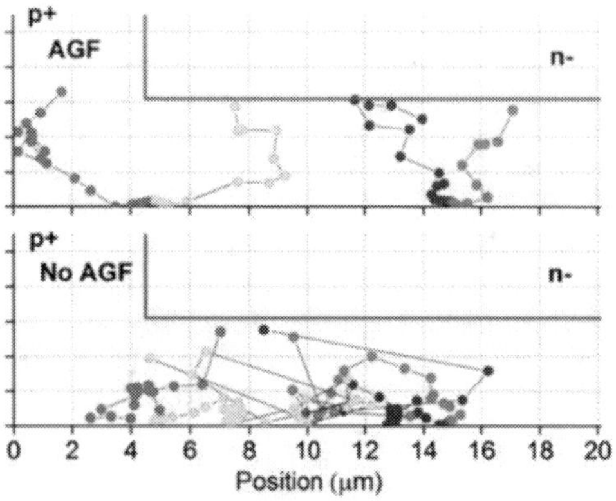

Figure 3. Four simulated electron tracks, showing particle positions at given time intervals, starting with initial photo-excitation at positions 4 and 5 respectively, with and without the presence of an AGF. The p-n junction is indicated in red.

The lower plot in figure 4 gives an example of the simulated times of arrival at the contacts for electrons and holes in a case where the APD gain $M = 81$. The upper plot shows the corresponding terminal current on the n-contact under the assumption of a constant terminal voltage. All electrons are collected within only 50 ps, whereas the holes take several ns to arrive at the contacts, i.e. it is the electrons that provide the fast response of the APD. Figure 5 shows a histogram of APD gain values (M) from 1000 simulations at position 1, with a mean value $M_{mean} = 54$, and a standard deviation of 23, resulting in a calculated excess noise factor $F \approx 1.2$ from the expression shown in figure 5.

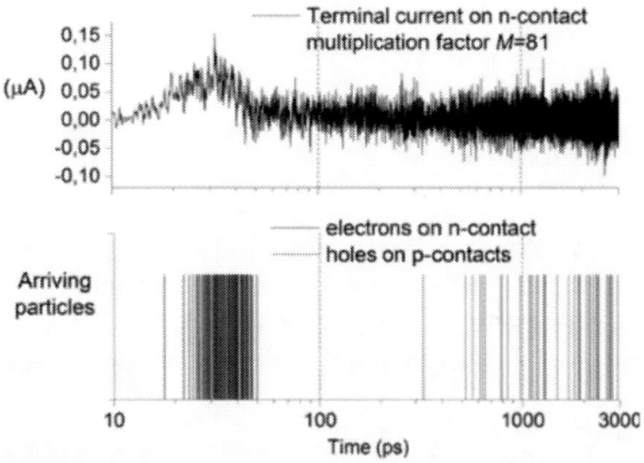

Figure 4. Terminal current plot and corresponding carrier time-of-arrival plot for an absorbed photon at position 1, with AGF and for $M = 81$.

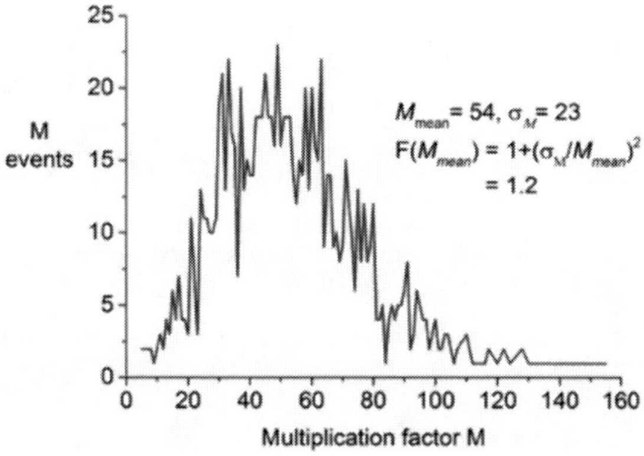

Figure 5. Histogram of electron multiplication events for photo-excitation at position 1.

Table 1. Simulation results for 7 V reverse bias at 77 K temperature. The response time, t_R (ns), has here been defined as the time when all excess electrons have reached their contact.

Position	M (AGF)	F (AGF)	t_R (AGF)	M (no AGF)	F (no AGF)	t_R (no AGF)
1	52	1.2	0.06	52	1.2	0.09
2	58	1.2	0.11	57	1.2	0.19
3	55	1.2	0.08	55	1.2	0.10
4	53	1.2	0.09	52	1.2	0.24
5	56	1.2	0.13	57	1.2	0.29
6	57	1.2	0.16	57	1.2	0.33

The main simulation results are summarized in Table 1. The APD gain is relatively independent of the excitation position, but a small increase is observed near the edges of the APD (positions 2, 3, 5, and 6), which has also been observed experimentally [8]. The excess noise factor is around 1.2 for all positions. It is evident that the electron AGF helps to speed up the response, not only for photons absorbed near the entrance face where the bare alloy field is strong (225 V/cm for electrons), but also for photons absorbed near the p-n junction where the alloy field is weak (50 V/cm for electrons). Response times are much shorter than electron lifetimes (typically 30 ns), so that virtually all photo-electrons will reach the multiplication region at the p-n junction.

5. Conclusions

We have seen that the dynamic response of a particular APD is indeed affected by alloy gradients. There were also notable positional differences in avalanche gain which apparently do not translate into positional variations in excess noise.

References

[1] Beck J D, Wan C-F, Kinch M A, and Robinson J E 2001, *Proc. SPIE* **4454**, 188
[2] Singh A, Srivastav V, Pal R 2011, *Optics & Laser Technology* **43**,1358
[3] Rothman J, Mollard L, Bosson S, Vojetta G, Foubert K, Gatti S, Bonnouvrier G, Salveti F, Kerlain A, Pacaud O 2012, *J. Electron. Mater.* **41**, 2928
[4] Bellotti E, Moresco M, and Bertazzi F 2011, *J. Electron. Mater.* **40**(8):1651-1656
[5] LoVecchio P, Wong K, Parodos T, Tobin S P, Hutchins M A, and Norton P W 2004, *Proc. SPIE* **5564**, 65
[6] Kraut E A, Vac J 1989, *Sci. Technol. A* **7** (2), 420
[7] Nasrallah S A-B, Mnasri S, Sfina N, Bouarissa N, Said M 2011, *J. Alloys and Compounds* **509**, 7677
[8] Reine M B, Marciniec J W, Wong K K, Parodos T, Mullarkey J D, Lamarre P A, Tobin S P, Minich R W, Gustavsen K A, Compton M, and Williams G M 2008, *J. Electron. Mater.* **37**, 1376
[9] Derelle S, Bernhardt S, Haı¨dar R, Deschamps J, Primot M B, Rothman J, Rommeluere S, and Gue´rineau N 2009, *J. Electron. Mater.* **38**, 1628
[10] Shockley W J 1938, *Appl. Phys.* Vol. **9**, 635
[11] Ramo S 1939, *Proceedings of the IRE* Vol. **27**, 584

Hydrodynamic Model for Silicon Carbide Semiconductors including crystal heating

O Muscato and V Di Stefano

Dipartimento di Matematica e Informatica, Università degli Studi di Catania, Viale A. Doria 6, 95125 Catania – Italy

E-mail: muscato@dmi.unict.it, vdistefano@dmi.unict.it

Abstract. A hydrodynamic model describing the electron transport in silicon carbide semiconductors, coupled with the heating of the crystal lattice is presented. It has been obtained by taking the moments of the coupled Boltzmann equations for the electrons and phonons, and by using the Maximum Entropy Principle in order to determine the transport coefficients and the constitutive equations. Simulation results in the bulk case are shown.

1. Introduction

Power semiconductor devices have attracted increasing attention as key components in a variety of power conversion units. Because of the mature technology of Si power devices currently employed in most applications, it is now difficult to achieve innovative breakthroughs in this field. Newly emerging semiconductors such as Silicon Carbide (SiC) are attractive for advanced power devices, owing to their superior physical properties such as their exceptional wide bandgap (three times more than Si), a high thermal conductivity (twice or triple that of Si), and a very high field breakdown (five times more than Si). Although these performances are very promising, SiC devices may suffer from severe self-heating effects which impose a limitation on both the output power and the power density of the devices. Self heating results in a higher lattice temperature in the transistor channel which can significantly deteriorate the current-voltage characteristics because of the reduction in the device parameters such as mobility and electron saturation velocity. For these reasons, electro-thermal simulations are mandatory in order to predict the behaviour of such devices.

If we want to go beyond the drift-diffusion equation scheme, valid in the linear irreversible thermodynamic regime, it is mandatory to explore the microscopic details of self-heating process.

The natural framework to deal with electro-thermal transport is given by a set of Boltzmann-like equations (called Bloch-Boltzmann-Peierls equation, hereafter BBP) which describe, from the kinetic point of view, the transport of electrons and phonons, coupled to the Poisson equation [1]. To solve numerically this system is an hard task, because it forms a set of partial integro-differential equations. One possibility is to develop, for such material, an Electro-thermal Monte Carlo (ETMC) scheme at expenses of huge computational efforts [2–11]. Another alternative is to take the moments of the BBPs obtaining a set of hydrodynamic-like equations. The main drawback is how to model the constitutive equations for the higher-order moments (i.e. the fluxes) as well as for the production terms, i.e., moments of the collisional operator. These problems have successfully been tackled for Si by means of the Maximum Entropy Principle (hereafter MEP) of extended thermodynamics [12]. Such an approach has successfully been used for simulating the electron transport [13–18] and the electro-thermal one in silicon devices [19–22]. Constructing such macroscopic model for SiC is the main aim of the present paper.

2. The Extended Hydrodynamic model

Over 200 crystal varieties of SiC, which are different by their atomic arrangement, are known. The 4H-SiC has a hexagonal lattice, and we have considered the valleys around the minima at the symmetry point M of the two lowest conduction bands. Due to the crystal symmetries, for each conduction band there are three equivalent valleys [23]. Then, for each of the two conduction bands, a Boltzmann-like kinetic equation for the unknown distribution function f_A must be considered. The phonon branches which mainly contribute to the transport phenomena in 4H-SiC are the polar and non-polar optical ones (in the Einstein approximation) and the acoustic one (in the Debye approximation). Again, for each η-th phonon branch, a Boltzmann-like kinetic equation for the unknown distribution function g_η must be added.

The BBP equations for phonons and electrons are coupled by the respective collisional operators, which depend on the scattering mechanisms. Here we have considered the electron scattering with ionized impurities (intravalley, Brooks and Herring model), acoustic (intravalley, elastic approximation), polar optical (intravalley, inelastic) and non-polar optical (*intervalley*, inelastic) phonons. The main complicacy introduced in such material is due to the *intervalley* scattering, which leaves the electron into a different valley after the collision with a phonon.

By multiplying the BBP kinetic equation describing the electron flow by suitable weight functions, one can obtain balance equations for the macroscopic quantities associated to the electron flow (called moments) in the A-th valley [24]

$$\frac{\partial n_A}{\partial t} + \frac{\partial (n_A V_A^i)}{\partial x^i} = n_A C_{n_A} \quad , \quad \frac{\partial (n_A V_A^i)}{\partial t} + \frac{\partial (n_A U_A^{ij})}{\partial x^j} + \frac{e n_A}{m_A^*} E_i = n_A C_{V_A}^i \tag{1}$$

$$\frac{\partial (n_A W_A)}{\partial t} + \frac{\partial (n_A S_A^i)}{\partial x^i} + e n_A V_A^i E_i = n_A C_{W_A} \tag{2}$$

$$\frac{\partial (n_A S_A^i)}{\partial t} + \frac{\partial (n_A F_A^{ij})}{\partial x^j} + e n_A E_j U_A^{ij} + \frac{e n_A W_A}{m_A^*} E^i = n_A C_{S_A}^i \tag{3}$$

where n_A, V_A^i, W_A, S_A^i are the electron density, average velocity, average energy and average energy-flux respectively, and $C_{n_A}, C_{V_A}^i, C_{W_A}, C_{S_A}^i$ the corresponding productions; U_A^{ij} and F_A^{ij} are the higher-order fluxes, and E_j the electric field. Similarly, one obtains the balance equations for the phonons

$$\frac{\partial W_p}{\partial t} = \sum_A P_p(f_A) \quad , \quad \frac{\partial W_n}{\partial t} = \sum_{AB} P_n(f_A, f_B) \tag{4}$$

$$\frac{\partial W_{ac}}{\partial t} + \frac{\partial Q^i}{\partial x^i} = \sum_A P_{ac}(f_A) \quad , \quad \frac{\partial Q^i}{\partial t} + v_s^2 \frac{\partial N^{ij}}{\partial x^j} = v_s^2 \sum_A P_{ac}^i(f_A) \tag{5}$$

where v_s is the sound velocity, W_{ac}, W_p, W_n are the energy densities for the acoustic, polar optical, non-polar optical phonons respectively, P_{ac}^i, P_p, P_n the corresponding productions, and Q^i the acoustic phonon energy-flux density. In the above system of PDEs, we have to determine the higher-order fluxes (i.e., $U_A^{ij}, F_A^{ij}, N^{ij}$) as well as the production terms (i.e., the right-hand-side). According to the Maximum Entropy Principle, if a certain number of moments is known, then the distribution functions which can be used for evaluating the unknown moments are those that extremize the entropy functional, under the constraint that those functions reproduce the known moments. By performing a suitable expansion around the thermal equilibrium, we obtain a hyperbolic system of PDEs, where the higher-order fluxes and the production terms are determined functions of the moments $n_A, V_A^i, W_A^i, S_A^i, W_p, W_n, W_{ac}, Q^i$. We observe that, in this scheme, no Fourier law relating the electron/phonon heat flow to the temperature gradient has been invoked because the energy-fluxes S_A^i, Q^i are considered as new variables.

3. The bulk case

The bulk case is represented by a SiC semiconductor with a uniform doping. If homogeneous initial conditions are taken, the above moment system reduces to a system of ordinary differential equations with the time as the only independent variable, and an electric field frozen into the sample. The parameters used in the simulation have been taken form [23]. For the initial conditions, we have taken the corresponding equilibrium values. In figure 1 we plot the electron density for the two conduction bands versus the simulation time, for an electric field of 450 kV/cm. The inversion of two electron populations begins at long times. In figure 2 we plot the electron velocity for the two conduction bands and their average (weighed with their densities). The velocity overshoot phenomenon is clearly observed. Finally, in figures 3, 4 the average electron energies and phonon temperatures are shown, respectively. The lattice temperature is defined as $T_L = (c_{ac}T_{ac} + c_pT_p + c_nT_n)/(c_{ac} + c_p + c_n)$ where c_{ac}, c_p, c_n are the volumetric specific heat for the respective phonon branches. Since the (bulk) semiconductor is supposed to be infinite without any boundary condition, and the lattice is not considered as a thermal reservoir, the heat produced by Joule effect inside the device cannot be dissipated and we expect a general overheating of the sample. The overheating of the sample has been predicted as function of the simulation time, and the contribution to the heat transport of the different phonon branches has been highlighted. The figure 4 shows a significant variation of the lattice temperature for a simulation time of $\simeq$ 100 ps. Since the transfer of energy from the electrons to phonons occurs in a time scale of $\simeq$ 0.3 ps., the energy transfer from the electrons to phonons can be assumed to occur instantaneously, when compared to the heat transport time scale. For this reason, in the simulation, one can use for the phonon flow a time step 100 times larger than the time step used for solving the electron flow (1)-(3), saving CPU time.

The simulation of real devices, where hot electron effects can play an important role in the energy dissipation phenomenon, using this hydrodynamic model will be the subject of the next researches.

Acknowledgments

This work has been supported by the research project AMBITION POWER (MIUR PON Ricerca e Competitività 2007-2013 Regioni Convergenza).

References

[1] Ziman J 1967 *Electrons and Phonons* (Oxford: Clarendon Press)
[2] Muscato O 2000 *COMPEL* **19** 812
[3] Muscato O and Wagner W 2005 *COMPEL* **24** 1351
[4] Muscato O, Wagner W and Di Stefano V 2010 *ESAIM: M2AN* **44** 1049
[5] Muscato O, Wagner W and Di Stefano V 2011 *Kinetic and Related Models* **4** 809
[6] Muscato O and Di Stefano V 2011 *COMPEL* **30** 519
[7] Sadi T, Kelsall R, Pilgrim N, Thobel J L and Dessene F 2012 *J. Comp. Electr.* **11** 118
[8] Muscato O, Di Stefano V and Wagner W 2013 *Comp. Math. with Applications* **65** 520
[9] Raleva K and Vasileska D 2013 *J. Comp. Electr.* **12** 601
[10] Nghiem T T T, Saint-Martin J and Dollfus P 2014 *J. Appl. Phys.* **116** 074514
[11] Muscato O, Wagner W and Di Stefano V 2014 *COMPEL* **33** 1198

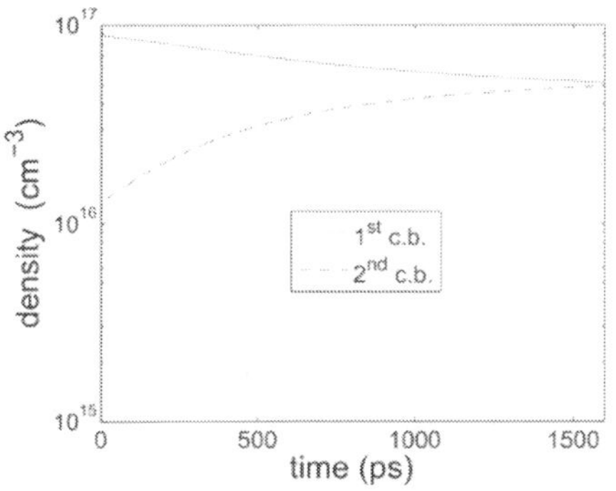

Figure 1. The electron density for the 1st, 2nd conduction band versus the simulation time.

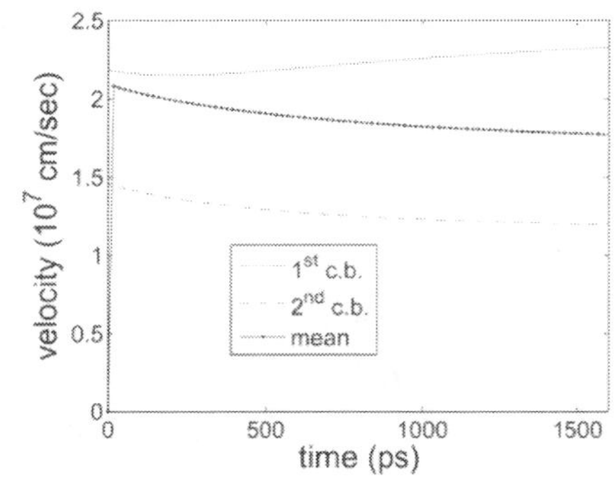

Figure 2. The electron velocity for the 1st, 2nd conduction band versus the simulation time, and their average.

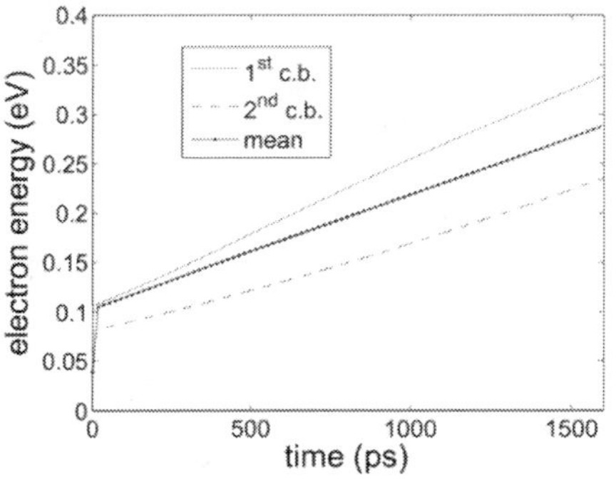

Figure 3. The electron energy for the 1st, 2nd conduction band versus the simulation time, and their average.

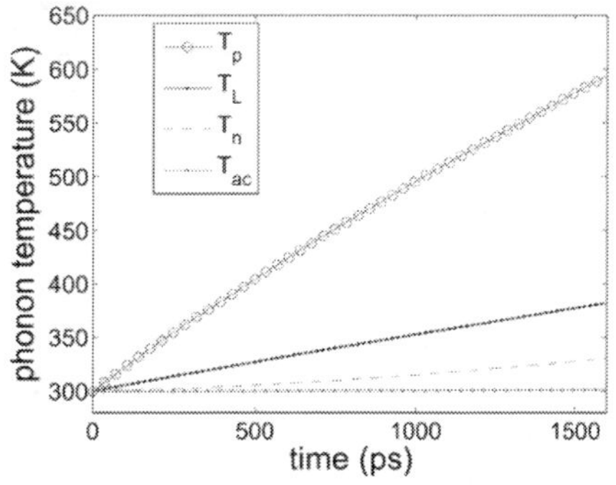

Figure 4. The phonon temperatures versus the simulation time, and the lattice temperature T_L.

[12] Jou D, Casas-Vázquez J and Lebon G 2001 *Extended irreversible thermodynamics* (Berlin: Springer-Verlag) p 517

[13] Muscato O, Pidatella R M and Fischetti M V 1998 *VLSI Design* **6** 247

[14] Anile A M, Muscato O and Romano V 2000 *VLSI Design* **10** 335

[15] Muscato O and Di Stefano V 2008 *J. Appl. Phys.* **104** 124501

[16] Muscato O and Di Stefano V 2012 *J. Comp. Electr.* **11** 45

[17] Di Stefano V and Muscato O 2012 *Acta Appl. Math.* **121** 225

[18] Muscato O and Di Stefano V 2014 *Cont. Mech. Therm.* **26** 197

[19] Muscato O and Di Stefano V 2011 *J. Phys. A:Math. Theor.* **44** 105501

[20] Muscato O and Di Stefano V 2011 *J. Stat. Phys.* **144** 171

[21] Muscato O and Di Stefano V 2011 *J. Appl. Phys.* **110** 093706

[22] Muscato O and Di Stefano V 2013 *Semicond. Sci. Technol.* **28** 025021

[23] Pennington G and Goldsman N 2004 *J. Appl. Phys.* **95** 4223

[24] Muscato O and Di Stefano V 2015 Electro-thermal transport in silicon carbide semiconductors via a hydrodynamic model (*Preprint* University of Catania)

Enhancement of thermoelectric efficiency in non-uniform semiconductor nanowires

K. H. Park[1,2], **P. N. Martin**[1,2], **and U. Ravaioli**[1,2]

[1]Department of Electrical and Computer Engineering,
[2]Beckman Institute for Advanced Technology and Science,
University of Illinois at Urbana-Champaign, Urbana, IL 61801, USA

E-mail: kpark39@illinois.edu

Abstract. Thermoelectric efficiency has been limited to a non-practical range because of the electronic material properties that are not easily controlled. Therefore, limiting thermal transport has been the best approach to enhance efficiency, and efforts to reduce the thermal conductivity even further have been extensively reported. In this work, we investigate thermoelectric transport coefficients of uniformly straight and sinusoidally undulated nanowires and compare the resulting thermoelectric figure-of-merit. The improved efficiency should suggest another strategy to effectively control thermal transport in semiconductor nanostructures.

1. Introduction

With widely growing interest in thermoelectric energy conversion, semiconductor nanowires have attracted great attention as possible candidates due to their potential for controlling both electronic and thermal transport. Recently, nanowires with diameters comparable to the phonon-mean-free-path have been studied intensively and have achieved improved thermoelectric efficiency, dictated by the dimensionless thermoelectric figure-of-merit ($ZT = S^2\sigma T/\kappa$), where S is the Seebeck coefficient, σ is the electrical conductivity, and κ is the thermal conductivity [1].

For semiconductor materials, it is very challenging to adjust the electrical power factor ($S^2\sigma$) to a desired level due to the interplay between S and σ [2]. However, the lattice thermal conductivity which has a significant contribution to the total thermal conductivity can be relatively easily manipulated by scaling, mainly owing to the phonon scattering that develops before scattering of electrons occurs in small-scale structures. The goal to limit thermal transport in ways that mildly affect electronic transport has been the most pursued strategy in thermoelectric engineering, an approach labeled as the 'phonon-glass electron-crystal' strategy [3,4].

In this study, we investigate thermoelectric properties of silicon nanowires oriented along [100] with non-uniform shape and compare their electronic and thermal transport with uniform shape nanowires, expecting an enhancement of the figure-of-merit ZT. In addition to the 1D nanowire structure in which phonons are confined in the transverse directions, non-uniformity will perturb phonon transport through the system [5]. Especially, sinusoidally undulated nanowires (sine NW) will be studied, as schematically depicted in figure 1, which can possibly realize a phonon blockade more efficiently without degrading much of the electronic transport. The length, width, undulation height, and the wavelength of the sinusoid are 108.6 nm, 4.34 nm, 4.34 nm, and 54.3 nm, respectively. The dimensions of the structure have been chosen as multiples of a lattice constant of silicon (0.543 nm) as molecular dynamics simulation requires exact locations of atoms. The width of the wire should effectively confine both electrons and phonons in the transverse directions, while only phonons will be disturbed by the corrugated surface during transport in the longitudinal direction because of their much larger mean-free-path (a few tens of nanometers for phonons in heavily doped silicon compared to a few nanometers for electrons) [6].

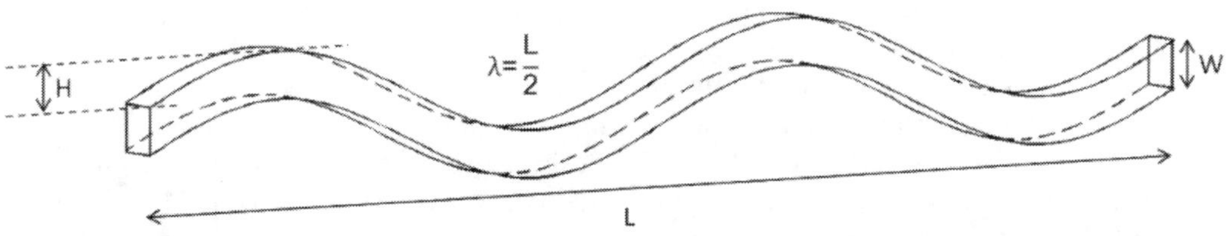

Figure 1. A schematic of a sinusoidal silicon nanowire with a square cross-section. The structural parameters are the length (L), width (W), the undulation height (H), and the wavelength of the sinusoidal wave (λ).

2. Approach

Even though the geometrical modification of nanostructures has been studied in the past, their work focused only on the degradation of thermal transport [3,7,8]. To examine trade-offs of non-uniform nanowires, studies on both electron and phonon transport should be conducted to find the maximum ZT. Loss of electronic transport due to the geometry variation can be examined by comparing an energy dependent electrical quantum conductance of the sine NWs with that of straight and uniform NWs (straight NW), employing a Recursive Green Function (RGF) approach [5,9]. The 3D conduction channel is modeled by a 1D tight-binding chain, and the quantum nature is accounted for by solving the Schrödinger equation in a series of 2D cross sections in the transverse direction. The total transmission coefficient $T(E) = \sum_{m,n}|t_{mn}|^2$ at energy E can be computed by summing the transmission probability t_{mn} over transverse sites (m,n) of each cross section [5]. $T(E)$ is then used to calculate the thermoelectric transport coefficients defined by solutions of the Boltzmann transport equation with the Landauer formalism. The electrical conductance (G) and the Seebeck coefficient (S) are expressed as [10,11]

$$G = \frac{2q^2}{h}\int_{-\infty}^{\infty} T(E)\left(-\frac{\partial f}{\partial E}\right) dE \ (1/\Omega), \tag{1}$$

$$S = -\frac{1}{qT}\frac{\int_{-\infty}^{\infty} T(E)(E-E_F)\left(-\frac{\partial f}{\partial E}\right)dE}{\int_{-\infty}^{\infty} T(E)\left(-\frac{\partial f}{\partial E}\right)dE} \ (\mathrm{V/K}). \tag{2}$$

The thermal conductivity of the sinusoidal NWs can be computed with molecular dynamics (MD) at 300 K. We adopt the reverse non-equilibrium MD method, which imposes a heat flux first and measures the resulting temperature gradient, performed with the Stillinger-Weber interatomic potential. The MD simulation should reveal a decrease in thermal conductivity with increasing number of undulations and for larger undulation heights in the sine NWs. There must be some variants on the transport coefficients due to the strain effect if the sine NW were fabricated or bent through mechanical stress. Previous work reported enhanced electrical conductivity under bending strain [12] and reduced thermal conductivity due to uniform strain and point defects of silicon nanowires [13], which may even enhance the thermoelectric figure-of-merit. We do not consider the strain effect in this study and leave our reference structure to be oriented in [100] direction. If we were to do the actual fabrication, this structure can be realized by the photolithography process as described in [14].

Journal of Physics: Conference Series **647** (2015) 012053 doi:10.1088/1742-6596/647/1/012053

3. Results

The quantum transmission of the straight and sine NWs is shown and compared in figure 2(a). The straight NW exhibits a step-function-like transmission as it represents quantum conductance at a corresponding energy level while the sinusoidal undulations directly affect the resistivity particularly at high energy. However, the undulations have little effect on low-energy transport, which leads this structure to be potentially useful of thermoelectric applications. Previous work on nanowire transport study has proven that the scattering at non-uniform surface intensifies at high energy because more conduction channels become available and efficiently couple to each other [15]. The parameters S, G, and S^2G of the wires calculated by taking into account the transmission coefficients, shown in figure 2(b-d). By definition, conductance can be converted into conductivity by multiplying L/A where A is the cross-section area. The computed S and G are within the range of experimentally measured data [3,4], but a direct comparison would not be meaningful as we consider electron scattering only due to the geometrical fluctuation and ignore otherwise. As expected, the sine NW becomes less conductive than the straight NW, reflecting the lowered transmission coefficient, and the S of the sine NW increases compared to that of the straight NW due to interplay with the conductance. Most importantly, S^2G, which is a direct measurement of the electronic effect on the ZT, decreases with the undulations above room temperature region, but the drop is not significant.

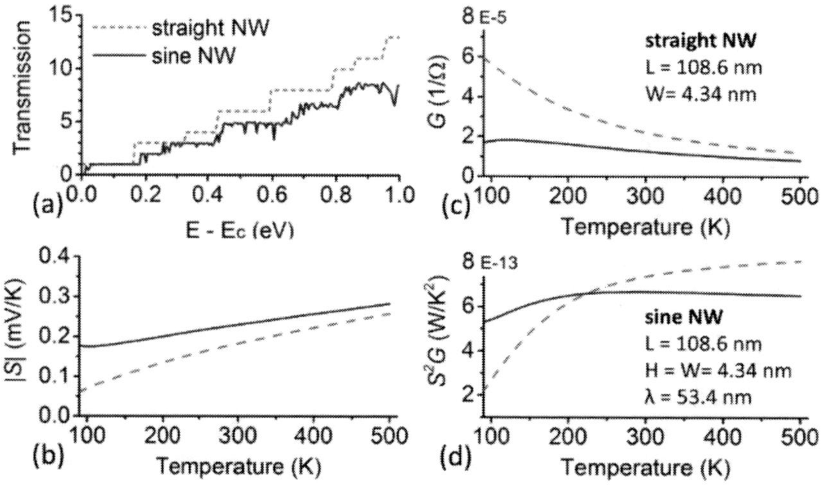

Figure 2. Comparisons of (a) electron transmission coefficient at 300 K, (b) Seebeck coefficient, (c) electrical conductance, and (d) power factor of the straight (dashed) and sine (solid line) NWs. Transmission is computed as a function of energy above the first conduction energy (E_c).

The structures under consideration have been relaxed through isothermal-isobaric and microcanonical ensemble processes for a total of 2 ns with the simulation time step of 1.0 fs, and the actual MD simulation has been performed for 2 ns. Unlike the electronic properties, κ can contribute to a great enhancement of ZT with the phonon confinement effect in low dimensional structures. This is shown in figure 3(a) where the thermal transport through the sine NW is indeed disrupted by the non-uniform geometry. Based on the calculated transport coefficients, the thermoelectric efficiency measurement ZT is computed and compared in figure 3(b) between the straight and sine NWs. The overall improvement for ZT of the sine NW is about 10 to 20 percent higher than that of the straight NW around 300 K.

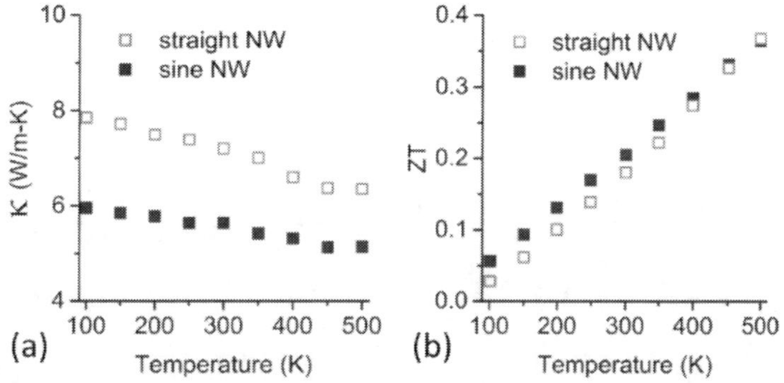

Figure 3. (a) Lattice thermal conductivity of straight and sine NWs calculated by reverse non-equilibrium MD simulation. (b) Comparison of the thermoelectric figure-of-merit ZT between straight and sine NWs.

4. Conclusion

We have investigated the thermoelectric efficiency of uniform and non-uniform shaped silicon nanowires. With the introduction of a sinusoidal-geometry variation, both electron and phonon transport are degraded, but an enhanced ZT may be achieved because there is a stronger reduction of thermal conductivity than of the electronic power factor, attributed to the large difference between electron and phonon mean-free-paths. Although the improvement obtained in this work was only around 10 to 20 percent, better enhancement can be realized by modulating structural parameters such as the cross section area, wire length, and undulation parameters. Further work will be to study benefits and downsides of the parameter modification and to find the limits of the corrugation. As we consider geometrical fluctuation as the sole perturbation to the nanowires, this strategy can be extended easily to any materials and structures.

References

[1] Chen G 2005 *Nanoscale Energy Transport and Conversion* (Oxford)
[2] Park K H, Mohamed M, Aksamija Z and Ravaioli U 2015 *J. Appl. Phys.* **117** 015103
[3] Hochbaum A I, Chen R, Delgado R D, Liang W, Garnett E C, Najarian M, Majumdar A and Yang P 2008 *Nature* **451** 163–7
[4] Boukai A I, Bunimovich Y, Tahir-Kheli J, Yu J-K, Goddard W A and Heath J R 2008 *Nature* **451** 168–71
[5] Martin P N and Ravaioli U 2009 *J. Phys. Conf. Ser.* **193** 012009
[6] Bux S K, Blair R G, Gogna P K, Lee H, Chen G, Dresselhaus M S, Kaner R B and Fleurial J-P 2009 *Adv. Funct. Mater.* **19** 2445–52
[7] Martin P N, Aksamija Z, Pop E and Ravaioli U 2009 *Phys. Rev. Lett.* **102** 1–4
[8] Martin P N, Aksamija Z, Pop E and Ravaioli U 2010 *Nano Lett.* **10** 1120–4
[9] Sols F, Macucci M, Ravaioli U and Hess K 1989 *J. Appl. Phys.* **66** 3892
[10] Nag B R 1980 *Electron Transport in Compound Semiconductors* (New York: Springer)
[11] Kim R, Datta S and Lundstrom M S 2009 *J. Appl. Phys.* **105** 034506
[12] Shao R, Gao P and Zheng K 2015 *Nanotechnology* **26** 265703
[13] Murphy K F, Piccione B, Zanjani M B, Lukes J R and Gianola D S 2014 *Nano Lett.* **14** 3785–92
[14] Abramova V V., Slesarev A S and Tour J M 2015 *Nano Lett.* 150331185249004
[15] Wang J, Polizzi E, Ghosh A, Datta S and Lundstrom M 2005 *Appl. Phys. Lett.* **87** 043101

Goldstone Mode Induced by Skyrmions and Collective Modes in Disordered Quantum Hall Crystals

I Kanazawa

epartment of Ph sics, Tok o akugei ni ersit , Koganeishi, Tok o 1 4 1, apan

mail `kanazawa@u-gakugei.ac.jp`

Abstract. e ha e proposed the e ecti e agrangian densit for the sk rmion like solitons around the lling factor $\nu \sim 1$, and ha e introduced the massless gauge eld oldstone mode induced the hedgehog like solitons. e ha e discussed the sk rmion glass eha iour.

1. Introduction

From eld theor method 1,2 , a ne t pe of topological e citation, Sk rmion, is predicted to occur around lling factor, $1/m$, ith odd integer m. Charged e citations in the uantum Hall ferromagnet are sk rmions. A sk rmion is a spin te ture, rotating se eral spins simultaneousl , and electric charge $\pm e/m$. A fractional charge e citation is possi le, ecause it is a coherent e citation. A sk rmion is a topological soliton spread o er t o energ le els. Sk rmion e citations can e detected e perimentall measuring their spin. An e perimental measurement 3 of the spin polari ation has een made around ν 1. The spin polari ation is o ser ed oth for $\nu \geq 1$ and for $\nu \leq 1$. It is a clear signal of sk rmion e citations. Recentl the present author has proposed that in uasi 2 1 dimensional uantum antiferromagnet the doped hole induces magnetic disordered hedgehog like soliton, hich is composed of the doped hole and the cloud of S 2 ang Mills elds ith the spin disorder around the hole, using the gauge in ariant e ecti e agrangian ith spontaneous s mmetr reaking 4 . In this stud , e tending the theoretical formula, e introduce the gauge in ariant e ecti e agrangian densit for the sk rmions around the lling factor 1, and present the massless gauge eld oldstone mode induced the Sk rmion. The massless gauge eld introduces the long range interaction, hich ill pla an important role in the Sk rme cr stal . Taking into account the correlation e ect among hedgehog like solitons, e ill discuss the mechanism of the long rela ation of the sk rmion like soliton d namics. In addition, e suggest the presence of the sk rmion glass.

2. A model system

The si e of the hedgehog like soliton is determined a competition et een Coulom energ and the eeman energ . That is, hen charge is spread o er a ider domain of radius $\sim R_C$, the energ ecomes smaller. n the other hand, the num er of re ersed spins is decreased, as increased the eeman energ . Taking account of that a hedgehog like soliton in planar geometr is similar to the 3 nonlinear sigma model the S 2 e citation 1 , e think that the

Journal of Physics: Conference Series **647** (2015) 012054 doi:10.1088/1742-6596/647/1/012054

pertur ing gauge elds A^a_μ introduced the hedgehog like soliton ha e a local S 2 s mmetr . Furthermore, it is assumed that S 2 gauge elds A^a_μ are spontaneousl roken through the Anderson Higgs mechanism in a a similar to the reaking of the uantum hall ferromagnetic s mmetr around the hedgehog like soliton. e set the s mmetr reaking $\langle \ |\phi_a| \ \rangle \quad \langle \ , \ , \mu\rangle$ of the ose eld ϕ_a in the agrangian densit as follo s,

$$
\begin{aligned}
L \quad & \frac{1}{2}\left(\partial_i S^j - g_1 \varepsilon_{\hat{i}\hat{j}k} A^a_i S^k\right)^2 \\
& -\frac{1}{4}\left(\partial_\nu A^a_\mu - \partial_\mu A^a_\nu \quad g_3 \varepsilon_{abc} A^b_\mu A^c_\nu\right)^2 \\
& \frac{1}{2}\left(\partial_\mu \phi_a - g_4 \varepsilon_{abc} A^b_\mu \phi_c\right)^2 - \lambda^2 \left(\phi_a \phi_a - \mu^2\right)^2 .
\end{aligned} \qquad 1
$$

In order to introduce the hedgehog like soliton ith a si e, R_C, e add the fourth term in . 1 , hich depends on the eeman energ and Coulom energ . $\hat{j}$ corresponds to the direction of the e ternal magnetic eld. hen the soliton is formed, e set the s mmetr reaking $\langle \ |\phi_a| \ \rangle \quad \langle \ , \ , \mu\rangle$. n the other hand, hen the anti soliton is formed, e set the s mmetr reaking $\langle \ |\phi_a| \ \rangle \quad \langle \ , \ , -\mu\rangle$. Then, e can o tain the e ecti e agrangian densit , $\mathcal{L}_{eff}$, at small introducing of hedgehog like solitons around the lling factor $\nu \quad 1$. The alue, $\mu \quad \langle \ |\phi_3| \ \rangle$, of the s mmetr reaking depends strongl on the eeman energ and the Coulom energ . That is, hen the eeman energ ecomes large, the alue of μ ecomes larger. n the other hand, hen the Coulom energ ecome larger, the alue of μ ecomes smaller.

$$
\begin{aligned}
\mathcal{L}_{eff} \quad & \frac{1}{2}\left(\partial_i S^j - g_1 \varepsilon_{\hat{i}\hat{j}k} A^a_i S^k\right)^2 \\
& -\frac{1}{4}\left(\partial_\nu A^a_\mu - \partial_\mu A^a_\nu \quad g_3 \varepsilon_{abc} A^b_\mu A^c_\nu\right)^2 \\
& \frac{1}{2}\left(\partial_\mu \phi_a - g_4 \varepsilon_{abc} A^b_\mu \phi_c\right)^2 \\
& \frac{1}{2} m_1^2 \left[\ A^1_\mu \ ^2 \quad A^2_\mu \ ^2 \right] \quad m_1 \left[A^1_\mu \partial_\mu \phi_2 - A^2_\mu \partial_\mu \phi_1\right] \\
& g_4 m_1 \left\{ \phi_4 \left[\ A^1_\mu \ ^2 \quad A^2_\mu \ ^2 \right] - A^3_\mu \left[\phi_1 A^1_\mu \quad \phi_2 A^2_\mu\right] \right\} \\
& -\frac{m_2^2}{2} \ \phi_4 \ ^2 - \frac{m_2^2 g_4}{2m_1} \phi_4 \ \phi_a \ ^2 - \frac{m_2^2 g_4^2}{m_1^2} \ \phi_a \phi_a \ ^2 ,
\end{aligned} \qquad 2
$$

here S^j is the spin parameter, $m_1 \quad \mu \cdot g_4$, and $m_2 \quad 2 \ 2 \ ^{1/2} \lambda \cdot \mu$. The e ecti e agrangian descri es t o massi e ector elds A^1_μ and A^2_μ, and one massless U 1 gauge eld A^3_μ. ecause masses of A^1_μ and A^2_μ are formed through the Higgs mechanism introducing the hedgehog like soliton, the elds A^1_μ and A^2_μ e ist around the hedgehog like soliton ithin the length of $\sim 1/m_2 \equiv R_C$. From the rst term in . 2 , the spin component parallel to $\hat{j}$, hich corresponds to the direction of the e ternal magnetic eld, is induced strongl around the hedgehog like soliton. The densit $\rho_{\hat{j}}$ for the spin component parallel to $\hat{j}$ around the hedgehog like soliton is gi en appro imatel

$$
\rho_{\hat{j}} \ r \ \sim \frac{1}{\pi} \frac{R_C^2}{\left[r^2 \quad R_C^2\right]^2} \cdot \frac{R_C^2}{\left[r^2 \quad R_C^2\right]} \qquad 3
$$

For the e ecti e g romagnetic ratio, g, $\ll 1$, the si e, $R_C \sim 1/m_1$, of the hedgehog like soliton is estimated Sondhi et al. 2 as

$$
\left(\frac{R_C}{l}\right)^3 \propto \frac{l}{a} \ g \ | \ |\ln g| \ ^{-1} \qquad 4
$$

here l and a are the andau length and ohr radius, respecti el . o e can de ne the topological num er q for e cited hedgehog like solitons as follo s,

$$q \quad \frac{1}{2\pi}\int_{\sum} \mathrm{d}S_{\mu\nu}\left(\partial_\mu A^3_\nu - \partial_\nu A^3_\mu\right),$$

here $\sum$ is a sphere, hose radius is larger than $R_C \sim 1/m_2$. If a sphere $\sum$ surrounds completel one soliton, hose center position is r_i, the alue of q_i is 1. If a sphere $\sum$ surrounds completel one anti soliton, hose center position is r_i, the alue of q_i is -1. hen the hedgehog like soliton is located at the position r_i and $|r - r_i| \ll 1/m_2 \sim R_C$ is assumed, the oldstone gauge eld A^3_μ r, r_i at the position r is represented as A^3_μ r, r_i $\propto q_i/|r - r_i|$. Thus e can introduce the interaction, hich is mediated the massless gauge eld oldstone mode , et een the hedgehog like solitons at positions r_i and r_j as $V_{ij} \propto q_i \cdot q_j/|r_i - r_j|$, hich pla s an important role in the Sk rme cr stal . It has een suggested ,1 that the e ect of disorder is important on the ground state of a t o dimensional electron gas in the uantum Hall regime at lling factors slightl de iating from unit . This suggests strongl that the aggregation of hedgehog like solitons is mulch related to the long rela ation time of the spin. In addition, stripe shaped magnetic domains has een o ser ed 11 . The stripe shape of the domain means that long range frustration pla s an important role on the domain formation 12 . In order to discuss the spin d namics, e en isage an e ecti e hamiltonian, H, for the hedgehog like soliton, O $r_{\bar{i}}$, hich is introduced in e . 2 ,

$$H \quad -J \sum_{<\bar{i},\bar{j}>} \frac{S_{\bar{i}\cdot\bar{j}}}{|S_{\bar{i}}||S_{\bar{j}}|} O\ r_{\bar{i}}\ \cdot O\ r_{\bar{j}}$$
$$\frac{1}{2}K \sum_{\bar{i}\neq\bar{j}} \frac{O\ r_{\bar{i}}\ \cdot O\ r_{\bar{j}}}{|r_{\bar{i}} - r_{\bar{j}}|}$$

ith $J > K >$ and the rst sum taken onl o er nearest neigh or the distance et een each magnetic soliton is $\leq 2R_c$ and the second taken o er all pair $\bar{i}$ / $\bar{j}$ means $|r_{\bar{i}} - r_{\bar{j}}| \gg 2R_c$. $S_{\bar{i}} \equiv \sum_{i\in\pi R_c^2} S_i$. That is, $S_{\bar{i}}$ is the summation of the ferromagnetic spin, S_i, ithin $\sim \pi R_c^2$ $\bar{i}$ around the hedgehog like soliton at the site $r_{\bar{i}}$. $S_{\bar{i}}$ represents the e ecti e spin of the soliton O $r_{\bar{i}}$. The rst term corresponds to short range ferromagnetic ordering interaction and the second corresponds to long range frustration. If g_3 in e . 2 is assumed to e e ual to $\sqrt{\pi/K}$, here K is the long range interaction constant in the e ecti e Hamiltonian, e . . Taking into account this model along lines originall suggested Cha es et al. 13 , there are t o emergent length scales that are long compared to R_c. The rst of these is the correlation length ξ_0 hich go erns the uctuations in the a sence of frustration. The second length, $L_0 \propto \xi_0^{-1} \cdot K^{-\frac{1}{2}}$, structure is roken up due to frustration, i.e. the frustration limited domain si e 14 . The characteristic length ξ_0 and L_0 are dependence on temperature and num er of hedgehog like solitons. If O $r_{\bar{i}}$ is the local order parameter for the ferromagnetic soliton, this means that

$$< O\ r_{\bar{i}}\ \cdot O\ r_{\bar{j}}\ > \propto R_c|r_{\bar{i}} - r_{\bar{j}}|^{-1} \mathrm{e\ p} -|r_{\bar{i}} - r_{\bar{j}}|/\xi_0$$
$$m^2$$

for $\xi_0 \ll |r_{\bar{i}} - r_{\bar{j}}| \ll L_0$,

$$< O\ r_{\bar{i}} \cdot O\ r_{\bar{j}}\ > \propto L_0 m^2 |r_{\bar{i}} - r_{\bar{j}}|^{-1} \mathrm{e\ p} -|r_{\bar{i}} - r_{\bar{j}}|/L_0$$

for $L_0 \ll |r_{\vec{i}} - r_{\vec{j}}|$, here m is the e pectation alue of the local order parameter. Then e rite the free energ F of the ferromagnetic domain of characteristic si e L as

$$F \quad \delta L^2 - \phi L^3 \quad S \; L \; L^3.$$

The coe cient ϕ is a measure of the free energ densit gained aggregation of the hedgehog like solitons. The domain all surface tension $\delta \propto \; \xi_0 \; ^{-2}$ is positi e. The strain coe cient $S \; L$ is $\sim S_0 L^2$ for L small compared to the radius of cur ature of the ideal space. The rela ation of a local order parameter in a nite s stem occurs most e cientl through the creation and mo ement of defect alls. The creation of such a defect all is proportional to δL^2. In this s stem, domains are special ecause reduction in the range of the local order parameter reduces the strain. So e e pect the collecti e acti ation free energ , $E_{col} \; L \; \propto \delta L^2 - m S_0 L^5$.

o , e shall consider the long rela ation time, hich can e descri ed in terms of a normali ed rela ation function, $f \; t$, for hich f 1. The collecti e eha ior can e descri ed in terms of domains ith a si e distri ution $\rho \; L$ and a most pro a le si e, $L \quad L_0 \gg R_c$. e introduce the rela ation function ithin each domain as e ponential ith a deca time dependent upon an acti ation free energ $E_{col} \; L \quad 1$ as follo s,

$$f \; t \qquad \int_0^\infty dL L^2 \rho \; L \; \text{e p}\{- \; t/\tau_\infty \; \text{e p} -E_{col} \; L \; /T \;\},$$

here τ_∞ is taken to e a T independent parameter.

3. Conclusion

e ha e introduced the gauge in ariant e ecti e agrangian densit for the sk rmion like solitons around the lling factor $\nu \sim 1$ and ha e introduced the massless gauge eld oldstone modes . e ha e discussed the mechanism of the long rela ation of sk rmion like solitons.

References

1 ee H and Kane C 1 *Phy. Rev. Lett.* **64** 1313
2 Sondhi S et al. 1 3 *Phy. Rev.* **B47** 1 41
3 arret S et al. 1 *Phy. Rev. Lett* **74** 112
4 Kana a a I 2 1 *Physica* **C 357-360** 14
Kana a a I 2 1 *Inter. J.Modern Phys* **B 15** 4 13
Kana a a I 2 *J. Phys Chem. Solid* **66** 13
Kana a a I 2 13 *J. Low Temp. Phy.* **172** 1
re et al. 1 *Phy. Rev. Lett.* **75** 2 2
Maude K et al. 1 *Phy. Rev. Lett.* **77** 4 4
1 eder een A and a aro V 1 *Phy. Rev. Lett* **82** 4
11 Shono T et al. 2 *Appl. Phys. Lett.* **77** 13 3
12 Coniglio A et al. 1 1 *Phy. Rev.* **B44** 12
13 Cha es et al. 1 *Physica* **A 225** 12
14 Ki elson S A et al. 1 4 *J. Chem. Phys.* **101** 231
1 Ki elson et al. 1 *Prog. Theor. Phys. Suppl.* **126** 2

Impact of a superlattice on electrical properties of AlGaN/GaN/sapphire 2DEG structures

V Jakštas[1], V Janonis[1], A Bičiūnas[1], R Aleksiejūnas[2], A Kadys[2], T Malinauskas[2] and I Kašalynas[1]

[1] Center for Physical Sciences and Technology, A. Gostauto 11, LT-01108 Vilnius, Lithuania
[2] Institute of Applied Research, Vilnius University, Sauletekio 9-III, LT-10222 Vilnius, Lithuania

E-mail: vytautas.jakstas@ftmc.lt, irmantas.kasalynas@ftmc.lt

Abstract. In this work we demonstrate the impact of the AlGaN/GaN superlattice (SL) on electrical properties of AlGaN/GaN heterostructures grown by MOCVD on sapphire substrate. Schottky diodes and high electron mobility transistors were processed of the AlGaN/GaN heterostructures with and without the SL. It was demonstrated that the presence of the SL increased the breakdown voltage and improved the saturation current and transistor transfer characteristics performance.

1. Introduction

AlGaN/GaN heterostructures are promising material for the high-frequency and high-power solid state electronic devices due to the unique physical properties of GaN, i.e. wide bandgap, high breakdown field, electron mobility, and good thermal stability and conductivity [1]. During the last decade a huge interest of scientist and technologist community to this material was shown towards the growth of higher quality heterostructures and the development of new electronic devices. One of expected is an efficient terahertz (THz) emitter development for THz spectroscopy and imaging applications at room temperature [2].

Successful designs require reliable electrical performance of AlGaN/GaN 2DEG structures, which in turn depends on the quality of the grown buffer and epitaxial layers. AlGaN/GaN heterostructures are usually grown on a foreign substrates, such as sapphire, Si, and SiC due to absence of the GaN substrates and requirement to maintain low budget. However, foreign substrates cause high threading dislocation density (TDD) in grown heterostructure layers. One of the possibilities to reduce the TDD is to insert a superlattice (SL) between the layers.

In this work we characterize differently grown 2DEG AlGaN/GaN heterostructures by deploying a Ni/Au contacts and investigating their electrical performance in temperature range 300-124 K. Temperature dependence of the I-V characteristics and C-V characteristics were investigated by measuring the Schottky barrier height, the ideality factor and the leakage current dependence on temperature for heterostructure grown on sapphire substrate with and without the SL.

2. Fabrication of samples

The AlGaN/GaN layers were grown by the metalorganic chemical vapour deposition (MOCVD) method providing two VU393 and VU394 samples with different heterostructures design on 2" diameter sapphire substrate. The first sample was consisted of n-type unintentionally-doped GaN layer, the AlN spacer and the $Al_xGa_{1-x}N$ layer. The main difference for the sample VU394 was that the SL consisted of 5 pairs of AlGaN/GaN layers and an additional GaN buffer was grown beneath the AlN spacer. The thickness and the molar concentration of Al in the top $Al_xGa_{1-x}N$ layer for the samples VU393 and VU394 were found of about 21 nm and 0.11, and 18 nm and 0.26, respectively.

The self-consistent design of the Schottky diodes and HEMTs were taken with the aim to reduce a complexity of production. A standard ultraviolet (UV) photolithography and e-beam thin films deposition system was used to obtain the electric contacts. The Ohmic contacts were formed of the Ti/Al/Ni/Au stack annealed at 830 °C in N_2 ambient for 30 s. The conditions of the Rapid Thermal Annealing (RTA) was experimentally optimised by using circular transmission line method (TLM) and determining the specific contact resistance [3]. The contact resistance and the specific contact resistivity were as low as 7.0 Ω and 3.4×10^{-5} Ω·cm^2, respectively.

The metals Ni/Au were processed afterward to form the Schottky contacts and finalize the fabrication process. SEM images of the electronic devices are shown in the figure 1. No mesa etching and no surface passivation were used. The proposed design with the minimized technological steps is very attractive for the fast and cheap prototyping of the electronic components.

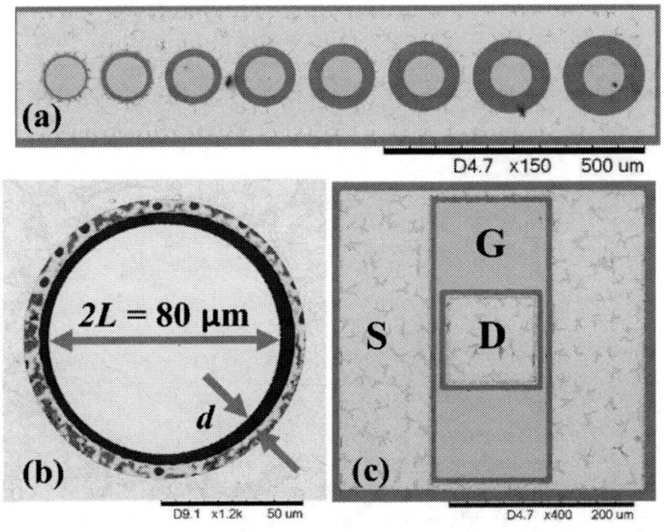

Figure 1. SEM images of (a) an array of the Schottky diodes with the distance d between the inner and the outer contacts varying from 5 µm to 40 µm, (b) enlarged single Schottky diode with $d = 5$ µm and the area of the central contact $S = 5.027\times10^5$ cm^2 that was left constant for all Schottky diodes, and (c) the HEMT with a square-ring-shape gate (G), here 'S' and 'D' denotes the source and drain electrodes, respectively. Gate length and width was 5.5 µm and 2×100 µm, respectively, the distance between G and S was 2.5 µm and the distance between S and D was 13 µm.

3. Results

3.1. Performance of the Schottky diodes

Measured I-V characteristics of the Schottky diodes at the applied direct current (DC) bias from -2 V to +2 V are shown in figure 2. The maximum forward current was found strongly depending over the distance d: at +2 V it changed from 146 A/cm^2 and 140 A/cm^2 when $d = 5$ µm to 396 A/cm^2 and 309 A/cm^2 when $d = 40$ µm for the samples VU393 and VU394, respectively. The leakage current for the diodes processed of VU394 sample has increased significantly above -1 V as it is seen in figure 2. The effect was explained by the different TDD measured with the atomic force microscope; for the samples VU393 and VU394 the TDD was of 1.6×10^8 cm^{-2} and 1.5×10^9 cm^{-2}, respectively. It was observed that the SL contribution was the other way than expected. However, the on-off ratio of all the VU394 Schottky diodes was almost 10^5, indicating that the increase of the leakage current was not critical for most applications.

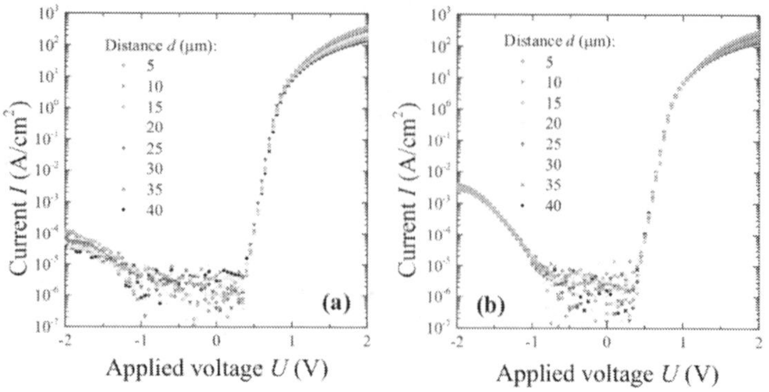

Figure 2. I-V characteristics of the samples (a) VU393 and (b) VU394 for the series of the Schottky diodes with varying d. The measured current was normalised to the area of the Schottky contact S.

Figure 3 (a) shows the I-V temperature (I-V-T) characteristics measured in the temperature range from room temperature (RT) up to 124 K. Strong dependence on the temperature was observed in the forward direction of the I-V characteristics. The threshold voltage of the Schottky diode increased with the decreasing temperature, however, the leakage current remained almost unaffected. Calculated Schottky barrier height Φ_b and ideality factor n were changing in a wide range from 1.09 eV and 1.2 at RT to 0.73 eV and 3.9 at 124 K temperature. Measured results demonstrated linear Φ_b dependence over n as it is shown in the figure 3 (b). This effect was attributed to the lateral inhomogeneities of the Schottky diodes [4]. The linear extrapolation of the data up to $n = 1$ provided the barrier height of about 1.1 eV.

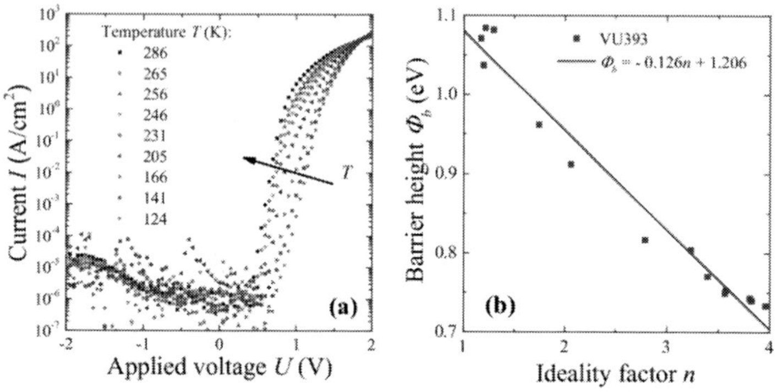

Figure 3. (a) The I-V-T characteristics of the sample VU393 with the measured current normalised to the area of the Schottky contact S and (b) the barrier height dependence on the ideality factor of the same Schottky diode.

The test for the breakdown voltage was done with the Schottky diodes processed of the VU393 and VU394 samples. The VU393 diodes sustained electric fields not higher than 35 kV/cm. However, the diodes of the sample VU394 did not undergo breakdown up to the applied voltage of -210 V corresponding breakdown field higher than 70 kV/cm. Since the TDD was 10 times higher in the sample with the SL, the increase of the breakdown voltage shows that the Schottky diodes undergo a breakdown in the sapphire and GaN interface.

3.2. Performance of the HEMTs

Figure 4 shows the DC characteristics of the HEMTs normalised to the gate width. The output characteristics were acquired at applied drain voltages from 0 V to +7 V. Maximum drain current density was obtained up to 188 mA/mm and 302 mA/mm for the samples VU393 and VU394, respectively. It was followed by the decrease of R_{ON} from 18 Ω to 10 Ω for the same samples indicating a significant improvement with the presence of the SL. In addition, the dependence of the leakage current over applied drain voltage, which was visible in the transfer characteristic of the samples VU393, vanished in the sample VU394 and the absolute values slightly decreased. Finally, the transconductance characteristics were measured steeper and its maximum value was obtained higher and improved from 121 mS/mm to 152 mS/mm level. This comparative analysis of the DC characteristics shows the positive impact of the SL to the performance of the HEMTs.

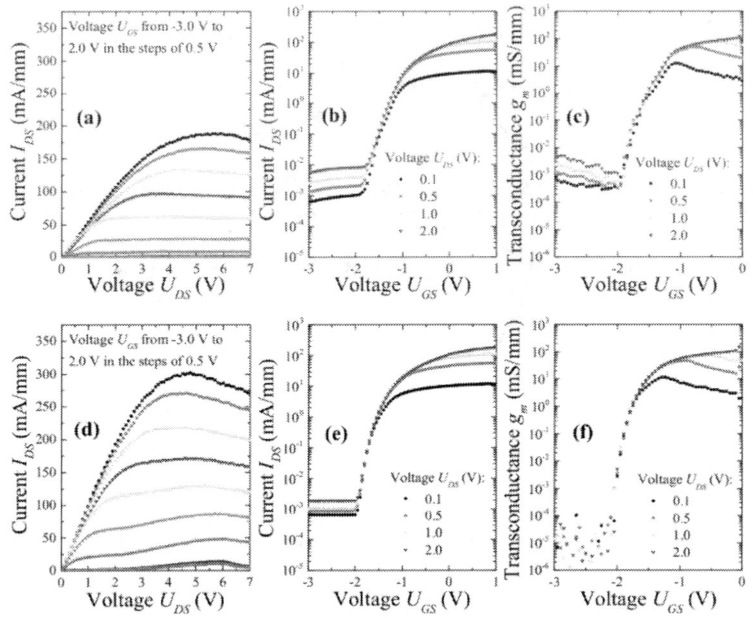

Figure 4. DC characteristics of the HEMT made of VU393 (parts a, b, and c) and of VU394 (parts d, e, and f) samples: (a, d) output characteristics at gate bias from +2 V to -3 V in the steps of 0.5 V, (b, e) transfer characteristics at the selected drain voltages, and (c, f) transconductance characteristics at the same drain voltages.

4. Conclusion

In summary, we have developed unpassivated surface Schottky diodes of the AlGaN/GaN structures grown with and without the SL on sapphire substrate. The results have shown higher leakage current of the Schottky diodes made of the sample with the SL, providing on-off current switching ratio up to 10^5. Despite higher TDD in the sample with SL, devices demonstrated much higher breakdown voltage. The comparison of the HEMTs made of the samples with and without the SL has shown improved parameters of the HEMT: $I_{DS\ MAX}$ has increased 1.6 times, the R_{ON} has decreased 1.8 times, $g_{m\ MAX}$ has increased 1.3 times, and the I_{DS} dependence over U_{DS} below U_{th} was not significant. The overall results have suggested a need to optimize the growth conditions of the SLs in order to overcome issue of increase of the TDD and further improve the performance of the electronic devices.

References

[1] Meng F, Zhang J, Zhou H, Ma J, Xue J, Dang L, Zhang L, Lu M, Ai S, Li X and Hao Y 2012 Transport characteristics of AlGaN/GaN/AlGaN double heterostructures with high electron mobility *J. Appl. Phys.* **112** 1–7

[2] Otsuji T, Watanabe T, Boubanga Tombet S A, Satou A, Ryzhii V, Popov V and Knap W 2014 Emission and detection of terahertz radiation using two-dimensional plasmons in semiconductor nanoheterostructures for nondestructive evaluations *Opt. Eng.* **53** 031206

[3] Jakštas V, Kašalynas I, Šimkienė I, Strazdienė V, Prystawko P and Leszczynski M 2014 Schottky diodes and high electron mobility transistors of 2DEG AlGaN/GaN structures on sapphire substrate *Lith. J. Phys.* **54** 227–32

[4] Tekeli Z, Altndal Ş, Çakmak M, Özçelik S, Çalişkan D and Özbay E 2007 The behavior of the I-V-T characteristics of inhomogeneous (Ni/Au) - Al 0.3Ga0.7N/AlN/GaN heterostructures at high temperatures *J. Appl. Phys.* **102** 0–8

Monte Carlo model for the analysis and development of III-V Tunnel-FETs and Impact Ionization-MOSFETs

V. Talbo[1,*], J. Mateos[1], T. González[1], Y. Lechaux[2], N. Wichmann[2], S. Bollaert[2] and B. G. Vasallo[1]

[1]Applied Physics Department, Salamanca University,Plaza de la Merced, s/n Salamanca 37008, Spain

[2]Institute of Electronics, Microelectronics and Nanotechnology, 59652 Villeneuve d'Ascq, France

[*]E-mail: vtalbo@usal.es

Abstract. Impact-ionization metal-oxide-semiconductor FETs (I-MOSFETs) are in competition with tunnel FETs (TFETs) in order to achieve the best behaviour for low power logic circuits. Concretely, III-V I-MOSFETs are being explored as promising devices due to the proper reliability, since the impact ionization events happen away from the gate oxide, and the high cutoff frequency, due to high electron mobility. To facilitate the design process from the physical point of view, a Monte Carlo (MC) model which includes both impact ionization and band-to-band tunnel is presented. Two ungated InGaAs and InAlAs/InGaAs 100 nm PIN diodes have been simulated. In both devices, the tunnel processes are more frequent than impact ionizations, so that they are found to be appropriate for TFET structures and not for I-MOSFETs. According to our simulations, other narrow bandgap candidates for the III-V heterostructure, such as InAs or GaSb, and/or PININ structures must be considered for a correct I-MOSFET design.

1. Introduction

Impact-ionization metal-oxide-semiconductor FETs (I-MOSFETs) have demonstrated to provide a subthreshold slope (SS) as low as ~4 mV/dec, one order of magnitude lower than those obtained with classical MOSFETs with a similar I_{on} [1]; however, the necessary source-to-drain voltage V_{DS} is still too large to be competitive with mainstream MOSFET technology. On the other side, III-V MOSFETs can work at V_{DS} lower than 0.5 V and deliver I_{on} currents near 1 A/mm [2], but with a large value of SS. Reliability is also an issue due to the degradation of the gate oxide. In order to overcome these weaknesses, III-V I-MOSFETs are being explored as promising devices for ultra-low power logical circuits. By means of an adequate heterostructure bandgap engineering, these devices should be able to improve both the reliability (by moving the impact ionization events away from the gate oxide) and the cutoff frequency (due to the higher mobility of III-V materials). The first option under study is InGaAs heterostructures because of the mature technology, but, as shown later, the tunnel effect appears for lower applied V_{DS} than impact ionization processes, thus making this behavior impossible as an I-MOSFET. In fact, tunnel-FETs (TFETs) [3] are direct opponents of I-MOSFETs for ultra-low SS digital applications. As an alternative, InAs-based heterostructures will have to be used for the fabrication of I-MOSFETs.

In order to support the development of III-V I-MOS transistors, and as an alternative to the test-and-error experimental procedure, this work reports the development of a Monte Carlo (MC) model, which incorporates impact ionization processes by means of the Keldysh approach [4,5] as well as band-to-band tunnel by means of the Wentzel-Kramers-Brillouin (WKB) method [6]. MC simulations are useful for the study of narrow bandgap III-V heterostructures from the physical point of view as well as ungated and gated PIN topologies. This model allows identifying the competition between the impact ionization and band-to-band tunneling at the onset of the conduction in InGaAs-based PIN diodes. As well, it can assist the design and optimization of I-MOSFET topologies in terms of I_{off} and I_{on} currents, V_{DS} and SS.

2. Physical Model

For the calculations we make use of an ensemble MC simulator self-consistently coupled with a 2D Poisson solver, which includes a detailed model for the hole transport, impact ionization processes and band-to-band tunnel, valid for the simulation of ungated $In_{0.53}Ga_{0.47}As$ and lattice matched InAlAs/InGaAs-based PIN diodes. The parameters for electrons in the involved materials can be found in Ref. [7]. The model used for hole dynamics is essential due to the P-region. A typical spherical and nonparabolic valence band structure is considered, including three sub-bands: heavy- and light-hole bands (HH and LH), degenerated at k=0 and characterized by a different curvature in k-space, and a third split-off band (SOH), in which the band warping is accounted for by the use of approximated overlap functions [9]. Ionized impurity, acoustic, polar and non-polar optical phonon scattering mechanisms are considered for holes [8,9]. The hole physical parameters used in the simulations are reported in [4,5]. To illustrate the hole transport in bulk InGaAs, Figure 1 presents (a) the mean velocity in the electric field direction and (b) the mean energy as a function of the electric field for different dopings.

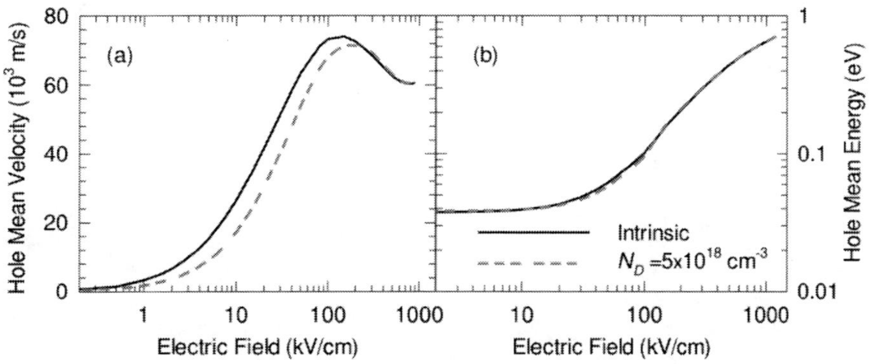

Figure 1. (a) Mean velocity in the electric field direction and (b) mean energy for holes in bulk InGaAs as a function of electric field for different P-type ionized impurity densities.

Impact ionization of electrons, which occurs in the Γ valley and leads to the generation of electron-hole pairs, is included in the MC simulator by using the Keldysh approach [10], where the probability per unit time of having an impact ionization event is given by $P(E)=S[(E- E_{th})/ E_{th}]^2$ if $E > E_{th}$, and $P(E)=0$ otherwise, E being the electron kinetic energy in the Γ valley, E_{th} the ionization threshold energy and S a measure of the softness or hardness of the threshold. E_{th} and S are considered as adjustable parameters to reproduce the ionization coefficients measured in bulk materials but they can be considered as adjustable parameters due to the large dispersion in the experimental measurements [11-14]. From each impact ionization occurrence, an electron in the Γ valley and a hole in the heavy-

Journal of Physics: Conference Series **647** (2015) 012056 doi:10.1088/1742-6596/647/1/012056

hole band emerge, while the electron originating the ionization process remains in the Γ valley. We have verified that hole impact ionization is negligible for the considered applied voltages [4].

Our MC simulator includes also band-to-band tunneling based on the WKB method [6], essential to reproduce the physical behavior of III-V PIN diodes in reverse bias conditions. In order to consider the injection from the P-region into the intrinsic layer of the PIN diode, we perform a discretization of the incident electrons energy and calculate the number of injected particles by tunnel effect at each time step (Δt=1 fs) in the whole energy range, as in Refs. [15-17]. The transmission coefficient through the energy barrier $qV(x)$ is calculated following the WKB model [6, 17], while the potential profile $V(x)$ is updated at each time step from the solution of the Poisson equation in the MC simulation. The Richardson constant A^* [15] can be considered as an adjustable parameter in the simulations in order to perform future experimental adjustments.

3. Results

Figure 2 shows (a) the I-V characteristics for reverse bias and (b) the position of tunnel events and impact ionization processes results provided by the MC model for a 100 nm PIN diode based on $In_{0.53}Ga_{0.47}As$, being the intrinsic reverse bias V=-3.0 V. In Figure 2(b) the energy bands have been included for clarity. The parameters considered for tunneling are of the order of those found for bulk [17] while the impact ionization probability has been exaggerated to check the possibility of being the dominant effect. Tunnel events take place along the intrinsic region, constrained by the shape of the energy bands. As well, impact ionization appears mainly at the N-side of the intrinsic region of the structure, where the electric field and then the electron energy reach the higher values. Tunnel effect is always the most significant mechanism, in fact, the impact ionization takes place just when there are enough electrons generated by tunnel effect in the P-side of the intrinsic region. Then, in the absence of tunneling, there is not any impact ionization event recorded in our simulations. Nevertheless, as can be seen in Figure 2(a), the value of I_{on} is enhanced by the presence of impact ionization mechanisms.

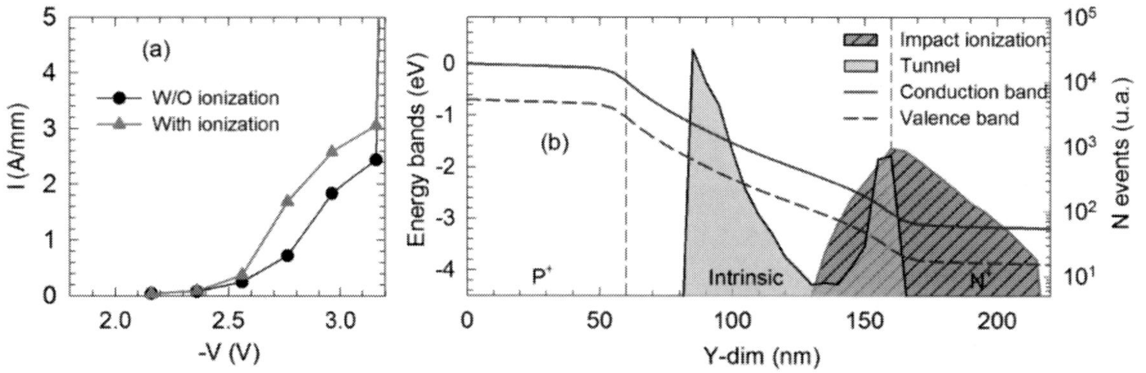

Figure 2. (a) I-V curves and (b) conduction and valence band energy bands, impact ionization and tunnel events for a InGaAs 100 nm PIN diode, being the reverse bias V=-3.0 V.

A 100 nm PIN diode based on the lattice-matched InAlAs/InGaAs heterostructure has been also qualitatively studied and simulation results have been obtained. The existence of impact ionization events, even if it is overestimated, does not lead to a significant enhancement in the I-V curves. For simplicity, Figure 3 presents (a) the I-V characteristics and (b) the position of just tunnel events in the absence of impact ionization. The reverse bias necessary for the achievement of I_{on} is of the order of that found for the InGaAs structure.

4. Conclusions

A MC model which incorporates impact ionization processes by means of the Keldysh approach and band-to-band tunnel by means of the WKB method has been developed for the study of narrow bandgap III-V heterostructures in order to support the development of III-V I-MOS transistors. The physical behavior of 100 nm PIN diodes based in InGaAs and InAlAs/InGaAs have been qualitatively analyzed to evaluate the competition between tunnel and impact ionization events. Even when the impact ionization is overestimated, the stronger physical process is the tunnel injection. Other narrow bandgap candidates for the III-V heterostructure, as InAs or GaSb, must be considered. As well, since the tunnel I_{on} of InGaAs homostructures can be improved when impact ionization takes place, designing PININ structures can also been considered for the development of III-V I-MOSFETs.

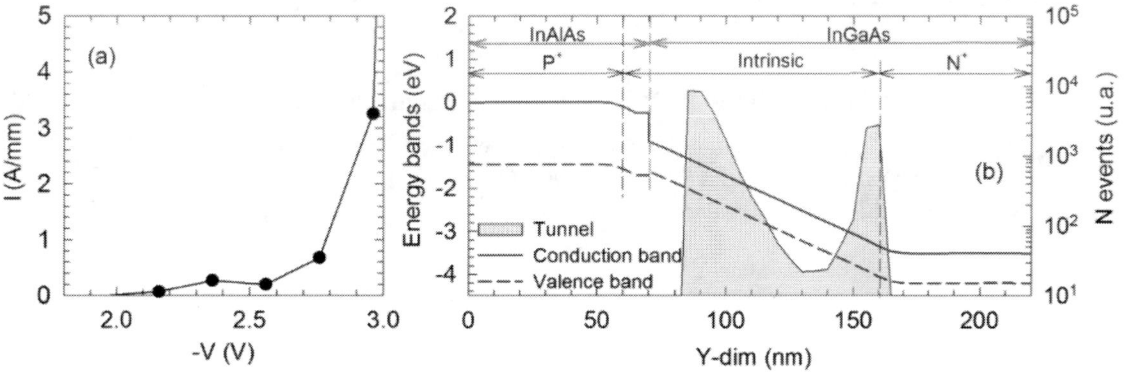

Figure 3. (a) I-V curves and (b) conduction and valence band energy bands, and tunnel events for a PIN diode based in the heterostructure InAlAs/InGaAs, being the reverse bias V=-3.0 V.

Acknowledgments
This work has been supported by the Ministerio de Economía y Competitividad-Dirección General de Investigación through the project TEC2013-41640-R, by the Consejería de Educación, Junta de Castilla y León, through Project SA052U13 and the Agence Nationale de la Recherche through the project ANR-13-JS03-0001-01.

References
[1] Q. Huang, R. Huang, Z. Wang, Z. Zhan and Y. Wang 2011 *Appl. Phys. Lett.* **99** 083507
[2] J. J. Gu, Y. Q. Wu and P. D. Ye 2011 *J. Appl. Phys.* **109** 053709
[3] W. Y. Choi, J. Y. Song, J. D. Lee, Y. J. Park and B.-G. Park 2005 *IEEE Inter. Electron devices Meeting (IEDM)* 955
[4] B. G. Vasallo, J. Mateos, D. Pardo and T. González 2003 *J. Appl. Phys.* **94** 4096
[5] B. G. Vasallo, J. Mateos, D. Pardo and T. González 2004 *J. Appl. Phys.* **95** 8271
[6] B. H. Bransden and C. J. Joachain 1987 *Introduction to Quantum Mechanics* New York: Longman Scientific & Technical
[7] J. Mateos, T. González, D. Pardo, V. Hoël and A. Cappy 1999 *Semicond. Sci. Technol.* **14** 864
[8] T. Brudevoll, T. A. Fjeldly, J. Baek and M. S. Shur 1990 *J. Appl. Phys.* **67** 7373
[9] M. Costato and L. Reggiani 1973 *Phys. Stat. Sol. (b)* **58** 471
[10] M. V. Fischetti 1991 *IEEE Trans. Electron Devices* **38** 634
[11] T. P. Pearsal 1980 *Appl. Phys. Lett.* **36** 218
[12] F. Osaka, T. Mikawa and T. Kaneda 1985 *IEEE J. Quantum Electron.* **21** 1326
[13] T. Kagawa, Y. Kawamura, H. Asai, M. Naganuma and O. Mikami 1989 *Appl. Phys. Lett.* **55** 993
[14] I. Watanable, T. Torikai, K. Matika, K. Fukushima and T. Uji 1990 *IEEE Electron Device Lett.* **11** 437
[15] D. Moro-Melgar, J. Mateos, T. González and B. G. Vasallo 2014 *J. Appl. Phys.* **116** 234502
[16] Sun L., Liu X. Y., Liu M., Du G. and Han R. 2003 *Semicond. Sci. Technol.* **18** 576
[17] Huang C.K., Zhang W. E. And Yang C. H. 1998 *IEEE Trans. on Electron Devices* **45** 4

Disorder effect on carrier mobility in Fullerene organic semiconductor

N Mendil[1], M Daoudi[2], Z Berkai[3] and A Belghachi[4]

[1,2,3,4] Laboratory of semiconductor device physics, University of Tahri Mohamed Bechar, P.O.B n°417 Bechar 08000, Algeria.
mendil_nesrine@yahoo.com

Abstract. The critical factor that limits the efficiencies of organic electronic devices is the low charge carrier mobility which is attributed to disorder in organic films. In this context, we have studied the effects of disorder on carrier mobility in organic Schottky diode of electrons for the fullerene (C_{60}). Our results show that the mobility is sensitive probes of structural phase transitions and order-disorder underlying C_{60}. Where it is one reason behind the low mobility which it take as value $1.4x10^{-2}$ $cm^2/V.s$ above critical temperature Tc =289K.

1. Introduction

Over the years, the world's industries and research organization have shown their interest in the renewable energy resources due to the limited fossil fuel reserves, which emit tons of carbon dioxide and other pollution every second. The efficient use of sunlight to produce the domestic electricity by means of solar cell devices is the perfect way to deal with the limitation of the reserve fossil fuel [1]. The recent growing demand for cost-effective renewable energy has generated considerable interest in the field of organic solar cells. Organic photovoltaic (OPV) cells have the potential to substantially lower costs for solar energy conversion, given their expected low materials and fabrication costs. Since fullerene (C_{60}) organic semiconductor was discovered, it has drawn the attention of physicists and chemists alike, which much work has been done on its physical and chemical properties.

At ambient conditions the C_{60} molecules are centered on sites of a face-centered-cubic (fcc) with a=1.42nm and energy gap about 1.5eV [2]. The order-disorder transition in C_{60} occurs around 289K at ambient pressure. At this transition freely rotating molecules get orientation ordered in a simple cubic lattice (sc) and below 90K the occupation of the nearly degenerate states does not change, so this temperature is identified with the quenching disorder transition [3-4].

In this work, we use a single active-layer OPV cell, where we have studied the effects of disorder on carrier mobility, conductivity and resistivity in C_{60} organic Schottky diode which it is the most frequently used as electron acceptor organic semiconductor due to its unique physical and chemical characteristics[5].

MENDIL Zoulikha Nesrine : mendil_nesrine@yahoo.com, phone number: (+213)0662012502

Journal of Physics: Conference Series **647** (2015) 012057 doi:10.1088/1742-6596/647/1/012057

2. Theoretical model

Charge transport mechanisms in organic semiconductors have a great importance since they are the basis for a new class of electronic and optoelectronic devices. In this context, the transport properties of charge carriers in the organic material have been investigated by using a mathematical simulation taking into account the effect of disorder, where the mobility (μ) varies depending on the temperature (T), the applied voltage (V) and the energetic disorder (δ), it can take 0.03eV (T<289K) and 0.06eV (T>289K) [6]. The charge carrier mobility is given by the following equation[7]:

$$\mu = \mu_{\infty} \exp\left[-\left(\frac{3\delta}{5K_BT}\right)^2 + 0.78\left(\left(\frac{\delta}{K_BT}\right)^{1.5} - 2\right)\sqrt{\frac{eaV}{\delta d}}\right] \quad (1)$$

where μ_{∞} = 0.03 cm^2/V.s, the mobility as the temperature is going to infinity and K_B is the Boltzmann constant [eV/K].

In the absence of traps the space charge limited the current density (J) can be written as a function of the applied voltage as[8]:

$$J = \frac{9}{8}\varepsilon_r\varepsilon_0\mu\frac{V^2}{d^3} \quad (2)$$

where ε_r=4 relative dielectric constant [7], d is the organic film thickness [cm].
The electrical resistivity of electron in C_{60} is expressed as follows:

$$\rho = \frac{V}{dJ} \quad (3)$$

3. Result and discussion

The temperature dependence of mobility is an important topic to understand the charge transport mechanism in organic materials, so in these figures 1 (a) and (b), we can see that the electron mobility decrease with increasing of temperature and lattice parameter.

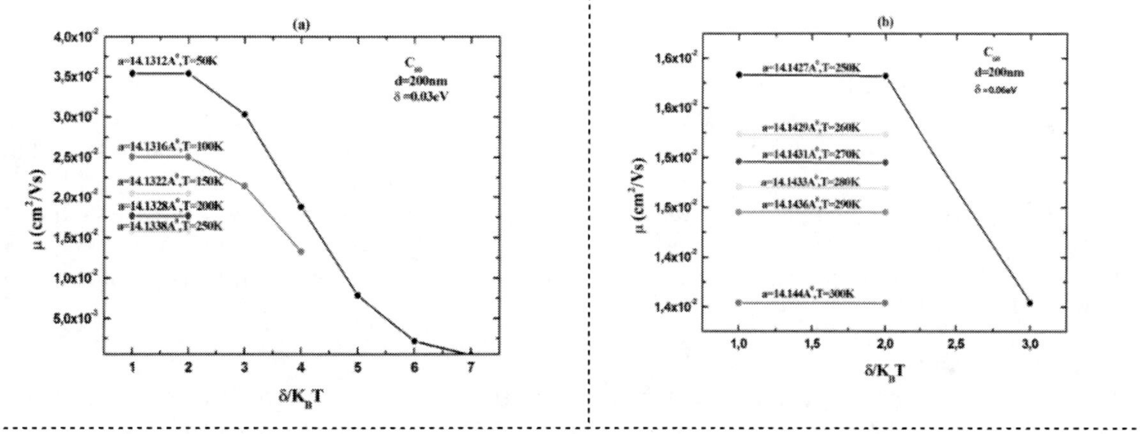

Figure 1. δ/K_BT dependence of electron mobility in C_{60}; for δ =0.03eV (a) and δ =0.06eV (b) with different temperature for diode thickness d=200nm.

In figure 1 (a), for the low temperature (T=50K - 100K) the electron mobility has a sufficient decrease from 3.5×10^{-2} cm^2/Vs up to 2.5×10^{-2} cm^2/Vs however when the temperature increase from 150K to 250K the electron mobility decrease from 2.1×10^{-2} cm2/Vs to 1.7×10^{-2} cm^2/Vs and it remains constant. In figure 1 (b), the mobility continues to decrease from 1.55×10^{-2} cm^2/Vs to 1.4×10^{-2} cm^2/Vs and it remains constant. If we compare the carrier mobility of amorphous silicon, which it is about

1 $cm^2V^{-1}s^{-1}$ [9] and that obtained by electron in C_{60} is about 1.5 $cm^2V^{-1}s^{-1}$, we can seem that C_{60} is one of the most useful electron-transport materials.

Figure 2 (a) and (b) show that the values of conductivity increase when $T \leq T_{C1}$=125 K in a remarkable way from 2.3×10^{-11}S/cm to 6.1×10^{-6}S/cm and the resistivity decrease from 4.33×10^{10} Ωcm to 1.6×10^{5} Ωcm exponentially, respectively, against at T> T_{C1} the conductivity and resistivity take a value of saturation σ =2. 5×10^{-5}S/cm and ρ = 4×10^{4} Ωcm until T= T_{C2}=250K. The conductivity increase with increasing of temperature, however the resistivity decrease in both cases of the energetic disorder (0.03eV and 0.06eV). In addition, the curves are divided into two parts due to the lattice parameter effect. We can also observe that at T_{C2}=250K σ =4. 7×10^{-6}S/cm and ρ= 2.126×10^{5} Ωcm , Fullerene crystal undergoes a first order phase transition from the fcc structure above T_{C2} to a sc structure belowT_{C2}.

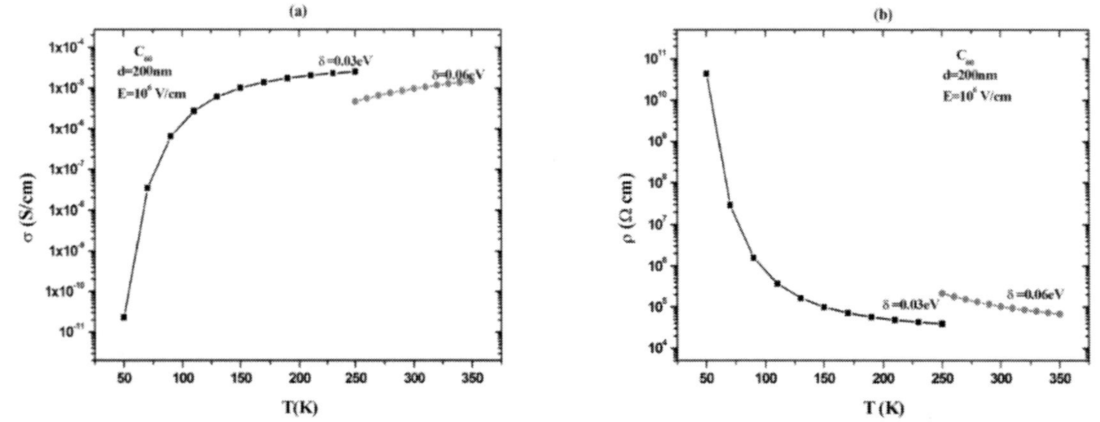

Figure 2. The variation of conductivity σ (a) and electrical resistivity ρ (b) of C_{60} layer for single carrier diode against Temperature.

The conductivity of the C_{60} single crystal is 17×10^{-8} $(\Omega.\ cm)^{-1}$ and the activation energy is 0.581eV. However, for the poly-crystalline films, the conductivity and activation energy are reported to be 10^{-6} to 10^{-8} $(\Omega.cm)^{-1}$ and 0.3-0.6eV, respectively[10]. For amorphous films, these values are within the intervals 10^{-7}-10^{-14} $(\Omega.cm)^{-1}$ and 0.5-1.1 eV. With improving the crystallinity of C_{60} films, their conductivity increases, and the activation energy fall [11]. So the conductivity of the crystalline C_{60} is obviously higher than that of the C_{60} amorphous films. So, we can conclude that C_{60} takes a behavior of poly-crystalline form when the temperature is in range of 50K-125K and amorphous form when T is in range of 125K-250K.

4. CONCLUSION

In this paper we have discussed the impact of energetic disorder on the charge carrier mobility, conductivity and resistivity. It was shown that the disorder is one of the key parameters to explain the transport in organic semiconductors. More specifically, increasing the disorder energy push the mobility decreases and it will be less temperature dependent.

ACKNOWLEDGEMENT

This work is partly supported by Algerian ministry of higher education and research (CNEPRU) project N° D03820140006

REFERENCES

[1] J Holmes, R Pandey, A A, Gunawan and K A Mkhoyan 2012 *Adv. Funct. Mater* **22** 617

[2] C W Chang, B C Regana,W Mickelson, R O Ritchie and A Zettl 2003 *Solid State Communications* **128** 359

[3] E A Katz, D Faiman, B Mishori, Y Shapira, A Isakina and M A Strzhemechn 2003 *J. Appl. Phys* **94** 7173

[4] E A Katz, D Faiman, S Shtutina, A P Isakina and K A Yagotintsev 2011 *Mat. Res. Soc. Symp. Proc.* 675

[5] J Y Hu, N N Niu, G Z Piao, Y Yang, Q Zhao, Y Yao, C Z Gu, C Q Jin and R C Yu 2012 *Carbon* 5458

[6] X Yu, Undergraduate Thesis 2012 *Monte Carlo simulation of charge transport in organic solar cells.* Department of Electrical and Computer Engineering, Duke University

[7] M Lenes, S W Shelton, A B Sieval, D F Kronholm, J C Hummelen, and P W M Blom 2009 *Adv. Funct. Mater* **19** 3002

[8] N Martinelli, PhD Thesis 2011 *Charge Transport in Organic Conjugated Materials: Impact of Lattice Dynamics and Coulombic Interaction Effects.* Université de Mons

[9] B Pevzner PhD Thesis 1995, *Transport and dielectric properties of thin fullerene (C_{60}) films.* Massachusetts Institute of Technology

[10] C Wen, J Li, K Kitazawa, T Aida, I Honma, H Komiyama, and K Yamada 1992 *Appl. Phys. Lett* **61** 2162

[11] J G Hou, A D Zhao, T Huang and S Lu 2014 *Encyclopedia of Nanoscience and Nanotechnology* **1** 409

Enhancement of radiative efficiencies of near-ultraviolet organic light-emitting diodes by localized surfacee Plasmon resonance enhancement effect

A Matsuoka[1], A. Emoto[2] and N. Ohtani[3]
Department of Electronics, Doshisha Univesrity,
3-1 Tatara-Miyakodani, Kyotanabe-shi, Kyoto 610-0321, Japan

E-mail: ohtani@mail.doshisha.ac.jp

Abstract. Near-UV emission organic light emitting diodes (OLEDs) containing silver (Ag) nanoparticles (NPs) were fabricated to improve their radiative efficiencies by the localized surface plasmon resonance (LSPR) enhancement effect. The samples studied were simple bi-layer structures consisting of an active layer and a hole-transporting layer (HTL). Poly[(9,9-dihexylfluoren-2,7-diyl)-co-(1,4-dimethyl-benzene-2,5-diyl)] (PF-DMB) was used as a UV emission material, whose emission peaking-wavelength is about 395 nm. Ag NPs were inserted between ITO and HTL, whose absorption peaking wavelength is corresponding to about 395 nm. Thus, the electroluminescence (EL) intensity of UV OLEDs containing Ag NPs was enhanced to about 2.5 times as large as that without Ag NP by the LSPR enhancement effect.

1. Introduction

Most of the research topics on organic light-emitting diodes (OLEDs) concentrate on the OLEDs operating in the visible light regime, because they are applied to flat panel displays and white light sources [1]. On the other hand, some research on OLEDs operating in the ultraviolet regime were reported [2, 3]. However, their radiative efficiencies are not excellent because high efficient carrier injection into large-bandgap emissive materials is difficult. In addition, they were fabricated using a vacuum evaporator because they consisted of low-weight molecular materials. Thus, it is easy to fabricate multiple-layer structures containing electron- and hole-transporting layers plus electron- and hole-injection layers for improving carrier injection into the wide-bandgap emissive materials. On the other hand, polymer-based OLEDs are fabricated using wet-process. It has an advantage to reduce the production-cost lower than that using vacuum evaporators. However, it is very difficult to fabricate multiple-layer structure OLEDs. Recently, localized surface plasmon resonance (LSPR) enhancement effect was energetically studied because LSPR was effective to improve the radiative efficiencies of LEDs [4-7]. LSPR is known that if the nanostructure is smaller than the wavelength of light and the free electrons can be displaced from the lattice of positive ions and collectively oscillate in resonance with the light. LSPR is also used in solar cells to enhance the energy conversion efficiencies [8].

In this research, near-UV emission polymer-based OLEDs containing silver (Ag) nanoparticles (NPs) were fabricated to improve their radiative efficiencies by the LSPR enhancement effect. Ag NPs have the potential to improve UV radiative efficiencies from OLEDs because the LSPR wavelength of

[1] To whom any correspondence should be addressed.
[2] To whom any correspondence should be addressed.
[3] To whom any correspondence should be addressed.

Ag NPs depends on the diameter of Ag NPs, thus LSPR wavelength can be tuned to the near-ultraviolet and visible regions [7].

2. Experimental

2.1 Extinction spectra of Ag NPs

Ag NPs were deposited on indium-tin-oxcide (ITO)-coated glass substrates using 3-(Aminopropyl) triethoxysilane (APTES) [9]. The diameter of Ag NPs was 10 nm. Figure 1 shows atomic force microscope (AFM) images of the two sample surfaces ; one contains only APTE, while another contains APTES and Ag NPs. The averaged surface roughness of the sample without APTES was less than 1 nm, while that of the sample with APTES was about 10 nm. This indicates that Ag NPs were uniformly deposited on the sample surface using APTES because the averaged roughess is equal to the diameter of Ag NPs. Figure 2 show extinction spectra of the two samples ; one contains APTES and Ag NPs, another contains only APTES. It is very clear that the absorption peak at around 390-nm was caused by Ag NPs in the sample containing Ag NPs. Thus, we concluded that Ag NPs were uniformly deposited on the ITO electrode and revealed absorption peaking wavelength at around 390 nm.

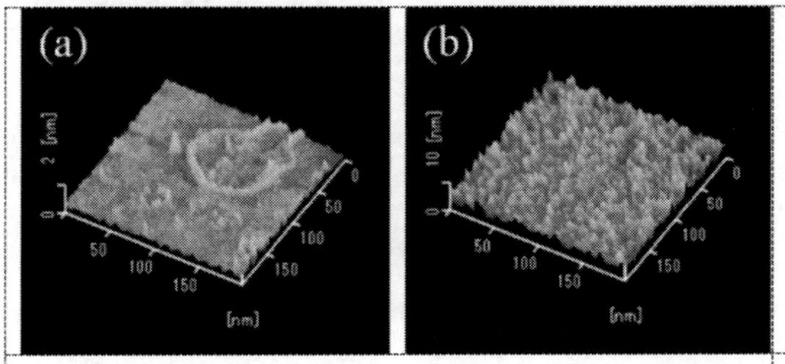

Figure 1. AFM images of the sample surfaces. (a) The sample contains only APTES, while (b) contains APTES and Ag NPs. The averaged surface roughness of (b) corresponds to the diameter of Ag NPs

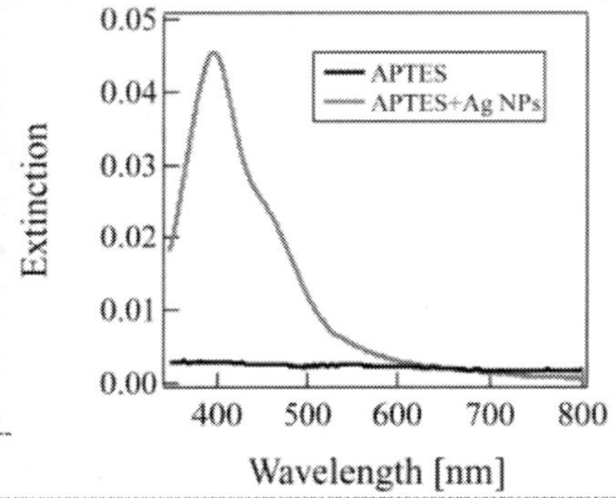

Figure 2. Extinction spectra of samples. Red line denotes the sample containing APTES and Ag NPs, while black line is only APTES

2.2 Fabrication of OLEDs

An emissive layer and a hole-transporting layer (HTL) were fabricated on the Ag NPs sheet by using spin-coating method. Thus, the samples studied were simple bi-layer structures. The following two polymer emissive materials were used in each OLED: Poly[(9,9-dihexylfluoren-2,7-diyl)-co-(1,4-dimethyl-benzene-2,5-diyl)] (PF-DMB) reveals a UV emission at 395 nm, while poly[(m-phyenylenevinylene)-alt-(2,5-dihexyioxy-p-phenylenevinylene)] (PPV) reveals a green emission at around 500 nm. This enables us to evaluate the LSPR enhancement effect because the emission of PF-DMB corresponds to the extinction peak of Ag NPs. Thus, the UV emission of PF-DMB should be enhanced by LSPR, while the emission of PPV should not be changed. In addition, poly(3,4-ethylenedioxythiophene)-polystyrenesulfonic acid (PEDOT:PSS) was used as a hole-transporting material. Finally, aluminium (Al) was vacuum deposited on the emissive layer as cathode metal. Consequently, two kinds of OLEDs were fabricated as follows and compared their emissions affected by LSPR enhancement effect.

Sample A ITO/PEDOT:PSS/PF-DMB/Al,

Sample B ITO/PEDOT:PSS/PPV/Al.

Ag NPs were inserted between ITO and PEDOT:PSS layers. To evaluate the LSPR enhancement effect originating from Ag NPs, we fabricated the samples A and B without Ag NPs. Note that Ag NPs are known to suffer oxidation, which leads to degradation of the LSPR enhancement effect in OLEDs. However, OLEDs are usually sealed not to expose to the air because organic materials are easily destroyed due to the oxidation and moisture. Thus, Ag NPs in OLEDs are also protected against the oxidant effect by the sealing effect. Consequently, the oxidant effect has little affect on Ag NPs.

3. Results and discussion

Figure 3(a) shows the injection current density dependence of lumnance of sample A. It is very clear that the luminance of the sample containing Ag NPs rapidly increased when the injection current was more than 220 mA/cm^2. Figure 3(b) shows EL spectra of the sample A when the injection current density was 270 mA/cm^2. UV-EL signals from PF-DMB were clearly observed in the both samples. In addition, the shape of both EL spectra is very similar. These results indicate that the OLEDs operating in the near UV regime were successfully fabricated. In this case, it is worthy of notice that the EL intensity of UV OLEDs containing Ag NPs was enhanced to about 2.5 times as large as that without Ag NPs. Consequently, this leads to an increase of the external quantum efficiency from $2.27 \times 10^{-4}\%$ to $5.42 \times 10^{-4}\%$ at the current injection of 270 mA/cm^2. Thus, the near UV EL emission was enhanced by the LSPR enhancement effect originating from Ag NPs.

In order to verify the above result, EL properties of sample B were shown in Fig. 4. Sample B should emit green light from PPV, which does not correspond to the extinction wavelength of Ag NPs. The luminance-current density properties shown in Fig. 4(a) reveal that the luminance did not drastically increase in the sample containing Ag NPs. In addition, EL spectra when the injection current density was 370 mA/cm^2 were not affected by the LSPR enhancement effect originating from Ag NPs as shown in Fig. 4(b). These results clearly demonstrate that the green emission was not enhanced by the LSPR enhancement effect. Thus, we conclude that the near UV emission was successfully enhanced by the LSPR enhancement effect because the extinction peaking-wavelength of Ag NPs corresponds to the emission wavelength of PF-DMB.

4. Conclusions

The localized surface plasmon resonance (LSPR) enhancement effect was applied to the near UV emission from polymer-based OLEDs for enhancing the near UV emissions. The extinction peaking-wavelength of Ag NPs whose diameter is about 10 nm was tuned to the near UV regime. As a result, the EL intensity of the UV OLEDs containing the Ag NPs was enhanced about 2.5 times as large as that without Ag NPs. This demonstrates that the LSPR enhancement effect is simple and useful method to enhance the UV emission from polymer-based OLEDs because the high-efficient carrier injection into wide-bandgap emissive materials is difficult in the polymer-based OLEDs.

References

[1] Ikawa S, Yagi S, Maeda T, Nakazumi H 2012 *Phys. Stat. Sol. (c)* **9** 2553
[2] Mikami A, Mizuno Y Takeda T 2008 *Digest of Tech. papers –SID Int. Symp.* **39** 215
[3] Chao T C, Lin, Y T, Yang C Y, Hung T S, Chou H C, Wu C C, Wong K T 2005 *Advanced Materials* **17** 992
[4] Gu X, Qiu T, Zhang W, Chu P K 2011 *Nanoscale Research Letters* **6** 1
[5] Sung J, Kim B, Choi C, Lee M, Park S, Lee E, Beom-Hoan O 2009 *Microelectron. Eng.* **86**, 1120
[6] Fujiki A, Uemura T, Zettu N, Akai-Kasaya M, Saito A, Kuwahara Y 2010 *Appl. Phys. Lett.* **96** 043307
[7] Rycenga M, Cobley C M, Zeng J, Li W, Moran C H, Zhang Q, Qin D, Xia Y 2011 *Chem. Rev.* **111** 3669
[8] Temple T L, Mahanara G D K, Reehal H S, and Bagnall D M 2009 *Solar Energy Materials and Solar Cells* **17** 1978
[9] Masitas R A, Khachian I V, Bill B L, Zamborini F P 2014 *Langmuir* **30** 13075

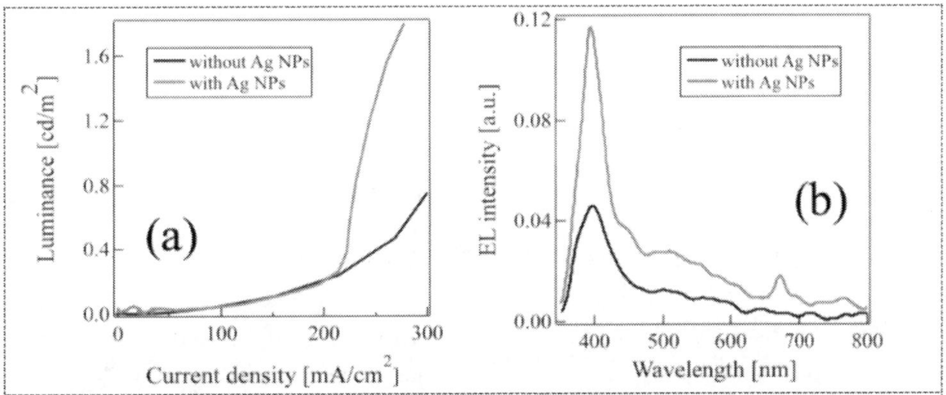

Figure 3. (a) Injection current density dependence of luminance of sample A. Red line denotes the sample containing Ag NPs, while black line denotes the sample not containing Ag NPs. (b) EL spectra of sample A when the same current density of 270 mA/cm^2 was injected. Red line denotes the sample containing Ag NPs, while black line denotes the sample not containing Ag NPs.

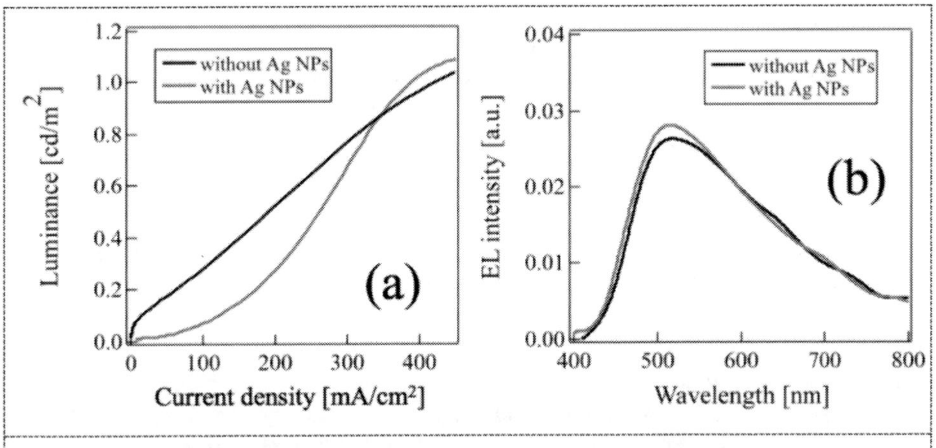

Figure 4. (a) Injection current density dependence of luminance of sample B. Red line denotes the sample containing Ag NPs, while black line denotes the sample not containing Ag NPs. (b) EL spectra of sample B when the same current density of 370 mA/cm^2 was injected. Red line denotes the sample containing Ag NPs, while black line denotes the sample not containing Ag NPs.

Journal of Physics: Conference Series **647** (2015) 012059 doi:10.1088/1742-6596/647/1/012059

Monte Carlo Study on Anomalous Carrier Diffusion in Inhomogeneous Semiconductors

N. Mori[1], R. J. A. Hill[2], A. Patanè[2], and L. Eaves[2]

[1] Division of Electrical, Electronic and Information Engineering, Osaka University, Osaka 565-0871, Japan

[2] School of Physics and Astronomy, University of Nottingham, Nottingham NG7 2RD, United Kingdom

E-mail: nobuya.mori@eei.eng.osaka-u.ac.jp

Abstract. We perform ensemble Monte Carlo simulations of electron diffusion in high mobility inhomogeneous InAs layers. Electrons move ballistically for short times while moving diffusively for sufficiently long times. We find that electrons show anomalous diffusion in the intermediate time domain. Our study suggests that electrons in inhomogeneous InAs could be used to experimentally explore generalized random walk phenomena, which, some studies assert, also occur naturally in the motion of animal foraging paths.

1. Introduction

Inhomogeneous semiconductors display distinctive electric and magnetic properties [1]. We recently demonstrated how linear magnetoresistance arises from the stochastic behavior of the electronic cycloidal trajectories around low-mobility islands in high-mobility inhomogeneous InAs epilayers [2, 3]. In an engineered inhomogeneous optical medium, Barthelemy *et al.* [4] demonstrated that light waves perform Lévy flight, a particular class of generalized random walk. Generalized random walk phenomena have been observed in nature and studied widely, ranging from electron [5,6], photon [4,7], and phonon [8,9] motion in solids and gases to animal foraging paths [10–12]. Here we present Monte Carlo studies demonstrating the possibility to experimentally explore generalized random walks in inhomogeneous InAs. We adopt the Monte Carlo method because it can handle arbitrary inhomogeneous patterns more easily than analytical models [13]. In addition, it can be easily extended to study the influence of high magnetic field [2, 3].

2. Calculation Method

We consider an inhomogeneous high-mobility InAs layer containing regions with low-mobility (low-μ regions) covering a fractional area f the remaining regions are called high-μ regions. The random spatial profile of the low-μ regions is generated from the power spectrum of the autocorrelation function $\langle\Delta(\boldsymbol{r})\Delta(\boldsymbol{r}')\rangle = \Delta^2 \exp(-|\boldsymbol{r} - \boldsymbol{r}'|^2/\Lambda^2)$. Here $\boldsymbol{r} = (x, y)$ is the two-dimensional in-plane vector and Λ is the correlation length. The low-μ regions $R_{\rm low}$ are defined according to the relation $R_{\rm low} = \{\boldsymbol{r}|\Delta(\boldsymbol{r}) < \Delta_{\rm th}\}$ where the threshold $\Delta_{\rm th}$ determines the low-μ coverage $f = R_{\rm low}/(R_{\rm low} + R_{\rm high})$. The scattering rate in the low-μ regions is assumed to be equal to $W_{\rm i} + W_{\rm e}$ and in the high-μ regions to $W_{\rm i}$. Here $W_{\rm i}$ is the weak inelastic scattering

Journal of Physics: Conference Series **647** (2015) 012059 doi:10.1088/1742-6596/647/1/012059

rate and W_e the strong elastic scattering rate. We assume that electrons relax to thermal equilibrium at $T = 300\,\mathrm{K}$ after inelastic scattering, while the momentum vector $\boldsymbol{k}$ is randomized isotropically after elastic scattering. The non-parabolicity of the electron energy dispersion of InAs is modeled as $\varepsilon(\boldsymbol{k})[1 + \alpha\varepsilon(\boldsymbol{k})] = \hbar^2 k^2/2m_e$ with $m_e = 0.023\,m_0$ and $\alpha = 2.2\,\mathrm{eV}^{-1}$. Note that we consider zero-field diffusion processes in a sample with low carrier density and the non-parabolicity plays a minor role. We performed ensemble Monte Carlo simulations to obtain the mean square displacement (MSD) at zero applied electric and magnetic fields:

$$\mathrm{MSD} = \frac{1}{N}\sum_{i=1}^{N} |\boldsymbol{r}_i(t) - \boldsymbol{r}_i(0)|^2. \tag{1}$$

Here N is the number of electrons. In the following simulations, the typical number of electrons is $N \sim 100000$.

3. Results and Discussion

Figure 1 shows examples of randomly generated patterns of low-μ regions with $\Lambda = 0.5\mu\mathrm{m}$. For $f < 0.5$, low-μ regions (colored regions) are isolated from each other and look like islands in the surrounding high-μ regions, while for $f > 0.5$, high-μ regions (white regions) look like islands. In spite of the fact that the low-μ coverage of the actual samples reported in Ref. [2] is $f \sim 0.05$, we mainly focus on high-f samples with $f = 0.8$, since we expect that relatively long range

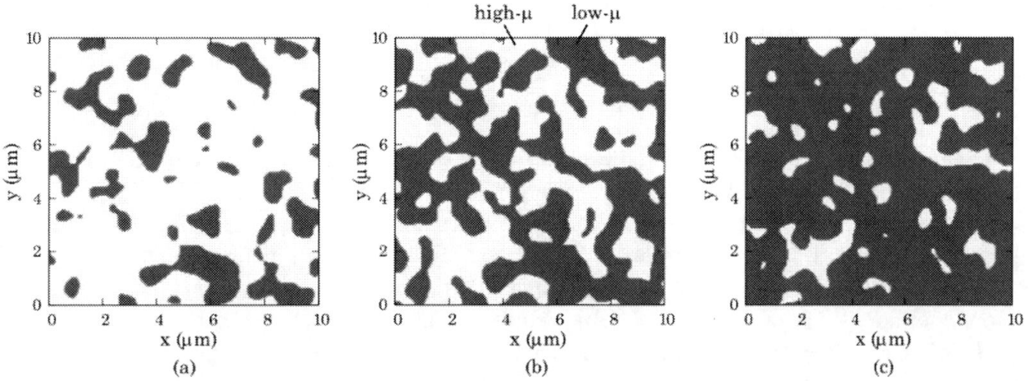

Figure 1. Examples of the generated distribution of low/high mobility islands for $\Lambda = 0.5\,\mu\mathrm{m}$ with the low-μ coverages (a) $f = 0.2$, (b) $f = 0.5$, and (c) $f = 0.8$. Colored areas represent the low-μ region.

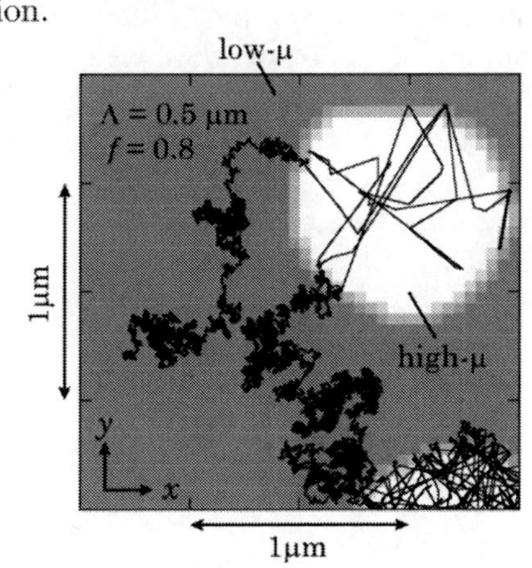

Figure 2. Example of a simulated trajectory in an inhomogeneous sample with $\Lambda = 0.5\,\mu\mathrm{m}$ and $f = 0.8$.

ballistic "jumps" occur when electrons enter the high-μ islands, which may lead to anomalous diffusion. Figure 2 shows an example of an electron trajectory in a sample with $f = 0.8$ and $\Lambda = 0.5\,\mu$m. The scattering rates are set to $W_i = 1.5 \times 10^{12}\,\mathrm{s}^{-1}$ and $W_e = 2.5 \times 10^{14}\,\mathrm{s}^{-1}$. Note the similarity between the trajectory and the possible Lévy motion of birds and animals – see for example Fig. 4 of Ref. [10].

Figure 3 shows the MSD for an inhomogeneous sample (solid line) compared with that for a homogeneous sample (dashed line). For the homogeneous sample, we assume that the scattering rate is given by an area-weighted average of the scattering rates in the low-μ and high-μ regions $W_i + fW_e$ $(= (W_i + W_e)f + W_i(1 - f))$. It can be seen that electrons move ballistically (MSD $\propto t^2$) for short times $t \lesssim 10$ fs while move diffusively (MSD $\propto t$) for sufficiently long times $t \gtrsim 1$ ns. In the intermediate time domain, electrons in the inhomogeneous sample show anomalous diffusion MSD $\propto t^{1.65}$ (super-diffusive) for 10 fs $\lesssim t \lesssim$ 1 ps and MSD $\propto t^{0.7}$ (sub-diffusive) for 1 ps $\lesssim t \lesssim$ 1 ns.

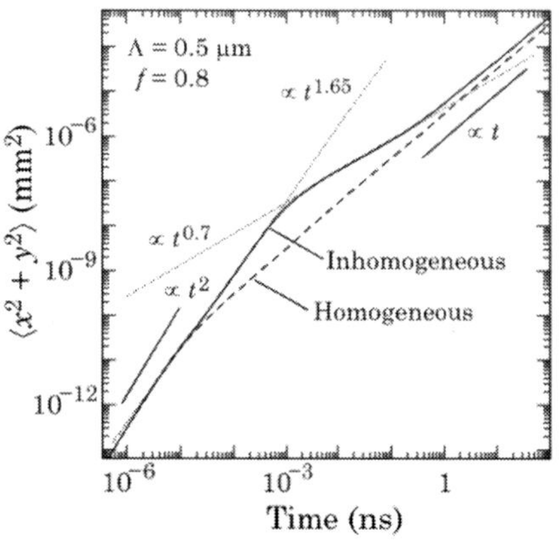

Figure 3. Mean square displacement as a function of t for an inhomogeneous sample (solid line) and a homogeneous sample (dashed line).

The transition from ballistic transport to anomalous diffusion should occur at the shortest scattering time $t_0 = (W_e + W_i)^{-1} \approx W_e^{-1} = 4$ fs. On the other hand, one can expect that the transition from anomalous to normal diffusion occurs at a time t_∞ when the long-time diffusion length becomes equal to the correlation length Λ. The crossover condition from anomalous to normal diffusion, therefore, may be written as $(Dt_\infty)^{1/2} = \Lambda$ or $t_\infty = \Lambda^2/D$. Here D is a diffusion coefficient given by $D = \ell^2/2\tau$ with ℓ being a mean free path and τ the scattering time. By assuming $\ell = v_{\mathrm{th}}\tau$, $\tau = W_e^{-1}$, and $v_{\mathrm{th}} = (2kT/m_e)^{1/2}$, we obtain

$$t_\infty = \frac{m_e W_e \Lambda^2}{kT}. \tag{2}$$

To check this relation, we calculate the MSD for inhomogeneous samples with various correlation lengths Λ (see Fig. 4). In addition, to show the anomalous behavior more clearly, figure 5 plots the MSD divided by t, which corresponds to the long-time diffusion coefficient as $t \to \infty$. The down-arrows represent t_∞ given by Eq. (2). It can be seen that the transition from the anomalous to the normal diffusion can be well described by Eq. (2).

We finally consider the transition from the super-diffusive to the sub-diffusive regime. As suggested in Ref. [4], the transition occurs at a time $t_{\mathrm{trans}} = d_{\mathrm{max}}/v$, where d_{max} is the greatest step length and v is the velocity of the random walker. In the present case, t_{trans} may be written as

$$t_{\mathrm{trans}} = \frac{\Lambda}{v_{\mathrm{th}}}. \tag{3}$$

The up-arrows in Fig. 5 represent t_{trans} given by Eq. (3). We find reasonable agreement between the Monte Carlo results and Eq. (3).

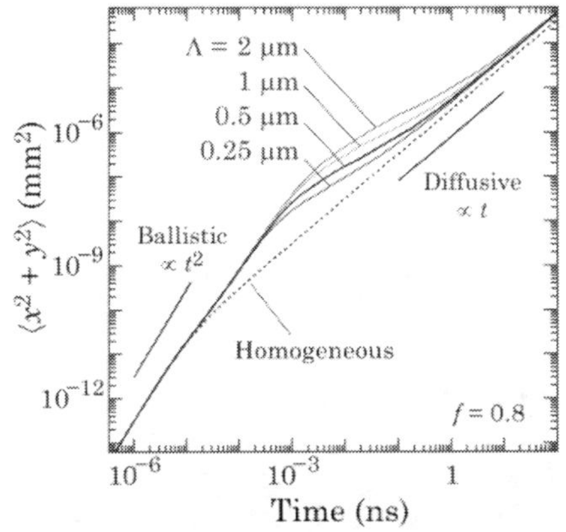

Figure 4. Mean square displacement (MSD) as a function of t for inhomogeneous samples with $\Lambda = 2$ (red), 1 (green), 0.5 (black), and $0.25\,\mu$m (blue). Dashed line shows MSD for a homogeneous samples.

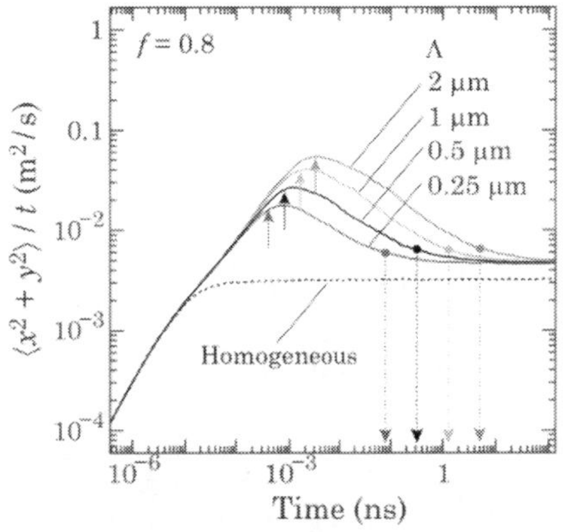

Figure 5. MSD divided by t for inhomogeneous samples with various correlation lengths Λ. Down-arrows (up-arrows) represent t_∞ ($t_{\rm trans}$) given by Eq. (2) (Eq. (3)) for $\Lambda = 2$ (red), 1 (green), 0.5 (black), and $0.25\,\mu$m (blue).

In this study, we considered inhomogeneous InAs. Other types of semiconductors, including alloys (e.g. InGaAs), may show similar behavior and are worth investigating. The anomalous diffusion presented here could be explored by measuring sample-size dependence of the carrier transmission (or conductance) to compare generalized random behavior which appears in a wide range of natural phenomena.

4. Conclusion

We have performed ensemble Monte Carlo simulations of electrons in inhomogeneous InAs layers. We find that electrons show anomalous diffusion in the intermediate time domain $t_0 \lesssim t \lesssim t_\infty$. Our study suggests that inhomogeneous InAs systems could be used to experimentally explore generalized random walk phenomena, which, some studies assert, also occur naturally in the motion of animal foraging paths.

References

[1] Shik A Y 1995 *Electronic Properties of Inhomogeneous Semiconductors* (London: Gordon and Breach Publishers)
[2] Kozlova N V, Mori N, Makarovsky O, Eaves L, Zhuang Q D, Krier A, and Patanè A 2012 *Nat. Commun.* **3** 1097
[3] Patanè A, Mori N, Makarovsky O, Kozlova N V, Pettinari G, Eaves L, Zhuang Q D, Krier A, Polimeni A, and Capizzi M 2013 *18th EDISON* (Matsue, Japan) Tu2-2
[4] Barthelemy P, Bertolotti J, and Wiersma D S 2008 *Nature* **453** 49
[5] Leadbeater M, Falko V I, and Lambert C J 1998 *Phys. Rev. Lett.* **81** 1274
[6] Luryi S, Semyonov O, Subashiev A, and Chen Z 2012 *Phys. Rev. B* **86** 201201(R)
[7] Mercadier N, Guerin W, Chevrollier M, and Kaiser R 2009 *Nat. Phys.* **5** 602
[8] Vermeersch B, Carrete J, Mingo N, and Shakouri A 2015 *Phys. Rev. B* **91** 085202
[9] Vermeersch B, Mohammed A M S, Pernot G, Koh Y R, and Shakouri A 2015 *Phys. Rev. B* **91** 085203
[10] Viswanathan G M , Afanasyev V, Buldyrev S V, Murphy E J, Prince P A, and Stanley H E 1996 *Nature* **381** 413
[11] Edwards A M, Phillips R A, Watkins N W, Freeman M P, Murphy E J, Afanasyev V, Buldyrev S V, da Luz M G E, Raposo E P, Stanley H E, and Viswanathan G M 2007 *Nature* **449** 1044
[12] Travis J 2007 *Science* **318** 742
[13] Revathi S and Balakrishnan V 1993 *J. Phys. A: Math. Gen.* ***26*** *5661*

Does the Finite Size of Electrons Affect Quantum Noise in Electronic Devices?

E. Colomés, D. Marian and X. Oriols

Departament d'Enginyeria Electrònica, Universitat Autònoma de Barcelona, Bellaterra 08193, Spain

E-mail: xavier.oriols@uab.es

Abstract. Quantum transport is commonly studied with the use of quasi-particle infinite-extended states. This leads to a powerful formalism, the scattering-states theory, able to capture in compact formulas quantities of interest, such as average current, noise, etc.. However, when investigating the spatial size-dependence of quasi-particle wave packets in quantum noise with exchange and tunneling, unexpected new terms appear in the quantum noise expression. For this purpose, the two particle transmission and reflection probabilities for two initial one-particle wave packets (with opposite central momentums) spatially localized at each side of a potential barrier are studied. After the interaction, each wave packet splits into a transmitted and a reflected component. It can be shown that the probability of detecting two (identically injected) electrons at the same side of the barrier is different from zero in very common (single or double barrier) scenarios. This originates an increase of quantum noise which cannot be obtained through the scattering states formalism.

1. Introduction

The scattering theory [1] is a useful formalism for quantum transport, allowing to obtain important quantities in a compact form (current, noise...). This theory uses as initial state for the electron quasi-particle infinite-extended states. Regarding this last feature, many reasonable questions appear: *Is it completely justified to choose as the initial state of an electron an infinite-extended state? Or at least, is it always mandatory? Does this initial state provide always the correct results? Can a finite spatially-extended initial state be chosen?* In fact, some relevant scattering experiments (Hong-Ou-Mandel kind) [2, 3] exhibit an increase of quantum noise compared to the expected results from the scattering-states theory. Here, we give theoretical support to these experiments, explaining its origin from the finite size of electrons.

With this motivation, we have recently discussed in Ref.[4] the size-dependence scattering probabilities when using as initial state a quasi-particle finite wave packet. We analyze scattering events when two quasi-particle wave packets impinge on a potential barrier from each side simultaneously (Fig.1) with the same energy but opposite momentum. We introduce the exchange interaction using as wave function the antisymmetric two-particle Slater determinant. Commonly, one would expect that after the interaction with the barrier (when the wave packet has split in a reflected and a transmitted part), only possibilities of finding both electrons at each side are possible $\mathcal{P}_{\mathcal{LR}}$ (Figs.1 (a) and (b)). The reason is that Pauli principle forbids two fermions being at the same position with the same state [5], and this is effectively the case when considering mono-energetic initial scattering states. However, if the initial states are not infinitely-extended, the reflected and transmitted states may not be equal and therefore, the cases where both electrons are found at the same place $\mathcal{P}_{\mathcal{LL}}$ and $\mathcal{P}_{\mathcal{RR}}$ (Fig.1 (c) and (d) respectively) become now possible.

In this conference, we go beyond the results of Ref.[4] and we explore which are the consequences of these new possibilities when analyzing quantum noise.

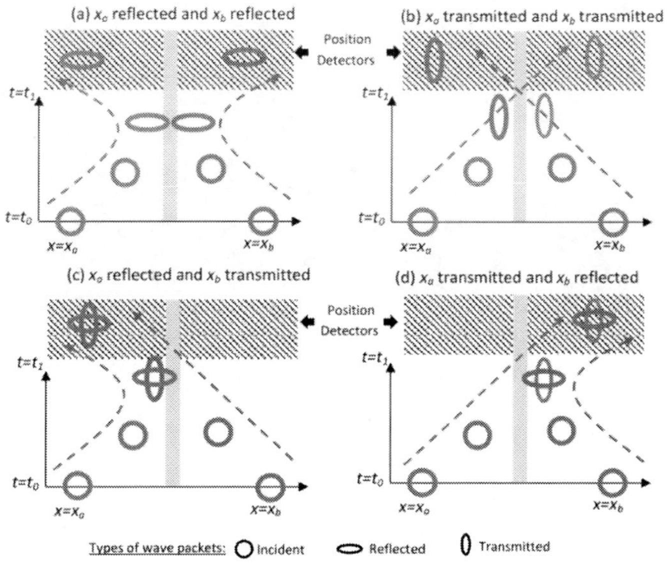

Figure 1. Two identically injected wave packets from the left x_a and from the right x_b of a scattering barrier. Solid regions represent the barrier region and shaded regions represent the particle detectors. (a) and (b) each particle is detected on a different side of the barrier at final time t_1 when the interaction with the barrier has almost finished. (c) and (d) both particles are detected on the same side of the barrier.

2. Two-particle probabilities

In order to study the size-dependence of quasi-particle wave packets in quantum noise with exchange and tunneling, we compute the probability of detecting two (quasi-)electrons at the same side of the barrier ($\mathcal{P}_{\mathcal{LL}}$ and $\mathcal{P}_{\mathcal{RR}}$) from the antisymmetric (Slater determinant) wave function (Φ) solution of the time-dependent Schrödinger equation in the configuration space. Contrarily to what is normally accepted [1], we notice (the explicit derivation can be seen in [4]) that in general, these probabilities ($\mathcal{P}_{\mathcal{LL}}$ and $\mathcal{P}_{\mathcal{RR}}$) are not zero:

$$\mathcal{P}_{\mathcal{LL}} = \int_{-\infty}^{0} dx_1 \int_{-\infty}^{0} dx_2 \ |\Phi|^2 = R_a T_b - |I_{a,b}^{r,t}|^2, \tag{1}$$

where R_a and T_b are the a-wave packet reflection and b-wave packet transmission coefficients respectively. The last term $|I_{a,b}^{r,t}|^2$ accounts for the overlapping among the different wave packets. An important feature is that depending on this term, which in turn depends on the wave packet size, we can obtain two particular limits: I) the results for infinite-extended scattering-states if the overlapping is maximum $\mathcal{P}_{\mathcal{LL}} = 0$ and II) the results for classical distinguishable particles if there is no overlapping $\mathcal{P}_{\mathcal{LL}} = R_a T_b$ (orthogonal states) [4].

As an example, we present here the case of a single barrier using two Gaussian wave packets as the quasi-particle states, more cases are exposed in [4]. As we can see in Fig.2, in general the probability fluctuates, as said above, between $\mathcal{P}_{\mathcal{LL}} = 0$ and $\mathcal{P}_{\mathcal{LL}} = R_a T_b$.

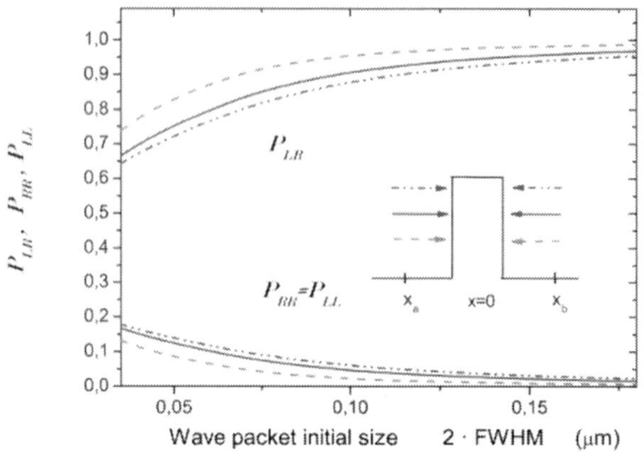

Figure 2. Probability of detecting two (quasi-)electrons at the same side of the barrier ($\mathcal{P}_{\mathcal{LL}}$ and $\mathcal{P}_{\mathcal{RR}}$) and one at each side ($\mathcal{P}_{\mathcal{LR}}$) for three energies, $E_{T=1/2} = 45meV$ (blue solid line), E_1=55meV and E_3=35meV, depending on the wave packet initial size, expressed through its Full Width at Half Maximum (FWHM). We observe that $\mathcal{P}_{\mathcal{LL}}$ or $\mathcal{P}_{\mathcal{RR}}$ are maximum for small wave packets and zero for large ones, while $\mathcal{P}_{\mathcal{LR}}$ reaches the maximum values for large wave packets.

3. Quantum noise in one- or two-particle processes

This simple analysis that we have developed has a quite surprising and far reaching consequence on the quantum noise. At high temperatures, due to the Fermi distribution (f_i, where i is the a or b reservoir), the processes which account for noise are mainly due to single or double scattering processes as the ones showed previously. The non-zero probability $\mathcal{P}_{\mathcal{LL}}$ leads to a modification of the conventional power spectral density $\langle S\rangle$ in these quasi-particles processes [1]. The (ensemble) power spectral density of the current fluctuations at zero frequency, $\langle S\rangle$ can be defined from:

$$\langle S\rangle = lim_{t_d\to\infty} 2q^2 \frac{\langle N^2\rangle_{t_d} - \langle N\rangle_{t_d}^2}{t_d}, \tag{2}$$

with

$$\langle N^2\rangle_{t_d} = \sum_{N=-\infty}^{N=\infty} P(N,t_d)N^2\ ,\ \langle N\rangle_{t_d} = \sum_{N=-\infty}^{N=\infty} P(N,t_d)N, \tag{3}$$

where we define $P(N, t_d)$ as the probability that N particles have been transmitted through the barrier during the time t_d. We consider positive N when the transmission is from left to right and negative otherwise. No net transmission of particles means $N = 0$.

Finally, from Eq.(2) and table 1, we achieve straightforwardly the quantum noise expression:

$$\langle S\rangle = \frac{4q^2}{h}\int_0^\infty dE\ \{T[f_a(1-f_a)+f_b(1-f_b)] + T(1-T)(f_a-f_b)^2 + 2\mathcal{P}_{\mathcal{LL}} f_a f_b\}. \tag{4}$$

The great difference between expression (4) and the well-known result of Büttiker [6] for a two-terminal mesoscopic device is the last term $2\mathcal{P}_{\mathcal{LL}} f_a f_b$. As commented before, $\mathcal{P}_{\mathcal{LL}}$ is zero for infinite-extended wave packets (i.e. scattering-states). In this way, the Büttiker expression can be straightforwardly obtained. However, the soundness and importance of our expression relies in the fact that, in general, quantum noise is increased due to the new scattering possibilities reported in Fig.1.

b \ a		Injection: Transmitted	Injection: Reflected	No injection
Injection	Transmitted	$\frac{\mathcal{P}_{\mathcal{LR}}}{2} f_a f_b$	$\mathcal{P}_{\mathcal{LL}} f_a f_b$	$T(1-f_a)f_b$
		0	-1	-1
	Reflected	$\mathcal{P}_{\mathcal{RR}} f_a f_b$	$\frac{\mathcal{P}_{\mathcal{LR}}}{2} f_a f_b$	$R(1-f_a)f_b$
		1	0	0
No injection		$T f_a(1-f_b)$	$R f_a(1-f_b)$	$(1-f_a)(1-f_b)$
		1	0	0

Table 1. Probability $P(N, t_d)$ (upper) that N (lower) electrons with energy E have effectively been transmitted from left to right reservoir during the time interval t_d.

4. Quantum noise in a multi-particle scenario

Regarding Eq.(2) one could call into question if the noise expression accomplishes the fluctuation-dissipation theorem, which states that at zero temperature noise is zero [7, 8]. At this temperature, due to the Fermi distribution, all possible states are occupied ($f_i = 1$). At first sight, our expression does not fulfil the theorem. Nevertheless, we must remember that at this temperature all the configuration space is fulfilled (below the Fermi energy) and then we cannot consider our one- or two-particle processes, we must consider a multi-particle scenario. As we show in [4], in the multi-particle scenario the new terms tend to zero as the configuration space is fulfilled, because overlapping among different wave packets is maximum and thus the fluctuation-dissipation theorem is perfectly accomplished.

5. Conclusions

In this conference we present the consequences on quantum noise of considering reasonable localized initial states and not the usual infinite-extended plane waves. For this purpose, we study the scattering possibilities (seen in Fig.1 and previously discussed in [4]) in a two-particle potential barrier scenario. It is shown that, contrarily to what it is assumed, if the transmitted and reflected components are not equal, both electrons can be found at the same side of the barrier (see Fig.2).

This fact carries an important consequence because there are new additional sources of noise. This feature is reflected in the quantum noise expression (2) with an additional term. We emphasize that within this expression, the well-known Büttiker expression is obtained as a particular limit when the wave packet size tends to infinite. This work give theoretical support to relevant and surprising experiments which were non well understood until now [2, 3].

Acknowledgments

This work has been partially supported by the "Ministerio de Ciencia e Innovación" through the Spanish Project TEC2012-31330 and by the Grant agreement no: 604391 of the Flagship initiative "Graphene-Based Revolutions in ICT and Beyond".

6. References

[1] Blanter Y M and Büttiker M 2000 *Phys. Rep.* **336** 1
[2] Bocquillon E, et al. 2013 *Science* **339** 1054
[3] Liu R C, Odom B, Yamamoto Y and Tarucha S 1988 *Nature* **391** 263
[4] Marian D, Colomés E and Oriols X 2015 *J. Phys. Condens. Matter* **27** 245302
[5] Pauli W 1925, Z. *Physik* **31** 765
[6] Büttiker M 1990 *Phys. Rev. Lett.* **65** 2901
[7] Nyquist H 1928 *Phys. Rev.* **32** 110
[8] Johnson J 1928 *Phys. Rev.* **32** 97

Journal of Physics: Conference Series **647** (2015) 012061 doi:10.1088/1742-6596/647/1/012061

Shot-noise suppression effects in InGaAs planar diodes at room temperature

Ó García-Pérez[1], H Sánchez-Martín, J Mateos[1], S Pérez[1], A Westlund[2], J Grahn[2] and T González[1]

[1] Department of Applied Physics, Universidad de Salamanca, Salamanca 37008, Spain
[2] Department of Microtechnology and Nanoscience, Chalmers University of Technology, Gothenburg SE-412 96, Sweden

E-mail: ogarcia@usal.es

Abstract. In this work, the noise characteristics of InGaAs planar diodes are studied. The presence of a recessed region originates a barrier in the potential profile, which can modulate the passage of ballistic carriers along the structure. This effect, in turn, may lead to suppressed levels of noise with respect to the full Poissonian value due to Coulomb interaction. With the aim of evidencing such phenomenon, the noise properties of a set of devices with different dimensions have been measured at room temperature. Some evidence of potential shot-noise suppression is observed in the results, but the undesired effect of resistive contacts and accesses has been found to be a limiting factor to quantify the suppression accurately.

1. Introduction

Noise is usually considered as unwanted fluctuations that appear in any electronic system and distort the processing of information signals. Nevertheless, studies on electronic noise can provide valuable information to better understand the electron transport mechanisms in semiconductor devices. For example, carrier interaction in mesoscopic devices can regulate its propagation statistics, which in turn may lead to suppressed levels of noise [1].

The effect of Pauli principle and/or Coulomb interaction on the carrier statistics in ballistic media, and how it may result in noise levels lower than the full shot noise value ($2qI$), is a field of interest for physicists and electronic engineers [2]-[3]. Pauli principle may have effect on the injecting statistics of the carriers at the contacts, whereas Coulomb interaction may affect the carrier propagation statistics through the device. Therefore, the current spectral density S_I of the device can be written as $S_I=F2qI$, where q is the electron charge, I is the bias current, and F is defined as the Fano factor, which quantifies the aforementioned suppression (or enhancement, which is also possible when positive correlations between carriers exist).

Since shot noise has its origin in the granularity of charge, it plays an important role when lengths are significantly scaled down. Therefore, understanding and characterizing the physics associated to the noise suppression is of great interest. However, the experimental evidence of such effects is not easy to obtain, mainly at room temperature (RT). The aim of this work is to identify suppression effects from the noise measurement of nanoscale InGaAs planar diodes at RT.

2. Device structure and Monte Carlo simulations

The geometry of the diodes, depicted in figure 1, is equivalent to the one of an ungated HEMT with a recess in between the drain and source accesses [4]. The recess-to-source and recess-to-drain distances will be denoted as L_S and L_D, and the length of the slot will be denoted as L_R. The presence of the recess, by removing part of the upper InGaAs layer (cap), allows focusing the electric field in that region. Under this condition, and considering the high mobility in the channel, quasi-ballistic transport of carriers can be expected under the recess even at RT.

In a first step, a set of devices has been simulated using the Monte Carlo particle method. The simulated potential profiles of a device with L_S=200 nm, L_R=160 nm, L_D=300 nm and a cap layer of 4 nm for different drain-to-source voltages are shown in figure 2. As it can be observed, an energy barrier appears under the recess generated from the quasi-ballistic electrons, which is expected to suppress the shot noise due to Coulomb correlations, as it was described in the previous section. The barrier height decreases with increasing bias, but it does not disappear. This is because above around 0.7 V, the value at which inter-valley mechanisms are activated in drain region [4], the increments of voltage are absorbed near the drain contact while the potential profile in the recess remains almost unaltered.

The simulated and measured I-V curves of a set of devices with varying L_R (and L_D=550 nm) and varying L_D (and L_R=160 nm) are represented in figure 3. The cap layer is 4 nm thick, L_S=200 nm, and a contact resistance of $2R_C$=0.35 **Ω**·mm has been added to the simulations for the direct comparison with the measurements. In all these devices the behavior in the simulated potential profile is equivalent to the case shown in figure 2 and described above.

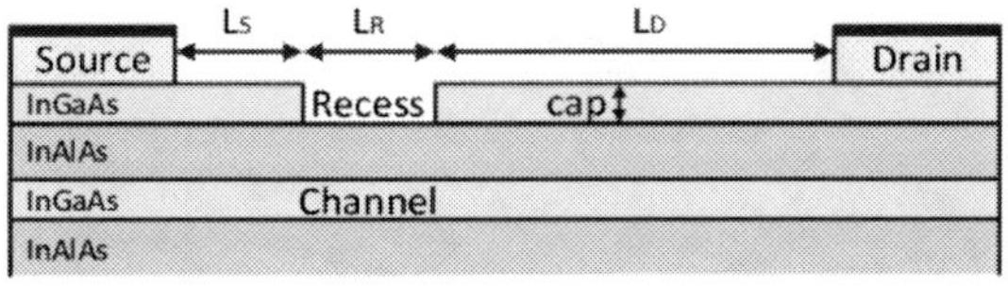

Figure 1. Geometry of the recessed planar diodes.

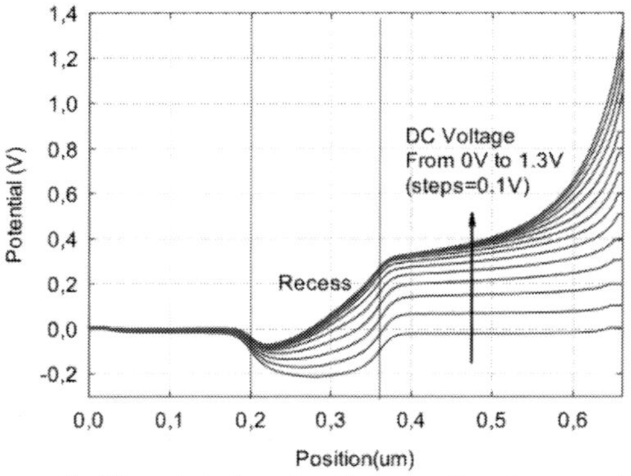

Figure 2. Simulated potential profile along a device with L_S=200 nm, L_R=160 nm and L_D=300 nm.

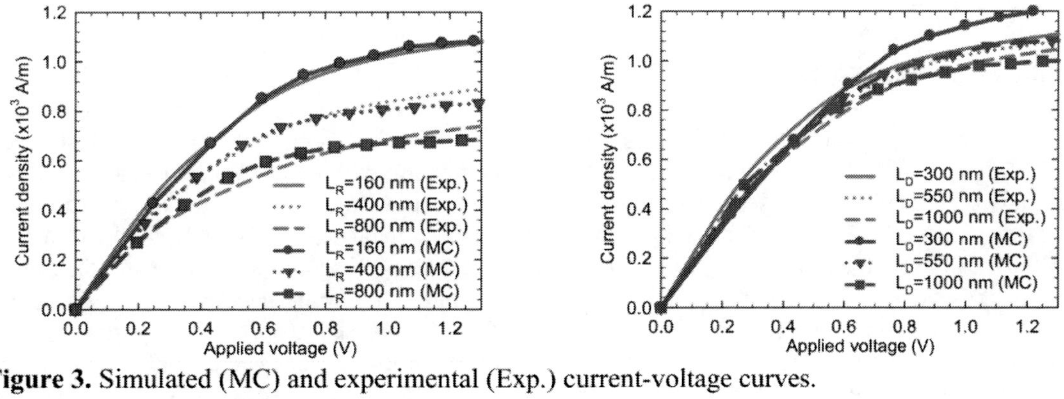

Figure 3. Simulated (MC) and experimental (Exp.) current-voltage curves.

3. Experimental results

For the measurement of the noise spectral density of the diodes, a PNA-X N5244A with Option 029 from *Agilent Technologies* has been used. This equipment allows the simultaneous acquisition of the noise power density and the complex impedance of a device with higher levels of accuracy and sensitivity. With both magnitudes, the conversion into current spectral density is straightforward. For each bias point, the spectral density value has been calculated averaging data between 20-30 GHz, ensuring that we are in the high-frequency white-noise plateau above $1/f$ noise.

The noise measurements were performed in the set of devices whose I-V curves were presented in figure 3, with: varying L_R (and L_D=550 nm), and varying L_D (and L_R=160 nm). The source access length is L_S=200 nm, the cap layer is 4 nm thick and the width is 10 µm. Additionally, a non-recessed device of 1300 nm length and 10 µm width has been characterized for comparison. The current spectral density S_I as a function of the bias current is shown in figures 4(a) and (c). When the current begins to saturate, there is a significant increase of S_I, which is not observed in the non-recessed device. The flat noise density of the non-recessed device indicates that thermal noise dominates in that case, due to the diffusive nature of the transport in the long structure. Also, as it is observed from the plots, the measured values of S_I are well below the full shot noise value $2qI$. Consequently, the values of F remain lower than one [figures 4(b) and (d)].

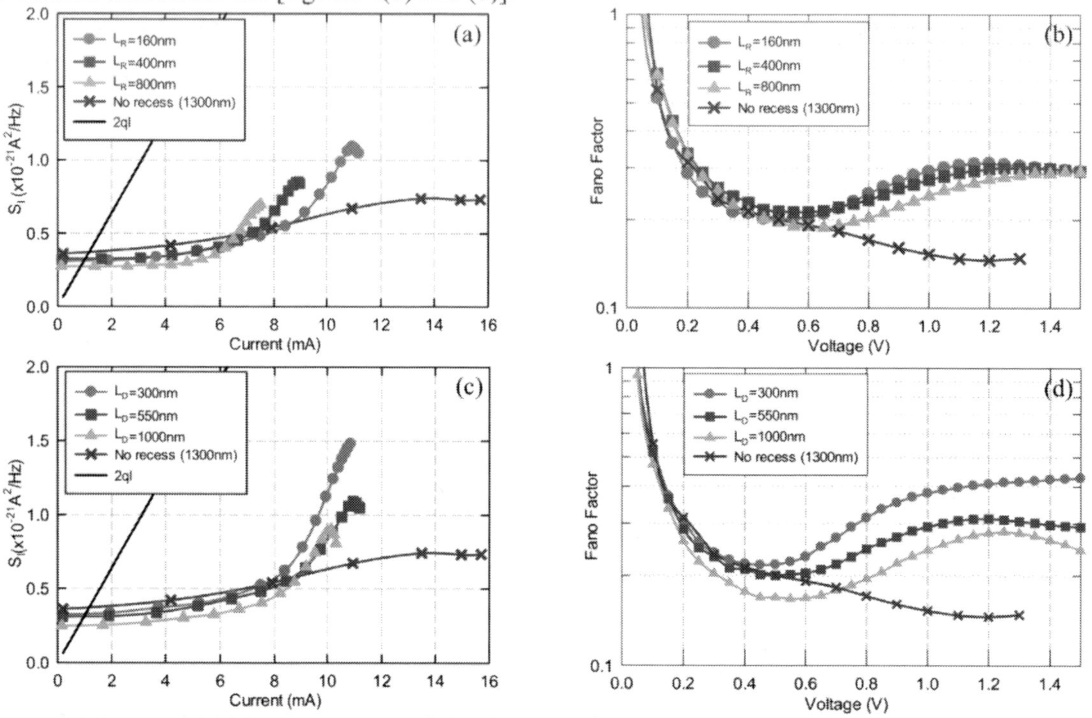

Figure 4. Measured (a)(c) current spectral density, and (b)(d) Fano factor.

4. Noise model and discussion

In order to determine and quantify the noise suppression effects directly attributed to the barrier under the recess, the noise contributions from the drain/source access regions and the contacts should be properly eliminated from the raw measurements presented in figure 4. For this purpose, a simple noise model has been developed based on the scheme shown in figure 5. The first module represents the contact resistance R_C=35 Ω, and its noise contribution is assumed to be thermal $S_C=4kT_0/R_C$, being k the constant of Boltzmann and T_0=300 K. The second module represents the zone of interest, i.e. the region of the barrier under the recess, and its noise contribution S_R is modeled as $S_R=F2qI$, assuming two different scenarios F=1 (full shot noise) and F=0 (fully suppressed noise) for comparison. Finally, the third module represents the drain access region, as a non-linear resistance $R_D(V)$, and its noise contribution is assumed to be thermal as $S_D=4kT_0/R_D(0)$. This strong assumption, which is not completely correct, allows to simplify the noise model and, as we will show later, it does not affect the main conclusions of this work. The values of R_R and R_D, which are bias dependent, are determined from the potential profiles and currents obtained from the Monte Carlo simulations. The resistance of the source region is considered negligible according to the potential profiles shown in figure 2. Finally, the resulting current spectral density S_I can be calculated following the formula shown in figure 5.

With the model described above, the theoretical values of S_I and F have been calculated for the same set of devices considered for the experiments in section 3, and the resulting plots are shown in figure 6. The calculated results reasonably reproduce the shape of the measured results in figure 4. As it was mentioned before, two extreme conditions are assumed in the recess (F=0 and F=1). However, its effect on the overall noise response is weak. Moreover, any change in the noise of the recess region is indistinguishable above 0.7 V due to the predominance of the drain resistance. Therefore, the signature of shot noise suppression would only be visible from the total noise at intermediate voltages. Nevertheless, even in those cases accurately quantifying the suppression is still challenging. Devices with reduced access and drain resistances should be fabricated in order to obtain more conclusive results (apart from complex low temperature measurements, where the thermal noise of the contacts and drain region would be lowered).

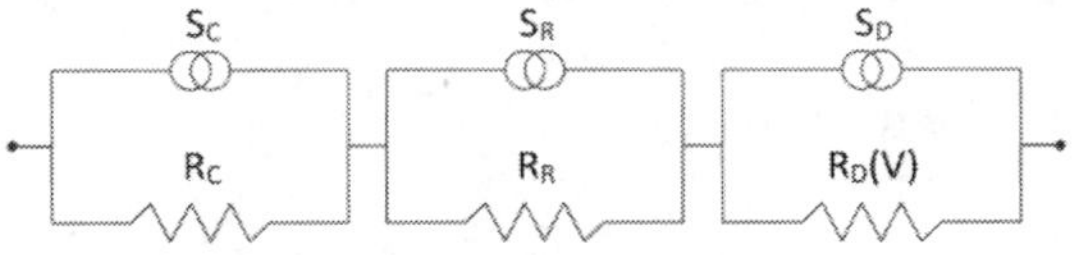

$$S_I = \frac{S_C R_C^2}{(R_C + R_R + R_D)^2} + \frac{S_R R_R^2}{(R_C + R_R + R_D)^2} + \frac{S_D R_D^2}{(R_C + R_R + R_D)^2}$$

Figure 5. Proposed noise model for the recessed diodes.

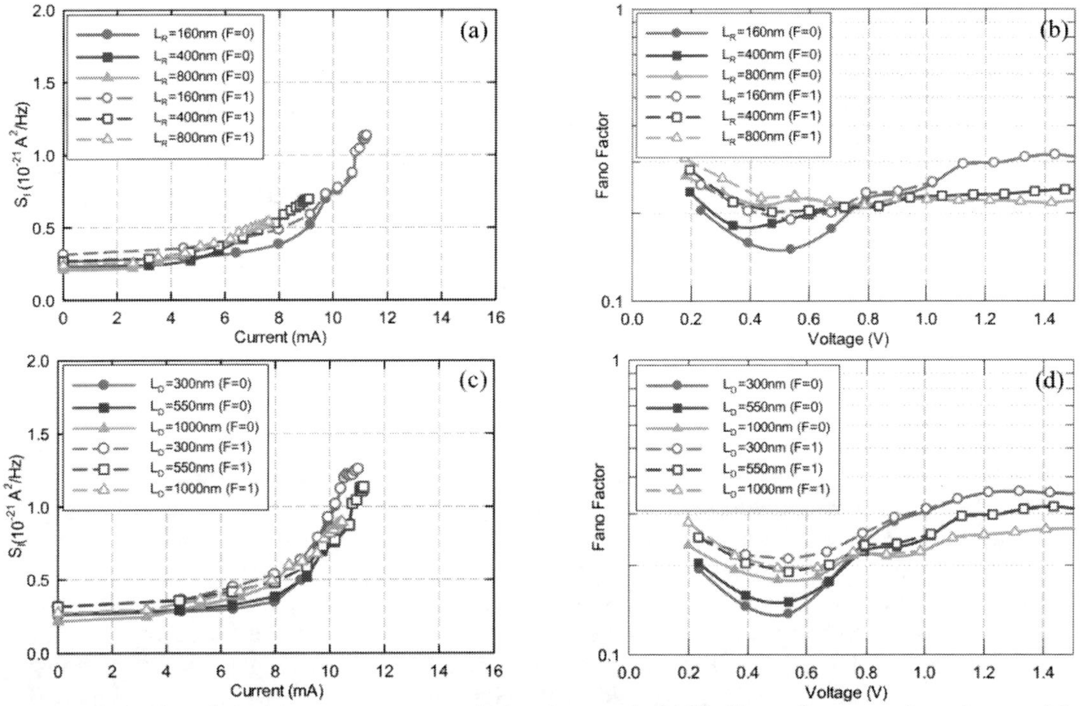

Figure 6. Calculated (a)(c) current spectral density, and (b)(d) Fano factor, using the model of figure 5.

Acknowledgements
This work has been partially supported by the Spanish MINECO through project TEC2013-41640-R, by the Consejería de la JCyL through project SA052U13, and by the European Social Fund (ESF).

References

[1] Landauer R 1998 *Nature* **392** 658-659
[2] Gomila C, Cantalapiedra I, Gonzalez T and Reggiani L 2002 *Phys. Rev. B* **66** 075302
[3] Naveh Y, Korotkov A N and Likharev K K 1999 *Phys. Rev. B* **60** 2169-2172
[4] Perez S, Gonzalez T, Pardo D and Mateos J 2008 *J. Appl. Phys.* **103** 094516

Improvement of interfacial and electrical properties of Al_2O_3/ n-$Ga_{0.47}In_{0.53}As$ for III-V impact ionization MOSFETs

Y Lechaux[1], A Fadjie[1], S Bollaert[1], V Talbo[2], J Mateos[2], T González[2], B G Vasallo[2] and N Wichmann[1]
[1]Institute of Electronics, Microelectronics and Nanotechnology, University of Lille 1, UMR 8520, 59652 Villeneuve d'Ascq, France
[2]Applied Physics Department, Salamanca University, Plaza de la Merced, s/n Salamanca 37008, Spain

E-mail: yoann.lechaux@ed.univ-lille1.fr

Abstract. In this work, Metal – Oxide – Semiconductor Capacitors (MOSCaps) based on *Al_2O_3/ n-$Ga_{0.47}In_{0.53}As$* interface have been studied. In order to have high MOSFETs performance, it is necessary to improve the semiconductor – oxide interface quality. It is observed that the $(NH_4)_2S$ passivation shows lower interface trap density in the order of $6\times10^{11}cm^{-2}.eV^{-1}$. Also, it is observed that O_2 plasma densification after a passivation in a NH_4OH solution improves the electrical behaviour of the charge control. Low interface trap density in the order of $1\times10^{12}cm^{-2}.eV^{-1}$ was obtained for different treatments presented in this work.

1. Introduction

During the past few years, the size reduction of the Metal – Oxide – Semiconductor Field Effect Transistor (MOSFET) was the key for the development of performant Integrated Circuits (ICs). One of the main issues is the reduction of power consumption related to the lowering of supply voltage V_{DD}. Today the decrease of V_{DD} is limited by the subthreshold swing SS. New device architectures are being investigated in order to have SS less than the 60mV/dec limit. I-MOSFET (Impact ionization MOSFET) is one candidate to obtain SS of few mV/dec [1]. Silicon I-MOSFET showed steep subthreshold swing in the order of 5mV/dec [2] with I_{ON} current in the range of 400mA/mm [2] however it requires large V_{DD}. III-V materials and engineering of bandgap energy is envisaged to reduce V_{DD}. In III-V I-MOSFET, gate oxide with high quality is an issue. Large majority of works used Al_2O_3 oxide deposited by ALD (Atomic Layer Deposition) to have low interface trap density D_{it} in the order of $10^{12}cm^{-2}.eV^{-1}$ [3] and very few report less D_{it} [4]. Recently, paper [5] reports an O_2 plasma treatment of the Al_2O_3 oxide to decrease the formation of oxide during the post-deposition annealing (PDA). In this paper, we present improvement of interfacial and electrical properties of Al_2O_3/ n-$Ga_{0.47}In_{0.53}As$ using O_2 plasma treatment during the formation of the oxide by ALD.

2. Device structure and experimental details

MOSCAPs were realised from a structure of n-doped $Ga_{0.47}In_{0.53}As$ lattice matched on InP wafer as shown in figure 1. The structure is made of a high n-doped GaInAs layer (N: $1\times10^{19}cm^{-3}$) to have a low resistance n-type ohmic contact and a low n-doped GaInAs layer (N: $2\times10^{16}cm^{-3}$) close to the oxide – semiconductor interface. The first step of fabrication process consist on cleaning, degreasing

Journal of Physics: Conference Series **647** (2015) 012062 doi:10.1088/1742-6596/647/1/012062

the surface and passivation before oxide growth by ALD. Then the thermal Al_2O_3 is deposited at 250°C from trimethylaluminium (TMA) and water (H_2O) precursors. Three samples were prepared. For the sample I, the native oxide was etched with a HCl solution diluted in deionized water (EDI) with a volume ratio of 1:3 during 2min. Then a 5% dilute $(NH_4)_2S$ solution was used to passivate the surface during 10min [3]-[6]. For the sample II, a 4% dilute NH_4OH solution was used to passivate the surface during 10min. Then 4nm Al_2O_3 was deposited by ALD at 250°C. For a third sample III, an O_2 plasma was used after 2nm oxide deposition on a sample treated with a dilute NH_4OH solution. Then 2nm oxide was deposited to have an oxide thickness of 4nm. On each sample, a PDA at 600°C during 1min under N_2H_2 atmosphere was used to stabilize the oxide. All treatments are resumed in table 1. The gate contact was realised with the evaporation of a sequence of Ni/ Au (100nm/ 250nm). The source n-type ohmic contact was realised with the evaporation of a sequence of Ti/ Pt/ Au on the highly n-doped GaInAs layer. The cross section of MOS gate stack was observed by the means of transmission electron microscopy showed in figure 2. We can observe in the middle the 4nm of non-densified Al_2O_3 surrounded by GaInAs at left and Ni metal at right.

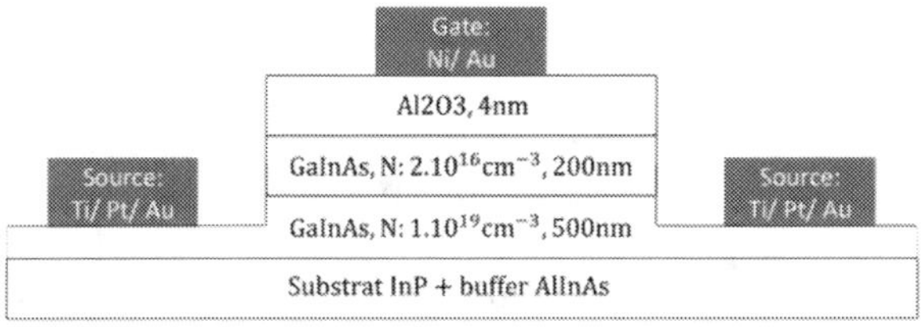

Figure 1. Structure of MOSCaps device at the end of the fabrication process.

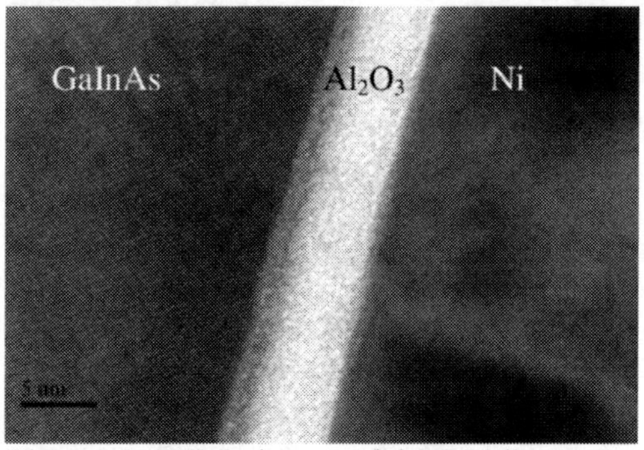

Figure 2. TEM picture of the MOS structure, we can see in the middle of Al_2O_3 with a uniform thickness of 4nm.

Table 1. Summary of different treatments and oxide deposition.

Sample	Surface treatments	Oxide Thickness (nm)	Densification	Post Deposition Annealing (°C)
I	HCl + $(NH_4)_2S$	4	no	600
II	NH_4OH	4	no	600
III	NH_4OH	4	yes	600

3. Results and discussion

3.1. Passivation effect

We compared dilute $(NH_4)_2S$ solution to passivate the surface of the sample I and dilute NH_4OH solution to passivate the surface of the sample II. To that purpose, capacitance – voltage (C – V) measurements were done at ambient temperature in function of gate voltage V_{gs} for various frequencies of the dynamic signal applied on the gate (figure 3). In figure 3(a), we represent C-V measurements for sample I and C-V measurements for sample II in figure 3(b). We can notice in both samples a high frequency dispersion in accumulation between V_{gs}=1V and V_{gs}=2V due to leakage current [7]. The sample I treated with the $(NH_4)_2S$ solution has showed around depletion, between V_{gs}=-0.75V and V_{gs}=0.25V, a lower frequency dispersion in comparison with sample II treated with the NH_4OH solution, indicating a lower interface trap density. However the identical stretch-out between V_{gs}=-0.75V and V_{gs}=0.25V on both samples has showed no significant modification in charge control. Furthermore, we can notice a "bump" in inversion, between V_{gs}=-2V and V_{gs}=-0.5V which indicates inversion assisted by interface trap states [8].

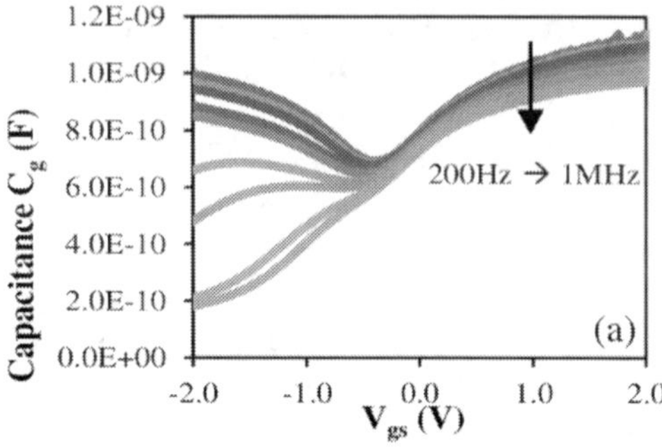

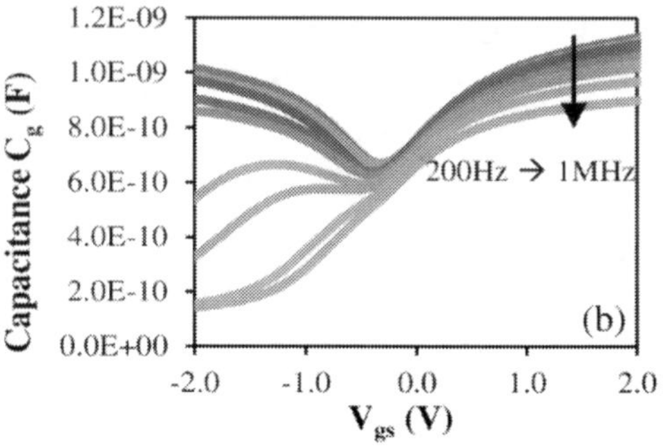

Figure 3. C – V characteristics of the MOS structure Al_2O_3/ n-GaInAs for sample I passivated with $(NH_4)_2S$ solution (a), and for sample II passivated with NH_4OH solution (b).

3.2. Oxide densification effect

We studied the densification effect using O_2 plasma in the sample III passivated with a dilute NH_4OH solution to improve the oxide – semiconductor interface quality. The post deposition O_2 plasma was used on the structure Al_2O_3/ Ge by Zhang et al. in order to improve GeO_x oxide [9]-[10], they succeed to have low value of D_{it} in the range of $10^{11}cm^{-2}.ev^{-1}$. In our case, the treatment consists on a densification of the oxide near interface by the mean of an O_2 plasma across 2nm of oxide. This treatment decreases the quantity of mobile charges in the oxide and fills dangling bonds after oxide deposition. Then 2nm of Al_2O_3 were added to have a total thickness of 4nm. C – V characteristics at ambient temperature in function of V_{gs} are showed in figure 4. We can notice at low frequency the minimum capacitance value is lower for the sample III than for sample II even though both are passivated with NH_4OH solution. Furthermore, the small stretch-out between V_{gs}=-0.5V and V_{gs}=0.5V for sample III indicates that the O_2 plasma improves the charge control by the gate and might improve the subthreshold swing SS of I-MOSFETs. And as for sample II, the large frequency dispersion in accumulation could be due to the high leakage current [7].

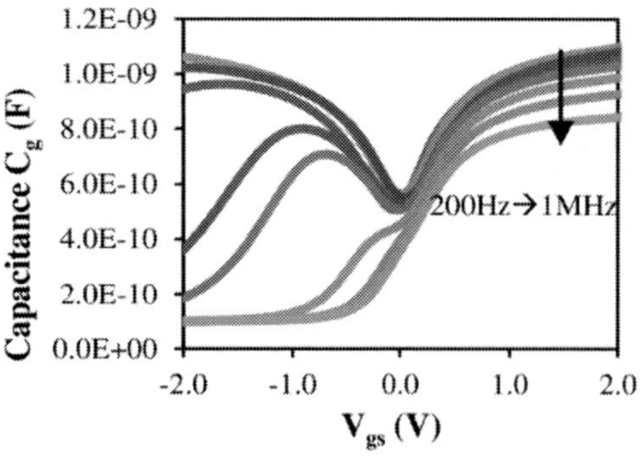

Figure 4. Capacitance measurements of the MOS structure Al_2O_3/ n-GaInAs with an oxide densification treatment (sample III).

3.3. Interface trap density

We extracted the interface trap density calculated by High frequency – Low frequency (HF-LF) method for each sample shown in figure 5 in function of gate – source voltage V_{gs}. All samples show D_{it} with the same order of magnitude, in the range of $1\times10^{12}cm^{-2}.eV^{-1}$. However, we can notice a low D_{it} of $6.5\times10^{11}cm^{-2}.eV^{-1}$ for the sample I passivated with $(NH_4)_2S$ solution compared to D_{it} of about $9.2\times10^{11}cm^{-2}.eV^{-1}$ and $1.5\times10^{12}cm^{-2}.eV^{-1}$ for samples II and III respectively. But the extraction of D_{it} in samples II and III may not be accurate because of the large frequency dispersion in accumulation.

Moreover, we have extracted flatband voltage V_{FB} in order to calculate the charge density Q_f in the oxide from $Q_f=(V_{FB}-\Phi_{MS})/C_{OX}$ (Table 2) where C_{OX} is oxide capacitance and Φ_{MS} ideal metal-semiconductor workfunction difference. We observed a decrease of charge density for sample III with densification and a shift of flatband voltage to its theoretical value (1).

$$\Phi_{MS} = \Phi_M - X - \frac{E_g}{2.q} - \frac{K.T}{q}.\ln(\frac{N_D}{n_i}) \quad (1)$$

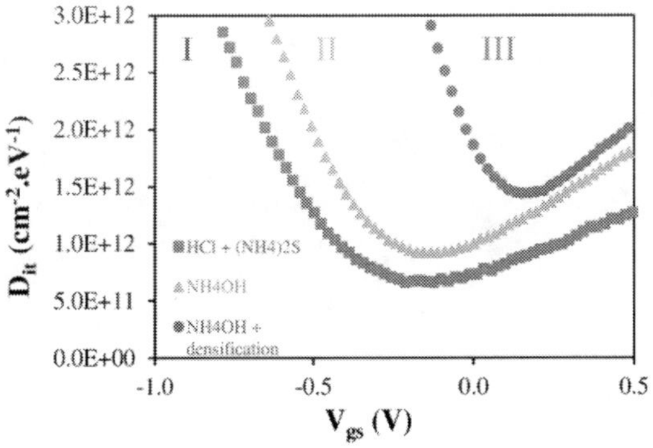

Figure 5. Extraction of the interface trap density D_{it} in function of the applied voltage V_{gs} from HF – LF method.

	HCl + $(NH_4)_2S$	NH_4OH	NH_4OH + densif.	Theory
Qf (x10^{11}cm^{-2})	64	45.1	1.14	-
V_{FB} (V)	-1,27	-0,96	-0,15	-0.13

Table 2. Calculated oxide charge density in function of the flatband voltage for each sample.

4. Conclusion
In summary, we studied effect of surface treatment on Al_2O_3/ n-$Ga_{0.47}In_{0.53}As$ interface. We have demonstrated a low D_{it} of about 6.5×10^{11}cm^{-2}.eV^{-1} for sample treated with dilute $(NH_4)_2S$ solution as surface treatment before oxide deposition. We have reported improvement in charge control by the gate using an O_2 plasma after oxide deposition, even if we obtained large interface trap density. This process needs to be optimized in order to build I-MOSFETs using higher dielectric permittivity oxide like HfO_2.

5. Acknowledgments
This work has been supported by the Agence Nationale de la Recherche through the project ANR-13-JS03-0001.

6. References
[1] Gopalakrishnan K, Griffin P B and Plummer J D 2002 IEEE IEDM 289
[2] Choi W Y, Song J Y, Lee J D, Park Y J and Park B-G 2005 IEEE IEDM 955
[3] Lin H-C,Wang W-E, Brammertz G, Meuris M and Heyns M 2009 *Microelectron. Eng.* **86** 1554
[4] Shin H S, Yum J H, Johnson D W, Harris H R, Hudnall T W, Oh J, Kirsch P, Wang W-E, Bielawski C W, Banerjee S K, Lee J C and Lee H D 2013 *Appl. Phys. Lett.* **103** 223504
[5] Fadjie-Djomkam A-B, Lechaux Y, Pastorek M, Ridaoui M, Wichmann N and Bollaert S 2015 EMC Session S:9
[6] O'Connor E, Brennan B, Djara V, Cherkaoui K, Monaghan S, Newcomb S B, Contreras R, Milojevic M, Hughes G, Pemble M E, Wallace R M and Hurley P K 2011 *J. Appl. Phys.* **109** 024101
[7] Yao B, Fang Z B, Zhu Y Y, Ji T and He G 2012 *Appl. Phys. Lett.* **100** 222903
[8] Trinh H D, Chang E Y, Wu P W, Wong Y Y, Chang C T, Hsieh Y F, Yu C C, Nguyen H Q, Lin Y C, Lin K L and Hudait M K 2010 *Appl. Phys. Lett.* **97** 042903
[9] Zhang R, Huang P C, Lin J C, Takenaka M and Takagi S 2013 *Appl. Phys. Lett.* **102** 081603
[10] Zhang R, Iwasaki T, Taoka N, Takenaka M and Takagi S 2011 *Appl. Phys. Lett.* **98** 112902

Journal of Physics: Conference Series **647** (2015) 012063 doi:10.1088/1742-6596/647/1/012063

Quasi-particle current in planar Majorana nanowires

Javier Osca and Llorenç Serra

Institut de Física Interdisciplinària i de Sistemes Complexos IFISC (CSIC-UIB), E-07122 Palma de Mallorca, Spain

Departament de Física, Universitat de les Illes Balears, E-07122 Palma de Mallorca, Spain

E-mail: javier@ifisc.uib-csic.es

Abstract. We calculate the local quasi-particle current of a Majorana state in a planar hybrid (superconductor-semiconductor) nanowire. In absence of perpendicular components of the magnetic field the current flows in circular trajectories without a preferred orientation. On the other hand, when a perpendicular component of the magnetic field is present the quasi-particle current circulates surrounding the Majorana density peak with an orientation established by the magnetic field.

1. Introduction

Majorana modes appear in quasi-1D wires as effectively charge-less, zero-energy eigensolutions. They arise from the splitting, through a phase transition, of bulk electronic states into pairs of quasi-particles on the wire ends, each one being its own antiparticle [1, 2, 3, 4]. These phase transitions are known to occur at particular values of the magnetic field [5] giving rise to characteristic phase diagrams. Recently, we have investigated the role of the magnetic orbital motion on the physics of Majorana states. In Ref. [6] it is shown that in a planar nanowire the main effect of the orbital motion is to change the Majorana phase boundaries. In general, stationary Majorana states can sustain non-vanishing local quasi-particle currents. Furthermore, these currents are altered by the kinetic orbital motion caused by the off plane components of the magnetic field. In this proceedings article we extend the work of Ref. [6] studying the quasi-particle current present in planar Majorana nanowires with and without components of the magnetic field perpendicular to the nanowire surface.

2. Physical model

Majoranas can be obtained in nanowires due the combined effects of s-wave superconductivity, Rashba interaction and an external magnetic field. We consider a nanowire where the electronic motion is restricted to the $\hat{x}$ (longitudinal) and $\hat{y}$ (transverse) directions in presence of these three effects. The homogeneous magnetic field points in an arbitrary direction and the edges are modeled as infinite square well potentials in the longitudinal and transverse directions (see Fig. 1a). Therefore the nanowire physics is described by a Hamiltonian of the Bogoliubov-de Gennes kind

$$\begin{aligned}\mathcal{H}_{BdG} = &\left(\frac{p_x^2+p_y^2}{2m} + V(x,y) - \mu\right)\tau_z + \Delta_s\,\tau_x + \frac{\alpha}{\hbar}\,(p_x\sigma_y - p_y\sigma_x)\,\tau_z \\ &+ \Delta_B\,(\sin\theta\cos\phi\,\sigma_x + \sin\theta\sin\phi\,\sigma_y + \cos\theta\,\sigma_z)\ ,\end{aligned} \tag{1}$$

Journal of Physics: Conference Series **647** (2015) 012063 doi:10.1088/1742-6596/647/1/012063

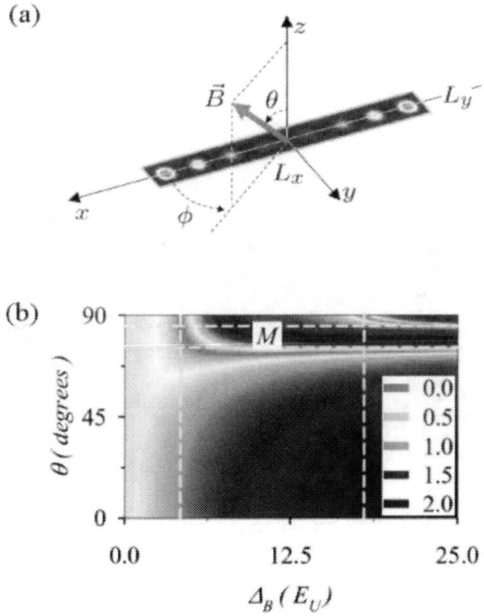

Figure 1. a) Schematic of a 2D planar nanowire showing the axis definitions. The density distribution of Majorana modes on the wire ends is qualitatively shown. b) Phase transition proximity measure with phase transitions (at zero value) in red, as a function of Δ_B and polar angle θ. The azimuthal angle remains $\phi = 0$. The results correspond to an InAs nanowire with $\alpha = 30\,\text{meVnm}$, $\Delta_s = 0.3\,\text{meV}$, gyro-magnetic factor $g = 15$, effective mass ratio $m^* = 0.033$, and $L_y = 150\,\text{nm}$ in a magnetic field range from zero to 6 T. Both figures taken from Ref. [6].

where the different terms are, in left to right order: kinetic energy, electrical potential V, chemical potential μ, the superconductor term with strength Δ_s, the Rashba spin orbit interaction term with coupling strength α and finally, the Zeeman term of a magnetic field of magnitude Δ_B and arbitrary polar and azimuthal angles $(\theta, \phi) \equiv \hat{n}$. The Pauli operators for isospin (particle-hole) are represented by $\tau_{x,y,z}$ while those for spin are $\sigma_{x,y,z}$. In a planar nanowire the perpendicular component of the magnetic field induces orbital motions of the nanowire quasi-particles through the substitution $\vec{p} \rightarrow \vec{p} + \frac{e}{c}\vec{A}(x, y)$ in the kinetic and Rashba terms (using the $e > 0$ convention). The additional orbital terms obtained with this substitution in Eq. (1) are given in Ref. [6].

The semi-infinite nanowire exact zero-energy solution can be obtained by means of the complex band structure method of Refs. [6, 7]. More specifically, the infinite nanowire eigensolutions are obtained numerically from the effective one dimensional Hamiltonian. Imposing the boundary condition to a superposition of infinite-nanowire solutions with different wave numbers yields the Majorana edge state. This method has the advantage of being computationally effective in comparison with direct diagonalization methods, even with fine spatial grids, and can also be used to obtain the phase diagram of a Majorana nanowire inside a tilted magnetic field as shown in Fig. 1b. In particular, a phase transition proximity measure is displayed where phase boundaries (zero values) appear as red lines, and the regions with a Majorana mode are labeled with an M while regions without a Majorana are not labeled. Dashed lines indicate the phase transition analytical limits

$$\Delta_{B,n}^{(c)} = \sqrt{\left(\mu - \epsilon_n + \frac{m\alpha^2}{2\hbar^2}\right)^2 + \Delta_s^2}\,, \tag{2}$$

and

$$\theta_n^{(c)} = \arccos\left(\frac{gm^*}{4(n - \frac{1}{2})}\right)\,, \tag{3}$$

where $n = 1, 2, 3, \ldots$. We refer to Ref. [6] for further details. The results are presented in characteristic units of the problem obtained by taking $\hbar$, m and the width of the nanowire

L_y as reference values. That is, our length and energy units are, respectively, $L_U \equiv L_y$ and $E_U \equiv \hbar^2/mL_y^2$.

3. Quasiparticle current

The quasi-particle continuity equation for the Hamiltonian of Eq. (1),

$$\frac{\partial\rho(x,y)}{\partial t} = -\nabla\cdot\vec{\jmath}(x,y)\,, \tag{4}$$

is obtained from the Majorana probability density $\rho(x,y) = \Psi(x,y)\Psi^*(x,y)$ and the time dependent Schrödinger equation $i\hbar\partial_t\Psi(x,y) = \mathcal{H}_{BdG}\Psi(x,y)$, where $\Psi(x,y)$ is defined as the four-component spinor $\Psi \equiv (\psi_{++},\psi_{-+},\psi_{+-},\psi_{--})^T$ in spin and isospin space. The rate of change of the Majorana density depends on the divergence of the quasi-particle current

$$\vec{\jmath}(x,y) = \Re\left[\,\Psi^*(x,y)\,\hat{v}\,\Psi(x,y)\,\right]\,, \tag{5}$$

that is directly proportional to the real part of the velocity operator $\hat{v} \equiv (\hat{v}_x,\hat{v}_y)$, where $\hat{v}_x = \partial\mathcal{H}/\partial p_x$ and $\hat{v}_y = \partial\mathcal{H}/\partial p_y$, in the same way as in Ref. [8]. For our particular Hamiltonian the x and y components of the velocity operator read

$$\hat{v}_x = -i\frac{\hbar}{m}\partial_x\tau_z + \frac{e}{mc}A_x(x,y) + \frac{\alpha}{\hbar}\sigma_y\tau_z\,, \tag{6}$$

$$\hat{v}_y = -i\frac{\hbar}{m}\partial_y\tau_z + \frac{e}{mc}A_y(x,y) - \frac{\alpha}{\hbar}\sigma_x\tau_z\,. \tag{7}$$

As the density is constant in time, the resulting current has zero divergence, $\nabla\cdot\vec{\jmath} = 0$. Equation (5) can be rewritten as the sum of a familiar quasi-particle current for non Rashba superconducting devices [9, 10], plus a Rashba current $\vec{\jmath}_{so}$

$$\vec{\jmath}(x,y) = \frac{\hbar}{m}\Im\left[\,\Psi^*(x,y)\,\nabla\tau_z\,\Psi(x,y)\,\right] + \frac{e}{mc}\,\rho(x,y)\,\vec{A}(x,y) + \vec{\jmath}_{so}(x,y)\,, \tag{8}$$

where

$$\vec{\jmath}_{so}(x,y) = \frac{\alpha}{\hbar}\Re\left[\,\Psi^*(x,y)\,(\sigma_y\hat{x} - \sigma_x\hat{y})\tau_z\,\Psi(x,y)\,\right]\,. \tag{9}$$

In this form it is possible to recover the usual expression for the current in absence of Rashba and superconductivity [9, 10] making the Rashba strength α zero and removing τ_z.

4. Results

We can see in Figs. 2a and 2b the quasi-particle current streamlines with the Majorana density $\rho(x,y)$ in the background, for two cases. In the first one (Fig. 2a) the magnetic field points along the longitudinal direction of the nanowire ($\theta = 90°$) while in the second one (Fig. 2b) the magnetic field has a component perpendicular to the surface of the nanowire ($\theta = 75°$). We notice that the current field is modified by the electronic orbital motion in the second case.

The main effect of the Majorana, in terms of quasi-particle transport, is to create circulating currents. For the longitudinally oriented magnetic field there is no preferred circulation orientation. As shown in Fig. 2c, the (z-component) rotational of the current has a maximum near the Majorana density peak and a minimum between the main and the secondary Majorana peaks. In the second case, with a perpendicular component of the magnetic field, a circulation orientation is enhanced over the other. We can see a strong current circulation around the Majorana density peak (see Fig. 2d) creating a maximum in the rotational near that point. This has similarities to what happens to the current near vortices in p-wave superconductors. On the other hand, the region of high circulating current around the Majorana peak is surrounded by regions of low or no current near the edges of the nanowire. This gives rise to minima in the value of the rotational.

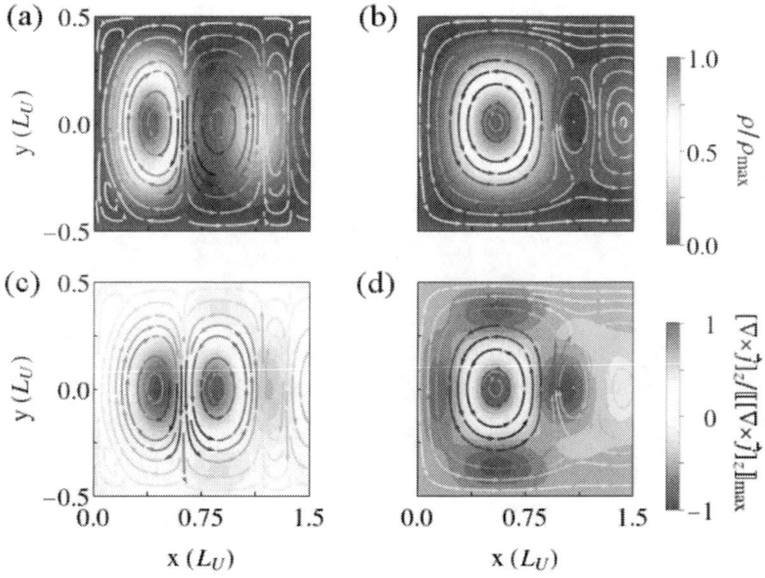

Figure 2. Current ($\vec{j}$) streamlines for longitudinal (a, c) and tilted field (b, d). The background in panels (a, b) shows the density, while in (c, d) it shows the z-component of the rotational $[\nabla \times \vec{j}]_z$. In each case the fields are normalized to their respective maximum values. The parameters are the same of Fig. 1. The streamlines are colored in a gray scale to show the regions of maximum current in black and the regions of zero current in white.

5. Conclusions

We have extended the study of Ref. [6] by calculating the local current associated with the Majorana state. We have shown that the main effect is the emergence of circulating currents. These currents follow closed trajectories without a preferred direction of circulation, in absence of a perpendicular component of the magnetic field. On the other hand, when a perpendicular component is present a quasi-particle current circulates around the Majorana peak in a vortex-like scenario. These quasi-particle currents may be difficult to measure experimentally in closed nanowires. However, in future work we will address the study of how these quasi-particle currents may change in open Majorana nanowires where they give a contribution to the nanowire conductance.

6. Acknowledgments

This work was funded by MINECO-Spain (grant FIS2011-23526), CAIB-Spain (Conselleria d'Educació, Cultura i Universitats) and FEDER. We hereby acknowledge the PhD grant provided by the University of the Balearic Islands.

References

[1] Alicea J 2012 *Rep. Prog. Phys.* **75** 076501
[2] Beenakker C W J 2013 *Annu. Rev. Condens. Matter Phys.* **4** 113
[3] Stanescu T D and Tewari S 2013 *J. Phys. Condens. Matter* **25** 233201
[4] Franz M 2013 *Nature Nanotechnology* **8** 149
[5] Lutchyn R M, Stanescu T D and Das Sarma S 2011 *Phys. Rev. Lett.* **106**(12) 127001
[6] Osca J and Serra L 2015 *Phys. Rev. B* **91**(23) 235417
[7] Serra L 2013 *Phys. Rev. B* **87** 075440
[8] Molenkamp L W, Schmidt G and Bauer G E W 2001 *Phys. Rev. B* **64**(12) 121202
[9] Vignale G and Rasolt M 1988 *Phys. Rev. B* **37**(18) 10685–10696
[10] Beenakker C 1992 *Low-Dimensional Electronic Systems* (*Springer Series in Solid-State Sciences* vol 111) ed Bauer G, Kuchar F and Heinrich H (Springer Berlin Heidelberg) pp 78–82 ISBN 978-3-642-84859-9

Journal of Physics: Conference Series **647** (2015) 012064 doi:10.1088/1742-6596/647/1/012064

Non-trivial Berry phase in the Cd_3As_2 3D Dirac semimetal

W. Desrat[1], C. Consejo[1], F. Teppe[1], S. Contreras[1], M. Marcinkiewicz[1], W. Knap[1,2], A. Nateprov[3] and, E. Arushanov[3]

[1] Laboratoire Charles Coulomb (L2C), UMR 5221 CNRS-Universit de Montpellier, Montpellier, France

[2] Institute of High Pressure Physics, Polish Academy of Sciences, Sokolowska 29/37 01-142 Warsaw, Poland

[3] Institute of Applied Physics, Academy of Sciences of Moldova, Chisinau MD 2028, Moldova

E-mail: wilfried.desrat@umontpellier.fr

Abstract. We present magnetotransport measurements performed on Cd_3As_2 samples. Our results confirm the existence of 3D Dirac fermions in this material with a π Berry phase. A metal-insulator transition driven by the magnetic field is also observed for $B < 1$ T. Finally the *in situ* rotation of the sample reveals that the Fermi surface is a weakly anisotropic ellipsoid.

1. Introduction

Cadmium arsenide (Cd_3As_2) is a semiconductor which has been extensively studied from the 60's to the 80's and was well-known for its high carrier mobility. Very recently, in 2013, Wang *et al.* computed the band structure of this material by first-principle calculations and showed that the conduction and valence bands cross along the $\Gamma - Z$ direction at finite $k_z = \pm 0.032$ $\AA^{-1}$ [1]. The presence of two Dirac cones makes Cd_3As_2 a robust 3D semimetal, protected by the crystal symmetry. ARPES measurements first revealed that the bulk band of Cd_3As_2 presented a linear dispersion along all three momentum directions [2]. This discovery has led to a quick renewed interest in this compound and resulted in an impressive amount of experimental breakthroughs in the space of two years : negative magnetoresistance due to chiral anomaly [3], transition to a superconducting state at GPa pressures [4], possible signatures of Weyl fermions due to the breaking of time reversal symmetry [5], etc. Here, we present preliminary results obtained by magnetotransport measurements on Cd_3As_2 samples. We confirm the existence of 3D Dirac fermions and show that the Fermi surface is not exactly spherical.

2. Temperature dependence of the longitudinal resistance

In this work, we focus on several samples cleaved from a Cd_3As_2 single crystal oriented along the [112] direction. The samples present a random shape with a non uniform thickness and are not polished intentionally. The main drawback is that we cannot access the bulk carrier concentration from the Hall voltage. However it is known that there is always a discrepancy between the Hall density and the concentration estimated from the Shubnikov-de Haas in these materials [6]. Thin gold wires are simply contacted with silver epoxy on top of visually flat surfaces obtained after cleavage. Four-wire measurements are carried out in either the van der

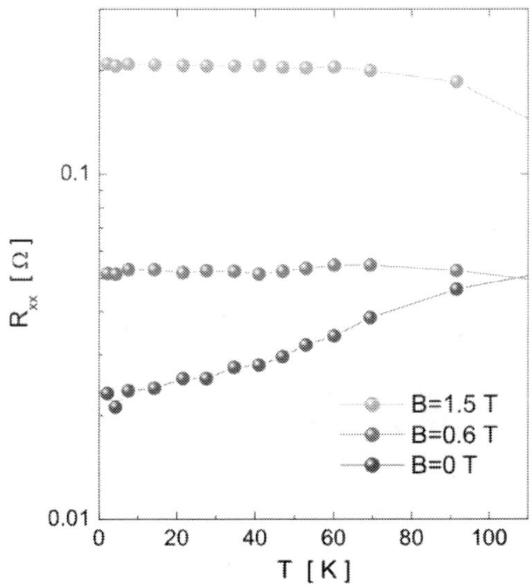

Figure 1. Longitudinal resistance R_{xx} measured as a function of the temperature for different magnetic field values. The metallic behavior observed at $B = 0$ T is replaced by an insulating T-dependence at $B = 1.5$ T. The sample undergoes a metal-insulator transition in between.

Pauw geometry or in a longitudinal configuration, when all four contacts are aligned. The latter is favored in order to obtain almost symmetric longitudinal resistance $R_{xx}(B)$ curves directly. We use a standard lock-in detection technique with a low frequency ac-current of 50 μA. The samples are placed inside a variable temperature insert surrounded by a superconducting coil of 13 T.

As the temperature is cooled down, the sample's resistivity at zero magnetic field exhibits the expected metallic behavior (Fig. 1). However we observe a metal-insulator transition as the magnetic field increases. At $B = 0.6$ T, the resistivity is almost independent on the temperature (for $T < 100$ K), and above this value, the sample behaves like an insulator. We underline that a metal-insulator transition has recently been reported in Cd_3As_2 induced by biasing a top gate [7]. Anyway the metal-insulator transition of Dirac fermions is still poorly understood and requires additional theoretical studies.

Figure 2 shows the magnetic field dependence of the longitudinal resistance R_{xx} from 0 to 13 T plotted for several temperatures up to $T = 153$ K. The resistance exhibits a clear positive magnetoresistance (MR) with well defined superimposed Shubnikov-de Haas (SdH) oscillations for the lowest temperatures. As T increases, the positive MR diminishes and the amplitude of the SdH oscillations fades out. The low field MR is parabolic, as illustrated by the open circles on the $T = 91$ K curve and becomes almost linear at high magnetic field (solid circles). The latter behavior has been observed in many topological insulators and 3D semimetals. The linear MR may arise in the presence of large mobility fluctuations in the sample or when the quantum limit is reached, and all electrons lie on the lowest Landau level.

The amplitude of the SdH oscillations normalized by the value at the lowest temperature $T = 2$ K is plotted vs T/B in Fig. 3. For a given magnetic field, the temperature dependence is given by the Lifshitz-Kosevitch formula, $R(T) \sim X/\sinh X$, where $X = 2\pi^2 k_B m^* T/B$. The equation fits the data nicely and provides an estimate of the effective mass of 0.04 m_0, which is in agreement with the previously reported values in this material [8].

3. Non-trivial Berry phase in Cd_3As_2

The analysis of the SdH oscillations can give access to another important property of the carriers. Indeed, the oscillating component of the magnetoresistance is given by $\Delta\rho \propto \cos(2\pi(B_f/B + 0.5 - \phi_B/2\pi))$, where B_f is the SdH oscillations frequency and ϕ_B the Berry phase. The precise determination of the phase of the oscillations is fundamental because it

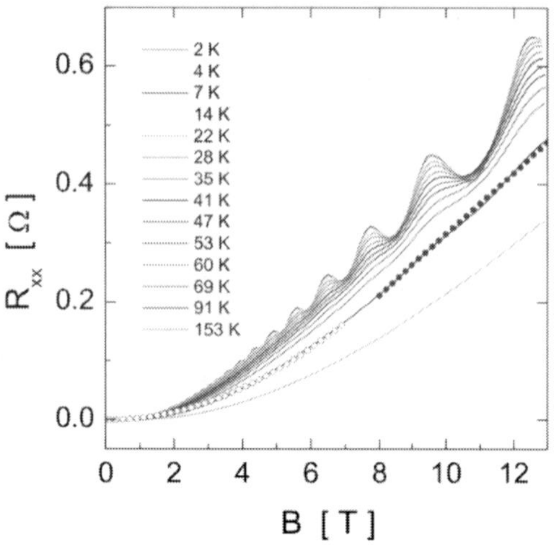

Figure 2. Magnetic field dependence of the longitudinal resistance for several temperatures. Open and closed circles show B-parabolic and B-linear dependences at low and high fields respectively.

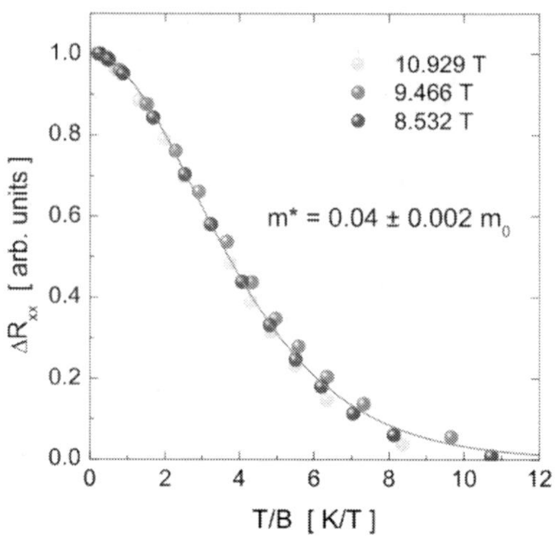

Figure 3. Normalized amplitude of the SdH oscillations plotted vs T/B. The solid line is a fit by the Lifshitz-Kosevitch formula which leads to an effective mass estimation of $m^* = 0.04\,m_0$.

allows to discriminate the nature of the carriers. Classical massive fermions and relativistic Dirac fermions do not exhibit the same Berry phase. For low carrier density samples, the Fermi energy is close to the Dirac point and Dirac fermions are expected with a typical π Berry phase signature. The phase is estimated from the intercept at the origin of a Landau plot as shown in the inset of Fig. 4. The inverse field of the oscillations minima and maxima is plotted as a function of the Landau level index (integer and half-integer values, respectively). Here, the intercept is equal to 0.55, i.e. $\phi_B = 1.1\pi$. This result indicates clearly a non-trivial Berry phase and the existence of 3D Dirac relativistic fermions in our Cd_3As_2 samples.

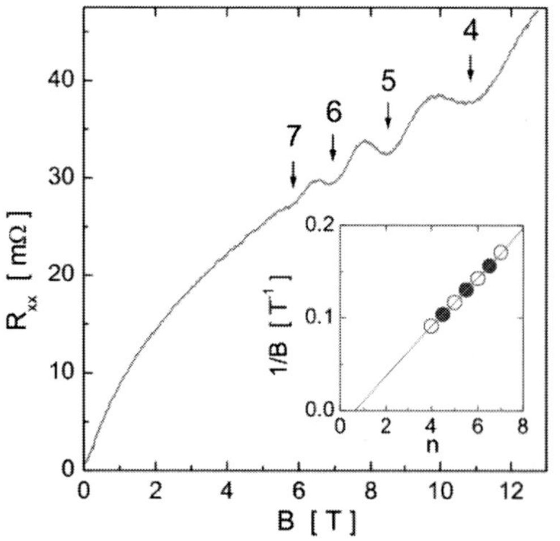

Figure 4. Longitudinal resistance vs the magnetic field at $T = 2$ K. Inset : inverse field of R_{xx} min and max (open/closed circles) vs the Landau level index.

4. Anisotropic Fermi surface

Finally we measured the magnetoresistance as a function of the tilt angle by an *in situ* rotation of the samples in respect to the magnetic field. Figure 5 plots the SdH frequency, i.e. the fundamental field B_f, vs the tilt angle. The frequency is not constant but varies between 37 and 43 T. This slight variation is the signature of 3D carriers, since 2D carriers are sensitive to the perpendicular component of the magnetic field only. The data should diverge when the surface becomes parallel to the applied field at 0° and 180°, which is not the case. Furthermore, the variation of the SdH frequency indicates that the cross section of the Fermi surface perpendicular to the magnetic field is not constant vs the rotation. It means that the Fermi surface is not a sphere but a weakly anisotropic ellipsoid, which confirms the conclusions of previous works [9]. Unfortunately we have not determined the in-plane crystallographic axes at this stage and cannot reconstruct the Fermi surface precisely. The 37 – 43 T range of the SdH frequency leads to an average Fermi radius varying between 0.033 and 0.036 $Å^{-1}$. We deduce a carrier concentration of $\sim 1.5 \times 10^{18}$ cm^{-3} in our samples.

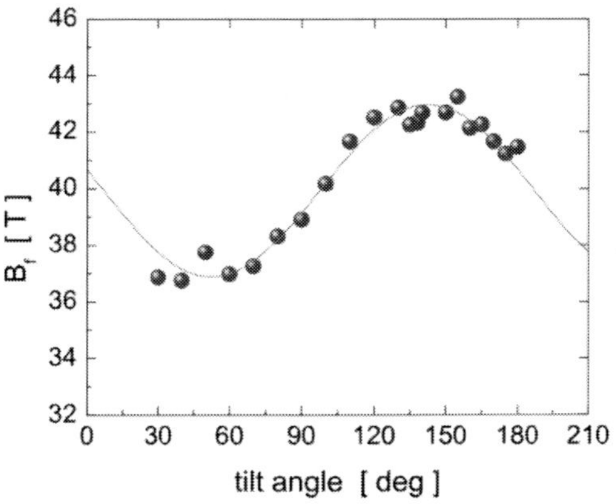

Figure 5. Fundamental field B_f as a function of the tilt angle. The behavior reveals a 3D Fermi surface with the shape of a slightly anisotropic ellipsoid (solid line is a cosine function). The 90° angle corresponds to the magnetic field normal to the sample's surface.

5. Conclusion

Transport measurements carried out at low temperature show unambiguous signatures of 3D relativistic Dirac fermions in Cd_3As_2 single crystals. A distinct Berry phase of π is estimated from the Landau plot. Additionnally, we confirm the weak anisotropy of the Fermi ellipsoidal surface.

References

[1] Wang Z, Weng H, Wu Q, Dai X and Fang Z 2013 *Phys. Rev. B* **88**(12) 125427
[2] Neupane M, Xu S Y, Sankar R, Alidoust N, Bian G, Liu C, Belopolski I, Chang T R, Jeng H T, Lin H, Bansil A, Chou F and Hasan M Z 2014 *Nat Commun* **5**
[3] Zhang C, Zhang E, Liu Y, Chen Z G, Liang S, Cao J, Yuan X, Tang L, Li Q, Gu T, Wu Y, Zou J and Xiu F 2015 (*Preprint* arXiv:1504.07698)
[4] He L, Jia Y, Zhang S, Hong X, Jin C and Li S 2015 (*Preprint* arXiv:1502.02509)
[5] Jeon S, Zhou B B, Gyenis A, Feldman B E, Kimchi I, Potter A C, Gibson Q D, Cava R J, Vishwanath A and Yazdani A 2014 *Nat Mater* **13** 851–856
[6] Narayanan A, Watson M D, Blake S F, Bruyant N, Drigo L, Chen Y L, Prabhakaran D, Yan B, Felser C, Kong T, Canfield P C and Coldea A I 2015 *Phys. Rev. Lett.* **114**(11) 117201
[7] Liu Y, Zhang C, Yuan X, Lei T, Wang C, He L, Che R and Xiu F 2014 (*Preprint* arXiv:1412.4380)
[8] He L P, Hong X C, Dong J K, Pan J, Zhang Z, Zhang J and Li S Y 2014 *Phys. Rev. Lett.* **113**(24) 246402
[9] Zhao Y, Liu H, Zhang C, Wang H, Wang J, Lin Z, Xing Y, Lu H, Liu J, Wang Y, Jia S, Xie X and Wang J 2014 (*Preprint* arXiv:1412.0330)

Journal of Physics: Conference Series **647** (2015) 012065 doi:10.1088/1742-6596/647/1/012065

The effect of interface roughness scattering on Si SOI FinFET with Ando's and extended Prange and Nee model

D Nagy[1], **M A Elmessary**[1,2], **M Aldegunde**[1,3] **and K Kalna**[1]

[1] ESDC, College of Engineering, Swansea University, Swansea SA2 8PP, Wales, UK
[2] Engineering Math & Physics Department, Faculty of Engineering, Mansoura University, Mansoura, Egypt

E-mail: D.Nagy.491828@swansea.ac.uk

Abstract. A new multi-subband interface roughness scattering (IRS) model is incorporated into a 3D Finite Element Monte Carlo simulator with 2D Schrödinger equation based quantum corrections. The model takes advantage of wavefunctions and energy levels obtained in solutions of Schrödinger equation on slices across the channel to calculate the respective form factors. The new 3D code is then used to forecast the performance of 10.7 nm gate length SOI Si FinFETs with rectangular cross-section and rounded corners. We found that the multi-subband IRS is much stronger at large electron kinetic energies resulting in a drive current of 600 mA/μm^2 that is 25% reduction when compared to 3D Ando model.

1. Introduction

Non-planar multi-gate transistors [1] are the leading solutions for sub-16 nm digital CMOS technology [2] scaled to ultimate limit thanks to superior electrostatic integrity and delivering a large ON-current. In this scaled process, interface roughness scattering (IRS) is considered to be one of the key limiting factors to carrier transport [3][4] because fabrication of the interface between semiconductor and channel in multi-gate nanoscaled devices is harder to control [5].
In this work, we have implemented a new multi-subband IRS model which uses wavefunctions and energy levels from our 3D finite element (FE) Monte Carlo (MC) toolbox [6] with 2D FE Schrödinger equation quantum corrections (SEQC) [7]. The wavefunctions and energy levels are used to calculate form factors entering the multi-subband scattering rate adapted to the FE mesh. The new multi-subband IRS model within the 3D FE MC is then applied to 10.7 nm gate length SOI FinFETs with rectangular cross-section (see Fig. 1), designed according to the ITRS guidelines [8], by the FE method and compared with a basic 3D Ando model.

2. Monte Carlo Code

In the past, a 3D FE ensemble MC simulation toolbox with density gradient (DG) [6] demonstrated an excellent agreement with experimental I_D-V_G characteristics of the 25 nm gate length SOI FinFET [9]. Later, the DG quantum corrections in MC toolbox has been replaced by SEQC [7] to predict transistor performance when in deep nanoscaled region [10].
A 3D FE mesh of the device in the 3D MC toolbox accommodates 2D FE planes perpendicular to the transport direction (see Fig. 2) [7]. The extracted 2D potential $V(y, z)$ from the 3D

[3] Present address: WCPM, School of Engineering, University of Warwick, Coventry CV4 7AL, England, UK

electrostatic potential $V(\mathbf{r})$ is used to solve the 2D Schrödinger equation:

$$-\frac{\hbar^2}{2}\nabla_\perp \cdot \left[(\mathbf{m}^*)^{-1} \cdot \nabla_\perp \psi(y,z)\right] + U(y,z)\psi(y,z) = E\psi(y,z), \tag{1}$$

where E is the energy, $(\mathbf{m}^*)^{-1}$ is the inverse effective mass tensor, $\psi(y,z)$ is the wavefunction penetrating the oxide, and $U(y,z) = -[qV(y,z) + \chi(y,z)]$ is the potential energy with $\chi(y,z)$ being the electron affinity. The eigenstates of the Schrödinger equation are used to calculate the 2D quantum density n_q. The quantum density is then interpolated onto the 3D simulation domain to obtain the quantum corrected potential which, in turn, is used to move particles. More details on the 3D FE MC toolbox can be found in Refs. [6] [7] [11].

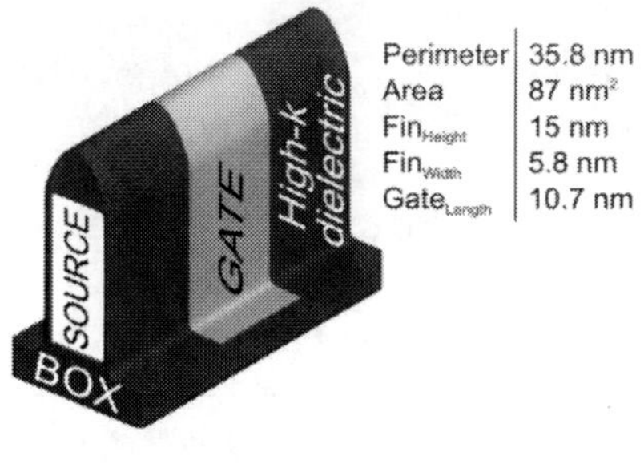

Figure 1. Schematic of the simulated Si SOI FinFET. The device has a Gaussian doping profile with σ of 3.45 nm in the transport direction and an effective oxide thickness (EOT) of 0.62 nm.

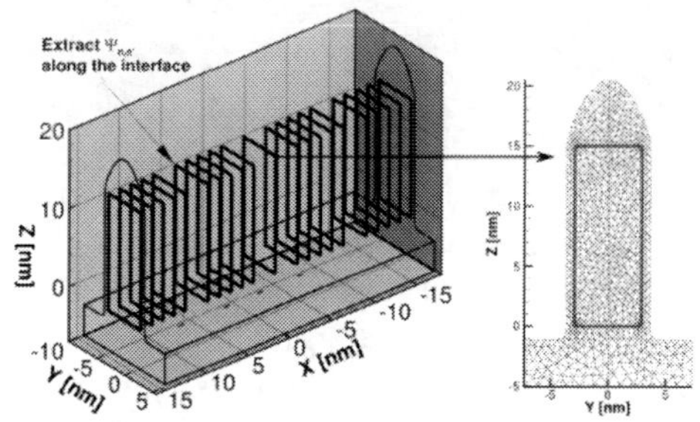

Figure 2. The interface slices along the transport direction (x-axis) that are also the position of the 2D FE slices used by the 2D Schrödinger solver.

3. Interface Roughness Scattering Model

Many IRS models require a power spectrum of the interface roughness commonly described by Gaussian and exponential forms [12]. We have chosen the latter as suggested in Ref. [13] given by:

$$C(\mathbf{q}) = \frac{\pi \Delta_{RMS}^2 \Lambda^2}{\sqrt{\left(1 + \frac{q^2\Lambda^2}{2}\right)^3}} \tag{2}$$

where Δ_{RMS} is the RMS height of the interface roughness, the Λ is the correlation length and the q is the wavevector. The value of Δ_{RMS} is 0.57 nm, taken from experimental data [14] and Λ is 1.7 nm [13].

3.1. Ando Model

The matrix element for Ando model is given by [15]:

$$\Gamma(\mathbf{k}) = \frac{m^* e^4}{\hbar^3 \epsilon_S^2} \frac{\Delta_{RMS}^2 \Lambda^2 E_\perp^2}{\sqrt{1 + k^2\Lambda^2}} \, \mathcal{E}\left(\frac{k\Lambda}{\sqrt{1 + k^2\Lambda^2}}\right) \tag{3}$$

where $E_\perp$ is the transverse electric field and $\mathcal{E}$ is the complete elliptic integral of the first kind. Fig. 4 shows the calculated maximum scattering rate for the IRS model using a rejection technique which takes into account the local electric field at the scattering event [16].

3.2. Extended Prange & Nee Model

The multi-subband EPN model is based on a 2D formulation of Prange & Nee scattering model for arbitrary paths [5]. Wavefunctions at each eigenstate are extracted along the interface for all the 21 slices (see Fig. 2), then the form factors are calculated by:

Journal of Physics: Conference Series **647** (2015) 012065 doi:10.1088/1742-6596/647/1/012065

$$f_{n,n'}(s) = \psi^*_{n,k}\psi_{n',k'}\Delta V \tag{4}$$

where n, n' are the initial and final energy levels, respectively, ΔV is the potential barrier between the Si channel and the high-κ dielectric. Finally the scattering rate is calculated by:

$$\frac{1}{\tau_{n,n'}(E)} = \frac{1}{2\hbar}\int_R |F_{n,n'}(q_\perp)|^2\; C(q)\; dq_\perp\; g_{1D,n'}(E)\; A \tag{5}$$

where the $F_{n,n'}(q_\perp)$ is a fast Fourier transform (FFT) of the form factors, $C(q)$ is the exponential power spectrum of the interface roughness, $g_{1D,n'}$ is the 1D density of states (DOS) and A is the area of the channel.

Fig. 3 shows the time-line of the simulation. A drift-diffusion (DD) simulation is run beforehand to acquire results for the subthreshold slope and also to provide an initial solution for the MC. To speed up the MC simulation, we run a pre-simulation using 3D Ando model followed by another pre-simulation using the multi-subband EPN model setting a maximum IRS rate for each energy. This rate is calculated as the maximum rate over all slices for the given energy. During the MC simulation itself, the instantaneous IRS rate, depending on the actual potential distribution in the device, is calculated to perform a rejection technique. The instantaneous IRS rate is interpolated from the nearest slice and assumes an exponential decay of the roughness potential strength as a function of distance from the interface [3]. Fig. 4 compares the maximum average scattering rates for both models. In case of the Ando model, the scattering rate reduces with increasing energy. For the multi-subband EPN model, the scattering has a maximum at the subband positions as expected. Note that the Ando model gives a higher probability of scattering below 0.15 eV while at a larger energy the trend is opposite.

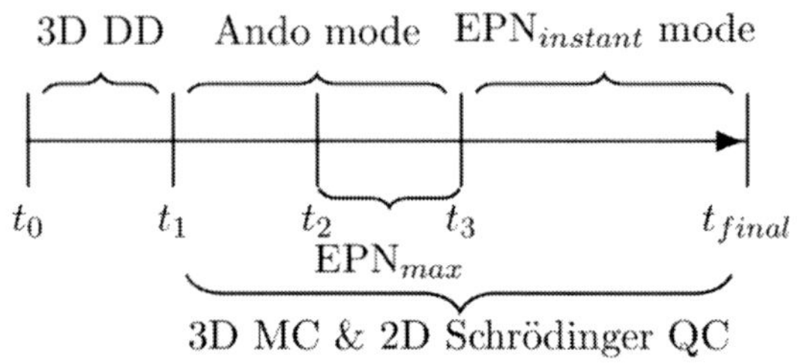

Figure 3. Time scale for the 3D FE MC with 2D Schrödinger QC toolbox.

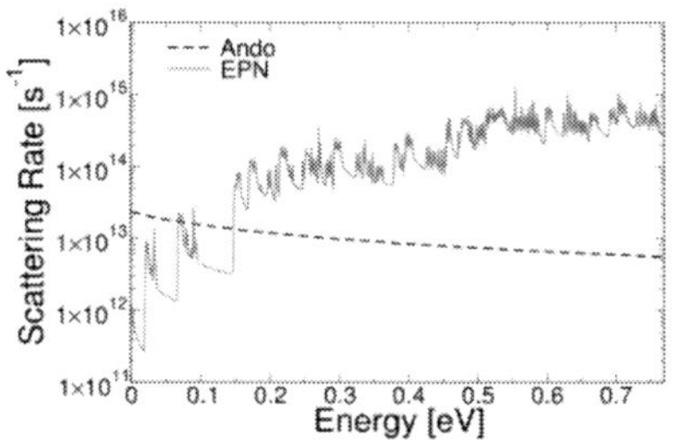

Figure 4. The maximum scattering rate from 3D (Ando) compared to multi-subbband (EPN) model. The EPN is for the 20 lowest electron subbands at V_D = 0.05 V and V_G = 0.0 V showed only up to 0.77 eV but calculated up to 3 eV.

4. Effect of the Interface Roughness

The new IRS model incorporated into the 3D SEQC FE MC toolbox and then applied to investigate n-channel SOI FinFET with a gate length of 10.7 nm and a ⟨100⟩ channel orientation designed following the ITRS guidelines [8]. We study a rectangular-like cross-section with rounded corners to demonstrate realistic nanoscale device geometry. Fig. 5 shows I_D-V_G characteristics, with the current normalized-to-area, comparing the IRS models at both low (V_D = 0.05 V) and high (V_D = 0.7 V) drain biases. At low drain bias, the current is slightly overestimated by the Ando model as the gate voltage increases above 0.4 V by about 8%. At high drain bias, the difference in overestimation of the current between the two IRS models largely increases to 25%. Table 1 lists the current at V_G-V_T = 0.7 V and subthreshold slope (SS) from drift-diffusion (DD) simulations. Note that drain induced barrier lowering (DIBL) obtained from MC is larger than DD since it accounts for quantum capacitance [10].

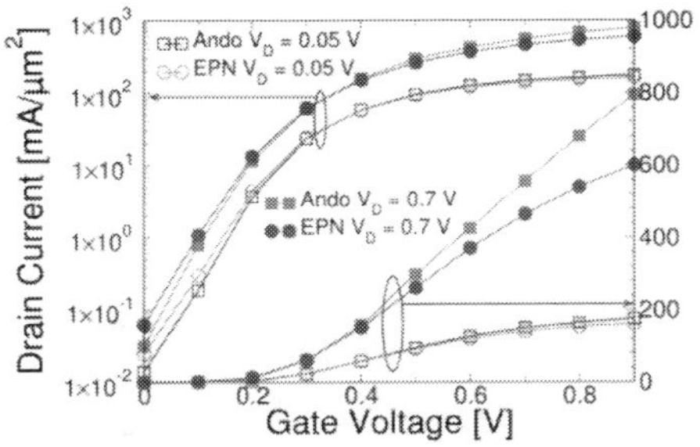

Figure 5. I_D-V_G characteristics at $V_D = 0.05$ V and 0.7 V using a normalised-to-area current for the 10.7 nm gate length FinFET comparing the 3D interface roughness model (Ando) with the multi-subband model (EPN).

Model	Ando	EPN
SS_{DD}^{LOW} [mV/dec]	70	
SS_{DD}^{HIGH} [mV/dec]	71	
$DIBL_{DD}$ [mV/V]	56	
$DIBL_{MC}$ [mV/V]	66	68
I_{MC}^{LOW} [mA/μm^2]	175	161
I_{MC}^{HIGH} [mA/μm^2]	796	600

Table 1. SS for drain biases of 0.05 V (LOW) and 0.7 V (HIGH) from the DD, DIBL from DD and from MC, and drive currents for normalized-to-area for 3D (Ando) and multi-subband (EPN) interface roughness models.

5. Conclusions

A new multi-subband EPN model has been incorporated into the 3D FE SEQC ensemble MC simulation toolbox for physically based modelling of nanoscale multi-gate transistors. The new 3D tool has been employed to study the effect of the IRS on I-V characteristics at low and high drain biases in the 10.7 nm gate length Si SOI FinFETs for a rectangular-like realistic cross-section. We have found that the multi-subband EPN IRS has a much stronger impact on non-planar devices for large electron kinetic energy (from above 0.18 eV) than the 3D Ando IRS. The 3D MC simulations that use the multi-subband EPN model predict a drive current of 161 mA/μm^2 at a low drain bias of 0.05 V. This is about 8% smaller when compared to the results obtained using the 3D Ando model, which is negligible difference. However, at a high drain bias of 0.7 V, the predicted drive current is 600 mA/μm^2, smaller by about 25%, which is noticeable reduction in the drain current. The IRS is thus essential for meaningful modelling of carrier transport in nanoscale multi-gate transistors.

References

[1] Huang X and *et al* 1999 *Electron Devices Meeting, 1999. IEDM '99. Technical Digest. International* pp 67–70
[2] Hook T 2012 *Custom Integrated Circuits Conference (CICC), 2012 IEEE* pp 1–7
[3] Jin S, Fischetti M V and Tang T w 2007 *J. Appl. Phys.* **102** 083715
[4] Wang J and *et al* 2005 *Appl. Phys. Lett.* **87** 043101
[5] Stanojevic Z and Kosina H 2013 *Simulation of Semiconductor Processes and Devices (SISPAD), 2013 International Conference on* pp 352–355
[6] Aldegunde M, Garcia-Loureiro A and Kalna K 2013 *IEEE T. Electron. Dev.* **60** 1561–1567
[7] Lindberg J and *et al* 2014 *IEEE T. Electron. Dev.* **61** 423–429
[8] ITRS(2012) International technology roadmap for semiconductors http://www.itrs.net/Links/2012ITRS/Home2012.htm
[9] Basker V and *et al* 2010 *VLSI Technology (VLSIT), 2010 Symposium on* pp 19–20
[10] Nagy D and *et al* 2015 *IEEE T. Nanotechnol.* **14** 93–100
[11] Aldegunde M, Seoane N, Garca-Loureiro A and Kalna K 2010 *Comput. Phys. Commun.* **181** 24 – 34
[12] Esseni D, Palestri P and Selmi L 2011 *Nanoscale MOS Transistors: Semi-Classical Transport and Applications* (Cambridge University Press)
[13] Goodnick S M and *et al* 1985 *Phys. Rev. B* **32**(12) 8171–8186
[14] Tang X and *et al* 2009 *IEEE T. Nanotechnol.* **8** 611–616
[15] Ferry D 2000 *Semiconductor Transport* (Taylor & Francis)
[16] Kalna K, Yang L and Asenov A 2005 *Solid-State Device Research Conference, 2005. ESSDERC Proceedings of 35th European* pp 169–172

Optimization of Ballistic Deflection Transistors by Monte Carlo Simulations

J-F Millithaler[1], I Iñiguez-de-la-Torre[2], J Mateos[2], T González[2], M Margala[1]
[1]Department of Electrical and Computer Engineering, University of Massachusetts, Lowell, MA 01854, USA
[2]Applied Physics Department, Salamanca University, Salamanca 37008, Spain

E-mail: jeanfrancois_millithaler@uml.edu

Abstract. This paper presents an optimization of the current-voltage characteristic of Ballistic Deflection Transistors. The implementation of an adequate surface charge model in a Monte Carlo tool shows a very good agreement with the available experimental data and allows us to predict the influence of different parameters, like temperature, channel and trench dimensions on the device output. These results are of importance for further use of this device in logical circuit applications.

1. Introduction

The Ballistic Deflection Transistor (BDT) [1] is a kind of semiconductor device that can drastically change the way to develop the electronics of tomorrow. This device associate the ballistic motion of electrons [2], offering very high speed performances, with a strategic geometry design allowing the steering of electrons, granting the opportunity to realize a complete logic gate [3-5]. The two symmetrical outputs can exhibit at the same time and for a given set of polarizations, one maximum of current and one minimum. In term of logical terms, this can be translated as logic high, or "1", and logic low, or "0", respectively. The purpose of this paper is to investigate major parameter dependencies in order to control the nonlinear behaviour of the BDT and provide the most useful output characteristic for logical applications.

2. Overview of the BDT and Monte Carlo Model

The BDT is an interesting device based on an InAlAs/InGaAs heterojunction, where a two-dimensional electron gas (2DEG) allows carriers to travel more than 2.5 times faster than in silicon. By using electron-beam lithography [6], the shape of the device can reach smaller dimensions than the electron mean free path (about 140 nm in InGaAs channel at 300 K) and provide ballistic transport at room temperature. The design of the BDT is presented in Fig. 1. It includes a grounded source, three drains (V_{TD}, V_{LD} and V_{RD}), where top drain is just a pull-up terminal, and two lateral gates biased in push-pull mode (V_{LG}=-V_{RG}). A negative bias applied on one gate will pinch off the corresponding drain channel. This will produce a steering effect, assisted by a triangular deflector, and the device will provide a nonlinear output characteristic [6, 7]. Fig. 2 shows a scanning electron microscope (SEM) image of a 300 nm wide channel BDT [6]. The simulated BDT has been directly designed over the SEM image, keeping all dimensions identical. Access regions have been removed for reducing the computation times and only the active region is simulated.

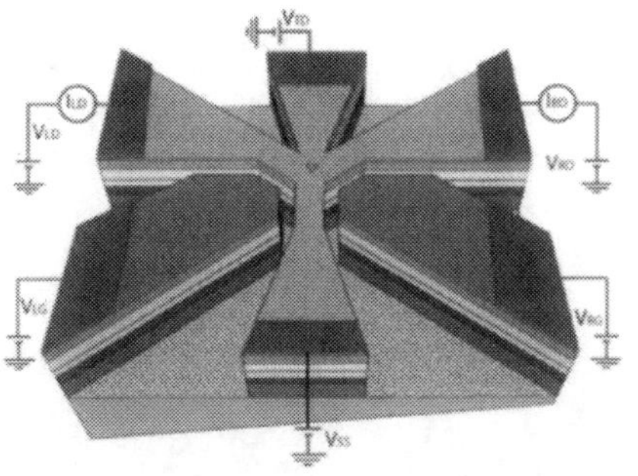

Figure 1: 3D topology of a typical BDT

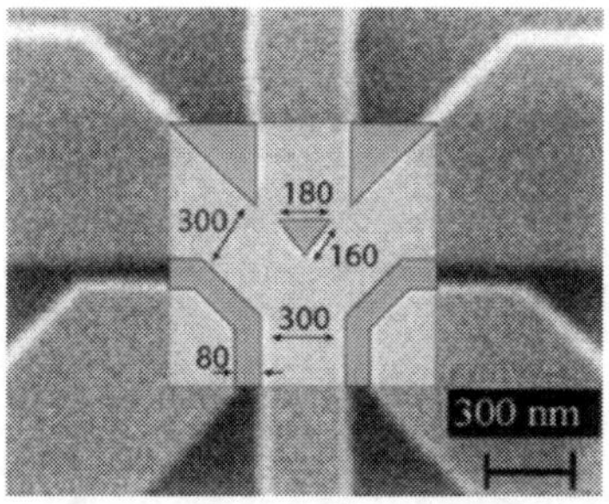

Figure 2: SEM image of BDT and sketch of simulated device. All dimensions are in nm.

The modelling tool is an ensemble MC simulator of the electron dynamics self-consistently coupled with a 2D Poisson solver using the finite differences approach [8]. A background doping $N_{db}=5\times10^{16}\,cm^{-3}$ is considered, the size of meshes is 5×5 nm and the time step is 1 fs. As devices are reaching nanometre size, the surface-volume ratio increase drastically and the effect of surface charges (at the interface semiconductor-dielectric), which led to a depletion mechanism in the channel, becomes crucial. The simulator can take care of these surface charges in two ways. The first way, so-called *constant charge model* (CCM), assume that all surface charges are bias and geometry independent. The value of σ in the MC code is a fitting parameter but adjusted via experimental measurements [9, 10]. The second way proposed is to set the surface charge value adaptive with the surrounding carrier concentration. The so-called *self-consistent charge model* (SCCM) allows the variation σ with the bias and position, increasing or decreasing its magnitude according the local electron density. More details about this model can be found in [11, 12].

3. Results

3.1 Effect of Surface Charges

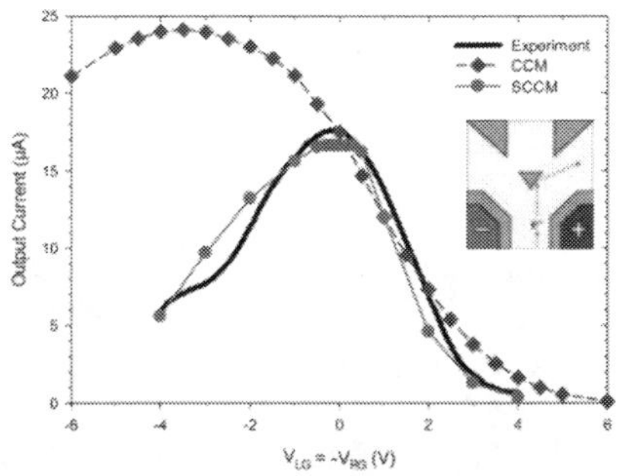

Figure 3: (a) Comparison of experimental, CCM and SCCM output current *vs.* gate voltage

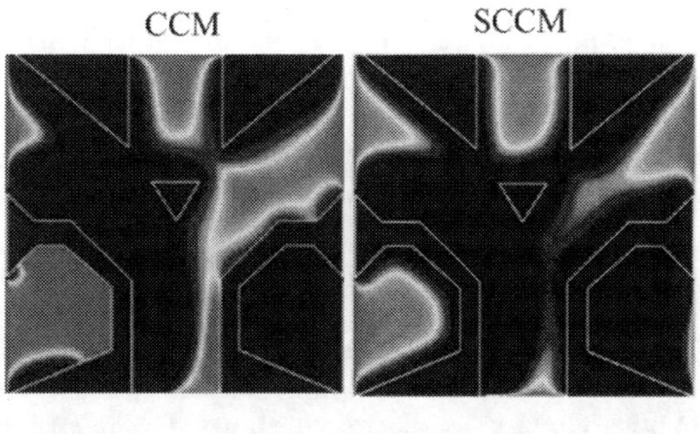

(b) 2D density maps of the BDT for the case of CCM and SCCM at a bias V_{LG}=-4 V

Fig. 3(a) presents the comparison of the experimental right drain current (left drain current is symmetric with respect to zero and is not traced for more clarity) *vs.* the left gate voltage [1] and the MC results obtained with the two presented models. The CCM is able to reproduce the experimental current in the vicinity of zero gate bias (V_{LG}=$-V_{RG}$=0.0 V) but clearly fails when the current should abruptly decrease for highly negative V_{LG}. On the other hand, the SCCM shows a remarkably good agreement with the measurements. The magnitude of the current and the sharp drop at both sides of the maximum are correctly reproduced. To help the reader to clarify the difference between the two

Journal of Physics: Conference Series **647** (2015) 012066 doi:10.1088/1742-6596/647/1/012066

models we are showing the 2D electron density map of the whole device in Fig. 3(b), for V_{LG}=-4 V, where the largest differences between the two models are found. Within the CCM the fixed value of the surface charge (σ=3.10^{15} m^{-2}) is not enough to completely deplete the source channel. Also, the electron concentration in the right drain channel is highly overestimated and as a consequence also the current. On the other case, the surface charges within the SCCM react to the rise in the amount of electrons in their vicinity, thus increasing in order to maintain a depletion region near the interfaces. The measured current decrease is then correctly reproduced.

3.2 Scaling down and temperature optimization

As the SCCM is the best model to correctly simulate the complex behaviour of the BDT, it is the one that will be used in the following. Fig. 4 shows the right drain current of the previously presented 300 nm wide channel BDT at room temperature and at 77 K. We observe, as expected, an increase of the maximum current from 16 μA up to 26 μA when T is decreased. Nevertheless, the bell-shape of the current is widened and the pinch-off is achieved for a higher value of the bias. Fig. 4 also shows the results for a smaller device, with the channel width as small as 100 nm. All the other dimensions have been scaled down proportionally and drain biases have been kept identical. The main result is that we can appreciate a sharper bell-shape of this current almost pinching-off the channel at -1.0 V. Therefore we demonstrate that scaling down the BDT will provide a much better gate control and an enhancement of the nonlinear output, making it suitable also for analog applications such as frequency doubling. Low temperature operation is also interesting, because for W_C=100 nm the maximum current increases a ×3 factor whereas it was only ×2 in the case of W_C=300 nm. Ballistic carriers are only affected by design device boundary thus low temperature shouldn't modify the output current. This means that even for the smallest device studied, the transport is not completely ballistic at room temperature, and a further downscaling is still necessary for reaching its optimum operation conditions.

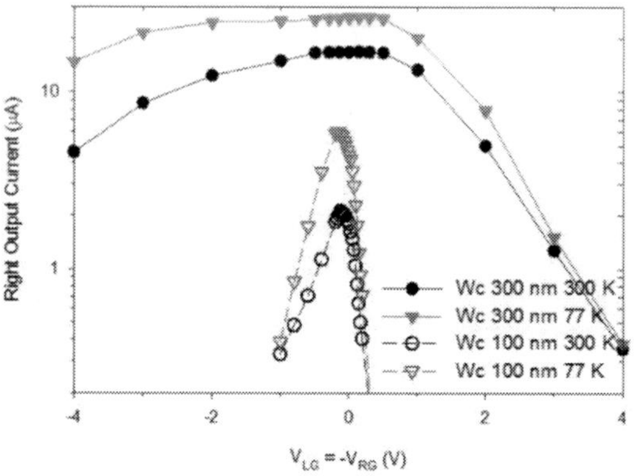

Figure 4: Simulations of right drain current for different channel widths and temperatures.

3.3 Trenches optimization

Controlling the trenches' width, W_t, is one of the biggest challenge in the BDT fabrication process because they are the key to provide the best performance of the device. In Fig. 5(a) we report the effect of W_t on the BDT's characteristics. Decreasing the width of the BDT trenches to 40 nm (the reference device, Fig. 2, has W_t=80 nm) narrows the bell-shape current, while increasing it to 120 nm has the opposite effect, as expected. However, the maximum current remains the same. On Fig. 6(a), the results for the smaller device, where W_C=100 nm, are plotted. As all dimensions have been scaled down proportionally, the reference W_t is now equal to 27 nm. Interestingly the variation of W_t has now more influence on the current peak magnitude but less on the pinch-off behaviour. The fabrication of a very small trench width might be a challenge, so that its filling with a high-k dielectric may provide a simpler solution and even enhanced performances. Up to now the air is used as dielectric, with ε=1 (obtained by simply digging the trenches onto the heterostructure). In the MC simulations the value of the permittivity must be modified in order to correctly capture the gate-to-channel coupling of the electric field through the semiconductor buffer. An intermediate value ε=10 (between the air, ε = 1, and the semiconductor, ε=13.88), provides the best fit [see Fig. 3(a)]. Figs. 5(b) and 6(b) show the output response when using ε=25 (corresponding to the case where trenches are filled with promising high-k dielectrics like ZrO_2 or HfO_2 [13]) and ε=70 (approaching the value of deionized water). In both cases, W_C=300 and 100 nm, the higher the value of the dielectric the stronger the gate control and the easier is to pinch-off the channel. Again, the effect is more pronounced in the smaller device, decreasing also the value of the current peak.

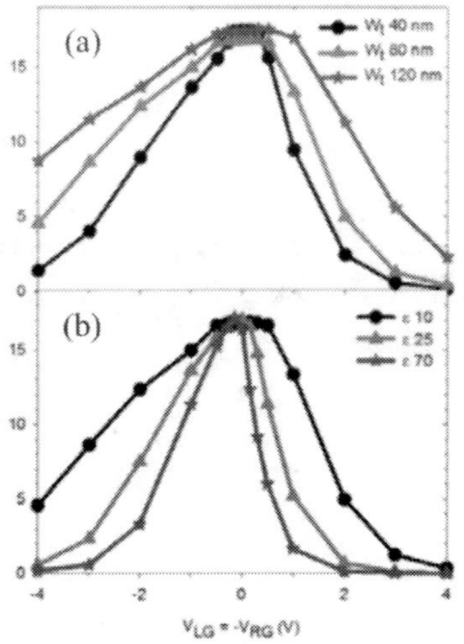

Figure 5: Output current for different W_t and ε, in the case W_C=300 nm

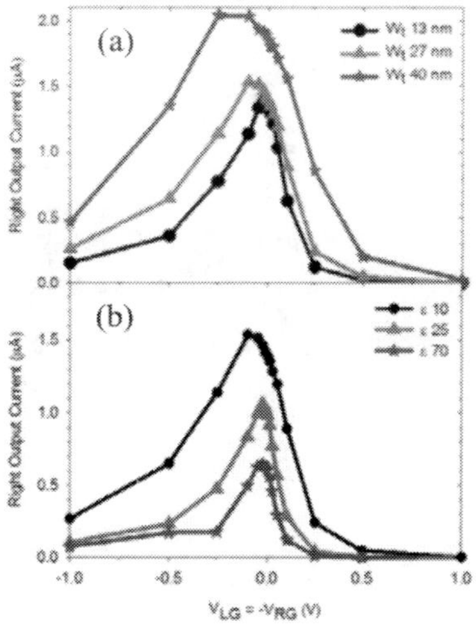

Figure 6: Output current for different W_t and ε, in the case W_C=100 nm

4. Conclusion

The use of a self-consistent surface charge model in our MC simulator has shown that the complex nonlinear behaviour of the BDT can be correctly simulated. The modelling of a real BDT provides very good qualitative and quantitative results and makes possible the optimization of the device. The effect of scaling down the channel and trench width and the permittivity of the dielectric filling the trenches have been studied and have shown that a better gate control can be reached. These results will be of importance for the possible use of BDTs in circuits for analog or logical applications

5. References

[1] V. Kaushal, I. Iñiguez-de-la-Torre and M. Margala, Solid-State Electronics 2011;56;120–129
[2] A. Song Encyclopedia Nanosci Nanotechnol 2004;9:371–89.
[3] HQ. Xu, I. Shorubalko, D. Wallin, I. Maximov, P. Omling, L. Samuelson L, IEEE Electron Dev Lett 2004;25:164.
[4] J. Sun, D. Wallin, I. Maximov, HQ. Xu, IEEE Electron Dev Lett 2008;29:540.
[5] CR. Müller, L. Worschech, P. Höpfner, S. Höfling, A.Forchel. IEEE Electron Dev Lett 2007;28:859.
[6] V. Kaushal, I. Iñiguez-de-la-Torre, H. Irie, G. Guarino, W. R. Donaldson, P. Ampadu, R. Sobolewski and M. Margala, IEEE Trans Nanotechnology 2010; 9:723
[7] Q. Diduck, H. Irie, and M. Margala, Int. J. High Speed Electron. Syst., 2009;19: 23–31
[8] J. Mateos, B. G. Vasallo, D. Pardo, T. González, J. S. Galloo, Y. Roelens, S. Bollaert and A. Cappy, Nanotechnology, 2003;14: 117
[9] J. S. Galloo, E. Pichonat, Y. Roelens, S. Bollaert, X. Wallart, A. Cappy, J. Mateos and T. González, 16th IPRM Conference Proceedings 2004
[10] T. González, B. G; Vasallo, D. Pardo and J. Mateos, Semicond. Sci. Technol. 2004;19:125–127
[11] I. Iñiguez-de-la-Torre, J. Mateos, T. González, D. Pardo, J. S. Galloo, S. Bollaert, Y. Roelens, and A. Cappy, Semicond. Sci. Technol. 2007;22: 663–670
[12] T. Sadi, F. Dessenne, and J.-L. Thobel, J. Appl. Phys. 2009;105: 053707
[13] Y. Wang, F. Zahid, J. Wang, and H. Guo, Phys. Rev. B 2012;85: 224110

Fabrication and characterization of multiband solar cells based on highly mismatched alloys

N López[1], A F Braña[1], C García Núñez[1], M J Hernández[1], M Cervera[1], M Martínez[1], K M Yu[2,3], W Walukiewicz[3], B J García[1]

[1]Grupo de Electrónica y Semiconductores, Departamento de Física Aplicada, Universidad Autónoma de Madrid, Spain.
[2]Department of Physics and Materials Science, City University of Hong Kong, China.
[3]Materials Sciences Division, Lawrence Berkeley National Laboratory, 1 Cyclotron Road, Berkeley, California 94720, USA.

E-mail: nair.lopez@uam.es

Abstract Multiband solar cells are one type of third generation photovoltaic devices in which an increase of the power conversion efficiency is achieved through the absorption of low energy photons while preserving a large band gap that determines the open circuit voltage. The ability to absorb photons from different parts of the solar spectrum originates from the presence of an intermediate energy band located within the band gap of the material. This intermediate band, acting as a stepping stone allows the absorption of low energy photons to transfer electrons from the valence band to the conduction band by a sequential two photons absorption process. It has been demonstrated that highly mismatched alloys offer a potential to be used as a model material system for practical realization of multiband solar cells. Dilute nitride $GaAs_{1-x}N_x$ highly mismatched alloy with low mole fraction of N is a prototypical multiband semiconductor with a well-defined intermediate band. Currently, we are using chemical beam epitaxy to synthesize dilute nitride highly mismatched alloys. The materials are characterized by a variety of structural and optical methods to optimize their properties for multiband photovoltaic devices.

1. Introduction

Recent attempts to increase the efficiency of solar cells beyond the Shockley and Queisser limit [1] have resulted in many novel device concepts. Among these the multiband solar cells (MSC) concept has been suggested and widely investigated [2,3]. The high power conversion efficiency of MSC arises from the increased photocurrent through absorption of low energy photons by the intermediate band (IB) while preserving a large band gap that determines the open circuit voltage. Several approaches have been proposed to implement the MSC device structure [2-4]. The existence of an IB has been demonstrated in a new class of alloys-highly mismatched alloys (HMAs) [5]. For example, substituting small amount of N atoms in GaAs splits the conduction band (E_C) of GaAs into two sub-bands: a low energy E_- and a higher energy E_+ bands [6-8]. The E_- band plays the role of the IB while the E_+ band becomes the E_C of the $GaAs_{1-x}N_x$ alloy.

The current work presents our study on the IB materials grown using a chemical beam epitaxy (CBE) system. The presence of an IB and the detection of the E_+ to E_- transition were studied using

Journal of Physics: Conference Series **647** (2015) 012067 doi:10.1088/1742-6596/647/1/012067

photoluminescence (PL) and photoreflectance (PR) techniques. The materials are characterized by a variety of structural and optical methods to optimize their performance for multiband photovoltaic devices. Additionally, the experimental results are analyzed with the band anticrossing model (BAC) showing an excellent agreement between experimental results and theoretical model.

2. Experimental

A CBE system has been used to grow the multiband HMA thin films under ultra-high vacuum conditions using gas precursors, which are in liquid phase at room temperature in equilibrium with their subatmospheric vapor pressure. The structure of the MSC consists of a p/n junction of the HMA layers sandwiched between a *p* or *n* blocking layers [10]. These blocking layers isolate the IB from the ohmic contacts so as to avoid the charge transport from the IB to the contacts. Therefore under solar irradiation three quasi-Fermi levels associated with the three bands of the material will form. This condition is necessary in order to preserve the open circuit voltage determined by the largest band gap of the material (E_+-E_V) [10]. Figure 1 shows a schematic of a MSC structure (inset) and its corresponding band diagram.

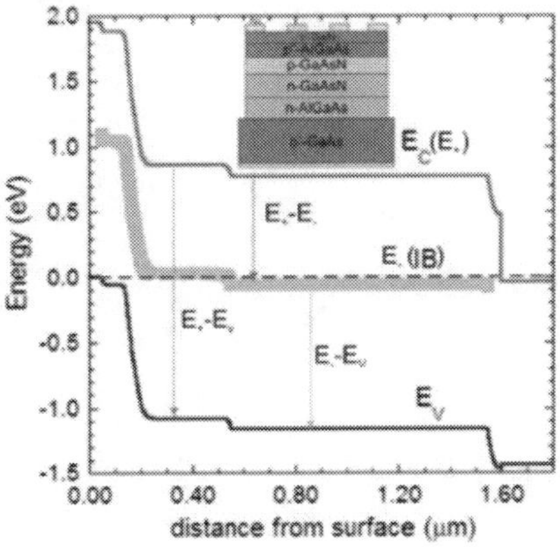

Figure 1. Structure and band diagram of a blocked intermediate band solar cell. The sub-band gap transitions from E_C (E_+) to the IB (E_-) are denoted as E_+-E_-, from E_- to E_V as E_--E_V and finally from E_+ to valence band (E_V) as E_+-E_V.

Dimethylhydrazine (DMHy), triethylgalium (TEGa) and tertiarybutylarsine (TBAs) precursors were used to grow $GaAs_{1-x}N_x$ layers by CBE. The gas precursor control system enables the control of fluxes of individual species on the substrate surface.

Fluxes of the different precursors determine both the growth rate as well as the alloy composition, which in turn determines the locations of the various band edges [11]. An alloy with an N mole fraction in the range x_N=0.02 to x_N=0.03 has the optimum band gap energies for an IB device [12]. To form a p/n junction Trimethyltin (TMSn) was used for *n*-type and carbon tetrabromide (CBr4) for *p*-type doping. Dopant levels have been chosen to optimize the performance of the device.

$Al_xGa_{1-x}As$ alloys were inserted as blocking layers during growth using tritertiary-bultylaluminum (TTBAl) as precursor for Al. This precursor is able to introduce high Al mole fractions [13]with a residual *p*-type doping level in the 10^{17} cm^{-3} range allowing to form $Al_xGa_{1-x}As$ layers with up x_{Al} = 0.45. A very thin *p*-GaAs capping layer was grown on the top of the structure to avoid oxidation of the $Al_xGa_{1-x}As$ layer.

3. Results

To obtain materials with the optimal band structure for a multiband device, growth parameters including substrate temperature, growth rate, fluxes of the various gas precursors were varied. N concentration of the samples was determined by measuring energies of interband transitions by PR and PL techniques. To obtain $GaAs_{1-x}N_x$ alloy with a band gap of close to 2 eV the N mole fraction should be x_N=0.023. For this composition the IB(E_-) is located at 1.17 eV above the E_V and 0.86 eV below E_+. The growth parameters were fixed at a growth rate of 0.5 μm/h, substrate temperature of 500 °C and a DMHy flux of 3 Torr. The energies of the E_+ and E_- are given by the BAC model (equation 1 [6, 7]).

$$E_{\pm} = \frac{(E_N + E_M) \pm \left[(E_N - E_M)^2 + 4V_{NM}^2 x\right]^{1/2}}{2} \quad (1)$$

where E_N is the energy of localized N level, E_M is the conduction band edge of the GaAs matrix and V_{NM}=2.7 eV is the coupling constant [6]. The E_-(k=0) and E_+(k=0) represents the intermediate and the conduction band edge, respectively. The E_V edge is not affected by the BAC and experiences only small downward shift of 20 meV per % of N content due to the E_V offset of GaAs and GaN. Figure 2 (a) shows the measured N mole fraction in $GaAs_{1-x}N_x$ layers as a function of the DMHy flux.

In order to calibrate the growth of $Al_xGa_{1-x}As$ blocking layers, $Al_xGa_{1-x}As$ layers were grown with a growth rate of 0.3 μm/h at a substrate temperature of 560 °C using different TTBAl fluxes; the composition of the grown layers was determined using spectroscopic ellipsometry (SE), PL and PR techniques. Figure 2 (b) shows the obtained dependence of the incorporated Al mole fraction as a function of the TTBAl flux used for different samples.

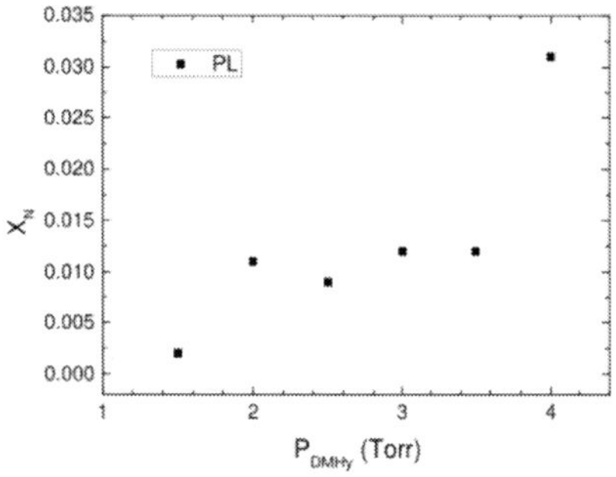

(a)

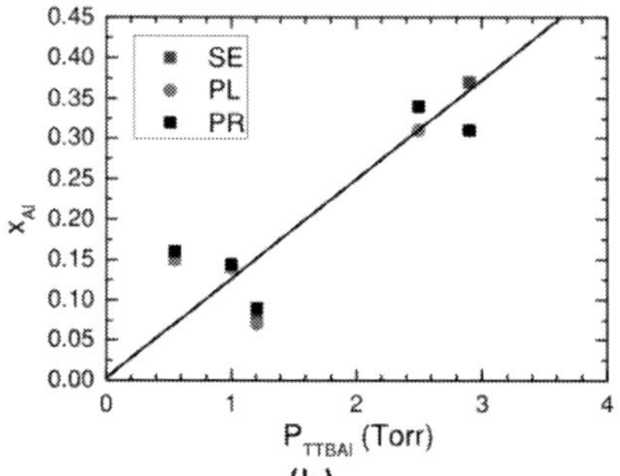

(b)

Figure 2. (a) N mole fraction in $GaAs_{1-x}N_x$ as a function of DMHy gas precursor flux at a growth rate of 0.5 μm/h and a substrate temperature of 500 °C. (b) Al mole fraction in $Al_xGa_{1-x}As$ as a function of TTBAl gas precursor flux at a growth rate of 0.3μm/h and a substrate temperature of 560 °C.

Finally, using the growth parameters determined from the calibration runs two blocked intermediate band (BIB) solar cells have been grown with different N content by varying the DMHy flux. In the dilute $GaA_{1-x}N_x$ alloy, the interaction between the localized N states and the extended states of the E_C of host GaAs matrix splits the E_C into two subbands. PL measurements on these samples demonstrated the presence of an IB(E_-) in both samples. As it is presented in Figure 3 (a) both transitions involving the IB are clearly observed for the sample grown with 3.75 Torr of DMHy. The low energy emission at 0.97 eV corresponds to the transitions between E_- and E_V whereas the higher energy peak at 1.21 eV can be attributed to the transition from E_+ to E_-; from this spectrum we observe that the E_- band is closer in energy to the E_V than to the $E_C(E_+)$. The N mole fraction for this sample is about x_N=0.05 which is significantly higher than the concentration of about x_N=0.03 estimated from the calibration samples. In addition, the PL spectrum shows a third peak at 1.37 eV which probably corresponds to a transition between E_- (IB) and the E_V of the AlGaAs blocking layer.

The PL spectra were analyzed using the BAC model. Figure 3 (b) shows the band gaps of $GaAs_{1-x}N_x$ as a function of the N mole fraction as calculated by the BAC model at low temperatures (12 K). The solid blue line corresponds to E_+ also denoted as the E_C of the $GaAs_{1-x}N_x$ alloy. The green line represents the energy level E_- that is equal to the energy separation E_-- E_V. Finally, the solid red line is the energy separation between E_+-E_-. The green-dash line and red-dash line are the PL energies measured for the sample grown with 3.75 Torr of DMHy pressure flux (red spectrum in figure 3a). Both PL transitions are in an excellent agreement with the transition energies predicted for a sample with x_N=0.05. Table 1 presents the theoretical calculation for the sub-band gaps and the PL experimental values. Also, as is observed in PL measurements, the E_- band is located closer to the E_V

than the conduction band (E_+). The BAC calculations show that the two sub-band gaps intersect for N content of about x_N=0.04.

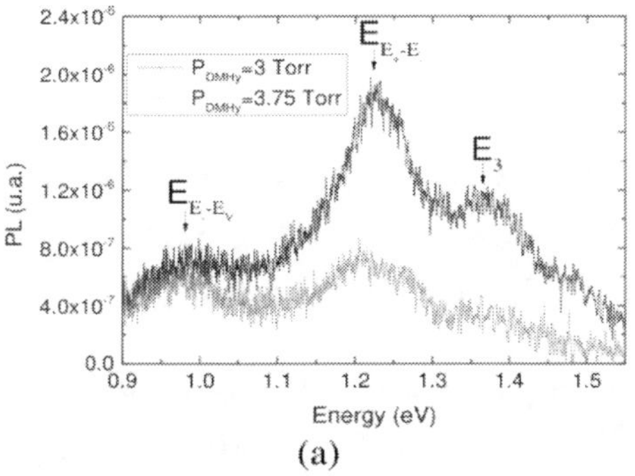

(a)

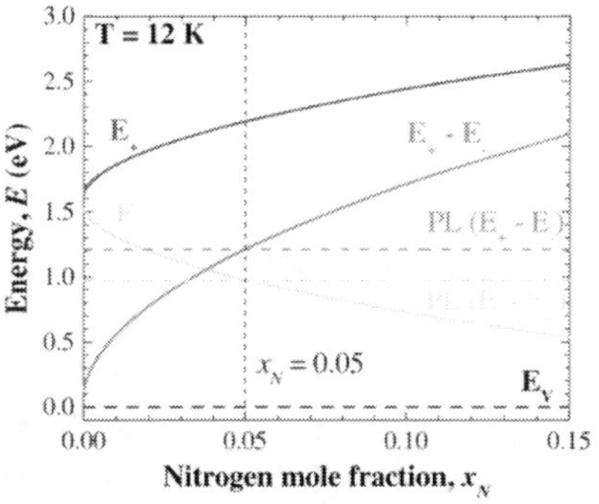

(b)

Figure 3. (a) Low temperature (12 K) photoluminescence measurements of samples grown with 3 Torr (black spectrum) and 3.75 Torr (red spectrum) of DMHy flux. (b) Theoretical calculations of the low temperature (12 K) sub-band gaps versus the N concentration using the BAC model are represented with solid lines. Dash lines correspond with the PL energy values.

Table 1. Theoretical and experimental values of the sub-band gaps for a sample with x = 0.05 of N.

	E_+-E_- (eV)	E_--E_V (eV)
BAC calculation	1.214	0.977
PL measurement	1.21	0.97

4. Conclusions

A chemical beam epitaxy system has been used to grow multiband solar cells based on a p/n junction of $GaAs_{1-x}N_x$ sandwiched between *p* and *n* blocking layers of $Al_xGa_{1-x}As$. The growth of $GaAs_{1-x}N_x$ and $Al_xGa_{1-x}As$ layers was optimized at 500 °C (0.5 μm/h) and 560 °C (0.3 μm/h) respectively. Two full structures have been grown with different N content in the $GaAs_{1-x}N_x$ layers by changing the DMHy flux used. These structures were characterized by PL measurements and studied by the BAC model. PL measurements have shown two emission peaks originating from transitions involving the IB, from E_+ to E_- and from E_- to E_V. The observed transition from E_+ to E_- demonstrates a significant optical coupling between those two bands. Comparison of the BAC model calculations with the experimental results allows for the determination of N concentration in the cell structures.

Acknowledgment

N. López acknowledges support of the European Commission by Marie Curie International Incoming Fellowship PIIF-GA-2012-326579. The authors also acknowledge the support from MINECO under TEC2013-48350-R project. W. Walukiewicz acknowledges support from the U.S. Department of Energy, Office of Science, Basic Energy Sciences, Materials Sciences and Engineering Division under Contract No. DE-AC02-05CH11231.

Reference

[1] W. Shockley and H. J. Queisser, *J. of Appl. Phys.* **32** (1961), 510-519.
[2] M. Wolf, *Proc. IRE* **48** (1960), 1246.
[3] A. Luque, et al., *Phys. Rev. Lett.* **78** (1997), 5014.
[4] N. López, et al., *J. of Solar Energy Engineering ASME* **129** (2007), 319-322.
[5] N. López, et al., *Phys. Rev. Lett.* **106** (2011), 028701.
[6] W. Shan, et al., *Phys. Rev. Lett.* **82** (1999), 1221.
[7] W. Walukiewicz, et al., *Phys. Rev. Lett.* **85** (2000), 1552.
[8] K. M. Yu, et al., *Phys. Rev. Lett.* **91** (2003), 246403.
[9] A. Luque et al., *Appl. Phys. Lett.* **87** (2005), 083505.
[10] N. López, et al., *26th EUPVSEC*-1BO.9.4 (2011), 97-100, 3-936338-27-2.
[11] K. M. Yu, et al., *J. Appl. Phys.* **106** (2009), 103709.
[12] L. Cuadra, et al., *IEEE Transactions on Electron Devices* **51** (2004), 1002-1007.
[13] A. Braña, et al., *18th European Molecular Beam Epitaxy Workshop*–TuP.13 (2015), 226.

Fabrication and Characterization of Fully Transparent ZnO Thin-Film Transistors and Self-Switching Nano-Diodes

Y. Sun[1], K. Ashida[1], S. Sasaki[1], M. Koyama[1], T. Maemoto[1], S. Sasa[1], S. Kasai[2], I. Iñiguez-de-la-Torre[3] and T. González[3]

[1]Nanomaterials Microdevices Research Center, Osaka Institute of Technology
5-16-1 Omiya, Asahi-ku, Osaka 535-8585, Japan
[2]Research Center for Integrated Quantum Electronics, Hokkaido University
Kita 8, Nishi 5, Kita-ku, Sapporo, Hokkaido 060-0808, Japan
[3]Departamento de Física Aplicada, Universidad de Salamanca,
Plaza de la Merced s/n, 37008 Salamanca, Spain

E-mail: d1d14301@st.oit.ac.jp

Abstract. Fully transparent zinc oxide (ZnO) based thin-film transistors (TFTs) and a new type of rectifiers calls self-switching nano-diodes (SSDs) were fabricated on glass substrates at room temperature by using low resistivity and transparent conducting Al-doped ZnO (AZO) thin-films. The deposition conditions of AZO thin-films were optimized with pulsed laser deposition (PLD). AZO thin-films on glass substrates were characterized and the transparency of 80 % and resistivity with 1.6×10^{-3} Ωcm were obtained of 50 nm thickness. Transparent ZnO-TFTs were fabricated on glass substrates by using AZO thin-films as electrodes. A ZnO-TFT with 2 μm long gate device exhibits a transconductance of 400 μS/mm and an ON/OFF ratio of 2.8×10^{7}. Transparent ZnO-SSDs were also fabricated by using ZnO based materials and clear diode-like characteristics were observed.

1. Introduction

Zinc Oxide (ZnO) have attracted attention as material suitable for use in transparent conductive films because of its wide band gap (E_g = 3.3 eV at 300 K), high electron mobility, rich in natural resources and low cost [1]. Therefore, ZnO can be used for a transparent electron devices on glass or flexible plastic substrates by a low temperature process. Thin-film transistors (TFTs) using ZnO thin-films for the channel layer are a famous application of it [2]. Additionally, low resistivity and transparency materials are required for the transparent conduction in electron devices. In recent years, indium tin oxide (ITO) has been widely applied as a low resistance, conductive material [3]. However, indium is toxic and it is a rare metal. On the other hand, ZnO based low resistance conductive materials are more advantageous than ITO. It is possible to reduce their resistivity by doping them with other n-type dopants, such as indium, gallium, and aluminum (Al) [4-6]. Out of those elements, aluminum has less poison and is rich in mineral resources. In this work, Al-Doped ZnO (AZO) thin-films deposited under optimum conditions were used as the electrodes of ZnO-TFTs.

On the other hand, development of radio frequency devices is active because the areas of goods management and security are employing electronic technology. New types of rectifiers for radio frequency identification (RFID) tags are studied by many groups. It is difficult to form high quality p-type materials from ZnO. As the result, it is difficult to fabricate a p-n junction diode by using ZnO films. One of the new types of rectifiers is called a self-switching nano-diode (SSD). SSD is a kind of rectifier without a p-n junction or a barrier structure so it is possible fabricate it with single material [7]. It has been reported that SSDs using compound semiconductor materials such as InGaAa and GaN [8, 9]. According to M. Y. Irshaid *et al.*, SSDs using ZnO thin-films on a glass substrate have demonstrated at least 51.5 MHz frequency [10]. In this work, fully transparent SSDs were fabricated by using ZnO thin-films as channel layers and AZO thin-films as electrodes.

2. Deposition and Characterization of Al-Doped ZnO

For a purpose of optimizing deposition conditions, AZO thin-films were grown on glass substrates (Corning Eagle XG) by pulsed laser deposition (PLD) at room temperature. A Nd:YAG laser (fourth harmonic, λ = 266 nm) is used for ablation of AZO ceramic target (ZnO-2 wt.% Al_2O_3). The laser repetition rate was 10 Hz and the laser pulse energy density is 2-3 J/cm^2. AZO thin-films with 50 nm thickness were grown under several different oxygen partial pressures (Po_2). The structure of these samples is shown in figure 2. The conditions of Po_2 were set in the range of 1×10^{-2} to 1×10^{-5} Torr, and without oxygen. The characteristics of AZO thin-films grown with different conditions were measured by optical transmission measurements, atomic force microscope (AFM), and Hall effect measurements.

Figure 1 shows the optical transmittance from 200 nm to 800 nm wavelength for the AZO films on glass substrates. The photograph in figure 1 shows the fabricated samples. The transmittance increases with increase of the Po_2 to 1×10^{-4} Torr. Higher transmittance was observed from a sample grown under 1×10^{-4} Torr of Po_2 and it was about 80 % in visible light region. The roughness of AZO thin-films surface was analyzed by an AFM. The roughness of root mean square (RMS) increases with increase in Po_2. The AZO films have a smooth surface with a value of RMS roughness below 1 nm when Po_2 is less than 1.0×10^{-4} Torr. Figure 2 shows the variation in carrier mobility, carrier density, and resistivity of the AZO films deposited under different conditions. Carrier mobility is not much changed even when Po_2 is changed. Carrier density is stable when Po_2 is lower than 1×10^{-4} Torr and its value is around 6-7$\times10^{20}$ cm^{-3}. However, it is decreased when Po_2 is higher than 1×10^{-4} Torr. Due to the carrier density, resistivity decreased at higher Po_2 but obtained about 2×10^{-3} Ωcm at lower Po_2 conditions. From those results, $Po_2 = 1\times10^{-4}$ Torr is a good condition for growing AZO thin-films as electrodes by PLD and using it to fabricate transparent electron devices.

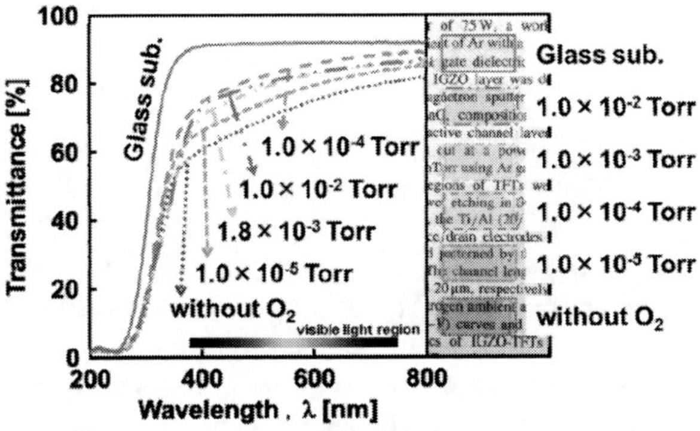

Figure 1. Optical transmittance for the AZO thin-films on glass and the photograph of text as seen through the samples.

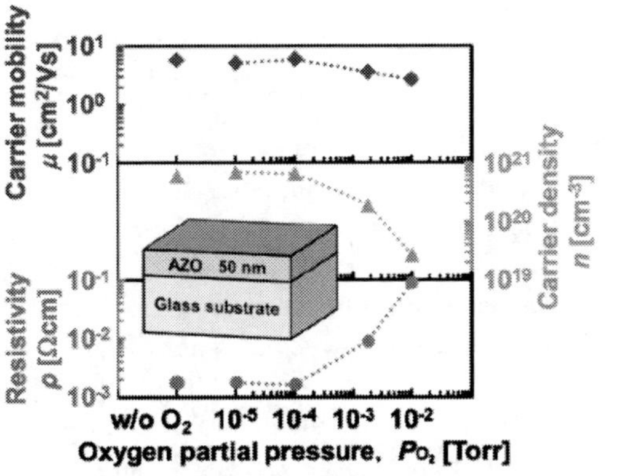

Figure 2. Carrier mobility, density, and resistivity of the AZO thin-films on glass and the structure of the samples.

3. Fabrication and Results of ZnO Thin-Films Transistors

In this work, transparent electron devices were fabricated with unheated process for application to flexible substrates. All the temperatures of the device processing were under 170 °C. Figure 3 shows a schematic diagram of a fabricated top-gate type ZnO-TFT. All layers of this TFT were grown by PLD. ZnO channel layer with a thickness of 40 nm was grown on a glass substrate under optimized Po_2 of 6×10^{-3} Torr. We previously reported that HfO_2 gate insulator layer grown at the high Po_2 is useful for a small hysteresis in the I_D-V_{GS} curves. The 50 nm thick HfO_2 gate insulator was deposited by PLD at Po_2 of 1×10^{-2} Torr. The electrodes were formed by 100 nm thick AZO thin-films at Po_2 of 1×10^{-4} Torr. A photograph shows ZnO-TFTs using metal for electrodes and using AZO in figure 4. From the comparison, the transparency of these two kinds of ZnO-TFTs are very different.

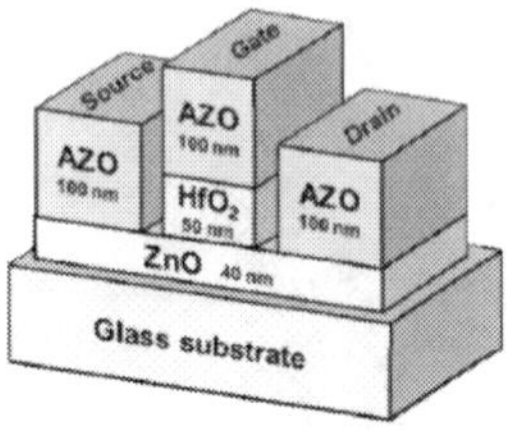

Figure 3. Schematic view of a top-gate ZnO-TFT fabricated in this work.

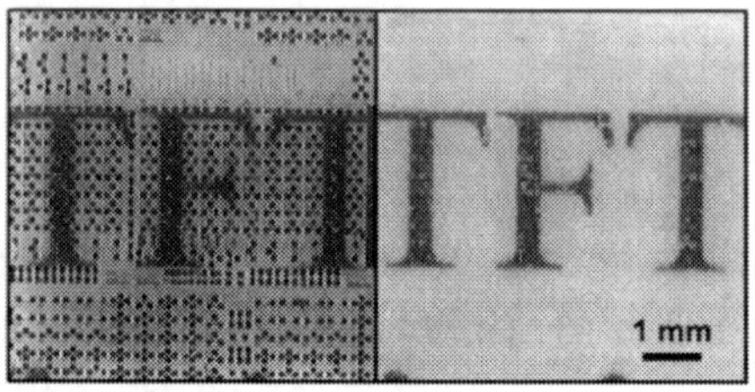

Figure 4. A photomicrograph of ZnO-TFTs using metal (left) or AZO (right) electrodes on glass.

A typical source-drain current (I_{DS})-voltage (V_{DS}) characteristics of a 2 μm gate length (L_G) device are illustrated in figure 5. The gate voltage (V_{GS}) was changed from -1 V to 7 V at 1 V steps. The device exhibits a maximum drain current of 1.4 mA/mm and a transconductance (g_m) of 400 μS/mm. Figure 6 shows the transfer characteristics (I_{DS}-V_{GS}) of the ZnO-TFT. V_{DS} was set to 8 V. The threshold voltage of this ZnO-TFT is 0.5 V and it has an ON/OFF ratio of 2.8×10^7, subthreshold voltage swing ~ 0.3 V/decade. From there results, we succeeded in fabricating transparent ZnO-TFTs reaching a practical level by using AZO thin-films as electrodes.

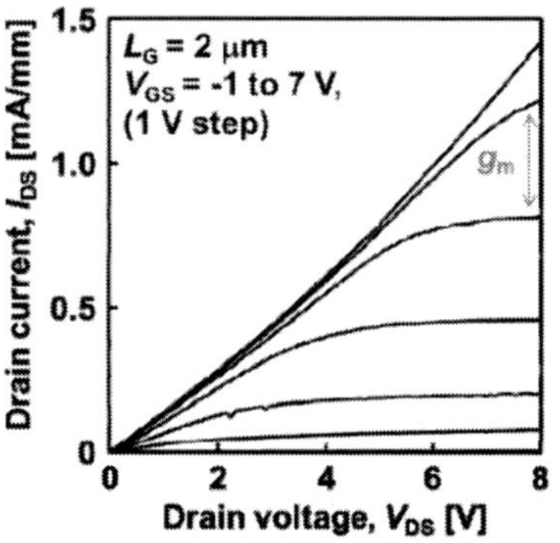

Figure 5. I_{DS}-V_{SD} characteristics at V_{GS} = 1 V intervals for a transparent ZnO-TFT.

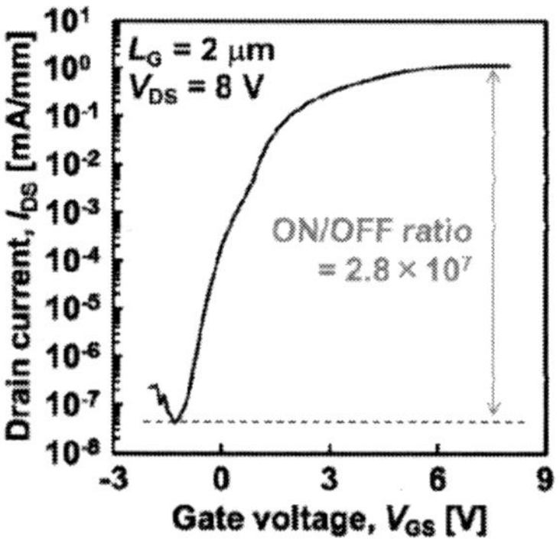

Figure 6. I_{DS}-V_{GS} characteristics with V_{DS} = 8 V of a transparent ZnO-TFT.

4. Fabrication and Results of ZnO Self-Switching Nano-Diodes

The fabrication process of ZnO-SSDs is very similar to that of ZnO-TFTs. A 40 nm thick ZnO thin-film is used as a channel layer. The carrier density of ZnO thin-films using in this work is 2.7×10^{18} cm^{-3} and carrier mobility of 4.4 $cm^2V^{-1}s^{-1}$ were obtained from Hall effect measurements. A simple structure of a ZnO-SSD shown at the inset of figure 7. Trenches on a 50 μm square ZnO channel layer were fabricated by electron beam lithography and wet chemical etching by using a diluted phosphoric acid. The L-shaped trenches were etched through the glass substrate to act as insulation. Ohmic electrodes were formed by 100 nm thick AZO thin-films.

Figure 7 shows the current-voltage (I-V) characteristics of a fabricated ZnO-SSD. The ZnO-SSD has a channel with length (L) of 1.8 μm and width (W) of 300 nm. Clear diode-like characteristics were observed. The ZnO-SSD was measured in the range of -40 V to 40 V. The SSD had a high breakdown voltage and the turn-on voltage was estimated to be about 24 V. Figure 8 shows a comparison of the sample's current in *I-V* characteristics for the device with and without ZnO-SSD. The current of the ZnO-SSD is lower because the area of SSD channel is smaller than normal channel. Rectification ratio (γ) was defined with a forward current (I_F, V = 40 V) divided by a reverse current (I_R, V = -40 V) and the γ of 2.6×10^1 was obtained from a fabricated ZnO-SSD. Fully transparent ZnO-SSDs are realized from achievements of fabricating transparent ZnO-TFTs.

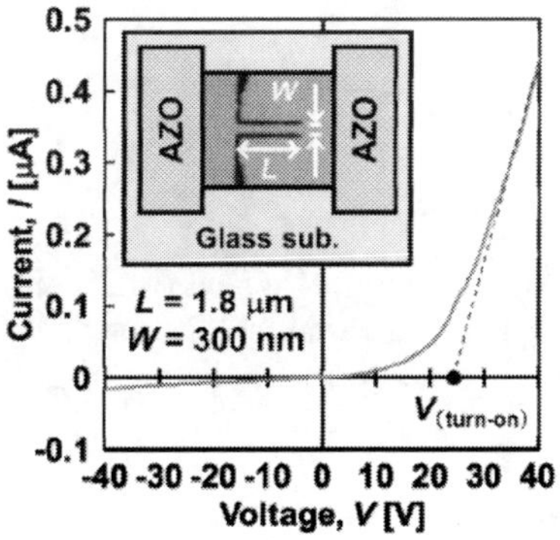

Figure 7. Schematic of a transparent ZnO-SSD fabricated on glass and its *I-V* characteristics.

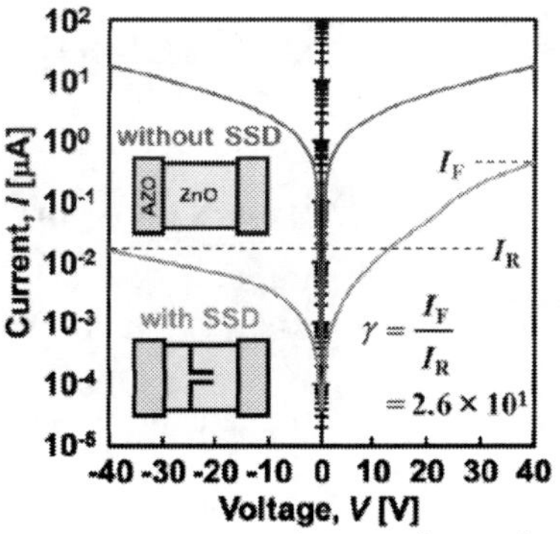

Figure 8. A comparison of performance between devices with SSD and without SSD.

5. Summary

Fully transparent ZnO-TFTs and SSDs were fabricated using AZO thin-films as transparent electrodes on glass substrate at room temperature. We optimized the deposition conditions to grow low resistance AZO thin-films with PLD. A 50 nm thick AZO thin-film on a glass substrate had resulted in a transmittance of 80 % in visible light region and a resistivity of 2×10^{-3} Ωcm. A transparent ZnO-TFT with the AZO thin-film as electrodes showed a clear pinch-off characteristic with an ON/OFF ratio of 2.8×10^7 and a g_m of 400 μS/mm. A new type of rectifier called SSDs with clear diode-like characteristics were fabricated using ZnO based transparent materials.

References

[1] R.D. Vispute, S. S. Hullavarad, D.E. Pugel, V. N. Kulkarni. S. Dhar, I. Takeuchi, and T. Venkatesan 2005 Wide Band Gap ZnO and ZnMgO Heterostructures for Future Optoelectronic Devices *Thin Films and Heterostructures for Oxide Electronics* ed Satishchandra B. Ogale (US: Springer) chapter 10 pp 301–330

[2] John F. Wager, Douglas A. Keszler and Rick E. Presley 2008 *Transparent Electronics* (US: Springer) pp 9-37

[3] Y. Shigesato 2011 In Based TCOs *Handbook of Transparent Conductors* ed D. Ginley, H. Hosono, D. C. Paine chapter 5 pp 149–160

[4] J.Park, J.Song, Y. Heo, J. Lee, J. Kim, W. Lim, L. Stafford, D. P. Norton and S. J. Pearton 2006 *J. Vac. Sci. Technol.* B **24** 2737

[5] S. Liang and X.Bi 2008 *J. Appl. Phys.* **104** 113533

[6] T. Minami, K. Okada, T. Miyata, J. Nomoto, Y. Hara and H. Abe 2009 *J. Vac. Sci. Technol.* A **27** 1006

[7] I. Iñiguez–de-la-Torre, J. Mateos, D. Pardo and T. González 2008 *J. Appl. Phys.* **103** 024502

[8] K. Y. Xu, G. Wang and A. M. Song 2008 *Appl. Phys. Lett.* **93** 233506

[9] A. Íñiguez-de-la-Torre, I. Íñiguez-de-la-Torre, J. Mateos, T. González, P. Sangaré, M. Faucher, B. Grimbert, V. Brandli, G. Ducournau and C. Gaquière 2012 *J. Appl. Phys.* **111** 113705

[10] M. Y. Irshaid, C. Balocco, Y. Luo, P. Bao, C. Brox-Nilsen1 and A. M. Song 2011 *Appl. Phys. Lett.* **99** 092101

High Efficiency Si Solar Cells Characterization Using Impedance Spectroscopy Analysis

A F Braña[1], E Forniés[2], N López[1], and B J García[1]

[1]Departmento de Física Aplicada, Universidad Autónoma de Madrid, Spain.
[2]Departamento de I+D. Aurinka PV Group, Rivas Vaciamadrid, Madrid, Spain

E-mail: alejandro.brana@uam.es

Abstract. Impedance Spectroscopy has been used to analyse commercial Si photovoltaic solar cells, to obtain information about minority carrier lifetimes, series and parallel resistances, and acceptor impurity densities. Silicon solar cells efficiencies ranging between 17 and 18% from different manufacturers have been analysed obtaining differences mainly in the electron lifetimes and doping densities. Relations between these parameters and DC curves are discussed.

1. Introduction

Silicon solar cells and panels still remain as the most used for the installation of photovoltaic power plants due to their reduced cost and mature fabrication process. New cells are produced with increasing efficiencies by means of the introduction of ameliorated processing techniques as well as improvements in material quality. Determining the characteristics of the performance of those solar cells is essential to improve their optimization.

Among other techniques, Impedance spectroscopy, as an AC characterization technique has been recognized as a valuable tool to obtain the transient response of charge carriers [1], which is not available when using only DC characterization. Bias dependent capacitance is one of the proposed techniques to obtain information on interface defects and on the carrier lifetime due to the fact that biasing or illuminating changes device dynamics, and that change becomes visible through capacitance variation.

The aim of this work is using impedance spectroscopy (IS) in commercial photovoltaic (PV) solar cells to compare their performances.

2. Experimental

PV cells used in this study were four commercial polycrystalline Si solar cells from different manufacturers, and with nominal efficiencies in the range of 17-18%, except sample C with 14%.

IV curves were obtained using a Cell tester from NPC Incorporated, model NPC 180 15 BPS. The system produces illumination from 0.5 to 1 sun (AM1.5G) and keeps a constant temperature of 25°C by means of a peltier cooler, avoiding heating effects due to illumination. Solar cell biasing was applied between -0.1V to 0.7V, to go from I_{sc} (0V) to slightly higher voltage values than V_{oc}.

Impedance measurements were performed using a HP4284 LCR meter from 20Hz to 1MHz with a small AC signal of 20mV (rms). Bias voltages ranged from -2V to 1V producing impedance curves. Measurement setup has been characterized to be able to simulate our AC networks taking into account the parasitic resistances and inductances at high frequencies. The obtained parameters were a resistance of 9Ω and an inductance of 5μH. Even they produced almost negligible effects when fitting the data to the equivalent circuits, those were taken into account in every sample.

Journal of Physics: Conference Series **647** (2015) 012069 doi:10.1088/1742-6596/647/1/012069

Silicon solar cells capacitance shows two main components as a function of bias [2]: a Mott–Schottky characteristic, due to the modulation of the Schottky barrier, at reverse and moderate forward bias, and a chemical capacitance [3], that increases exponentially for intense forward bias.

Two main processes are responsible of carrier extraction in crystalline solar cells: minority carrier diffusion and recombination [4]. In standard silicon solar cells, Fermi level splitting occurs at forward bias, being E_{Fn} the electron quasi-Fermi level, and E_{F0} as the equilibrium Fermi level. In this situation, we may define $qV_F = E_{Fn} - E_{F0}$ for forward biased p-n junctions where V_F is the potential associated with the splitting of the Fermi levels, related in our case to the accumulation of minority carriers with density n (electrons) [3]. Using Boltzmann statistics for the electron density the capacitance per unit area associated with the homogeneous accumulation in a layer of thickness L is given by

$$C_\mu = \frac{q^2 L n_0}{k_B T} e^{qV_F/\eta k_B T} \quad (1)$$

where n_0 is the minority carrier density in equilibrium, and η is the diode quality factor.

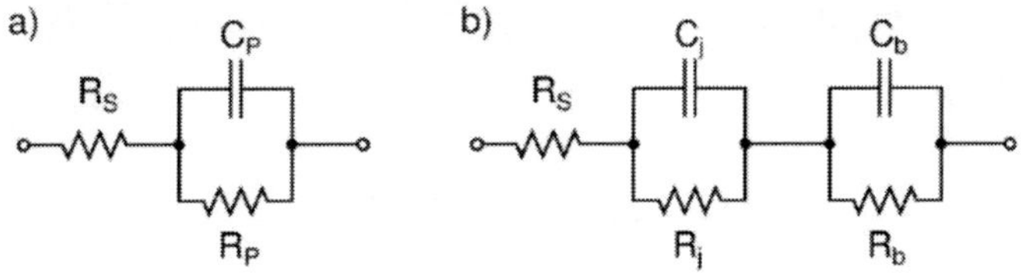

Figure 1. a) Equivalent circuit for IS. b) Equivalent circuit showing junction capacitance C_j as well as pp^+ back contact junction capacitance.

The studied solar cells are fabricated on p-type Si, with a highly P-doped emitter to form the n^+p junction and with a back surface field (BSF) at the back contact forming a pp^+ junction. Impedance spectra will be fitted using the equivalent circuits in figure 1. Part a) shows the simple circuit for a solar cell obtained in the LCR system. Figure 1b shows the equivalent circuit for the AC measurements taking into account both junctions, C_j and R_j for the n^+p junction, and C_b and R_b for the BSF junction. In our model, C_j stands for the chemical capacitance and R_j is the recombination resistance (R_{rec}) given by $R_{rec} = \tau_n / C_\mu$.

For reverse bias, Mott-Schottky behaviour is expected as an excess of majority carriers is present in the n^+p junction. Both the built-in potential V_{bi} and the carrier density N_A may be obtained through the well-known expression

$$C^{-2} = \frac{2(V_{bi} - V)}{A^2 q \varepsilon \varepsilon_0 N_A} \quad (2)$$

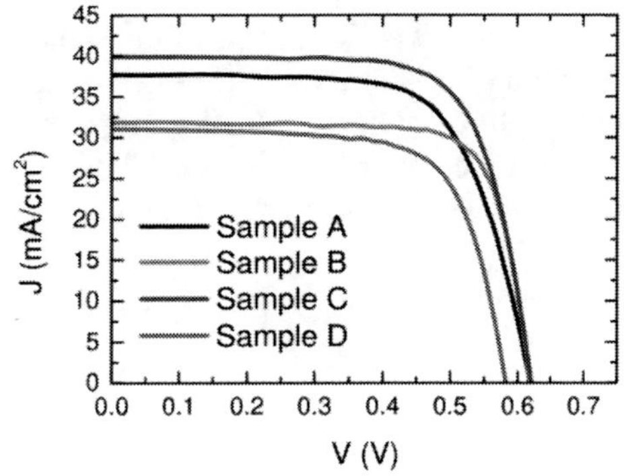

Figure 2. JV characteristics at 1 sun (AM1.5G, 25°C) of the four solar cells studied in this work.

3. Results and discussion

JV characteristics for the studied samples at 1 sun are shown in figure 1, showing differences mainly in the J_{sc} value, responsible for the differences in power conversion efficiency of the solar cells. Moreover, JV curves do not give enough information to find an explanation to the low V_{oc} value of sample D.

Further characterization of each sample comprises C-V measurements in the dark at different frequencies. 2kHz was chosen for the analysis from the C-f curves for a constant C region, and also far from the cutoff frequency. The experimental results show a linear behaviour of C^{-2} through a wide range of bias voltages, yielding both V_{bi} and N_A after fitting to equation (2). Fitting results are summarized in table 1.

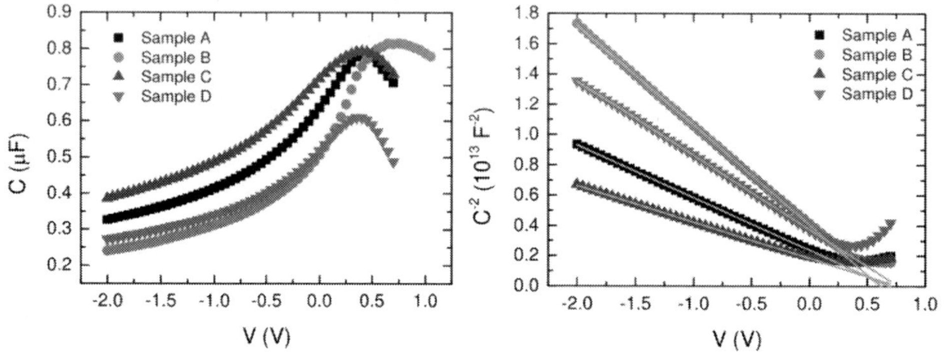

Figure 3. Solar cells dark CV characteristics at a frequency of 2 kHz, and corresponding Mott- Schottky plot with fitting lines to equation (2).

According to equation (1), the junction capacitance should increase as a function of forward bias. However, that is not the case after a given voltage in every analysed sample. It has been shown in classic Schottky diodes [5] that under moderate to large forward bias conditions, minority carrier injection can be limited depending on the barrier heights at the metal–semiconductor junction.

Table 1. CV and DC parameters for the studied cells

Sample	N_A (10^{16} cm^{-3})	V_{bi} (V)	J_{SC} (mA/cm^2)	V_{OC} (mV)
A	1.1	0.69	38.1	617
B	0.9	0.62	33.9	619
C	1.7	0.77	40.0	618
D	0.8	0.77	28.0	516

Next step in solar cell characterization was IS analysis of the samples. Figure 4 summarizes electrical impedance measurements in the complex plane for different voltage bias. After LCR measurements, complex impedance is obtained in the form $Z = Z' + iZ''$. Those values are plotted in a Nyquist plot producing arcs with different amplitudes for each voltage. Those semicircles are easily described by means of the equivalent circuit of figure 1a), and fittings to that model was performed using the *EIS Spectrum Analyser* software [6].

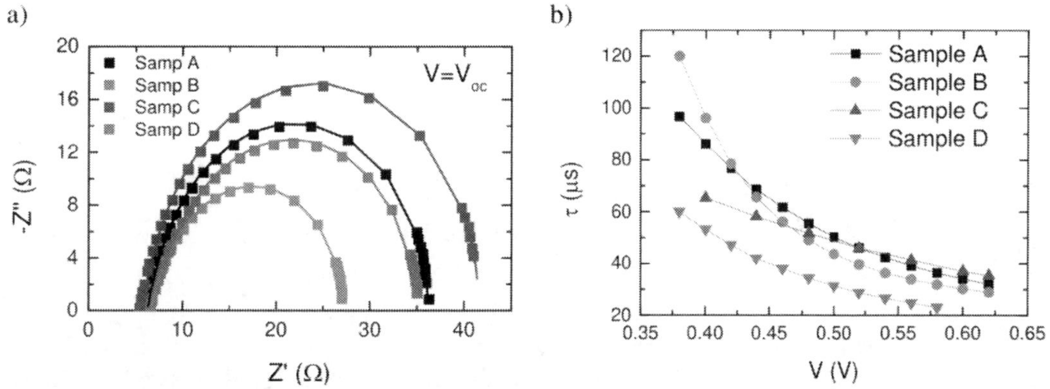

Figure 4. a) Nyquist plot in dark at a voltage equivalent to the V_{oc}. Solid lines are fits to the model of figure 1a) using EIS Spectrum Analyser [6]. b) Electron lifetimes obtained for each sample as a function of cell bias.

Series resistance in the model is the value of the starting point of the imaginary part of the impedance (Z''), which turns out to be very similar in all the cells, in close relation to de DC characterization curves. Moreover, semicircle width is the value of the recombination (junction) resistance and the maximum imaginary impedance corresponds to the product $C_P R_P$ from which the electron lifetime was calculated (figure 4b). The obtained lifetimes at (V_{oc}), are consistent with DC characterization. The contribution of the p–p+ junction is not detected in the impedance spectra for samples A, B and C, until a strong forward bias is applied (not shown here). This is due to the fact that at moderate forward bias, the voltage across the n+–p and p–p+ junctions compresses the space-charge layer, consequently reducing the junction width. However, sample D shows a remarkable asymmetry in the impedance spectrum indicating that BSF capacitance is playing an important role, mainly at high bias voltages.

4. Conclusions

IS analysis has been used as a technique to analyse solar cell behaviour. Different cell parameters, as series and parallel resistances, capacitance, minority carrier lifetime, acceptor impurities density have been obtained as function of applied bias. We showed that electron lifetimes follow DC characteristics in terms of short-circuit current density. It's also been shown that BSF needs to be taken into account to explain the impedance spectrum.

Acknowledgments

Financial support is acknowledge from Ministerio de Economía y Competitividad under project TEC2013-4835-R

References

[1] Anil-Kumar R, Suresh M S, Nagaraju J. 2001 *IEEE Trans. Electron Devices* **48** 2177
Yadav P, Tripathi B, Pandeya K, Kumar M 2014 *Phys.Chem.Chem.Phys.* **16** 15469
[2] Mora-Seró I, Luo Y, Garcia-Belmonte G, Bisquert J, Muñoz D, Voz C, Puigdollers J, Alcubilla R 2008 *Sol. Energy Mater. Sol. Cells* **92** 505.
[3] Bisquert J 2008 *Phys. Rev. B* **77** 235203
[4] Sze S M, 1981 *Physics of Semiconductor Devices*, John Wiley and Sons, New York
[5] Green M A, Shewchun J 1973 *Solid State Electron.* **16** 1141.
[6] Bondarenko A S, Ragoisha G A 2005 *Progress in Chemometrics Research*, Pomerantsev A. L., Ed.; Nova Science Publishers: New York, pp. 89–102 (the program is available online at http://www.abc.chemistry.bsu.by/vi/analyser/)

Towards a life-time-limited 8-octave-infrared photoconductive germanium detector

S G Pavlov [a,1], N Deßmann [b], A Pohl [b], N V Abrosimov [c], M Mittendorff [d,e], S Winnerl [d], R Kh Zhukavin [f], V V Tsyplenkov [f], D V Shengurov [f], V N Shastin [f,g], H-W Hübers [a,h]

[a] Institute of Optical Sensor Systems, German Aerospace Center (DLR), 12489 Berlin, Germany,
[b] Institut für Optik und Atomare Physik, Technische Universität Berlin, 10623 Berlin, Germany
[c] Leibniz-Institute for Crystal Growth, 12489 Berlin, Germany
[d] Helmholtz-Zentrum Dresden-Rossendorf, 01328 Dresden, Germany
[e] Technische Universität Dresden, 01069 Dresden, Germany
[f] Institute for Physics of Microstructures, Russian Academy of Sciences, 603950 Nizhny Novgorod, Russia
[g] Nizhny Novgorod State University, 603950 Nizhny Novgorod, Russia
[h] Humboldt-Universität zu Berlin, 12489 Berlin, Germany

E-mail: sergeij.pavlov@dle.de

Abstract. Ultrafast, ultra-broad-band photoconductive detector based on heavily doped and highly compensated germanium has been demonstrated. Such a material demonstrates optical sensitivity in the more than 8 octaves, in the infrared, from about 2 mm to about 8 μm. The spectral sensitivity peaks up between 2 THz and 2.5 THz and is slowly reduced towards lower and higher frequencies. The life times of free electrons/holes measured by a pump-probe technique approach a few tenths of picoseconds and remain almost independent on the optical input intensity and on the temperature of a detector in the operation range. During operation, a detector is cooled down to liquid helium temperature but has been approved to detect, with a reduced sensitivity, up to liquid nitrogen temperature. The response time is shorter than 200 ps that is significantly faster than previously reported times.

1. Introduction

Broad-band infrared detectors are requested for the in-situ diagnostic of power and quality of broad-band wavelength tunable pulsed infrared free electron lasers. Usually such lasers deliver picosecond pulses separated with typical temporal intervals of a few nanosecond (1/20/40 ns at the FELIX Facility in the Radboud University, Nijmegen, The Netherlands; 9.2 ns at the FEL Facility, the Institute of Scientific and Industrial Research, Osaka University, Japan; 16 ns at the CLIO Facility, the Universite Paris-Sud, Orsay, France; 77 ns at the FELBE Facility, Helmholz Zentrum Dresden-Rossendorf, Germany; 180 ns at the NovoFEL Facility, the Budker Institute for Nuclear Physics, Novosibirsk, Russia).

[1] S G Pavlov, sergeij.pavlov@dlr.de

Journal of Physics: Conference Series **647** (2015) 012070 doi:10.1088/1742-6596/647/1/012070

Germanium photoconductive detectors (GeD) are known for their very broad-band high spectral sensitivity in infrared [1]. The detector utilizes photoconductivity of free electron/holes optically excited from the ground state of dominating impurity. Since conventional applications of GeDs are in sensitive low-noise detection, they are usually from low doped (~ 10^{14} cm^{-3}) and low compensated crystals, standard dopants for p-type is gallium (p-Ge:Ga) and for n-type is antimony (n-Ge:Sb). The responsivity of such a material peaks at wavelengths around 100 μm and the long-wavelength cut-off is determined by the binding energy of the dopant to the valence band or the conduction band, respectively (figure 1, left). For time-resolved detection of optical pulses, significant reduction of conductivity recovering times is required. This can be achieved, for instance, by increasing of number of traps for free charge carriers. Such an effect can result from compensation of a dominant impurity in germanium by doping with centers causing opposite conductivity. As a rule, reduction of the recombination time of free charge carriers occurs proportionally to the number of attracting Coulomb centers [2]. Response time of ~2 ns in the far-infrared was observed for neutron transmutation doped high-compensated ($N_D/N_A = 0.32$-0.52) p-Ge:Ga:As:Se samples with a moderate acceptor (Ga) concentration of $N_A = (1$-$2)\times10^{15}$ cm^{-3} and a total compensating donor (As+Se) concentration of $N_D = (5.2$-$12)\times10^{14}$ cm^{-3} [3]. Similar response times have been obtained under special electrical bias of conventionally compensated GeD with $N_D/N_A < 0.7$ near impurity breakdown [4]. The speed of such detectors is set by technical limitations such as the bias circuit, the geometry of the detector crystal, and the electric field applied to the detector.

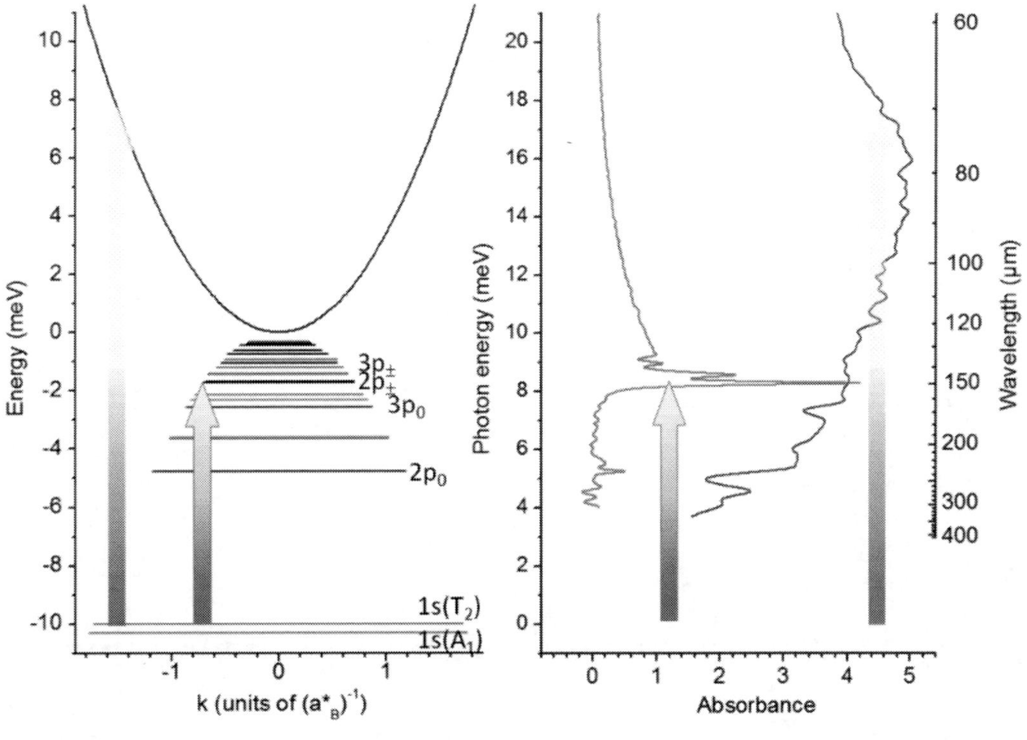

Figure 1. Left: Band structure of electronic states of germanium in vicinity of the bottom of the conduction band. Discrete line are ground ($1s(A_1)$) and excited levels (marked are few of odd-parity states serving as terminating states for the dipole allowed optical transitions from the ground state) of antimony donor atoms in the germanium bandgap. Right: Impurity absorption spectra of two germanium crystals: low doped low compensated n-Ge:Sb #471 (discrete intracenter optical transitions are clearly seen) and heavily doped highly compensated n-Ge:Sb:Ga #538 taken at low crystal temperature, 6 K. Vertical bold arrows indicate intracenter (bound-bound) and impurity-continuum (bound-free) optical transitions.

Journal of Physics: Conference Series **647** (2015) 012070 doi:10.1088/1742-6596/647/1/012070

2. Experimental

2.1. Germanium detectors

Dedicated doping of germanium crystals by dominating and compensating atoms, gallium and antimony, has been performed in specially designed gradient germanium crystals up to almost a critical concentration of metal-insulate transition for shallow impurities. Different samples with different level of compensation cut from these crystals have been characterized by absorption spectroscopy at ~6 K with a Fourier transform infrared spectrometer (FTIR) in the range from 4 to 400 μm. The spectra of heavily doped samples reveal an absorbance of up to five (absorption coefficient up to 100 cm^{-1}) (figure 2, right). Ohmic contacts have been prepared by evaporation of metal (Au+Sb for n-Ge and Al for p-Ge) on the opposite 2×0.5 mm^2 surfaces of the 4×2×0.5 mm^3 samples. The samples were attached to a cold finger of a liquid helium flow cryostat. Connections to the outer circuit were made by the coaxial stainless steel cables.

2.2. Fast detection

Detection of the ~10 ps infrared pulses at different wavelength has been carried out at the FELBE facility in the Helmholtz Zentrum Dresden-Rossendorf. We used a conventional DC biasing with a 7 ps rise time bias tee. Photocurrent has been directly fed into the input of a 20 GHz band oscilloscope, a DPO72004C model. Typical photoresponse on the FEL pulse is shown in figure 2. Dynamic range of detection exceeds 15 dB and was limited by a digital noise of the scope. Response signal dropped with increasing temperature but remained until liquid nitrogen temperature of 77 K [5].

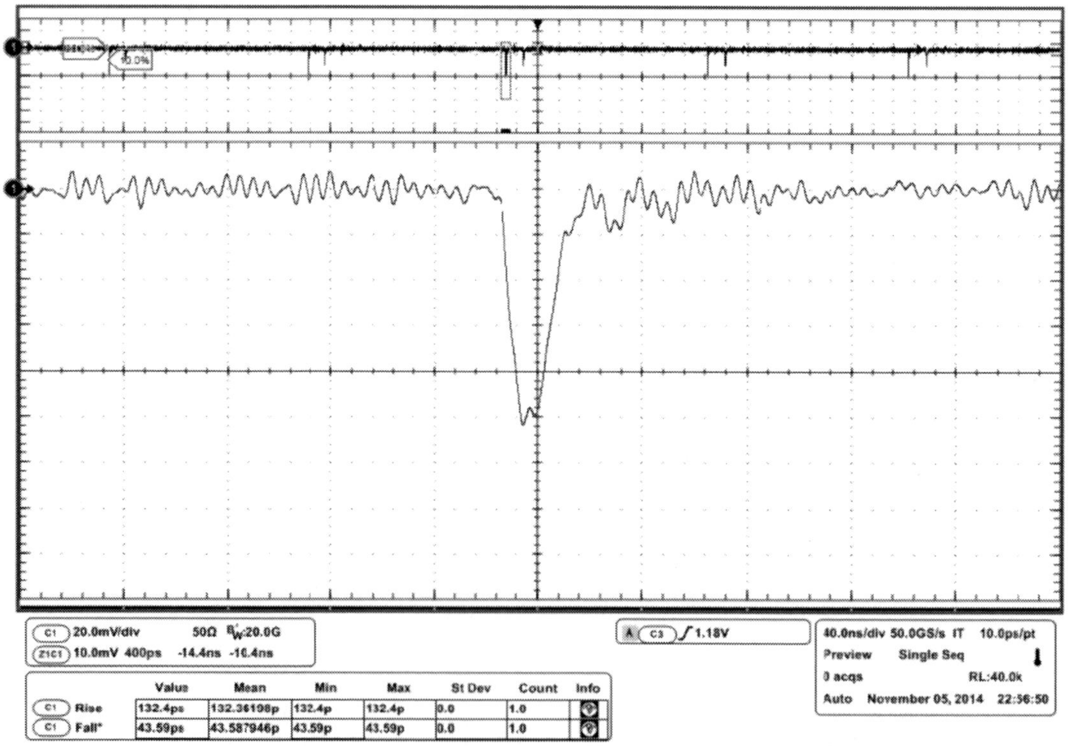

Figure 2. Response of the p-Ge:Ga:Sb # 539-4 detector on a FELBE pulse @ 19 μm taken with a fast scope, no amplifier. Upper signal (multi-pulse response) has a vertical resolution of 20 mV/division and time scale of 40 ns/division; the lower signal (single pulse): 10 mV/div and 400 ps/div, correspondingly. Rise time is 132 ps. FWHM is about 160 ps. The sample is at T ~ 10 K.

Both, uncompensated and compensated GeD resolve single FEL micropulses (separation is 77 ns). Uncompensated samples demonstrate significantly longer response, the shortest response times have been obtained for the heavily doped ((2-3)$\times10^{16}$ cm^{-3}) p-Ge:Ga:Sb samples with ~97% compensation.

2.3. Fast detection

Lifetimes of nonequilibrium electrons/holes have been measured by a pump-probe technique at the FELBE. These measurements show significant drop of recombination times (under 200 ps) in the samples with the large number of attracting recombination centers (figure 3) achieved by high level of compensation. Low compensated samples have lifetimes longer than 1 ns.

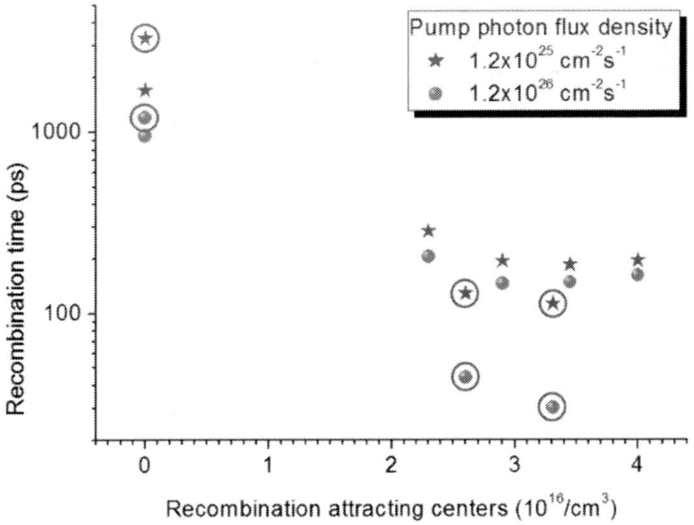

Figure 3. Dependence of the lifetime for p-Ge:Ga:Sb (data points shown in the circles) and n-Ge:Sb:Ga samples with a different concentration of domi-nant and compensating centers on a number of attracting centers for different pump photon flux densities. Pumping is in the state continuum (in the bottom of the conduction or valence bands, corres-pondingly). The samples were held at T ~ 10 K.

3. Conclusions

We have demonstrated ultrafast, ultra-broad-band (from millimeter waves to infrared) GeD based on heavily doped and highly compensated crystals. The response time of around 200 ps corresponds to the lifetime of nonequilibrium charge carriers as measured for the same semiconductors.

4. Acknowledgements

This research was supported by the German-Russian program "Research on technological advances of radiation sources of photons and neutrons based on accelerators and neutron sources in cooperation with research organizations and universities of the Federal Republic of Germany" (InTerFEL project, BMBF No. 05K2014 and Russian Ministry of Science and Education (ID No. RFMEFI61614X0008). We are grateful to P. Michel and the FELBE team for their dedicated support. N. D. and A. P. gratefully acknowledge support by the Helmholtz Research School on Security Technologies.

5. References

[1] Sclar N 1984 *Prog. Quant. Electr.* **9** 149
[2] Abakumov V N, Perel' V I and Yassievich I N 1977 *Sov. Phys. JETP* **45** 354
[3] Hegmann F A, Williams J B, Cole B, Sherwin M S, Beeman J W and Haller E E 2000 *Appl. Phys. Lett.* **76** 262
[4] Morozov S V, Marem'yanin K V, Erofeeva I V, Yablonskiy A N, Antonov A V, Gavrilenko L V, Rumyantsev V V and Gavrilenko V I 2010 *Semiconductors* **44** 1476
[5] Deßmann N, Pavlov S G, Pohl A, Abrosimov N V, Winnerl S, Mittendorff M, Zhukavin R Kh, Tsyplenkov V V, Shengurov D V, Shastin V N and Hübers H-W 2015 *Appl. Phys. Lett.* **106** 171109

Institute of Physics
Dirac House, Temple Back
Bristol BS1 6BE UK

ISSN: 1742-6588
ISBN 978-1-5108-1474-5

CURRAN ASSOCIATES INC.
proceedings.com